Walsh Series and Transforms

Mathematics and Its Applications (*Soviet Series*)

Volume 64

Walsh Series and Transforms

Theory and Applications

by

B. Golubov

Moscow Institute of Engineering,
Moscow, U.S.S.R.

A. Efimov

Moscow Institute of Engineering,
Moscow, U.S.S.R.

and

V. Skvortsov

Moscow State University,
Moscow, U.S.S.R.

KLUWER ACADEMIC PUBLISHERS

DORDRECHT / BOSTON / LONDON

Library of Congress Cataloging-in-Publication Data

Golubov, B. I. (Boris Ivanovich)
[Rfady i preobrazovanifa Uolsha. English]
Walsh series and transforms : theory and applications / by B. Golubov, A. Efimov, V. Skvortsov.
p. cm. -- (Mathematics and its applications. Soviet series ; v. 64)
Translation of: Rfady i preobrazovanifa Uolsha.
Includes index.
ISBN 0-7923-1100-0 (alk. paper)
1. Walsh functions. 2. Decomposition (Mathematics) I. Efimov, A. V. (Aleksandr Vasil'evich) II. Skvortsov, V. A. (Valentin Anatol'evich) III. Title. IV. Series: Mathematics and its applications (Kluwer Academic Publishers). Soviet series ; v. 64.
QA404.5.G6413 1991
515'.243--dc20 90-26705

ISBN 0-7923-1100-0

Published by Kluwer Academic Publishers,
P.O. Box 17, 3300 AA Dordrecht, The Netherlands.

Kluwer Academic Publishers incorporates
the publishing programmes of
D. Reidel, Martinus Nijhoff, Dr W. Junk and MTP Press.

Sold and distributed in the U.S.A. and Canada
by Kluwer Academic Publishers,
101 Philip Drive, Norwell, MA 02061, U.S.A.

In all other countries, sold and distributed
by Kluwer Academic Publishers Group,
P.O. Box 322, 3300 AH Dordrecht, The Netherlands.

Printed on acid-free paper

This is the translation of the work
РЯДЫ И ПРЕОБРАЗОВАНИЯ УОЛША
ТЕОРИЯ И ПРИМЕНЕНИЯ
Published by Nauka, Moscow, © 1987
Translated from the Russian by W. R. Wade

Printed in the Netherlands

SERIES EDITOR'S PREFACE

'Et moi, ..., si j'avait su comment en revenir, je n'y serais point allé.'

Jules Verne

The series is divergent; therefore we may be able to do something with it.

O. Heaviside

One service mathematics has rendered the human race. It has put common sense back where it belongs, on the topmost shelf next to the dusty canister labelled 'discarded non-sense'.

Eric T. Bell

Mathematics is a tool for thought. A highly necessary tool in a world where both feedback and non-linearities abound. Similarly, all kinds of parts of mathematics serve as tools for other parts and for other sciences.

Applying a simple rewriting rule to the quote on the right above one finds such statements as: 'One service topology has rendered mathematical physics ...'; 'One service logic has rendered computer science ...'; 'One service category theory has rendered mathematics ...'. All arguably true. And all statements obtainable this way form part of the raison d'être of this series.

This series, *Mathematics and Its Applications*, started in 1977. Now that over one hundred volumes have appeared it seems opportune to reexamine its scope. At the time I wrote

> "Growing specialization and diversification have brought a host of monographs and textbooks on increasingly specialized topics. However, the 'tree' of knowledge of mathematics and related fields does not grow only by putting forth new branches. It also happens, quite often in fact, that branches which were thought to be completely disparate are suddenly seen to be related. Further, the kind and level of sophistication of mathematics applied in various sciences has changed drastically in recent years: measure theory is used (non-trivially) in regional and theoretical economics; algebraic geometry interacts with physics; the Minkowsky lemma, coding theory and the structure of water meet one another in packing and covering theory; quantum fields, crystal defects and mathematical programming profit from homotopy theory; Lie algebras are relevant to filtering; and prediction and electrical engineering can use Stein spaces. And in addition to this there are such new emerging subdisciplines as 'experimental mathematics', 'CFD', 'completely integrable systems', 'chaos, synergetics and large-scale order', which are almost impossible to fit into the existing classification schemes. They draw upon widely different sections of mathematics."

By and large, all this still applies today. It is still true that at first sight mathematics seems rather fragmented and that to find, see, and exploit the deeper underlying interrelations more effort is needed and so are books that can help mathematicians and scientists do so. Accordingly MIA will continue to try to make such books available.

If anything, the description I gave in 1977 is now an understatement. To the examples of interaction areas one should add string theory where Riemann surfaces, algebraic geometry, modular functions, knots, quantum field theory, Kac-Moody algebras, monstrous moonshine (and more) all come together. And to the examples of things which can be usefully applied let me add the topic 'finite geometry'; a combination of words which sounds like it might not even exist, let alone be applicable. And yet it is being applied: to statistics via designs, to radar/sonar detection arrays (via finite projective planes), and to bus connections of VLSI chips (via difference sets). There seems to be no part of (so-called pure) mathematics that is not in immediate danger of being applied. And, accordingly, the applied mathematician needs to be aware of much more. Besides analysis and numerics, the traditional workhorses, he may need all kinds of combinatorics, algebra, probability, and so on.

In addition, the applied scientist needs to cope increasingly with the nonlinear world and the extra mathematical sophistication that this requires. For that is where the rewards are. Linear models are honest and a bit sad and depressing: proportional efforts and results. It is in the non-linear world that infinitesimal inputs may result in macroscopic outputs (or vice versa). To appreciate what I am hinting at: if electronics were linear we would have no fun with transistors and computers; we would have no TV; in fact you would not be reading these lines.

There is also no safety in ignoring such outlandish things as nonstandard analysis, superspace and anticommuting integration, p-adic and ultrametric space. All three have applications in both electrical engineering and physics. Once, complex numbers were equally outlandish, but they frequently proved the shortest path between 'real' results. Similarly, the first two topics named have already provided a number of 'wormhole' paths. There is no telling where all this is leading - fortunately.

Thus the original scope of the series, which for various (sound) reasons now comprises five subseries: white (Japan), yellow (China), red (USSR), blue (Eastern Europe), and green (everything else), still applies. It has been enlarged a bit to include books treating of the tools from one subdiscipline which are used in others. Thus the series still aims at books dealing with:

- a central concept which plays an important role in several different mathematical and/or scientific specialization areas;
- new applications of the results and ideas from one area of scientific endeavour into another;
- influences which the results, problems and concepts of one field of enquiry have, and have had, on the development of another.

Fourier series and the Fourier transform are of enormous importance in mathematics. They are based on the trigonometric orthogonal system of functions. However, this is but one orthonormal system and depending on the domain of interest other systems may be more useful, i.e. better adapted to the phenomena being studied or modeled. Examples are the Haar system and various systems based on wavelets. Another most important example is the Walsh system which is based on rectangular waves rather than sinusoidal ones. These appear to be preferred in a number of cases, for instance in signal processing.

There is no doubt about the importance of Walsh-Fourier series and transforms in harmonic analysis, in signal processing, in probability theory, in image processing, etc. It is therefore slightly surprising that no systematic treatment of the topic appeared before. However, here is one by a group of authors which have contributed significantly to the field.

The shortest path between two truths in the real domain passes through the complex domain.

J. Hadamard

La physique ne nous donne pas seulement l'occasion de résoudre des problèmes ... elle nous fait pressentir la solution.

H. Poincaré

Never lend books, for no one ever returns them; the only books I have in my library are books that other folk have lent me.

Anatole France

The function of an expert is not to be more right than other people, but to be wrong for more sophisticated reasons.

David Butler

Bussum, December 1990

Michiel Hazewinkel

TABLE OF CONTENTS

Chapter 1

WALSH FUNCTIONS AND THEIR GENERALIZATIONS

Chapter 2

WALSH-FOURIER SERIES BASIC PROPERTIES

Chapter 3

GENERAL WALSH SERIES AND FOURIER-STIELTJES SERIES QUESTIONS ON UNIQUENESS OF REPRESENTATION OF FUNCTIONS BY WALSH SERIES

Chapter 4

SUMMATION OF WALSH SERIES BY THE METHOD OF ARITHMETIC MEANS

Chapter 5

OPERATORS IN THE THEORY OF WALSH-FOURIER SERIES

Chapter 6

GENERALIZED MULTIPLICATIVE TRANSFORMS

Chapter 7

WALSH SERIES WITH MONOTONE DECREASING COEFFICIENTS

Chapter 8

LACUNARY SUBSYSTEMS OF THE WALSH SYSTEM

Chapter 9

DIVERGENT WALSH-FOURIER SERIES ALMOST EVERYWHERE CONVERGENCE OF WALSH-FOURIER SERIES OF $\mathbf{L}^2$ FUNCTIONS

Chapter 10

APPROXIMATIONS BY WALSH AND HAAR POLYNOMIALS

Chapter 11

APPLICATIONS OF MULTIPLICATIVE SERIES AND TRANSFORMS TO DIGITAL INFORMATION PROCESSING

Chapter 12

OTHER APPLICATIONS OF MULTIPLICATIVE FUNCTIONS AND TRANSFORMS

APPENDICES

PREFACE

The classical theory of Fourier series deals with the decomposition of functions into sinusoidal waves. Unlike these continuous waves, the Walsh functions are "rectangular waves". Such waves have been used frequently in the theory of signal transmission and it has turned out that in some cases these "waves" are preferred to the sinusoidal ones. In this book we give an introduction to the theory of decomposition of functions into Walsh series and into series with respect to the more general multiplicative systems. We also examine some applications of this theory.

The orthonormal system which is now called the Walsh system was introduced by the American mathematician J.L. Walsh in 1923. The development of the theory of Walsh series has been (and continues to be) strongly influenced by the classical theory of trigonometric series. Because of this it is inevitable to compare results on Walsh series to those on trigonometric series. There are many similarities between these theories, but there exist differences also. Much of this can be explained by modern abstract harmonic analysis, which studies orthonormal systems from the point of view of the structure of topological groups. This point of view leads in a natural way to a new domain of definition for the Walsh functions. As it is useful to consider the trigonometric functions "in complex form", i.e., defined on the circle group instead of the interval $[0, 2\pi)$, even so we shall see that it is convenient to define the Walsh system on a group which differs from the circle group in an essential way. This means that the Walsh system, and more generally each multiplicative system, provides an important model on which one can verify and illustrate many questions from abstract harmonic analysis. The Walsh system is also of great interest to anyone specializing in the theory of orthogonal systems since it is one of the simplest examples of a complete, bounded, orthonormal system. And, the so-called lacunary subsystems of the Walsh system play an essential role in probability theory.

In addition to progress made during the last 10-15 years in theoretical research on Walsh series, a number of works have been published which are concerned with applications of Walsh functions to scientific computing, coding theory, digital signal processing, e.t.c. Beginning in 1970 and repeating almost every year after that, there has been a conference in the United States of America which dealt with some aspect of applications of Walsh functions and has generated a collection of articles on the subject.

In 1969 and 1977, H. Harmuth published two monographs which were translated into Russian in 1977 and 1980 ([9], [10] [1]). Although they contained much material on the applications of Walsh functions, the mathematical theory was almost completely missing. Thus the techniques implimented there were presented without an adequate theoretical foundation. Another monograph [27] was devoted to applied questions. But until now, no book has been published[2], either in the Soviet Union or abroad, which contains an account of the theoretical foundations of Walsh series and Walsh transforms accessible to a broad spectrum of specialists in applied mathematics.

One of the main goals of this book is to remove this flaw to some extent. This book is intended for a wide audience of engineers, technical specialists and graduate students preparing for a career in applied mathematics. In addition to these, this book may also be of interest to graduate students in any of the mathematical sciences, since it can be used as an introduction to further study of Fourier analysis on groups. This book will give the reader access to the literature on Walsh series (a fairly complete survey of this literature up to 1970 appeared in the article of Balašov and Rubinšteĭn [[1]], and surveys of more recent research in the article of Wade [[4]] and in the closing pages of the monograph [1]) and, for those who wish to continue to questions in a more abstract setting, it will give an introduction to monographs [1], [24], [26], and others. Acquaintance with the theory of

[1] Numbers inside square brackets denote references which appear in the bibliography found at the end of this book. References to scholarly articles will contain the family name of the author and the number of the article as it appears in his list in the bibliography (for example, Efimov [[1]]). References without the author's name refer to monographs or textbooks which are listed separately in the bibliography (for example [1], [2]).

[2] Translator's note: A comprehensive monograph on the theory of Walsh series and transforms was published in 1990 by Adam Hilger (Institute of Physics) Publishing, Ltd, Bristol and New York. It is "Walsh Series: An Introduction to Dyadic Harmonic Analysis" by F. Schipp, W.R. Wade, and P. Simon, with assistance from J. Pàl.

Walsh series is also useful for the study of general questions in the theory of orthogonal series.

The first 10 chapters of this book deal with foundations of a theoretical nature, and Chapters 11 and 12 are connected with applications. Chapters 1 and 2 are fundamental for all that follows and by themselves are sufficient preparation for further study of both theoretical and applied material. Chapters 3-5 contain results concerning uniqueness of representation of functions by Walsh series, questions about summability and convergence in $\mathbf{L}^p$ of Walsh-Fourier series. This material is not used directly in the final chapters and on first reading those primarily interested in applications may restrict themselves to a passing acquaintance with these chapters. On the other hand, the concepts considered in Chapter 6 concerning multiplicative transforms are widely used in the last two chapters.

Chapters 7 and 8, where we consider Walsh series with monotone coefficients and lacunary series, contain only elementary information about these important classes of Walsh series whose theory has extensive connections with other closely related areas, in particular, as was mentioned above, with the theory of probability.

Chapter 9 is devoted to the specific questions of convergence and divergence of Walsh-Fourier series and is intended primarily for mathematicians. Here we give Hunt's proof of the Walsh analogue of Carleson's Theorem, about convergence of Fourier series of functions in the class $\mathbf{L}^2$, which in its basic features coincides with the trigonometric proof. However, many of the technical details are simpler in the Walsh case and this allows the reader to grasp more easily the basic ideas of the proof and will prepare him for further study of the trigonometric case.

In Chapter 10 we consider the problem of approximation by Walsh polynomials and by polynomials in the multiplicative systems. This problem is fundamental for many applications of the Walsh system. Finally, in Chapters 11 and 12 we examine methods for applying the Walsh system and its generalizations to digital information processing, to construct special computational devices, to digital filtering, and to digital holograms.

In order to aid the reader who is only acquainted with an undergraduate curriculum in mathematics, we include at the end of this book several Appendices containing background information about more advanced mathematical material which is used in the body of this book, namely, information about group theory, measure theory, the Lebesgue integral, and functional analysis. These appendices are followed by a commentary which includes brief remarks of a historical nature and references for sources of material which appears in each chapter. In view of the fact, as was mentioned above, that the theory of Walsh series already has some excellent and comprehensive survey articles our commentary gives further information only about the latest developments in this area.

Chapters 1-5 (except §1.5, §2.5, and §2.7) and Chapter 9 were written by V.A. Skvorcov, Chapters 7, 8, 10 (except §10.5) and §2.7 were written by B.I. Golubov, Chapters 6, 11, 12, and §10.5 by A.V. Efimov, §1.5 was written jointly by A.V. Efimov and V.A. Skvorcov, and §2.5 was written jointly by B.I. Golubov and V.A. Skvorcov.

The authors hope that this book will draw attention to the applicability of our subject and at the same time precipitate further theoretical investigations of solutions to applied problems.

The authors convey sincere thanks to B.F. Gapoškin and A.I. Rubinšteĭn, who read a manuscript version of this book and gave several valuable remarks which helped improve the presentation of this material.

The Authors

FOREWORD

The sections in each chapter have interior enumeration and theorems and formulae each have their own enumeration in each section. For example, §1.5 is the fifth section in Chapter 1, **2.3.5** is the fifth theorem in §2.3, and (2.3.5) is the fifth formula in §2.3. Similarly, **A5.2** is the second section of Appendix 5, and **A5.2.3** is the third theorem in **A5.2**. The beginning of a proof will be marked by "PROOF.", and the end of a proof will be marked with a " ∎".

The sign "≡" frequently denotes equal by definition.

The Lebesgue measure of a set A will be denoted by $\operatorname{mes} A$, but (and this is so especially when A is an interval Δ) we shall also use the shorter notation $|A|$.

We presume that the reader is familiar with the symbolism generally accepted from set theory, including the symbol $\emptyset$, which represents the empty set.

As usual, (a, b) denotes the open interval from a to b, $[a, b]$ the closed interval from a to b, and $[a, b)$ the half open interval from a to b which contains the point a but not the point b.

Other notation will be introduced as needed.

Chapter 1

WALSH FUNCTIONS AND THEIR GENERALIZATIONS

§1.1. The Walsh functions on the interval [0,1).

Consider the function defined on the half open unit interval $[0, 1)$ by

$$r_0(x) = \begin{cases} 1 & \text{for } x \in [0, 1/2) , \\ -1 & \text{for } x \in [1/2, 1) . \end{cases}$$

Extend it to the real line by periodicity of period 1 and set $r_k(x) \equiv r_0(2^k x)$ for $k = 0, 1, \dots$ and real x. The functions $r_k(x)$ are called the *Rademacher functions.*

It is evident from this definition that

$$r_{k+m}(x) = r_k(2^m x), \tag{1.1.1}$$

and

$$\int_{m/2^k}^{(m+1)/2^k} r_k(x)\, dx = 0 \tag{1.1.2}$$

for all integers $m,\ k \geq 0$. It is also clear that each $r_k(x)$ has period $1/2^k$, is constant (with constant value $+1$ or -1) on the dyadic intervals $[m/2^{k+1}, (m+1)/2^{k+1})$, $m = 0, \pm 1, \pm 2, \dots$, and although it has a jump discontinuity at each point of the type $m/2^{k+1}$, it is always continuous from the right. A graph of $r_k(x)$ on $[0, 1)$ for $k = 1$ can be found in Fig. 1.

The Rademacher functions are sometimes defined by

$$r_k(x) = \operatorname{sgn} \sin 2^{k+1} \pi x,$$

where

$$\operatorname{sgn} t = \begin{cases} 1 & \text{for } t > 0, \\ 0 & \text{for } t = 0, \\ -1 & \text{for } t < 0. \end{cases}$$

This definition differs only slightly from the one above. Namely, these r_k's are 0 at the jumps instead of being continuous from the right. We draw attention to this fact to emphasize that for us the Rademacher functions never assume the value 0.

The *Walsh system* $\{w_n(x)\}_{n=0}^{\infty}$ is obtained by taking all possible products of Rademacher functions. In connection with this we shall use the following enumeration of the Walsh system. (This enumeration is called the Paley enumeration; see

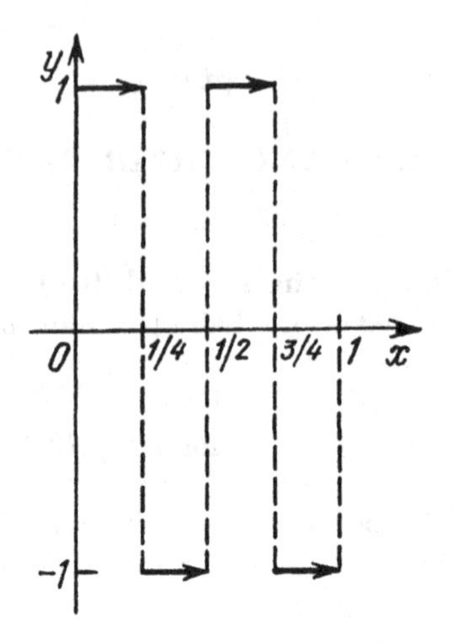

Figure 1.

the commentary on Chapter 1). Set $w_0(x) \equiv 1$. To define $w_n(x)$ for $n \geq 1$, represent the natural number n as a dyadic expansion, i.e., in the form

$$n = \sum_{i=0}^{k} \varepsilon_i 2^i, \tag{1.1.3}$$

where $\varepsilon_k = 1$ and $\varepsilon_i = 0$ or 1 for $i = 0, 1, \dots, k-1$. Such an n obviously satisfies $2^k \leq n < 2^{k+1}$, where $k = k(n)$. Set

$$w_n(x) = \prod_{i=0}^{k} (r_i(x))^{\varepsilon_i} = r_k(x) \prod_{i=0}^{k-1} (r_i(x))^{\varepsilon_i}. \tag{1.1.4}$$

In particular, each Walsh function $w_n(x)$, $n \geq 1$, takes on only the values 1 and -1, and is continuous from the right.

The definitions we have given for the Rademacher functions and the Walsh functions make sense on the entire real line. However, frequently the Walsh functions are considered only on the interval $[0, 1)$. This is the natural domain of definition, as will be seen in §1.2 where we shall give another definition of the Walsh functions.

Suppose $2^k \leq n < 2^{k+1}$. Notice by (1.1.3) that $n - 2^k = \sum_{i=0}^{k-1} \varepsilon_i 2^i$. It follows from (1.1.4) that

$$w_n(x) = r_k(x) w_{n-2^k}(x). \tag{1.1.5}$$

In particular, $r_k(x) = w_{2^k}(x)$.

We also notice that the following results are true.

1.1.1. *The finite product of integer powers of Rademacher functions is a Walsh function $w_\ell(x)$. Moreover, if k is the maximal index of the Rademacher functions* *appearing in this product, then $\ell < 2^{k+1}$, and $\ell = 0$ if and only if each r_i appears in this product with an even power.*

PROOF. Notice from definition that

$$r_i^m(x) = \begin{cases} 1 & \text{when } m \text{ is even,} \\ r_i(x) & \text{when } m \text{ is odd.} \end{cases}$$

Thus a finite product of powers of Rademacher functions which contains s odd powers reduces to a product of the form

$$r_{i_1}(x) \cdot r_{i_2}(x) \cdot \ldots \cdot r_{i_s}(x),$$

for certain indices $i_1 < i_2 < \cdots < i_s \le k$. It follows from (1.1.4) that this product is precisely the Walsh function $w_\ell(x)$ where $\ell = 2^{i_1} + 2^{i_2} + \cdots + 2^{i_s}$. It follows that $0 < \ell < 2^{i_s+1} \le 2^{k+1}$. If a finite product has no odd powers, then it is everywhere equal to 1 and coincides with the Walsh function $w_0(x)$. This finishes the proof of **1.1.1.** ∎

1.1.2. *The product of two Walsh functions $w_n(x)$ and $w_m(x)$ coincides identically with a third Walsh function $w_\ell(x)$, i.e., $w_n(x) \cdot w_m(x) = w_\ell(x)$. Moreover, if $m \le n < 2^{k+1}$, then $\ell < 2^{k+1}$, and $\ell = 0$ if and only if $n = m$.*

PROOF. By (1.1.4) the product of $w_n(x)$ and $w_m(x)$ is a product of powers of Rademacher functions whose indices do not exceed k. Hence the result follows from **1.1.1.** ∎

In §1.2 we shall be more precise about the relationship between the indices n, m and ℓ which appear in Theorem **1.1.2.**

It is evident from the definition that each Walsh function is constant on certain half-open dyadic intervals. We shall use a special notation for such intervals. Namely, we shall denote the *(dyadic) intervals of rank* $k \ge 0$ by

$$\Delta_m^{(k)} \equiv [m/2^k, (m+1)/2^k), \qquad 0 \le m \le 2^k - 1. \tag{1.1.6}$$

For convenience we also set $\Delta_0^{(0)} \equiv [0, 1)$.

We shall call the collection of all intervals of rank k the *dyadic* (or *binary*) *net of rank* k and denote it by $\mathcal{N}_k$. We shall occasionally refer to the intervals $\Delta_m^{(k)}$ as nodes of the net $\mathcal{N}_k$. Each of these nets provides a partition of the interval $[0, 1)$, namely

$$[0, 1) = \bigcup_{m=0}^{2^k-1} \Delta_m^{(k)}.$$

The interval of rank k which contains a point x will be denoted by $\Delta_{(x)}^{(k)}$. Of course for each point x the sequence $\Delta_{(x)}^{(k)}$ is a nested set of intervals which shrinks to the point x.

Notice that any two (dyadic) intervals are either disjoint or subsets of one another.

1.1.3. *For each $0 \le n < 2^{k+1}$, the Walsh function $w_n(x)$ is constant on the intervals $\Delta_m^{(k+1)}$, $0 \le m < 2^{k+1}$ and takes on the value 1 or -1. Moreover, $w_n(x) = 1$ for all $x \in \Delta_0^{(k+1)}$.*

PROOF. If $n < 2^{k+1}$ then the Rademacher functions appearing in the product (1.1.4) (which defines $w_n(x)$) have indices which do not exceed k. Thus it is clear that $w_n(x)$ is constant on each $\Delta_m^{(k+1)}$ and identically 1 on $\Delta_0^{(k+1)}$. ∎

Notice for $2^k \le n < 2^{k+1}$ that the intervals $\Delta_m^{(k+1)}$ are the largest dyadic intervals on which the function $w_n(x)$ is constant.

1.1.4. *If $2^k \le n < 2^{k+1}$ for some $k \ge 0$ then*

$$\int_{m/2^k}^{(m+1)/2^k} w_n(x)\,dx = 0 \tag{1.1.7}$$

for each $m = 0, 1, \dots 2^k - 1$.

PROOF. Write $w_n(x)$ as a product by (1.1.5). Notice by (1.1.3) that the function $w_{n-2^k}(x)$ is constant on the interval $\Delta_m^{(k)}$ with value equal to 1 or -1. In view of (1.1.2) it follows that

$$\int_{m/2^k}^{(m+1)/2^k} w_n(x)\,dx = \pm \int_{m/2^k}^{(m+1)/2^k} r_k(x)\,dx = 0.$$

This proves (1.1.7). ∎

It is immediate from (1.1.7) that

$$\int_0^1 w_n(x)\,dx = \begin{cases} 1 & \text{for } n = 0, \\ 0 & \text{for } n \ge 1. \end{cases} \tag{1.1.8}$$

1.1.5. *The Walsh system satisfies the orthogonality condition*

$$\int_0^1 w_n(x) \cdot w_m(x)\,dx = \begin{cases} 1 & \text{for } n = m, \\ 0 & \text{for } n \ne m, \end{cases}$$

i.e., the Walsh system forms an orthonormal system on $[0, 1)$.

PROOF. This fact follows immediately from **1.1.2** and (1.1.8). ∎

By using periodic extensions of the Walsh functions to the whole real line, it is easy to see that

$$w_{n2^m}(x) = w_n(2^m x). \tag{1.1.9}$$

Indeed, write

$$n2^m = 2^m \sum_{i=0}^{k} \varepsilon_i 2^i = \sum_{i=0}^{k} \varepsilon_i 2^{i+m}$$

and combine (1.1.4) with (1.1.1).

An important consequence of (1.1.9) is that

(1.1.10)

$$\begin{aligned}\int_{\Delta_j^{(m)}} w_{n2^m}(x) w_{p2^m}(x)\,dx &= \int_{\Delta_j^{(m)}} w_n(2^m x) w_p(2^m x)\,dx \\ &= 2^{-m} \int_j^{j+1} w_n(x) w_p(x)\,dx \\ &= \begin{cases} 2^{-m} & \text{for } n = p, \\ 0 & \text{for } n \neq p. \end{cases}\end{aligned}$$

§1.2. The Walsh system on the group.

In this section we introduce another method of generating the Walsh system which changes the domain of definition.

It turns out that a better and more natural domain on which to define the Walsh functions is given by the set of sequences whose entries are either 0 or 1, namely, sequences of the form

(1.2.1) $$\overset{*}{x} = \{x_0, x_1, x_2, \dots, x_j, \dots\},$$

where $x_j = 0$ or 1 for $j = 0, 1, \dots$. Each such sequence gives rise to a series

(1.2.2) $$\sum_{j=0}^{\infty} x_j \cdot 2^{-j-1},$$

which is the dyadic expansion of some point x in the interval [0,1], namely,

(1.2.3) $$x = \sum_{j=0}^{\infty} x_j 2^{-j-1}.$$

Notice that the correspondence between the set of sequences of the form (1.2.1) and the set of series of the form (1.2.2) is evidently 1-1 but the correspondence between series of the form (1.2.2) and points in the interval [0,1] which are represented by the formula (1.2.3) is not 1-1 because every dyadic rational has two expansions of the form (1.2.2) : one which is finite and one which is infinite (with x_j identically 1 for j large).

Because of this, the usual interval [0,1] is not a suitable model for a geometric interpretation of the set of sequences (1.2.1). In stead, we use the so-called modified interval $[0,1]^*$, which consists of expressions of the form (1.2.2) but not their sums. One can think of the modified interval $[0,1]^*$ as the usual interval in which each interior dyadic rational x has been split into two points, a *left* point $x-0$ which is the infinite dyadic expansion of x, and a *right* point $x+0$ which is the finite expansion of x. The dyadic rationals 0 and 1 are not split because they each have only one expansion of the form (1.2.2), the expansion of 0 being the expansion in which every coefficient is zero.

We shall now show that the set of sequences (1.2.1) can be made into a commutative group. (For the definition of a group, see **A1.1.**) Define an algebraic operation $\oplus$, which we call addition, by the following process. The *sum* of two sequences $\overset{*}{x} = \{x_j\}_{j=0}^{\infty}$ and $\overset{*}{y} = \{y_j\}_{j=0}^{\infty}$ is the sequence $\overset{*}{z} = \{z_j\}_{j=0}^{\infty}$ given by

$$\overset{*}{z} = \overset{*}{x} \oplus \overset{*}{y} = \{x_j \oplus y_j\}_{j=0}^{\infty} \tag{1.2.4}$$

where

$$x_j \oplus y_j = \begin{cases} 0 & \text{for } x_j + y_j = 0 \text{ or } 2, \\ 1 & \text{for } x_j + y_j = 1, \end{cases}$$

i.e., the operation $\oplus$ is coordinate addition of two sequences modulo 2.

It is obvious that this operation is associative and commutative. The inverse operation $\ominus$, defined by $\overset{*}{z} \equiv \overset{*}{x} \ominus \overset{*}{y}$ if and only if $\overset{*}{z} \oplus \overset{*}{y} = \overset{*}{x}$, evidently coincides with the operation $\oplus$. In particular, if $\overset{*}{0}$ represents the sequence all of whose entries are zero, then $\overset{*}{0}$ is the zero element of the group and $\overset{*}{y} \oplus \overset{*}{y} = \overset{*}{0}$ for all sequences $\overset{*}{y}$.

The *commutative group* whose elements are sequences of the form (1.2.1) and whose addition is the operation $\oplus$ defined by the formula (1.2.4) will be denoted by G.

Since there is a 1-1 correspondence between the sequences (1.2.1) and the modified interval $[0,1]^*$, the operation $\oplus$ can be carried over in a natural way to $[0,1]^*$, making it a group as well.

The group structure of G gives a convenient domain of definition for the Walsh functions, whose properties are intimately connected with this structure.

We shall use $\{\overset{*}{w}_n(\overset{*}{x})\}_{n=0}^{\infty}$ to denote the Walsh system whose domain is the group G or the modified interval $[0,1]^*$. This system is defined in the following way. Let n be a natural number whose dyadic expansion has the form (1.1.3) and let $\overset{*}{x}$ be an element of the group G of the form (1.2.1). Set

$$\overset{*}{w}_n(\overset{*}{x}) = (-1)^{\sum_{i=0}^{k} \varepsilon_i x_i}. \tag{1.2.5}$$

For $n = 2^i$ we see that $\overset{*}{w}_{2^i}(\overset{*}{x}) = (-1)^{x_i}$. Naturally, this subsequence of the Walsh system will be called the Rademacher system on the group G and will be denoted by $\{\overset{*}{r}_i(\overset{*}{x})\}$. Using this notation we see that (1.2.5) is analogous to (1.1.4), namely,

$$\overset{*}{w}_n(\overset{*}{x}) = \prod_{i=0}^{k}((-1)^{x_i})^{\varepsilon_i} = \prod_{i=0}^{k}(\overset{*}{r}_i(\overset{*}{x}))^{\varepsilon_i}. \tag{1.2.6}$$

It is clear from definition (1.2.5) that for each $n < 2^{k+1}$, the function $\overset{*}{w}_n(\overset{*}{x})$ is constant with value 1 or -1 on sets of the form $\{\overset{*}{x} : x_j = x_j^0,\ j = 0, 1, \ldots, k\}$, for each choice of fixed coordinates $\{x_0^0, x_1^0, \ldots, x_k^0\}$. These sets can be indexed by integers $0 \le m < 2^{k+1}$ in the following way. For each such m let the x_j^0's be determined by

$$m = \sum_{j=0}^{k} x_j^0 2^{k-j}, \tag{1.2.7}$$

and set

$$\overset{*}{\Delta}{}_m^{(k+1)} \equiv \{\overset{*}{x} : x_j = x_j^0,\ j = 0, 1, \ldots, k\}. \tag{1.2.8}$$

We shall presently see that this notation consistent with the notation introduced in §1.1.

Consider the transformation $\lambda : G \to [0,1]$ which takes each sequence $\overset{*}{x}$ of the form (1.2.1) to the number $x = \lambda(\overset{*}{x})$ which satisfies (1.2.3). Thus $\overset{*}{x}$ is the sequence of coefficients of the dyadic expansion of $\lambda(\overset{*}{x})$. The transformation λ is well-defined but, as we remarked above, is not 1-1. We shall examine what happens to the sets (1.2.8) under the transformation λ, i.e., look at the set of points x of the form (1.2.3) which are images under λ of sequences from the set (1.2.8). Clearly, for such points x,

$$\sum_{j=0}^{k} x_j^0 2^{-j-1} \le x \le \sum_{j=0}^{k} x_j^0 2^{-j-1} + \sum_{j=k+1}^{\infty} 2^{-j-1} = \sum_{j=0}^{k} x_j^0 2^{-j-1} + 2^{-k-1}.$$

But (1.2.7) implies

$$\sum_{j=0}^{k} x_j^0 2^{-j-1} = \left(\sum_{j=0}^{k} x_j^0 2^{k-j} \right) 2^{-k-1} = m/2^{k+1},$$

i.e., each such point satisfies the inequality

$$m/2^{k+1} \le x \le (m+1)/2^{k+1}.$$

It follows that each element of the set $\overset{*}{\Delta}{}_m^{(k+1)}$ is a sequence of coefficients of dyadic expansions of points from $[m/2^{k+1}, (m+1)/2^{k+1}]$, the closure of the interval $\Delta_m^{(k+1)}$. This shows that the notation introduced in (1.2.8) is consistent with the notation introduced in (1.1.16). Moreover, one can interpret $\overset{*}{\Delta}{}_m^{(k+1)}$ geometrically as the subset of the modified interval $[0,1]^*$ which differs from the usual interval

$[m/2^{k+1}, (m+1)/2^{k+1}]$ in that its interior dyadic rationals have been split into two points and its endpoints are understood to be $(m/2^{k+1})+0$ and $((m+1)/2^{k+1})-0$. Notice that as interpreted the intervals $\overset{*}{\Delta}{}_{m-1}^{(k+1)}$ and $\overset{*}{\Delta}{}_{m}^{(k+1)}$ have no common points, since the right endpoint of the first interval is $(m/2^{k+1})-0$ and the left endpoint of the second interval is $(m/2^{k+1})+0$.

It is easy to verify that the sets $\overset{*}{\Delta}{}_0^{(k)}$, $k=0,1,\ldots$, $(\overset{*}{\Delta}{}_0^{(0)} \equiv G)$, are subgroups of G and for each fixed k, that the sets $\overset{*}{\Delta}{}_m^{(k)}$, $m=0,1,\ldots,2^k-1$, exhaust the cosets (see **A1.2**) of the subgroup $\overset{*}{\Delta}{}_0^{(k)}$ in the group G.

The following identity is important to the theory of Walsh series[1]:

$$\overset{*}{w}_n(\overset{*}{x} \oplus \overset{*}{y}) = \overset{*}{w}_n(\overset{*}{x})\overset{*}{w}_n(\overset{*}{y}), \qquad n=0,1,\ldots. \tag{1.2.9}$$

It reveals a connection between the properties of the Walsh functions as defined on the group G, and the group structure of G. Its proof follows easily from the definitions of the operation $\oplus$ and the Walsh functions. Indeed, it is clear from (1.2.4) that the usual sum x_i+y_i and the sum $x_i \oplus y_i$ are either both even or both odd. This same observation holds for the sums $\sum_{i=0}^k \varepsilon_i(x_i+y_i)$ and $\sum_{i=0}^k \varepsilon_i(x_i \oplus y_i)$, where ε_i are defined by the decomposition (1.1.3). It remains to apply (1.2.5).

In order to formulate the next property of the Walsh functions, we introduce an operation on the non-negative integers similar to the operation $\oplus$ on the group G. For this, represent each non-negative integer n as an element of the group by setting

$$\overset{*}{n} = \{\varepsilon_0, \varepsilon_1, \ldots, \varepsilon_k, 0, 0, \ldots\},$$

where the coefficients ε_j are determined by the equation (1.1.3). Let G_0 represent the collection of sequences in G which contain only finitely many non-zero entries, i.e., sequences which from some point on are identically zero. Clearly, the map $n \to \overset{*}{n}$ is a 1-1 transformation from the collection of non-negative integers onto the subgroup G_0. Let $\overset{*}{m}$ be the element of the subgroup G_0 which corresponds to m. Then $\overset{*}{n} \oplus \overset{*}{m}$ belongs to G_0 and, under this transformation, has a non-negative integer preimage which we shall denote by $n \oplus m$. Thus to obtain the dyadic sum of two numbers n and m, take their dyadic expansions and add the dyadic coefficients coordinatewise modulo 2.

As an analogue of (1.2.9) we obtain

$$\overset{*}{w}_n(\overset{*}{x})\overset{*}{w}_m(\overset{*}{x}) = \overset{*}{w}_{n\oplus m}(\overset{*}{x}), \qquad x \in G. \tag{1.2.10}$$

By using the Walsh functions as defined on the group G we can give a new definition of the Walsh functions on the unit interval which is equivalent to that

[1] Identity (1.2.9) shows that $\overset{*}{w}_n(\overset{*}{x})$ are the *characters* of the group G.

given in §1.1. For this we look at the connection between the half-open interval [0,1) and the group G in more detail. We mentioned above that the transformation from G to [0,1) fails to be 1-1 only at dyadic rational points, because each dyadic rational in (0,1) has two dyadic expansions. If we agree to use only the finite expansion of these dyadic rationals, then each point $x \in [0,1)$ uniquely determines the sequence (1.2.1). Thus we can define a transformation $g : [0,1) \to G$ by the formula

$$g(x) \equiv \overset{*}{x} \equiv \{x_0, x_1, \dots, x_j, \dots\}, \tag{1.2.11}$$

where x_j is determined by the formula (1.2.3), with the convention about dyadic rationals which we just agreed upon. Let G' be that subset of G obtained by removing all sequences which are identically 1 from some point on. It is clear that $g(x) \neq g(y)$ for $x \neq y$, so the transformation g is a 1-1 map from the half open interval [0,1) onto G'. Hence we can define the inverse transformation g^{-1} on G' which apparently coincides with the restriction to G' of the map λ defined above.

Earlier, when we showed that the transformation λ takes sets of the form $\overset{*}{\Delta}{}_m^{(k+1)}$ to intervals, we in fact showed that the finite dyadic expansion of the point $m/2^{(k+1)}$ has the form $\sum_{j=0}^{k} x_j^0 2^{-j-1}$, where the coefficients x_j^0 are defined by (1.2.7). It is easy to check that the dyadic expansions of the interior points of the interval $[m/2^{k+1}, (m+1)/2^{k+1}) = \Delta_m^{(k+1)}$ has coefficients x_j which coincide with x_j^0 for $0 \le j \le k$. Consequently, the transformation g takes the interval $\Delta_m^{(k+1)}$ to a subset of $\overset{*}{\Delta}{}_m^{(k+1)}$. In fact, the image of $\Delta_m^{(k+1)}$ under g is precisely the set $\overset{*}{\Delta}{}_m^{(k+1)} \bigcap G'$.

It is now possible to give the following definition of the Walsh functions on [0,1):

$$w_n(x) = \overset{*}{w}_n(g(x)) = (-1)^{\sum_{i=0}^{k} \varepsilon_i x_i}, \qquad n = 0, 1, \dots, \tag{1.2.12}$$

where $\{\varepsilon_i\}$ is determined by (1.1.3) and $\{x_i\}$ is the sequence of coefficients from the dyadic expansion of x with the convention that the finite expansion is used when x is a dyadic rational.

We shall show that this definition is equivalent to the definition given by (1.1.4). For this, notice first that by (1.1.4) and (1.2.6) it is enough to prove

$$r_i(x) = \overset{*}{r}_i(g(x)) = (-1)^{x_i} \tag{1.2.13}$$

for $i = 0, 1, \dots$.

To prove this last identity, notice that the coordinate x_i of the sequence $g(x)$ (see (1.2.11)) is fixed as x ranges over an interval of the form $[m/2^{i+1}, (m+1)/2^{i+1})$, since as was noticed above, g takes such an interval to a subset of $\overset{*}{\Delta}{}_m^{(i+1)}$. Moreover, the value of x_i (see (1.2.8)) coincides with x_i^0 from the expansion of m in (1.2.7), where k has been replaced by i. Since the coefficients x_i^0 alternate between 0 and 1 as m ranges from 0 to $2^{i+1} - 1$, it follows that the values of $(-1)^{x_i}$ alternate between 1 and -1 as x moves through the intervals of rank i. Hence the description above of r_i coincides with the definition given in §1.1. This completes the proof of identities (1.2.13) and (1.2.12). In particular, it is natural to restrict the domain of the Walsh functions of a real variable to the half open interval [0,1).

By using the transformation g one can transfer the group operation $\oplus$ from G or from the modified interval $[0,1)^*$ to the usual unit interval in the following way. First recall that the map g is only defined on the subgroup G', which consists of sequences which are not identically 1 from some point on. Set

$$x \oplus y = g^{-1}(g(x) \oplus g(y)), \qquad g(x) \oplus g(y) \in G'. \tag{1.2.14}$$

Thus to find the sum $x \oplus y$ it is necessary to look at the sequences $\overset{*}{x}$ and $\overset{*}{y}$ of the form (1.2.1), obtained from the dyadic expansions (1.2.3) of the points x and y (where the finite expansion has been used when a dyadic rational is involved). Add them to obtain $\overset{*}{x} \oplus \overset{*}{y}$ and if this sequence does not terminate in 1's, use the transformation g^{-1} to take $\overset{*}{x} \oplus \overset{*}{y}$ to a point in [0,1). This point is the value of $x \oplus y$.

It is evident that the sum $x \oplus y$ is not defined for all pairs $x, y \in [0,1)$. This is done so that the operation $\oplus$ on [0,1) will preserve many of the important properties that are enjoyed by the group operation. For the theory of Walsh series the most significant of these properties is that the shift operation $G \to G \oplus \overset{*}{y}$ is a 1-1 transformation from the group G onto itself, and that for each fixed k it induces a permutation of the finite collection of cosets $\overset{*}{\Delta}{}_j^{(k)}$, $0 \le j \le 2^k - 1$. (See **A1.3**). Examining the corresponding shift operation on [0,1) (with respect to $\oplus$ as defined above), we shall see that if we neglect a countable subset then it too is a 1-1 transformation. Before we formulate the corresponding properties concerning cosets, we shall make several preliminary remarks.

Recall that the subset $G \setminus G'$ of the group G consists of all sequences of the form (1.2.1) whose entries are identically 1 from some point on. Hence $G \setminus G'$ is countable. If for some fixed $\overset{*}{y}$ we denote by $(G \setminus G') \oplus \overset{*}{y}$ the collection of all sums of the form $\overset{*}{z} \oplus \overset{*}{y}$, where $\overset{*}{z} \in G \setminus G'$, then it is also clear that $(G \setminus G') \oplus \overset{*}{y}$ is countable for each $\overset{*}{y} \in G$.

Define a set $I_y \equiv \lambda(G' \oplus g(y))$, where λ is the transformation defined above which takes the group G onto the unit interval [0,1]. Since $[0,1) \setminus I_y \subset \lambda(G \setminus (G' \oplus g(y))) = \lambda((G \setminus G') \oplus g(y))$ and this last set is the image under λ of a countable set, we see that the set $[0,1) \setminus I_y$ is also countable.

We are now prepared to formulate and prove the following result.

1.2.1. *For each fixed y the sum $x \oplus y$ is defined for all $x \in I_y \equiv \lambda(G' \oplus g(y))$, i.e., at all but countably many points in $[0,1)$. The transformation $x \to x \oplus y$ is a 1-1 transformation of the set I_y onto itself. Moreover,*

$$(\Delta_j^{(k)} \bigcap I_y) \oplus y = \Delta_{j_1}^{(k)} \bigcap I_y$$

for each dyadic interval $\Delta_j^{(k)}$, *where the integer* j_1 *depends on* y *and satisfies* $j_1 = j$ *if and only if* $y \in \Delta_0^{(k)}$. *If* y *is a dyadic rational then* $I_y = [0,1)$. *In particular, the shift operation* $x \to x \oplus y$ *maps* $[0,1)$ *onto itself when* y *is a dyadic rational.*

PROOF. Recall that the transformation g takes [0,1) onto G' and that on the set G', the transformation λ coincides with g^{-1}. Consequently, the transformation $\lambda = g^{-1}$ is 1-1 on the set $(G' \oplus g(y)) \bigcap G'$ and we have both $g(I_y) = (G' \oplus g(y)) \bigcap G'$ and $g(\Delta_j^{(k)} \bigcap I_y) = \overset{*}{\Delta}{}_j^{(k)} \bigcap (G' \oplus g(y)) \bigcap G'$. Furthermore, it is clear that

$$\left((G' \oplus g(y)) \bigcap G'\right) \oplus g(y) = G' \bigcap (G' \oplus g(y)).$$

It follows from (1.2.14) that the sum $x \oplus y$ is defined for all $x \in I_y$ and that the map $x \to x \oplus y$ is a 1-1 transformation from I_y onto itself.

It is well known (see **A1.3**) that on any group a shift preserves cosets. Thus a shift by $g(y)$ takes a coset $\overset{*}{\Delta}{}_j^{(k)}$ of the group G to another one, say $\overset{*}{\Delta}{}_{j_1}^{(k)}$. Moreover, it is clear that

$$\left(\overset{*}{\Delta}{}_j^{(k)} \bigcap (G' \oplus g(y)) \bigcap G'\right) \oplus g(y) = \overset{*}{\Delta}{}_{j_1}^{(k)} \bigcap G' \bigcap (G' \oplus g(y)).$$

This verifies the statements concerning the intervals $\Delta_j^{(k)}$.

It is evident that $\overset{*}{\Delta}{}_j^{(k)} \oplus g(y) = \overset{*}{\Delta}{}_j^{(k)}$ if and only if $g(y) \in \overset{*}{\Delta}{}_0^{(k)}$, i.e., $y \in \Delta_0^{(k)}$. Moreover, for each dyadic rational y the sequence $g(y)$ is finite so $G' \oplus g(y) = G'$. These observations complete the proof of **1.2.1**. ∎

It is now possible to obtain the following analogue of (1.2.9) for the Walsh functions $w_n(x)$:

1.2.2. *Let* $x, y \in [0,1)$. *If the sum* $x \oplus y$ *is defined then*

$$w_n(x \oplus y) = w_n(x)w_n(y) \tag{1.2.15}$$

for each $n = 0, 1, \ldots$. *Thus for a fixed* y *identity (1.2.15) holds for all but countably many points* $x \in [0,1)$, *and when* y *is a dyadic rational it holds for all* $x \in [0,1)$.

PROOF. Fix a natural number n. In view of **1.2.1** we need only prove (1.2.15) in the case when the sum $x \oplus y$ is defined. Suppose, then, that $g(x) \oplus g(y) \in G'$. Hence $g^{-1}(g(x) \oplus g(y))$ exists and we may apply definitions (1.2.14), (1.2.12), identity (1.2.9), and again definition (1.2.12). We obtain

$$\begin{aligned} w_n(x \oplus y) &= w_n(g^{-1}(g(x) \oplus g(y))) = \overset{*}{w}_n(gg^{-1}(g(x) \oplus g(y))) \\ &= \overset{*}{w}_n(g(x) \oplus g(y)) = \overset{*}{w}_n(g(x))\overset{*}{w}_n(g(y)) = w_n(x)w_n(y). \end{aligned}$$ ∎

It is now easy to see that

$$w_n\left(x \oplus \frac{1}{2^{k+1}}\right) = -w_n(x), \qquad 2^k \le n < 2^{k+1},\ x \in [0,1). \tag{1.2.16}$$

Indeed, write $n = 2^k + \sum_{i=0}^{k-1} \varepsilon_i 2^i$ and notice by definition (1.2.12) that

$$w_n(1/2^{k+1}) = (-1)^{1\cdot 1} = -1.$$

Since (1.2.15) implies

$$w_n\left(x \oplus \frac{1}{2^{k+1}}\right) = w_n(x) w_n\left(\frac{1}{2^{k+1}}\right),$$

it follows that (1.2.16) holds.

An immediate consequence of (1.2.12) is that an analogue of (1.2.10) holds without the kind of restriction on the domain which appeared in the statement of **1.2.2**. Namely, the following identity is true:

$$w_n(x) w_m(x) = w_{n\oplus m}(x), \qquad x \in [0,1). \tag{1.2.17}$$

This identity is a more precise version of **1.1.2**. In connection with this remark notice that if $m \le n < 2^{k+1}$ then $n \oplus m < 2^{k+1}$ and $n \oplus m = 0$ if and only if $n = m$.

1.2.3. *Let $\overset{*}{x}$ and $\overset{*}{y}$ be elements of the group G and λ be the transformation defined above which takes G onto $[0,1]$. Then*

$$|\lambda(\overset{*}{y}) - \lambda(\overset{*}{x})| \le \lambda(\overset{*}{y} \oplus \overset{*}{x}). \tag{1.2.18}$$

PROOF. Let $\overset{*}{x} = \{x_i\}_{i=0}^{\infty}$ and $\overset{*}{y} = \{y_i\}_{i=0}^{\infty}$. If $\lambda(\overset{*}{y}) = \lambda(\overset{*}{x})$ the inequality is obvious. By symmetry we may suppose that $\lambda(\overset{*}{y}) > \lambda(\overset{*}{x})$. Consider the sequence $\overset{*}{z} = \{z_i\}_{i=0}^{\infty}$ defined by

$$z_i = \begin{cases} y_i & \text{if } x_i = 0, \\ 1, & \text{if } x_i = 1. \end{cases}$$

Since for each i we have $z_i \ge y_i$, it is clear that

$$\lambda(\overset{*}{z}) = \sum_{i=0}^{\infty} z_i 2^{-i-1} \ge \sum_{i=0}^{\infty} y_i 2^{-i-1} = \lambda(\overset{*}{y}).$$

Furthermore, $z_i \ge x_i$ and $z_i - x_i = z_i \oplus x_i \le y_i \oplus x_i$ for each i. It follows that $\lambda(\overset{*}{z}) - \lambda(\overset{*}{x}) = \sum_{i=0}^{\infty} z_i 2^{-i-1} - \sum_{i=0}^{\infty} x_i 2^{-i-1} \le \sum_{i=0}^{\infty} (y_i \oplus x_i) 2^{-i-1} = \lambda(\overset{*}{y} \oplus \overset{*}{x})$. But

$\lambda(\overset{*}{y}) - \lambda(\overset{*}{x}) \le \lambda(\overset{*}{z}) - \lambda(\overset{*}{x})$. Consequently, $\lambda(\overset{*}{y}) - \lambda(\overset{*}{x}) \le \lambda(\overset{*}{y} \oplus \overset{*}{x})$ which verifies (1.2.18) in the case under consideration. ∎

The quantity $\rho_G^*(\overset{*}{x}, \overset{*}{y}) \equiv \lambda(\overset{*}{x} \oplus \overset{*}{y}) = \lambda(\overset{*}{x} \ominus \overset{*}{y})$ plays the role of a distance[2] between two elements $\overset{*}{x}$ and $\overset{*}{y}$ in G. Similarly, a new concept of distance between two points x and y in the unit interval [0,1) can be defined by

$$\rho^*(x, y) \equiv \lambda(g(x) \oplus g(y)). \tag{1.2.19}$$

If x and y are points for which the sum (1.2.14) is defined, then $\rho^*(x, y) = x \oplus y$.

Since $x = \lambda(g(x))$ and $y = \lambda(g(y))$ always hold, it follows directly from (1.2.18) that

$$|y - x| \le \rho^*(x, y) \qquad x, y \in [0, 1). \tag{1.2.20}$$

This inequality reveals the connection between ρ^* and the usual metric on the real line.

From definition (1.2.19) it is clear that the distance ρ^* is invariant under translation, namely, if $x \oplus z$ and $y \oplus z$ are defined then

$$\rho^*(x, y) = \rho^*(x \oplus z, y \oplus z). \tag{1.2.21}$$

1.2.4. *a) If $x, y \in \Delta_j^{(k)}$, then $\rho^*(x, y) \le 1/2^k$.*

b) If $\rho^*(x, y) \le 1/2^k$ then there exists a dyadic interval of the form $\Delta_j^{(k-1)}$ which contains both x and y.

PROOF. a) Since we have agreed to use the finite dyadic expansion for each dyadic rational, it is clear that if $x, y \in \Delta_j^{(k)}$ then the dyadic coefficients of x and y satisfy $x_i = y_i$ for $i = 0, 1, \dots, k-1$. Thus the sum $g(x) \oplus g(y)$ has the form $(0, 0, \dots, 0, z_k, z_{k+1}, \dots)$ which implies

$$\rho^*(x, y) = \lambda(g(x) \oplus g(y)) = \sum_{j=k}^{\infty} \frac{z_j}{2^{j+1}} \le \frac{1}{2^k}.$$

b) Suppose to the contrary that x and y belong to different dyadic intervals of rank $k - 1$. Then we can choose an integer $0 \le i \le k - 2$ such that the coefficients of 2^{-i-1} in the dyadic expansions of x and y are different. Consequently, the i-th element of the sequence $g(x) \oplus g(y)$ equals 1. This means that the dyadic expansion of the number $\lambda(g(x) \oplus g(y))$ contains the term 2^{-i-1}. Since $i \le k - 2$ it follows that $\rho^*(x, y) \ge 2^{-k+1} > 2^{-k}$. This establishes b). ∎

---

[2] That the function $\lambda(\overset{*}{x} \ominus \overset{*}{y})$ satisfies the triangle inequality follows directly from (1.2.18). After recognizing this, it is easy to verify that this function satisfies all the usual properties of a metric.

1.2.5. *If $y_n \to y$ in the usual sense and y is a dyadic irrational then $\rho^*(y_n, y) \to 0$.*

PROOF. For any k choose a natural number $n_0(k)$ such that $y,\ y_n \in \Delta_j^{(k)}$ for all $n \geq n_0$ and some integer j. It remains to apply **1.2.4** a). ∎

It is not difficult to verify that G (respectively, $[0,1)$) is a metric space under the distance ρ_G^* (respectively, ρ^*) (see **A2.1** and **A2.2**).

It is important to notice that on these metric spaces the Walsh functions are continuous. This is connected with the fact that each Walsh function has a jump discontinuity, in the classical sense, only at dyadic rationals. And, on the modified interval each dyadic rational is not only split into two pieces but these pieces are sufficiently far from each other as measured by the distance ρ_G^*; we leave it to the reader to verify this fact and provide the actual calculations. An similar situation prevails on the unit interval [0,1) for the distance ρ^*. In this case, the dyadic rationals are not split in two, but a pair of disjoint dyadic intervals are a positive distance from each other.

§1.3. Other definitions of the Walsh system. Its connection with the Haar system.

For each $0 \leq n < 2^{k+1}$ and $0 \leq m < 2^{k+1}$ denote by $w_{n,m}^{(k+1)}$ the constant value which the Walsh function $w_n(x)$ takes on the dyadic interval $\Delta_m^{(k+1)}$ (and which the Walsh function $\overset{*}{w}_n(\overset{*}{x})$ takes on the set $\overset{*}{\Delta}{}_m^{(k+1)}$). We shall examine the $2^{k+1} \times 2^{k+1}$ matrix $(w_{n,m}^{(k+1)})$. Our interest in this matrix stems from the fact that it completely determines the first 2^{k+1} Walsh functions.

1.3.1. *The matrix $(w_{n,m}^{(k+1)})$ is symmetric and orthogonal.*

PROOF. Since the elements of this matrix depend on k but k is fixed, we shall drop the superscript $(k+1)$ from the notation in this proof.

We first verify that $w_{n,m} = w_{m,n}$. Write n in the form (1.1.3) and write m in the form (1.2.7). Recall that the numbers x_j^0, $j = 0, 1, \ldots, k$, determine the set $\overset{*}{\Delta}{}_m^{(k+1)}$ (see (1.2.8)). We obtain from definition (1.2.12) that

$$w_{n,m} = (-1)^{\sum_{i=0}^{k} \varepsilon_i x_i^0}.$$

Reverse the roles of n and m. By (1.1.3), (1.2.7), and (1.2.8) we see that the set $\overset{*}{\Delta}{}_n^{(k+1)}$ is determined by the numbers $x_i = \varepsilon_{k-i}$ for $i = 0, 1, \ldots, k$. Moreover, for each i the coefficient of 2^i in the dyadic expansion of m of the form (1.1.3) is given by x_{k-i}^0. Therefore, we have by formula (1.2.12) that

$$w_{m,n} = (-1)^{\sum_{i=0}^{k} x_{k-i}^0 \varepsilon_{k-i}}.$$

This coincides with the formula for $w_{n,m}$ given above. In particular, the matrix $(w_{n,m}^{(k+1)})$ is symmetric.

Orthogonality of the matrix follows easily from orthogonality of the Walsh system $\{w_n(x)\}$. Indeed, by **1.1.3** we have

$$\int_0^1 w_n(x)w_\ell(x)\,dx = \sum_{m=0}^{2^{k+1}-1} \int_{\Delta_m^{(k+1)}} w_n(x)w_\ell(x)\,dx = \sum_{m=0}^{2^{k+1}-1} w_{n,m}^{(k+1)} w_{\ell,m}^{(k+1)} \cdot 2^{-k-1}.$$

Thus it follows from **1.1.5** that the matrix $(w_{n,m}^{(k+1)})$ is orthogonal. ∎

Theorem **1.3.1** allows us to prove the following useful result:

1.3.2. *Any function $P(x)$, which is constant on all dyadic intervals of the form $\Delta_m^{(k)}$, $0 \le m \le 2^k - 1$, can be represented in the form*

$$P(x) = \sum_{i=0}^{2^k-1} a_i w_i(x),$$

i.e., $P(x)$ is a Walsh polynomial whose non-zero coefficients have indices no greater than $2^k - 1$. Moreover, this representation of $P(x)$ is unique.

PROOF. Use the notation $w_{n,m}^{(k+1)}$ introduced above and let p_m denote the constant value which $P(x)$ assumes on the dyadic interval $\Delta_m^{(k)}$. We obtain the following system of 2^k equations in the unknowns a_i, $0 \le i \le 2^k - 1$:

$$\sum_{i=0}^{2^k-1} a_i w_{i,m}^{(k)} = p_m, \qquad 0 \le m \le 2^k - 1.$$

By Theorem **1.3.1**, the determinant of this system is non-zero. Hence the function $P(x)$ can be represented as promised, and this representation is unique among the Walsh polynomials of order no greater than $2^k - 1$.

Since by hypothesis the function $P(x)$ is also constant on any dyadic interval $\Delta^{(\ell)}$ of rank $\ell > k$, the remarks above remain true if ℓ is substituted for k. Hence the representation of $P(x)$ is unique among the class of Walsh polynomials of order no greater than $2^\ell - 1$. But this class contains the class of Walsh polynomials of order no greater than $2^k - 1$. Since the representations are unique in both classes, it follows that there is but one representation. In particular, the representation must be unique in the class of all Walsh polynomials. ∎

We shall now show that the matrix $(w_{n,m}^{(k+1)})$ can be constructed from the matrix $(w_{n,m}^{(k)})$. First, we establish the relationships

$$\begin{cases} w_{2n,m}^{(k+1)} = w_{2n+1,m}^{(k+1)} = w_{n,m}^{(k)}, \\ w_{2n,2^k+m}^{(k+1)} = -w_{2n+1,2^k+m}^{(k+1)} = w_{n,m}^{(k)} \end{cases} \tag{1.3.1}$$

for $0 \le n \le 2^k - 1$ and $0 \le m \le 2^k - 1$. If $n = \sum_{i=0}^{k-1} \varepsilon_i 2^i$, then $2n = \sum_{i=1}^{k} \varepsilon_{i-1} 2^i$ and $2n + 1 = 2^0 + \sum_{i=1}^{k} \varepsilon_{i-1} 2^i$. Furthermore, if m is the index of the interval $\Delta_m^{(k)}$, written in the form (1.2.7) with $k-1$ in place of k, then converting to intervals of rank $k+1$ we see that the expansions of m and $2^k + m$ can be written in the form $m = \sum_{i=1}^{k} x_{i-1}^0 2^{k-i}$, $2^k + m = 2^k + \sum_{i=1}^{k} x_{i-1}^0 2^{k-i}$. Applying formulae (1.2.8), (1.2.5), and (1.2.12) we obtain

$$w_{2n,m}^{(k+1)} = (-1)^{\sum_{i=1}^{k} \varepsilon_{i-1} x_{i-1}^0} = (-1)^{\sum_{i=0}^{k-1} \varepsilon_i x_i^0} = w_{n,m}^{(k)},$$

$$w_{2n,2^k+m}^{(k+1)} = (-1)^{0 \cdot 1 + \sum_{i=1}^{k} \varepsilon_{i-1} x_{i-1}^0} = w_{n,m}^{(k)},$$

$$w_{2n+1,m}^{(k+1)} = (-1)^{1 \cdot 0 + \sum_{i=1}^{k} \varepsilon_{i-1} x_{i-1}^0} = w_{n,m}^{(k)},$$

and

$$w_{2n+1,2^k+m}^{(k+1)} = (-1)^{1 \cdot 1 + \sum_{i=1}^{k} \varepsilon_{i-1} x_{i-1}^0} = -w_{n,m}^{(k)}.$$

This proves (1.3.1).

Since the matrices $(w_{n,m}^{(k)})$ and $(w_{n,m}^{(k+1)})$ are symmetric, (1.3.1) can be rewritten in the form

$$\text{(1.3.1')} \qquad \begin{cases} w_{n,2m}^{(k+1)} = w_{n,2m+1}^{(k+1)} = w_{n,m}^{(k)}, \\ w_{2^k+n,2m}^{(k+1)} = -w_{2^k+n,2m+1}^{(k+1)} = w_{n,m}^{(k)} \end{cases}$$

for $0 \le n \le 2^k - 1$, and $0 \le m \le 2^k - 1$.

The equations in (1.3.1) can be used to generate the matrices $(w_{n,m}^{(k)})$ recursively. To obtain the matrix $(w_{n,m}^{(k+1)})$ write each row of $(w_{n,m}^{(k)})$ twice, with the new copies under the old ones. This makes an intermediate $2^{k+1} \times 2^k$ matrix. To fill in the rest of the columns of $(w_{n,m}^{(k+1)})$, write a copy of each row of the intermediate matrix to the right of its row but multiply each element of the copied even rows by -1. To illustrate this process, we write the matrices of order 2×2, $2^2 \times 2^2$, $2^4 \times 2^4$:

$$\begin{pmatrix} 1 & 1 \\ 1 & -1 \end{pmatrix}, \begin{pmatrix} 1 & 1 & 1 & 1 \\ 1 & 1 & -1 & -1 \\ 1 & -1 & 1 & -1 \\ 1 & -1 & -1 & 1 \end{pmatrix}, \begin{pmatrix} 1 & 1 & 1 & 1 & 1 & 1 & 1 & 1 \\ 1 & 1 & 1 & 1 & -1 & -1 & -1 & -1 \\ 1 & 1 & -1 & -1 & 1 & 1 & -1 & -1 \\ 1 & 1 & -1 & -1 & -1 & -1 & 1 & 1 \\ 1 & -1 & 1 & -1 & 1 & -1 & 1 & -1 \\ 1 & -1 & 1 & -1 & -1 & 1 & -1 & 1 \\ 1 & -1 & -1 & 1 & 1 & -1 & -1 & 1 \\ 1 & -1 & -1 & 1 & -1 & 1 & 1 & -1 \end{pmatrix}.$$

Since these matrices are symmetric, the doubling process works equally well with columns. Thus it is easy to see that the values of the first 2^k Walsh functions, as prescribed by the matrices $(w_{n,m}^{(k)})$, $(w_{n,m}^{(k+1)})$, and all subsequent ones, are consistent with one another. Indeed, the first identity of (1.3.1') shows exactly that for $0 \leq n \leq 2^k - 1$ the values of the function $w_n(x)$ generated by matrix of order $2^{k+1} \times 2^{k+1}$ on each of the intervals $\Delta_{2m}^{(k+1)}$ and $\Delta_{2m+1}^{(k+1)}$ which make up the interval $\Delta_m^{(k)}$ coincide with the values of this function on the interval $\Delta_m^{(k)}$ as prescribed by the matrix of order $2^k \times 2^k$.

Thus it is clear that the collection of Walsh functions is uniquely determined by the sequence of matrices $(w_{n,m}^{(k)})_{k=0}^{\infty}$, where $(w_{0,0}^{(0)}) = (1)$. On the other hand equations (1.3.1) together with the agreement that the matrix of order 1×1 has the form $(w_{0,0}^{(0)}) = (1)$, completely determine the matrices $(w_{n,m}^{(k)})$, $k = 0, 1, \ldots$. This gives a *new, equivalent definition of the Walsh system*, not only on the interval [0,1), but also on the group G.

Finally, the matrices $(w_{n,m}^{(k)})$ allow us to establish a connection between the Walsh system and the Haar system, another system used widely in the theory of orthogonal expansions of functions. This will give us one more equivalent definition of the Walsh system on [0,1), this time in terms of Haar functions.

The *Haar system* $\{h_n(x)\}_{n=0}^{\infty}$ is defined on the interval [0,1) in the following way. For all $x \in [0, 1)$ set $h_0(x) = 1$. Write each integer $n \geq 1$ as $n = 2^k + m$, for some integers $k \geq 0$, $0 \leq m \leq 2^k - 1$, and set

$$h_n(x) = \begin{cases} 2^{k/2} & \text{for } x \in \Delta_{2m}^{(k+1)}, \\ -2^{k/2} & \text{for } x \in \Delta_{2m+1}^{(k+1)}, \\ 0 & \text{for } x \in [0, 1) \setminus \Delta_m^{(k)}. \end{cases} \tag{1.3.2}$$

(We notice that according to this definition, the Haar functions are continuous from the right at each point of discontinuity; this convention differs from that frequently found in the literature, where the Haar functions are defined at jump discontinuities to be the average of their left and right limits.)

Clearly, the Haar function with index $n = 2^k + m$, $0 \leq m \leq 2^k - 1$, is non-zero only on the dyadic interval $\Delta_m^{(k)}$, $m = 0, 1, \ldots, 2^k - 1$.

By using the matrices $(w_{n,m}^{(k)})$, we shall now show that the Walsh functions $w_i(x)$, $2^k \leq i \leq 2^{k+1} - 1$, can be written as a linear combination of Haar functions with indices in the same range, namely, that

$$w_{2^k+n} = 2^{-k/2} \sum_{m=0}^{2^k-1} w_{n,m}^{(k)} h_{2^k+m}, \qquad 0 \leq n \leq 2^k - 1. \tag{1.3.3}$$

It is easy to see that the sum on the right side of (1.3.3) is constant on the interval $\Delta_{2m}^{(k+1)}$ with value $w_{n,m}^{(k)}$, and constant on the interval $\Delta_{2m+1}^{(k+1)}$ with value $-w_{n,m}^{(k)}$. Hence the values of $w_{2^k+n}(x)$ for $0 \leq n \leq 2^k - 1$, as calculated by the formula (1.3.3), coincide with the true values that these functions must satisfy as described in the second formula of (1.3.1'). Thus definition (1.3.3) is equivalent to the one given above using the matrices $(w_{n,m}^{(k)})$. This establishes (1.3.3). Since it holds for any $k \geq 0$, it defines each Walsh function $w_i(x)$ for $i \geq 1$.

Equations (1.3.3) can be interpreted in the following way. The linear transformation induced by the matrix $(w_{n,m}^{(k)})$ takes the vector $(h_{2^k}, h_{2^k+1}, \dots, h_{2^{k+1}+1})$ to the vector $(w_{2^k}, w_{2^k+1}, \dots, w_{2^{k+1}+1})$. Since the matrix $(w_{n,m}^{(k)})$ is symmetric and orthogonal hence its own inverse, we see that (1.3.3) has the following analogue which defines Haar functions in terms of Walsh functions:

$$h_{2^k+n} = 2^{k/2} \sum_{m=0}^{2^k-1} w_{n,m}^{(k)} w_{2^k+m}, \qquad 0 \leq n \leq 2^k - 1. \tag{1.3.4}$$

Defining the Walsh system by using (1.3.3) is most useful in those cases when a certain property of Haar series corresponds to a similar one for Walsh series.

§1.4. Walsh series. The Dirichlet kernel.

By a Walsh series we shall mean a series of the form

$$\sum_{i=0}^{\infty} a_i w_i(x), \tag{1.4.1}$$

where the coefficients a_i are, by convention, real.

We shall isolate several properties enjoyed by the partial sums

$$S_n(x) = \sum_{i=0}^{n-1} a_i w_i(x) \tag{1.4.2}$$

of (1.4.1).

1.4.1. *For each n, $1 \leq n \leq 2^{k+1}$, the sum $S_n(x)$ has a constant value on the intervals $\Delta_m^{(k+1)}$, $0 \leq m < 2^{k+1}$.*

PROOF. This observation follows directly from **1.1.3.** ∎

For each positive integer $n \leq 2^k$, the constant value which a partial sum $S_n(x)$ assumes on an interval $\Delta_m^{(k)}$ will be denoted by $s_{n,m}^{(k)}$.

1.4.2. *For each interval $\Delta_m^{(k)} = \Delta_{2m}^{(k+1)} \bigcup \Delta_{2m+1}^{(k+1)}$ and each integer n satisfying $2^k < n \leq 2^{k+1}$, the equations*

$$s_{n,2m}^{(k+1)} + s_{n,2m+1}^{(k+1)} = 2 s_{2^k,m}^{(k)}, \tag{1.4.3}$$

and

$$\int_{\Delta_m^{(k)}} S_n(x)\, dx = \int_{\Delta_m^{(k)}} S_{2^k}(x)\, dx = |\Delta_m^{(k)}| \cdot s_{2^k,m}^{(k)} \tag{1.4.4}$$

hold.

PROOF. Write the sum $S_n(x)$ in the form

$$S_n(x) = S_{2^k}(x) + \sum_{i=2^k}^{n-1} a_i w_i(x).$$

Thus by (1.2.16) we have

$$S_n\left(x \oplus \frac{1}{2^{k+1}}\right) - S_{2^k}\left(x \oplus \frac{1}{2^{k+1}}\right) = -S_n(x) + S_{2^k}(x). \tag{1.4.5}$$

Using the definition of the operation $\oplus$ it is not difficult to see that if x belongs to one of the intervals $\Delta_{2m}^{(k+1)}$ or $\Delta_{2m+1}^{(k+1)}$, then $x \oplus 1/2^{k+1}$ belongs to the other one. Moreover, both these points belong to the same interval $\Delta_m^{(k)}$. Consequently, $S_{2^k}(x \oplus 1/2^{k+1}) = S_{2^k}(x)$ and we obtain from (1.4.5) that

$$S_n(x) + S_n\left(x \oplus \frac{1}{2^{k+1}}\right) = 2S_{2^k}(x). \tag{1.4.6}$$

In view of the notation introduced above, this identity verifies (1.4.3). We obtain (1.4.4) multiplying (1.4.3) by $|\Delta_{2m}^{(k)}|$ and recalling that $|\Delta_m^{(k)}| = 2|\Delta_{2m}^{(k)}|$. ∎

Apply (1.4.4) for $n = 2^{k+1}$ and iterate. We obtain

$$\int_{\Delta_m^{(k)}} S_{2^{k+\ell}}(x)\, dx = \int_{\Delta_m^{(k)}} S_{2^k}(x)\, dx, \qquad \ell \geq 0. \tag{1.4.7}$$

The partial sums of the series $\sum_i w_i(x)$ play such an important role in the theory of Walsh-Fourier series that they receive a special notation:

$$D_n(x) = \sum_{i=0}^{n-1} w_i(x). \tag{1.4.8}$$

Analogous to the trigonometric case, we shall refer to these partial sums as the *Dirichlet kernels for the Walsh system.*

The Dirichlet kernels satisfy the following properties.

1.4.3. *For $1 \leq n \leq 2^{k+1}$ the kernel $D_n(x)$ is constant on each interval $\Delta_m^{(k+1)}$, and*

$$D_n(x) = n \qquad \text{for } x \in \Delta_0^{(k+1)}. \tag{1.4.9}$$

PROOF. The first part of this result is a special case of **1.4.1**. Identity (1.4.9) follows directly from the fact that for $x \in \Delta_0^{(k+1)}$ and $0 \leq i < 2^{k+1}$, each Walsh function $w_i(x)$ is identically 1 (see **1.1.3**). ∎

Relationship (1.1.8) implies

$$\int_0^1 D_n(x)\,dx = 1, \qquad n = 1,2,\dots. \tag{1.4.10}$$

Let n be an integer written in the form $n = 2^k + m$ where $1 \le m \le 2^k$. Then

$$D_n(x) = D_{2^k}(x) + \sum_{i=2^k}^{2^k+m-1} w_i(x).$$

Applying (1.1.5) to each of the functions $w_i(x)$ for $2^k \le i < n-1 < 2^{k+1}$, we obtain

$$D_n(x) = D_{2^k}(x) + r_k(x) \sum_{i=2^k}^{2^k+m-1} w_{i-2^k}(x) = D_{2^k}(x) + r_k(x) \sum_{s=0}^{m-1} w_s(x).$$

Consequently, we have verified the formula

$$D_n(x) = D_{2^k}(x) + r_k(x) D_m(x) = D_{2^k}(x) + w_{2^k}(x) D_m(x), \tag{1.4.11}$$

for all $n = 2^k + m,\ 1 \le m \le 2^k$.

Substituting $m = 2^k$, we see that

$$D_{2^{k+1}}(x) = (1 + r_k(x)) D_{2^k}(x). \tag{1.4.12}$$

We shall now establish the identity

$$D_{2^k}(x) = \begin{cases} 2^k & \text{for } x \in \Delta_0^{(k)}, \\ 0 & \text{for } x \in [0,1) \setminus \Delta_0^{(k)}. \end{cases} \tag{1.4.13}$$

The proof is by induction on k. For $k = 0$ it is obvious. Suppose the formula holds for some $k \ge 0$. To obtain it for $k+1$, combine (1.4.12) with the facts that $\Delta_0^{(k)} = \Delta_0^{(k+1)} \bigcup \Delta_1^{(k+1)}$ and the Rademacher functions were defined so that

$$r_k(x) = \begin{cases} 1 & \text{for } x \in \Delta_0^{(k+1)}, \\ -1 & \text{for } x \in \Delta_1^{(k+1)}. \end{cases}$$

It follows that $D_{2^{k+1}}(x) = 2D_{2^k}(x) = 2^{k+1}$ for $x \in \Delta_0^{(k+1)}$ and $D_{2^{k+1}}(x) = 0$ for $x \in \Delta_1^{(k+1)}$. This verifies (1.4.13) for $k+1$. Hence (1.4.13) holds for all non-negative integers k.

The inequality

$$|D_n(x)| \le 2^{i-1}, \qquad x \in \Delta_1^{(i)},\ i = 1,2,\dots,\ n = 1,2,\dots \tag{1.4.14}$$

will play an important role for us. To prove this inequality, fix $i \geq 1$ and proceed by induction on k, where $2^k < n \leq 2^{k+1}$. Notice that the inequality is obvious for $n = 1 = 2^0$. Suppose that the inequality holds for $n \leq 2^k$ and let $2^k < n \leq 2^{k+1}$. Let $m = n - 2^k$ and notice that (1.4.11) holds.

We consider three cases: $i \leq k$, $i = k+1$, and $i > k+1$.

If $i \leq k$ then $\Delta_1^{(i)} \subset [0,1) \setminus \Delta_0^{(i)} \subset [0,1) \setminus \Delta_0^{(k)}$. Hence by (1.4.13), $D_{2^k}(x) = 0$ for $x \in \Delta_1^{(i)}$. In view of (1.4.11), it follows that $D_n(x) = r_k(x) D_m(x)$ for such points x. In particular,

$$(1.4.15) \qquad |D_n(x)| = |D_m(x)|, \qquad x \in \Delta_1^{(i)}$$

Since $m = n - 2^k \leq 2^k$, inequality (1.4.14) holds for $D_m(x)$ by the inductive hypothesis. Hence (1.4.15) implies (1.4.14) for $D_n(x)$ when $i \leq k$.

Passing to the case $i = k+1$, notice that $r_k(x) = -1$ for $x \in \Delta_1^{(k+1)}$. Recall also that $D_m(x) = m$ for $x \in \Delta_1^{(k+1)} \subset \Delta_0^{(k)}$ (see (1.4.9)). Thus it follows from (1.4.11) and (1.4.13) that $D_n(x) = 2^k - m$ for $x \in \Delta_1^{(k+1)}$. In particular, we have proved (1.4.14) in the case $i = k+1$.

It remains to examine the case $i > k+1$. In this situation $\Delta_1^{(i)} \subset \Delta_0^{(k+1)}$ and therefore by (1.4.9), $D_n(x) = n \leq 2^{k+1} \leq 2^{i-1}$. Thus inequality (1.4.14) holds in this case as well.

This completes the inductive step from k to $k+1$. Thus the proof of (1.4.14) is finished.

It is not difficult to see that inequality (1.4.14) implies

$$(1.4.16) \qquad |D_n(x)| < 1/x, \qquad x \in (0,1),\ n = 1, 2, \ldots.$$

Indeed, if $x \neq 0$ then choose an i such that $x \in \Delta_1^{(i)}$, i.e., $2^{-i} \leq x < 2^{-i+1}$ and apply (1.4.14) to obtain

$$|D_n(x)| \leq 2^{i-1} < 1/x.$$

§1.5. Multiplicative systems and their continual analogues.

The Walsh system is a special case of a more general class of function systems, the so-called multiplicative systems. We shall define these systems here by means of a direct generalization of the definitions in §1.2. As is the case for the Walsh system, these systems can be defined on the interval [0,1), extended to the whole real line by periodicity of period one, or defined on some compact group similar to the group G.

We begin with a description of a class of groups of interest to us.

Let

$$(1.5.1) \qquad \mathbf{P} \equiv \{p_1, p_2, \ldots, p_j, \ldots\}, \qquad p_j \geq 2,\ j \geq 1$$

be a fixed sequence of natural numbers. Using **P** we define a set of sequences of integers of the form

$$\overset{*}{x} = \{x_1, x_2, \ldots, x_j, \ldots\}, \qquad 0 \le x_j \le p_j - 1, \ j \ge 1. \tag{1.5.2}$$

This set becomes a group, which we shall denote by $G(\mathbf{P})$, if we use the binary operation

$$\overset{*}{x} \oplus \overset{*}{y} \equiv \{x_j \oplus y_j\}_{j=1}^{\infty}, \qquad x_j \oplus y_j = x_j + y_j \pmod{p_j}. \tag{1.5.3}$$

In the special case when the elements p_j of the sequence **P** are identically 2, then the group $G(\mathbf{P})$ coincides with the group G introduced in §1.2. Notice, however, that in contrast to the dyadic case, the operation $\ominus$ (which is the inverse of $\oplus$ on $G(\mathbf{P})$) is different from $\oplus$. In fact, this operation can be defined as coordinate subtraction modulo p_j, i.e.,

$$x_j \ominus y_j \equiv \begin{cases} x_j - y_j, & x_j \ge y_j, \\ p_j + x_j - y_j, & x_j < y_j. \end{cases}$$

We shall define a multiplicative system on the group $G(\mathbf{P})$ which will be indexed by the non-negative integers n. To do this we use the sequence **P** to write each non- negative integer n in **P**-adic form. This is a direct generalization of the dyadic expansion (1.1.3). First, set

$$m_0 = 1, \quad m_j = \prod_{s=1}^{j} p_s, \tag{1.5.4}$$

where p_s are the members of the sequence (1.5.1). Next, write each n in the form

$$n = \sum_{j=1}^{k} \alpha_j m_{j-1}, \qquad 0 \le \alpha_j \le p_j - 1, \ j = 1, 2, \ldots, k. \tag{1.5.5}$$

This will be called the **P**-adic expansion of n.

As in the dyadic case, each number n corresponds to an element $\overset{*}{n}$ of the group $G(\mathbf{P})$, namely, if n has **P**-adic expansion (1.5.5) then

$$\overset{*}{n} = \{\alpha_1, \alpha_2, \ldots, \alpha_k, 0, 0, \ldots\}.$$

This element is a finite sequence. As in §1.2, the map $n \to \overset{*}{n}$ allows us to transfer the group operation $\oplus$ from the group to the set of non-negative integers.

For each integer n with $\mathbf{P}$-adic expansion (1.5.5) and for sequence $\overset{*}{x}$ of the form (1.5.2), define the n-th function of the system $\{\overset{*}{\chi}_n(\overset{*}{x})\}_{n=0}^{\infty}$ by[3]

$$\overset{*}{\chi}_n(\overset{*}{x}) = \exp\left(2\pi i \sum_{j=1}^{\infty} \frac{\alpha_j x_j}{p_j}\right). \tag{1.5.6}$$

It is easy to see that these functions satisfy the equations[4]

$$\begin{cases} \overset{*}{\chi}_n(\overset{*}{x} \oplus \overset{*}{y}) = \overset{*}{\chi}_n(\overset{*}{x})\overset{*}{\chi}_n(\overset{*}{y}), \\ \overset{*}{\chi}_n(\overset{*}{x} \ominus \overset{*}{y}) = \overset{*}{\chi}_n(\overset{*}{x})\overline{\overset{*}{\chi}_n(\overset{*}{y})}, \end{cases} \tag{1.5.7}$$

and

$$\begin{cases} \overset{*}{\chi}_{n\oplus m}(\overset{*}{x}) = \overset{*}{\chi}_n(\overset{*}{x})\overset{*}{\chi}_m(\overset{*}{x}), \\ \overset{*}{\chi}_{n\ominus m}(\overset{*}{x}) = \overset{*}{\chi}_n(\overset{*}{x})\overline{\overset{*}{\chi}_m(\overset{*}{x})}, \end{cases} \tag{1.5.8}$$

analogous to equations (1.2.9) and (1.2.10) for the Walsh system. Notice that the system of functions (1.5.6) becomes the Walsh system $\{\overset{*}{w}_i\}$ in the particular case when $p_j = 2$ for all j.

As it was for the Walsh system, the domain of definition for the functions (1.5.6) can be transformed in a 1-1 fashion to a "modified" interval $[0,1]_{\mathbf{P}}^*$, or, if we relax the 1-1 condition on some countable set, can be transformed to the unit interval [0,1). The details are as follows. Notice that each sequence $\overset{*}{x}$ of the form (1.5.2) from the group $G(\mathbf{P})$ corresponds to a series

$$\sum_{j=1}^{\infty}(x_j/m_j), \tag{1.5.9}$$

where the m_j's are defined by equation (1.5.4). This series evidently converges and is the $\mathbf{P}$-adic expansion of some number x which equals the sum of this series. The transformation $\lambda_{\mathbf{P}} : \overset{*}{x} \to x = \sum_{j=1}^{\infty}(x_j/m_j)$ takes the group $G(\mathbf{P})$ onto the interval [0,1]. It is not 1-1 since each $\mathbf{P}$-adic rational has two expansions, a finite one and an infinite one. If we consider these two expansions as different points then, as in the dyadic case in §1.2, we obtain a "modified" interval $[0,1]_{\mathbf{P}}^*$, which gives a geometric interpretation of the group $G(\mathbf{P})$.

Corresponding to the system $\{\overset{*}{\chi}_n\}$ there is a multiplicative system of functions on the interval [0,1) defined analogously with the definition (1.2.12) which uses

[3] The systems described here, which in the literature are usually called Price systems (see [1], p. 68), do not exhaust the entire class of multiplicative systems.

[4] Hence the system $\{\overset{*}{\chi}_n(\overset{*}{x})\}_0^{\infty}$ is the character system for the group $G(\mathbf{P})$.

the transformation $g_{\mathbf{P}} : x \to \overset{*}{x}$. This transformation is defined so that each point $x \in [0,1)$ whose expansion is given by (1.5.9) corresponds to the sequence of the form (1.5.2) whose entries are the $\mathbf{P}$-adic coefficients of x. As before, we adhere to the convention that the expansion used for each $\mathbf{P}$-adic rational is the finite one. Thus for each n of the form (1.5.5) we have

$$\chi_n(x) = \overset{*}{\chi}_n(g_{\mathbf{P}}(x)) = \exp\left(2\pi i \sum_{j=1}^{\infty} \frac{\alpha_j x_j}{p_j}\right). \tag{1.5.10}$$

Let x be any real number. Notice that the $\mathbf{P}$-adic coefficients x_j of the expansion (1.5.9) of x can be computed by the formula

$$x_j = [x m_j] \pmod{p_j}, \qquad j \geq 1,$$

where for each real number a, $[a]$ represents the greatest integer in a. Moreover, for $\mathbf{P}$-adic rational points this formula gives the coefficients of the finite expansion. Similarly, the coefficients α_j of the expansion (1.5.5) of a natural number n can be computed by the formula

$$\alpha_j = \left[\frac{n}{m_{j-1}}\right] \pmod{p_j}, \qquad j \geq 1.$$

It is clear that the functions

$$\chi_{m_{j-1}}(x) = \exp\left(2\pi i \frac{x_j}{p_j}\right), \qquad j = 1, 2, \ldots, \tag{1.5.11}$$

play a role here analogous to that played by the Rademacher functions in the definition of the Walsh system. In particular, we can write each $\chi_n(x)$ in the form

$$\chi_n(x) = \prod_{j=1}^{k} (\chi_{m_{j-1}}(x))^{\alpha_j}. \tag{1.5.12}$$

It is easy to check for $n < m_k$ that the function $\chi_n(x)$ is constant on intervals of the form

$$\delta_r^{(k)} = \left[\frac{r}{m_k}, \frac{r+1}{m_k}\right), \qquad 0 \leq r \leq m_k - 1. \tag{1.5.13}$$

These intervals are $\mathbf{P}$-adic analogues of the dyadic intervals $\Delta_r^{(k)}$. On the group $G(\mathbf{P})$ or on the modified interval $[0,1]^*_{\mathbf{P}}$ they correspond to the sets

$$\overset{*}{\delta}{}_r^{(k)} = \{\overset{*}{x} : x_j = x_j^0,\ j = 1, 2, \ldots, k\},$$

where $r = \sum_{j=1}^{k} x_j^0 m_{k-j}$.

Notice once and for all that

$$\delta_r^{(k-1)} = \bigcup_{s=rp_k}^{(r+1)p_k-1} \delta_s^{(k)} \tag{1.5.14}$$

and that

$$x_k = s \pmod{p_k}, \qquad x \in \delta_s^{(k)}. \tag{1.5.15}$$

As in §1.2 we can transfer the operation $\oplus$ from the group $G(\mathbf{P})$ to the unit interval [0,1). Moreover, for each fixed y the identity

$$\chi_n(x \oplus y) = \chi_n(x)\chi_n(y) \tag{1.5.16}$$

holds for all but countably many points x in [0,1). (The proof of this fact is similar to that of **1.2.2.**)

We shall make several more observations about the system $\{\chi_n(x)\}$.

By using formula (1.5.11) and summing the resulting geometric series, it is not difficult to verify

$$\sum_{q=0}^{p_k-1} (\chi_{m_{k-1}}(x))^q = \begin{cases} p_k & \text{for } x \in \delta_{rp_k}^{(k)},\ r = 0, 1, \dots, m_{k-1}-1, \\ 0 & \text{for } x \in \delta_r^{(k-1)} \setminus \delta_{rp_k}^{(k)}. \end{cases} \tag{1.5.17}$$

The system $\{\chi_n(x)\}$ is orthonormal. This will follow from (1.5.8), (1.5.10) when we establish

$$\int_{\delta_r^{(k)}} \chi_n(x)\, dx = 0, \qquad m_{k-1} \le n < m_k,\ 0 \le r \le m_k - 1. \tag{1.5.18}$$

To prove (1.5.18) notice for $j < k$ that the function $\chi_{m_{j-1}}(x)$ is constant on $\delta_r^{(k-1)}$ and has modulus 1 there. Thus by (1.5.12) we have

$$\left| \int_{\delta_r^{(k-1)}} \chi_n(x)\, dx \right| = \left| \int_{\delta_r^{(k-1)}} (\chi_{m_{k-1}}(x))^{\alpha_k}\, dx \right|.$$

Since $\chi_{m_{k-1}}(x)$ is constant on the intervals $\delta_s^{(k)}$, we see by (1.5.14), (1.5.11), and (1.5.15) that this last integral reduces to the sum

$$\frac{1}{|\delta_s^{(k)}|} \sum_{q=0}^{p_k-1} \exp\left(2\pi i \frac{q}{p_k} \alpha_k\right) = 0.$$

This proves (1.5.18) and shows that the system $\{\chi_n(x)\}$ is orthonormal.

The Dirichlet kernels for the system $\{\chi_n(x)\}$ will share the same notation we used for the Walsh system:

$$D_n(x) = \sum_{s=0}^{n-1} \chi_s(x). \tag{1.5.19}$$

(No confusion will arise from this choice of notation. It will be clear from the context which system we are talking about.)

Analogous to (1.4.11), one obtains immediately from definitions (1.5.10) and (1.5.12) that

$$D_n(x) = D_{\alpha_k m_{k-1}}(x) + (\chi_{m_{k-1}}(x))^{\alpha_k} D_r(x),$$

for $n = \alpha_k m_{k-1} + r$, $1 \leq r \leq m_{k-1}$. Here, as before, we have set $D_0(x) \equiv 0$. Since

$$D_{\alpha_k m_{k-1}}(x) = \sum_{q=0}^{\alpha_k - 1} (\chi_{m_{k-1}}(x))^q D_{m_{k-1}}(x),$$

it follows that

$$D_n(x) = D_{m_{k-1}}(x) \sum_{q=0}^{\alpha_k - 1} (\chi_{m_{k-1}}(x))^q + (\chi_{m_{k-1}}(x))^{\alpha_k} D_r(x) \tag{1.5.20}$$

for $n = \alpha_k m_{k-1} + r$ and $1 \leq r \leq m_{k-1}$. In particular, by setting $\alpha_k = p_k - 1$ and $r = m_{k-1}$ we obtain

$$D_{m_k}(x) = D_{m_{k-1}}(x) \sum_{q=0}^{p_k - 1} (\chi_{m_{k-1}}(x))^q.$$

We have therefore by induction and (1.5.17) that

$$D_{m_k}(x) = \begin{cases} m_k & \text{for } x \in \delta_0^{(k)}, \\ 0 & \text{for } x \in [0,1) \setminus \delta_0^{(k)}. \end{cases} \tag{1.5.21}$$

We shall now pass to the construction of a continual analogue of multiplicative systems, that is a system whose index set is a continuum.

Notice that the system of functions (1.5.6) can be viewed as a single function of two variables, namely, $\overset{*}{\chi}(\overset{*}{x}, \overset{*}{n}) \equiv \chi_n(\overset{*}{x})$. From this point of view, the role of both variables in definition (1.5.6) is similar, and the only difference between them is that the second variable does not take values from the whole group $G(\mathbf{P})$, but only from the countable subgroup consisting of finite sequences.

We are interested in a generalization of the function $\overset{*}{\chi}(\overset{*}{x}, \overset{*}{n})$ in which the second variable ranges over a continuum of points instead of the discrete set of points in $G(\mathbf{P})$ consisting of finite sequences. This continuum of points must itself also be a group.

We begin by constructing a class of "continuum" groups whose cartesian products will form the region of definition of the function $\overset{*}{\chi}$. This function can be viewed as an extension of $\overset{*}{\chi}(\overset{*}{x}, \overset{*}{n})$ and thus a generalization of multiplicative systems.

Let $\mathcal{P}$ be an arbitrary doubly infinite sequence of natural numbers of the form

$$\mathcal{P} \equiv \{\dots, p_{-j}, \dots, p_{-1}, p_1, p_2, \dots, p_j, \dots\}, \tag{1.5.22}$$

where $p_j \geq 2$ for $j = \pm 1, \pm 2, \dots$. (For convenience we have omitted the index 0 here.) Define a group $G(\mathcal{P})$ as the set of sequences of the form

$$\overset{*}{x} = \{\dots, x_{-j}, \dots, x_{-1}, x_1, x_2, \dots, x_j, \dots\}, \tag{1.5.23}$$

where $0 \leq x_j \leq p_j - 1$ for $j = \pm 1, \pm 2, \dots$ and $x_{-j} = 0$ for $j > k(\overset{*}{x}) \geq 1$. Define a group operation $\oplus$ on $G(\mathcal{P})$ by using (1.5.3) with one difference: the index j takes on all integer values except zero. Clearly, the group $G(\mathcal{P})$ consists of sequences which are only infinite to the right.

Let $\mathcal{P}'$ represent the reverse sequence of $\mathcal{P}$, i.e.,

$$\mathcal{P}' \equiv \{\dots, p'_{-j}, \dots, p'_{-1}, p'_1, p'_2, \dots, p'_j, \dots\} \tag{1.5.24}$$

where

$$p'_j = p_{-j}, \qquad j = \pm 1, \pm 2, \dots. \tag{1.5.25}$$

Thus the group $G(\mathcal{P}')$ consists of sequences of the form

$$\overset{*}{x}' = \{\dots, x'_{-j}, \dots, x'_{-1}, x'_1, x'_2, \dots, x'_j, \dots\}, \tag{1.5.26}$$

where $0 \leq x'_j \leq p_{-j} - 1$ for $j = \pm 1, \pm 2, \dots$ and $x'_{-j} = 0$ for $j > k(\overset{*}{x}') \geq 1$.

Define a function[5] of two variables $\overset{*}{\chi}(\overset{*}{x}, \overset{*}{x}')$ for each $(\overset{*}{x}, \overset{*}{x}') \in G(\mathcal{P}) \times G(\mathcal{P}')$ by

$$\overset{*}{\chi}(\overset{*}{x}, \overset{*}{x}') = \exp\left\{2\pi i \left(\sum_{j=1}^{k(\overset{*}{x}')} \frac{x_j x'_{-j}}{p_j} + \sum_{j=1}^{k(\overset{*}{x})} \frac{x_{-j} x'_j}{p_{-j}}\right)\right\}. \tag{1.5.27}$$

[5] $\overset{*}{\chi}(\overset{*}{x}, \overset{*}{x}')$ as a function of the first variable is a character of the locally compact group $G(\mathcal{P})$.

Let

$$(1.5.28)\qquad m_0 \equiv 1,\ m_j \equiv \prod_{s=1}^{j} p_s,\ m_{-j} \equiv \prod_{s=1}^{j} p_{-s}, \qquad j = 1, 2, \ldots.$$

In view of (1.5.25) we have

$$m'_0 = m_0 = 1,\ m'_j \equiv \prod_{s=1}^{j} p'_s = m_{-j},\ m'_{-j} \equiv \prod_{s=1}^{j} p'_{-s} = m_j$$

for $j = 1, 2, \ldots$.

We shall describe transformations which take the groups $G(\mathcal{P})$ and $G(\mathcal{P}')$ into the set of real numbers. In contrast to the groups $G(\mathbf{P})$ which were by (1.5.9) identified with the unit interval, these groups $G(\mathcal{P})$ and $G(\mathcal{P}')$ will be identified with the positive real axis. To accomplish this, correspond each sequence (1.5.23) to a series of the form

$$(1.5.29)\qquad \sum_{j=1}^{k(\overset{*}{x})} x_{-j} m_{-j+1} + \sum_{j=1}^{\infty} \frac{x_j}{m_j},$$

and each sequence (1.5.26) to a series of the form

$$(1.5.30)\qquad \sum_{j=1}^{k(\overset{*}{x}')} x'_{-j} m_{j-1} + \sum_{j=1}^{\infty} \frac{x'_j}{m_{-j}}.$$

The transformation $\lambda_{\mathcal{P}}$ which takes an element $\overset{*}{x} \in G(\mathcal{P})$ to the sum of the series (1.5.29) (respectively, the transformation $\lambda_{\mathcal{P}'}$ which takes an element $\overset{*}{x}' \in G(\mathcal{P}')$ to the sum of the series (1.5.30)) is a map from the group $G(\mathcal{P})$ (respectively, $G(\mathcal{P}')$) onto the positive real axis $[0, \infty)$. Moreover, these transformations fail to be 1-1 only at $\mathcal{P}$-adic rationals (respectively, $\mathcal{P}'$-adic rationals). If, in analogy with the modified interval $[0, 1)^*_{\mathbf{P}}$, we form the modified rays $[0, \infty)^*_{\mathcal{P}}$ and $[0, \infty)^*_{\mathcal{P}'}$ in which each $\mathcal{P}$- adic rational (respectively, $\mathcal{P}'$-adic rational) has been split into two points, then the groups can be mapped in a 1-1 fashion onto these modified rays. Thus we obtain a suitable geometric model for the groups $G(\mathcal{P})$ and $G(\mathcal{P}')$.

The usual ray $[0, \infty)$ can be mapped in a 1-1 fashion into the groups $G(\mathcal{P})$ and $G(\mathcal{P}')$ by transformations $g_{\mathcal{P}}$ and $g_{\mathcal{P}'}$. These transformations are defined in the following way. Let $g_{\mathcal{P}} : x \to \{x_j\}_{\pm j=1}^{\infty}$ and $g_{\mathcal{P}'} : x' \to \{x'_j\}_{\pm j=1}^{\infty}$, where x_j and x'_j are the coefficients of the corresponding expansions: (1.5.29) for x, and (1.5.30) for x'. Here, as before, we agree to take the finite expansion when x is a $\mathcal{P}$-adic rational or x' is a $\mathcal{P}'$-adic rational.

These coefficients are determined by the equations

$$(1.5.31)\qquad \begin{cases} x_j = [xm_j] \pmod{p_j}, & x_{-j} = [x/m_{1-j}] \pmod{p_{-j}}, \\ x'_j = [xm_{-j}] \pmod{p_{-j}}, & x'_{-j} = [x'/m_{j-1}] \pmod{p_j} \end{cases}$$

for $j = 1, 2, \dots$, where $[a]$ represents the greatest integer in a.

Thus,

$$(1.5.32)\qquad \begin{cases} x = \sum_{j=1}^{k(x)} x_{-j}m_{1-j} + \sum_{j=1}^{\infty} x_j/m_j \equiv [x] + \{x\}, \\ x' = \sum_{j=1}^{k(x')} x'_{-j}m_{j-1} + \sum_{j=1}^{\infty} x'_j/m_{-j} \equiv [x'] + \{x'\}. \end{cases}$$

(Here, $\{a\}$ represents the fractional part of a number a.) Using this notation, we define a function on $[0,\infty) \times [0,\infty)$ by

$$(1.5.33)\quad \chi(x, x') = \overset{*}{\chi}(g_{\mathcal{P}}(x), g_{\mathcal{P}'}(x')) = \exp\left\{ 2\pi i \left(\sum_{j=1}^{k(x')} \frac{x_j x'_{-j}}{p_j} + \sum_{j=1}^{k(x)} \frac{x'_j x_{-j}}{p_{-j}} \right) \right\}.$$

We shall make a number of observations about the function $\chi(x, x')$ and the corresponding function $\overset{*}{\chi}(\overset{*}{x}, \overset{*}{x}')$.

1.5.1. $\chi(0, x') = \chi(x, 0) = |\chi(x, x')| = 1$. *Similarly, if* $0_{\mathcal{P}}$ *is the zero element of the group* $G(\mathcal{P})$ *and* $0_{\mathcal{P}'}$ *is the zero element of the group* $G(\mathcal{P}')$ *then* $\overset{*}{\chi}(0_{\mathcal{P}}, \overset{*}{x}') = \overset{*}{\chi}(x, 0_{\mathcal{P}'}) = |\overset{*}{\chi}(\overset{*}{x}, \overset{*}{x}')| = 1$.

1.5.2. *Let* $\mathcal{P}$ *be a sequence of the form (1.5.22) and consider the sequences* $\mathbf{P} = \{p_j\}_{j=1}^{\infty}$ *and* $\mathbf{P}' = \{p'_j\}_{j=1}^{\infty}$, *where* $p'_j = p_{-j}$. *If* $x' = n$ *for some non-negative integer* n *then* $\chi(x, n) = \chi_n(\{x\})_{(\mathbf{P})}$, *where* $(\chi_n(x))_{(\mathbf{P})}$ *is the multiplicative system of type (1.5.10) determined by the sequence* $\mathbf{P}$. *If* $x = n$ *for some non-negative integer* n *then* $\chi(n, x') = \chi_n(\{x'\})_{(\mathbf{P}')}$, *where* $(\chi_n(x))_{(\mathbf{P}')}$ *is the multiplicative system of type (1.5.10) determined by the sequence* $\mathbf{P}'$.

(Similar properties hold also for the function $\overset{*}{\chi}(\overset{*}{x}, \overset{*}{x}')$.)

PROOF. These facts follow directly from the definitions, since if the fractional part of one of the variables in (1.5.33) is zero then the value of the function does not depend on the integer part of the other variable. ∎

1.5.3. *Using the notation introduced in* **1.5.2**,

$$\chi(x, x') = \chi(\{x\}, [x']) \cdot \chi([x], \{x'\}) = \chi_{[x']}(\{x\})_{(\mathbf{P})} \cdot \chi_{[x]}(\{x'\})_{(\mathbf{P}')}.$$

(This identity shows that $\chi(x, x')$ is the *cross product* system of $\chi_n(\{x'\})_{\mathbf{P}'}$ and $\chi_m(\{x\})_{\mathbf{P}}$. (See Vilenkin, Zotikov [1].)

PROOF. To verify this identity it is enough to write (1.5.33) in the form

$$\chi(x, x') = \exp\left(2\pi i \sum_{j=1}^{k(x')} \frac{x_j x'_{-j}}{p_j} \right) \cdot \exp\left(2\pi i \sum_{j=1}^{k(x)} \frac{x'_j x_{-j}}{p_{-j}} \right)$$

and combine (1.5.32) with **1.5.2**. ∎

1.5.4. *Let $k \geq 1$. For each fixed $x' < m_k$ (where m_k is defined by (1.5.28)), the function $\chi(x, x')$ is constant in x on intervals of the form (1.5.13), $r = 0, 1, \ldots$. For each fixed $x < m_{-k}$, the function $\chi(x, x')$ is constant in x' on intervals of the form*

$$\delta'^{(k)}_r = [r/m_{-k}, (r+1)/m_{-k}), \qquad r = 0, 1, \ldots.$$

PROOF. Fix x' and observe that the second factor in the statement of **1.5.3** is constant in x on the interval $[[x], [x]+1)$. Similarly, the first factor is by **1.5.2** constant on each $\delta^{(k)}_r$. This proves the first half of **1.5.4**. The second half is proved in a similar way. ∎

1.5.5. *The function $\overset{*}{\chi}(\overset{*}{x}, \overset{*}{x}')$ is multiplicative in each of its variables, i.e.,*

$$\overset{*}{\chi}(\overset{*}{x}, \overset{*}{y})\overset{*}{\chi}(\overset{*}{x}, \overset{*}{z}) = \chi(\overset{*}{x}, \overset{*}{y} \oplus \overset{*}{z}), \quad \overset{*}{\chi}(\overset{*}{x}, \overset{*}{y})\overset{*}{\chi}(\overset{*}{z}, \overset{*}{y}) = \chi(\overset{*}{x} \oplus \overset{*}{z}, \overset{*}{y}),$$

and

$$\overset{*}{\chi}(\overset{*}{x}, \overset{*}{y})\overline{\overset{*}{\chi}(\overset{*}{x}, \overset{*}{z})} = \chi(\overset{*}{x}, \overset{*}{y} \ominus \overset{*}{z}), \quad \overset{*}{\chi}(\overset{*}{x}, \overset{*}{y})\overline{\overset{*}{\chi}(\overset{*}{z}, \overset{*}{y})} = \chi(\overset{*}{x} \ominus \overset{*}{z}, \overset{*}{y}).$$

PROOF. This result follows directly from definition (1.5.27). ∎

Notice that these identities generalize the identities (1.5.7) and (1.5.8).

An analogue of **1.5.5** holds for the function $\chi(x, x')$ with the same reservations concerning the domain that were imposed for equation (1.5.16). Specifically, for each fixed x and y

$$\chi(x, y)\chi(z, y) = \chi(x \oplus z, y), \quad \chi(x, y)\overline{\chi(z, y)} = \chi(x \ominus z, y) \tag{1.5.34}$$

hold for all but countably many z in $[0, \infty)$.

For certain applications involving the multiplicative function $\chi(x, x')$, it is important to impose an additional symmetry assumption on the sequence (1.5.22), namely that

$$p_{-j} = p_j, \qquad j = 1, 2, \ldots. \tag{1.5.35}$$

In this case the group $G(\mathcal{P}')$ coincides with the group $G(\mathcal{P})$, and clearly both groups are completely determined by the sequence $\mathbf{P} = (p_1, p_2, \ldots, p_j, \ldots)$. In addition to properties **1.5.1** - **1.5.5**, the corresponding functions $\overset{*}{\chi}(\overset{*}{x}, \overset{*}{x}')$ and $\chi(x, x')$ also satisfy the following symmetry conditions:

$$\overset{*}{\chi}(\overset{*}{x}, \overset{*}{x}') = \overset{*}{\chi}(\overset{*}{x}', \overset{*}{x}), \quad \chi(x, x') = \chi(x', x) \tag{1.5.36}$$

As an analogue of Dirichlet kernels for the multiplicative function $\chi(x, x')$, we introduce the kernel

$$D_{x'}(x) \equiv D(x, \check{x}') \equiv \int_0^{x'} \chi(x, t)\, dt. \tag{1.5.37}$$

(The sign ˘ appears above the variable over which we integrate.)

The connection between the Dirichlet kernel $D(x, \breve{x}')$ and the kernel $D_n(x)_{(\mathbf{P})}$ defined in (1.5.19) is given by the following formula:

$$D(x, \breve{x}') = \begin{cases} D_{[x']}(x)_{(\mathbf{P})} + \{x'\}\chi_{[x']}(x)_{(\mathbf{P})} & \text{for } 0 \leq x < 1, \\ \chi_{[x']}(\{x\})_{(\mathbf{P})} \int_0^{\{x'\}} \chi_{[x]}(t)_{(\mathbf{P}')}\, dt & \text{for } x \geq 1. \end{cases} \tag{1.5.38}$$

We shall prove this formula. For $0 \leq x < 1$ we use **1.5.3** and **1.5.1** to obtain

$$\begin{aligned} D(x, \breve{x}') &= \int_0^{x'} \chi_{[t]}(\{x\})_{(\mathbf{P})} \chi_{[x]}(\{t\})_{(\mathbf{P}')}\, dt \\ &= \int_0^{[x']} \chi_{[t]}(x)_{(\mathbf{P})}\, dt + \int_{[x']}^{[x']+\{x'\}} \chi_{[t]}(x)_{(\mathbf{P})}\, dt \\ &= \sum_{n=0}^{[x']-1} \left(\int_n^{n+1} \chi_n(x)_{(\mathbf{P})}\, dt \right) + \{x'\}\chi_{[x']}(x)_{(\mathbf{P})} \\ &= D_{[x']}(x)_{(\mathbf{P})} + \{x'\}\chi_{[x']}(x)_{(\mathbf{P})}. \end{aligned}$$

On the other hand, let $x \geq 1$. Since

$$\int_n^{n+1} \chi_k(\{t\})_{(\mathbf{P}')}\, dt = 0$$

for $k = 1, 2, \ldots,\ n = 0, 1, 2, \ldots,$ we have

$$\begin{aligned} D(x, \breve{x}') &= \int_0^{[x']} \chi_{[t]}(\{x\})_{(\mathbf{P})} \chi_{[x]}(\{t\})_{(\mathbf{P}')}\, dt \\ &\qquad + \int_{[x']}^{[x']+\{x'\}} \chi_{[t]}(\{x\})_{(\mathbf{P})} \chi_{[x]}(\{t\})_{(\mathbf{P}')}\, dt \\ &= \sum_{n=0}^{[x']-1} \left(\int_n^{n+1} \chi_n(\{x\})_{(\mathbf{P})} \chi_{[x]}(\{t\})_{(\mathbf{P}')}\, dt \right) \\ &\qquad + \int_{[x']}^{[x']+\{x'\}} \chi_{[x']}(\{x\})_{(\mathbf{P})} \chi_{[x]}(\{t\})_{(\mathbf{P}')}\, dt \\ &= \chi_{[x']}(\{x\})_{(\mathbf{P})} \int_{[x']}^{[x']+\{x'\}} \chi_{[x]}(\{t\})_{(\mathbf{P}')}\, dt \\ &= \chi_{[x']}(\{x\})_{(\mathbf{P})} \int_0^{\{x'\}} \chi_{[x]}(t)_{(\mathbf{P}')}\, dt. \end{aligned}$$

This proves formula (1.5.38).

If we interchange the roles of x and x' we obtain another generalization of the Dirichlet kernels for the function $\chi(x, x')$:

$$D_x(x') \equiv D(\check{x}, x') \equiv \int_0^x \chi(t, x')\, dt, \tag{1.5.39}$$

Here x plays the role of an index and x' that of a variable. For this kernel, the formula analogous to (1.5.38) turns out to be:

$$D(\check{x}, x') = \begin{cases} D_{[x]}(x')_{(\mathbf{P}')} + \{x\}\chi_{[x]}(x')_{(\mathbf{P}')} & \text{for } 0 \le x' < 1, \\ \chi_{[x]}(\{x'\})_{(\mathbf{P}')} \int_0^{\{x\}} \chi_{[x']}(t)_{(\mathbf{P})}\, dt & \text{for } x' \ge 1, \end{cases}$$

where $D_{[x]}(x')_{(\mathbf{P}')}$ is the kernel of the form (1.5.19) for the system $\{\chi_n(x')\}_{(\mathbf{P}')}$.

We close this chapter with the following theorem.

1.5.6. *For each $k \ge 0$ the systems*

$$\left\{\psi_{\nu k}(y) \equiv \frac{1}{\sqrt{m_k}}\chi\left(\frac{\nu}{m_k}, y\right)\right\}_{\nu=0}^{\infty}$$

and

$$\left\{\phi_{\nu k}(x) \equiv \frac{1}{\sqrt{m_{-k}}}\chi\left(x, \frac{\nu}{m_{-k}}\right)\right\}_{\nu=0}^{\infty}$$

are uniformly bounded and orthonormal on the respective intervals $[0, m_k)$ and $[0, m_{-k})$.

PROOF. We shall verify the result for the first system. The proof for the second system is similar.

Clearly,

$$\int_0^{m_k} \chi\left(\frac{\nu}{m_k}, y\right) \overline{\chi\left(\frac{\nu}{m_k}, y\right)}\, dy = \int_0^{m_k} \left|\chi\left(\frac{\nu}{m_k}, y\right)\right|^2 dy = m_k$$

for each ν. If $\nu \ne \mu$ then by the second equation in (1.5.34) and identities (1.5.37) and (1.5.38) we have

$$\begin{aligned} \int_0^{m_k} \chi\left(\frac{\nu}{m_k}, y\right) \overline{\chi\left(\frac{\mu}{m_k}, y\right)}\, dy &= \int_0^{m_k} \chi\left(\frac{\nu}{m_k} \ominus \frac{\mu}{m_k}, y\right) dy \\ &= D\left(\frac{\nu}{m_k} \ominus \frac{\mu}{m_k}, \check{m}_k\right) \\ &= \begin{cases} D_{m_k}(\nu/m_k \ominus \mu/m_k)_{(\mathbf{P})} & \text{for } 0 \le \nu/m_k \ominus \mu/m_k < 1, \\ 0 & \text{for } \nu/m_k \ominus \mu/m_k \ge 1. \end{cases} \end{aligned}$$

But $\nu \neq \mu$ so $\nu/m_k \ominus \mu/m_k \geq 1/m_k$. In particular, $\nu/m_k \ominus \mu/m_k \notin \delta_0^k$. Thus by (1.5.21) $D_{m_k}(\nu/m_k \ominus \mu/m_k) = 0$. Consequently,

$$\int_0^{m_k} \chi\left(\frac{\nu}{m_k}, y\right) \overline{\chi\left(\frac{\mu}{m_k}, y\right)}\, dy = 0, \qquad \nu \neq \mu.$$

Since the uniform boundedness of this system is obvious, the proof of this theorem is complete. ▮

Chapter 2

WALSH-FOURIER SERIES. BASIC PROPERTIES.

Within the collection of all Walsh series, Walsh-Fourier series play a crucial role. These are the series of the form (1.4.1) whose coefficients are given by the formula

$$a_i = \int_0^1 f(t)w_i(t)\,dt,$$

for some integrable function f. From this definition it is clear that the concept of a Fourier series is intimately connected with the theory of measures and integrals.

For the case considered in §1.2, namely the system $\{\overset{*}{w}_n(\overset{*}{x})\}$ defined on the group G, the formula for the Fourier coefficients is similar:

$$a_i = \int_G f(\overset{*}{t})\overset{*}{w}_i(\overset{*}{t})\,d\mu.$$

Here the μ represents Haar measure for the group G (see **A3.5**) and the integral is meant in the sense of Lebesgue. This integral satisfies a fundamental property which is extremely important for the theory of Fourier series, namely it is translation invariant, i.e.,

$$\int_G f(\overset{*}{t} \oplus \overset{*}{x})\,d\mu(\overset{*}{t}) = \int_G f(\overset{*}{t})\,d\mu(\overset{*}{t})$$

for any integrable function f and any element $\overset{*}{x} \in G$. This property, on which many results of Fourier series are based, once again emphasizes that the most natural domain on which to define the Walsh functions is the group G. The theory of Walsh- Fourier series is more elegant on the group G than on the real line or the unit interval because results can be formulated on the group with less restrictions than are necessary to formulate the same results on the unit interval. This is due in part to the fact that the unit interval is not a group under the operation $\oplus$. But, the theory in the group setting requires mastery of certain techniques including the concept of Haar measure, and other closely related ideas. In connection with this, the language of the theory of Walsh-Fourier series on the group may seem at first unnecessarily abstract and perhaps unusual for the uninitiated reader who is interested most of all in the applications.

For this reason we have decided to set forth here the foundations of the theory of Walsh series on the more intuitive version of the Walsh system which is defined on the unit interval [0,1). The reader who has mastered the concepts mentioned above is urged while studying these results about series in the system $\{w_n(x)\}$ to constantly have in view the parallel results about series in the system $\{\overset{*}{w}_n(\overset{*}{x})\}$. In this light, he should interpret $w_n(x)$ as $\overset{*}{w}_n(\overset{*}{x})$, the dyadic intervals $\Delta_j^{(k)}$ as the equivalence classes $\overset{*}{\Delta}{}_j^{(k)}$ of the group G, and all integrals as integrals with respect to Haar measure on G.

In those few cases when translation to the language of the group is not merely mechanical, it will be made sufficiently clear.

§2.1. Elementary properties of Walsh-Fourier series. Formulae for partial sums.

As we mentioned above, the Walsh-Fourier series of a function f, Lebesgue integrable on [0,1), is a series of the form (1.4.1) whose coefficients are given by

$$a_i = \int_0^1 f(t)w_i(t)\,dt. \tag{2.1.1}$$

We notice at once that every uniformly convergent Walsh series must be a Walsh-Fourier series. This follows from the fact that a uniformly convergent series can be integrated term by term and the fact that the Walsh system is orthonormal (see Theorem **1.1.5**). In particular, any polynomial can be viewed as the Fourier series of its sum.

We shall presently see that on the unit interval the Lebesgue integral is *translation invariant* with respect to the operation $\oplus$, i.e.,

$$\int_0^1 f(t \oplus x)\,dt = \int_0^1 f(t)\,dt. \tag{2.1.2}$$

This property plays an important role in the study of Fourier series. (We note for each fixed x that the integrand on the left side of (2.1.2) fails to be defined at countably many points t (see §1.2). However, countable sets are of Lebesgue measure zero and the Lebesgue integral cannot distinguish between functions which differ only on a set of measure zero.)

To prove (2.1.2) we begin by showing that Lebesgue measure is translation invariant under the operation $\oplus$, i.e., that for any Lebesgue measurable set $E \subset [0,1)$ and any $x \in [0,1)$ the set $E \oplus x \equiv \{t \oplus x : t \in E\}$ is measurable and

$$\operatorname{mes}(E \oplus x) = \operatorname{mes}(E). \tag{2.1.3}$$

(In the group case this property is built into the construction of Haar measure.) Recall from Theorem **1.2.1** that for each positive integer k, the map $\Delta^{(k)} \to \Delta^{(k)} \oplus x$ is a permutation on the collection of dyadic intervals of rank k. Thus if E is a union of non-overlapping dyadic intervals then $E \oplus x$ is also a union of non-overlapping dyadic intervals (except countably many points) and thus its measure does not change. Since any open set can be written as a countable union of dyadic intervals,

(2.1.3) holds for any open subset of [0,1). In view of the definition of measurable sets (see **A3.2.3**) it follows that (2.1.3) holds for all measurable sets E.

It is now clear that (2.1.2) holds for characteristic functions of any measurable set $E \subset [0,1)$ since

$$\int_0^1 \chi_E(t)\,dt = \text{mes}(E).$$

Hence (2.1.2) holds for all simple functions, i.e., for functions of the form $\sum_{i=1}^{\infty} c_i \chi_{E_i}$ where E_i are non- overlapping and measurable. Finally, since the Lebesgue integral is defined in terms of limits of integrals of simple functions (see **A4.2**), we conclude that (2.1.2) holds for any integrable function.

Before we isolate several fundamental properties enjoyed by Walsh- Fourier series, we introduce additional notation.

Analogous to the trigonometric case, the Fourier coefficients (2.1.1) will frequently be denoted by $\widehat{f}(i)$. Thus we shall write the Walsh-Fourier series of a function f in the form

$$\sum_{i=0}^{\infty} \widehat{f}(i) w_i(x), \tag{2.1.4}$$

where

$$\widehat{f}(i) = \int_0^1 f(t) w_i(t)\,dt. \tag{2.1.5}$$

We shall use the symbol $\widehat{f}$ to represent the sequence $\{\widehat{f}(i)\}_{i=0}^{\infty}$ of Fourier coefficients of a function f, i.e., the map $i \to \widehat{f}(i)$ which takes the set of non-negative integers into the set of real numbers. We shall also consider $\widehat{f}$ as the image of a function f under the map $f \to \widehat{f}$ defined on the set of functions integrable on [0,1). We shall call this transformation the *Walsh-Fourier transformation* for the interval $\widehat{[0,1)}$. Frequently the transformation $f \to \widehat{f}$ will be denoted simply by the symbol $\widehat{f}$.

We point out several properties which this transformation satisfies.

2.1.1. *The transformation $\widehat{f}$ is linear, i.e., if f, g are integrable functions and α, β are real numbers then*

$$(\alpha f + \beta g)\widehat{\ } = \alpha \widehat{f} + \beta \widehat{g}.$$

PROOF. This property follows immediately from definition (2.1.5). ∎

2.1.2. *If $f_a(t) \equiv f(t \oplus a)$ then*

$$\widehat{f_a}(i) = w_n(a)\widehat{f}(i).$$

PROOF. Since the integral is translation invariant under $\oplus$ and since $(t\oplus a)\oplus a = t$, we see by identity (1.2.15) that

$$\begin{aligned}\widehat{f_a}(i) &= \int_0^1 f(t\oplus a)w_i(t)\,dt \\ &= \int_0^1 f(t)w_i(t\oplus a)\,dt \\ &= w_i(a)\int_0^1 f(t)w_i(t)\,dt \\ &= w_i(a)\widehat{f}(i). \quad \blacksquare\end{aligned}$$

Proposition **2.1.1** shows that under the Walsh-Fourier transformation, the sum of two functions corresponds to the sum of their Fourier coefficients. It is natural to ask whether there is an operation of two functions which corresponds to the product of their Fourier coefficients. In other words, given two integrable functions f and g can we find a third function ϕ such that $\widehat{\phi}(i) = \widehat{f}(i)\cdot\widehat{g}(i)$ for all $i = 0, 1, 2, \ldots$. By looking at polynomials, it is easy to see that such a function ϕ will not be obtained by multiplying the functions f and g. However, such a function can be obtained from a generalized product $*$ which is called the *convolution*. The convolution of two functions f and g is defined by the identity

$$\phi(x) = (f*g)(x) \equiv \int_0^1 f(t\oplus x)g(t)\,dt. \tag{2.1.6}$$

It is not at all obvious that the integral in (2.1.6) exists and defines an integrable function. We shall show that this function is defined at least for almost every x in [0,1) (i.e., for all x except those in some set of measure zero), is Lebesgue integrable on [0,1), and satisfies the inequality

$$\int_0^1 |(f*g)(x)|\,dx \le \int_0^1 |f(x)|\,dx \int_0^1 |g(x)|\,dx. \tag{2.1.7}$$

If we denote the norm of the space $\mathbf{L}([0,1))$ of integrable functions by $\|\cdot\|_1$ (see **A5.2**), then (2.1.7) can be written in the form

$$\|f*g\|_1 \le \|f\|_1\,\|g\|_1.$$

Since

$$\int_0^1 |(f*g)(x)|\,dx \le \int_0^1\int_0^1 |f(t\oplus x)g(t)|\,dt\,dx = \int_0^1 |(|f|*|g|)(x)|\,dx, \tag{2.1.8}$$

and since each integrable function is almost everywhere finite, it suffices to show that the integral $\int_0^1 (|f| * |g|)(x)\, dx$ is finite. But by Fubini's Theorem (see **A4.4.4**) translation invariance of the Lebesgue integral we see that

$$\begin{aligned}\int_0^1 \int_0^1 |f(t \oplus x)|\, |g(t)|\, dt\, dx &= \int_0^1 \int_0^1 |f(t \oplus x)|\, |g(t)|\, dx\, dt \\ &= \int_0^1 |g(t)| \left(\int_0^1 |f(t \oplus x)|\, dx \right) dt \\ &= \int_0^1 |g(t)| \left(\int_0^1 |f(x)|\, dx \right) dt \\ &= \int_0^1 |f(x)|\, dx \int_0^1 |g(t)|\, dt.\end{aligned}$$

Hence the integral in question is finite. Moreover, if we combine this inequality with (2.1.8) we conclude that (2.1.7) holds.

We shall now prove that if $\phi(x) = (f * g)(x)$ then

$$\widehat{\phi}(i) = \widehat{f}(i)\widehat{g}(i), \qquad i = 0, 1, 2, \ldots. \tag{2.1.9}$$

This identity can be obtained by several applications of Fubini's Theorem (see **A4.4.4**). To apply Fubini's Theorem it is necessary to show, as we did above, that the integrand is absolutely integrable on the square $[0,1) \times [0,1)$. Thus by Fubini's Theorem, (1.2.15), and translation invariance of the integral we obtain

$$\begin{aligned}\widehat{\phi}(i) &= \int_0^1 (f * g)(x) w_i(x)\, dx \\ &= \int_0^1 w_i(x) \int_0^1 f(t \oplus x) g(t)\, dt\, dx \\ &= \int_0^1 \int_0^1 w_i(x) f(t \oplus x) g(t)\, dx\, dt \\ &= \int_0^1 \left(\int_0^1 f(t \oplus x) w_i(t \oplus x) g(t) w_i(t)\, dx \right) dt \\ &= \int_0^1 g(t) w_i(t) \left(\int_0^1 f(t \oplus x) w_i(t \oplus x)\, dx \right) dt \\ &= \widehat{f}(i) \int_0^1 g(t) w_i(t)\, dt \\ &= \widehat{f}(i)\widehat{g}(i).\end{aligned}$$

We shall denote the partial sums of the Walsh-Fourier series of a function f by $S_n(x, f)$ or more briefly by $S_n(x)$. We shall find an integral expression for these partial sums. Indeed, substitute definition (2.1.5) into the expression for the partial sums of a Walsh-Fourier series (2.1.4). By (1.2.15) we obtain

$$S_n(x,f) = \sum_{i=0}^{n-1} \int_0^1 f(t)w_i(t)w_i(x)\,dt = \int_0^1 f(t)\left(\sum_{i=0}^{n-1} w_i(t \oplus x)\right)\,dt.$$

By the definition of the Dirichlet kernels (see (1.4.8)) and translation invariance of the integral, we arrive at the formula

$$S_n(x,f) = \int_0^1 f(t)D_n(x \oplus t)\,dt = \int_0^1 f(x \oplus t)D_n(t)\,dt \tag{2.1.10}$$

for $n = 1, 2, \ldots$. One can also arrive at this formula by starting with the definition of convolution and applying (2.1.9).

The partial sums of Walsh-Fourier series of order 2^k are of fundamental importance. For such partial sums it is clear by (2.1.10) and (1.4.13) that

$$\begin{aligned} S_{2^k}(x,f) &= \int_0^1 f(x \oplus t)D_{2^k}(t)\,dt \\ &= 2^k \int_{\Delta_0^{(k)}} f(x \oplus t)\,dt = \frac{1}{|\Delta_0^{(k)}|}\int_{\Delta_0^{(k)}} f(x \oplus t)\,dt \end{aligned}$$

for any $k \geq 0$ and $x \in [0,1)$. If $x \in \Delta_j^{(k)}$, $t \in \Delta_0^{(k)}$, and $x \oplus t$ is defined, then by Theorem **1.2.1** we have $x \oplus t \in \Delta_j^{(k)}$ (the countable set of values t where the sum $x \oplus t$ is not defined does not affect the integration above.) We conclude that

$$S_{2^k}(x,f) = \frac{1}{|\Delta_j^{(k)}|}\int_{\Delta_j^{(k)}} f(t)\,dt, \qquad x \in \Delta_j^{(k)}. \tag{2.1.11}$$

It is useful to notice that identity (2.1.11) is a characterization of Walsh-Fourier series. Indeed, if the partial sums S_{2^k} of a series (1.4.1) satisfies

$$S_{2^k}(x) = \frac{1}{|\Delta_j^{(k)}|}\int_{\Delta_j^{(k)}} f(t)\,dt \qquad x \in \Delta_j^{(k)},\ j = 0, 1, 2, \ldots, 2^k - 1, \tag{2.1.12}$$

for some $f \in \mathbf{L}([0,1))$ and all $k \geq 0$, then the series (1.4.1) is the Fourier series of the function f.

To prove this notice first that each coefficient a_i of the series (1.4.1) can be viewed as a Fourier coefficient of any partial sum of this series $S_n(x)$ for $n > i$. Next, fix i and choose k such that $i < 2^k$. It follows that

$$a_i = \int_0^1 S_{2^k}(t)w_i(t)\,dt. \tag{2.1.13}$$

By **1.1.3** the function $w_i(x)$ is constant on each interval $\Delta_j^{(k)}$ and we denote this constant value by $w_{i,j}$. Moreover, the sum $S_{2^k}(x)$ is also constant on each $\Delta_j^{(k)}$ and we denote this constant value by $s_{2^k,j}$. Consequently, we have by (2.1.13) and (2.1.12) that

$$\begin{aligned} a_i &= \sum_{j=0}^{2^k-1} \int_{\Delta_j^{(k)}} S_{2^k}(t) w_i(t)\, dt \\ &= \sum_{j=0}^{2^k-1} w_{i,j} s_{2^k,j} |\Delta_j^{(k)}| \\ &= \sum_{j=0}^{2^k-1} w_{i,j} \int_{\Delta_j^{(k)}} f(t)\, dt \\ &= \sum_{j=0}^{2^k-1} \int_{\Delta_j^{(k)}} f(t) w_i(t)\, dt = \widehat{f}(i). \end{aligned}$$

We have proved the following theorem.

2.1.3. *A Walsh series (1.4.1) is the Walsh-Fourier series of some function f, integrable on $[0,1)$, if and only if all partial sums of this series of order 2^k satisfy identity* (2.1.12).

§2.2. The Lebesgue constants.

The *Lebesgue constants* for the Walsh system are defined by

$$L_n \equiv \int_0^1 |D_n(t)|\, dt, \tag{2.2.1}$$

for $n = 1, 2, \ldots$. As we shall see below, these constants play an essential role in the study of questions concerning convergence of Walsh-Fourier series.

Notice by (1.4.13) that

$$L_{2^k} = 1. \tag{2.2.2}$$

Thus some subsequence of the Lebesgue constants is bounded. We shall show that there are other subsequences which grow without bound. Never the less, the following shows that $\log_2 n$ is an upper bound for the growth of any subsequence of Lebesgue constants.

2.2.1. *The Lebesgue constants satisfy the inequality*

$$L_n \le \log_2 n, \qquad n \ge 4. \tag{2.2.3}$$

PROOF. For the proof suppose that $2^k \le n \le 2^{k+1}$ for some $k \ge 2$. Recall that $[0,1) = \Delta_0^{(k)} \cup \left(\cup_{i=1}^k \Delta_1^{(i)}\right)$ and $|\Delta_1^{(i)}| = 2^{-i}$. Hence by estimate (1.4.14) and identity (1.4.11) we have

$$L_n = \int_{\Delta_0^{(k)}} |D_n(t)|\,dt + \sum_{i=1}^{k} \int_{\Delta_1^{(i)}} |D_n(t)|\,dt$$
$$\le 1 + \sum_{i=1}^{k} 2^{i-1} \cdot 2^{-i} = 1 + \frac{k}{2}.$$

Since $2^k \le n$ implies $k \le \log_2 n$, it follows that $L_n \le 1 + (\log_2 n)/2$. In particular, (2.2.3) holds for $n \ge 4$. ∎

The following proposition shows that the order of the upper estimate in (2.2.3) is exact, i.e., for some subsequence of natural numbers $\{n_k\}$ the sequence L_{n_k} grows like $\log_2 n_k$. Moreover, we shall see that this subsequence can be chosen so that $2^k \le n_k < 2^{k+1}$ for each $k \ge 0$. In fact, we shall show that one such subsequence is given by

$$n_{2s} = \sum_{i=0}^{s} 2^{2i}, \quad n_{2s+1} = \sum_{i=0}^{s} 2^{2i+1}, \qquad s = 0,1,2,\dots. \tag{2.2.4}$$

2.2.2. *If natural numbers n_k are defined by (2.2.4) then*

$$L_{n_k} > \frac{1}{2}\left(\frac{k}{2}+1\right) > \frac{1}{4}\log_2 n_k, \qquad k = 0,1,2,\dots. \tag{2.2.5}$$

PROOF. Since

$$n_{2s} = \frac{2^{2s+2}-1}{3} < \frac{4}{3}2^{2s}, \quad n_{2s+1} = 2\frac{2^{2s+2}-1}{3} < \frac{4}{3}2^{2s+1},$$

we have

$$n_k < \frac{4}{3}2^k, \qquad k = 0,1,2,\dots. \tag{2.2.6}$$

Consequently,

$$|D_{n_k}(t)| < \frac{4}{3}2^k, \qquad t \in [0,1),\ k = 0,1,2,\dots. \tag{2.2.7}$$

To prove (2.2.5) we show first that

$$\int_{2^{-k-2}}^{1} |D_{n_k}(t)|\,dt \ge \frac{1}{2}\left(\frac{k}{2}+1\right), \qquad k = 0,1,2,\dots. \tag{2.2.8}$$

We prove (2.2.8) for the case $k = 2s$ by induction on s, i.e., we shall show that

$$\int_{2^{-2s-2}}^{1} |D_{n_{2s}}(t)|\,dt \geq \frac{1}{2}(s+1), \qquad s = 0, 1, 2, \dots. \tag{2.2.9}$$

For $s = 0$ (2.2.9) is obvious, since $n_0 = 1$ and

$$\int_{1/4}^{1} |D_1(t)|\,dt = \frac{3}{4} > \frac{1}{2}.$$

For the general inductive step, suppose that

$$\int_{2^{-2s}}^{1} |D_{n_{2(s-1)}}(t)|\,dt \geq \frac{1}{2}s \tag{2.2.10}$$

for some integer $s \geq 1$. By (1.4.13) we know that $|D_{2^{2s}}(t)| = 2^{2s}$ for $t \in \Delta_0^{(2s)}$. On the other hand, (2.2.7) implies $|D_{n_{2(s-1)}}(t)| < 2^{2s}/3$ for all $t \in [0, 1)$. Thus

$$|D_{2^{2s}}(t)| - |D_{n_{2(s-1)}}(t)| > \frac{2}{3}2^{2s}, \qquad t \in \Delta_0^{(2s)}. \tag{2.2.11}$$

But (2.2.4) implies

$$n_{2s} = 2^{2s} + n_{2(s-1)}. \tag{2.2.12}$$

Hence applying (1.4.11) in conjunction with (2.2.11) we obtain

$$|D_{n_{2s}}(t)| \geq |D_{2^{2s}}(t)| - |D_{n_{2(s-1)}}(t)| > \frac{2}{3}2^{2s}, \qquad t \in \Delta_0^{(2s)}.$$

Bearing in mind that $(2^{-2s-2}, 2^{-2s}) \subset \Delta_0^{(2s)}$, we come to the inequality

$$\int_{2^{-2s-2}}^{2^{-2s}} |D_{n_{2s}}(t)|\,dt > \frac{3}{4}2^{-2s} \cdot \frac{2}{3} \cdot 2^{2s} = \frac{1}{2}. \tag{2.2.13}$$

Another application of (1.4.11) in conjunction with (1.4.13) and (2.2.10) yields

$$\int_{2^{-2s}}^{1} |D_{n_{2s}}(t)|\,dt = \int_{2^{-2s}}^{1} |D_{n_{2(s-1)}}(t)|\,dt > \frac{1}{2}s. \tag{2.2.14}$$

Combining inequalities (2.2.13) and (2.2.14), we obtain (2.2.9)

For the case $k = 2s + 1$ ($s = 0, 1, \dots$) we must show

$$\int_{2^{-2s-3}}^{1} |D_{n_{2s+1}}(t)|\,dt \geq \frac{1}{2}\left(s + \frac{1}{2} + 1\right).$$

This inequality holds for $s = 0$ since in this case $n_{2s+1} = n_1 = 2$ and

$$\int_{1/8}^{1} |D_2(t)|\,dt = 2 \cdot \frac{3}{8} = \frac{3}{4}.$$

The rest of the proof is similar to the case when s is even.

Thus (2.2.8) is established.

Notice by (2.2.6) that $\log_2 n_k < k+1$. Thus we conclude by (2.2.8) that inequality (2.2.5) holds as required. ∎

§2.3 Moduli of continuity of functions and uniform convergence of Walsh-Fourier series.

To study the question of uniform convergence of a Walsh-Fourier series of some function f it is natural to estimate the difference $S_n(x,f) - f(x)$. According to (2.1.10) and (1.4.10), this difference can be written in the form

$$S_n(x,f) - f(x) = \int_0^1 (f(x \oplus t) - f(x))D_n(t)\,dt. \tag{2.3.1}$$

By combining (2.1.11) with the obvious identity $f(x) = (1/|\Delta_j^{(k)}|)\int_{\Delta_j^{(k)}} f(x)\,dt$, we see that for partial sums of order 2^k, formula (2.3.1) has the form

$$S_{2^k}(x,f) - f(x) = \frac{1}{|\Delta_j^{(k)}|}\int_{\Delta_j^{(k)}} (f(t) - f(x))\,dt \qquad x \in \Delta_j^{(k)}. \tag{2.3.2}$$

This leads us to the following theorem:

2.3.1. *If f is a function continuous on the interval $[0,1]$ then the subsequence of partial sums $\{S_{2^k}(x,f)\}$ of the Walsh-Fourier series of f converges to f uniformly on $[0,1)$.*

PROOF. Let $\varepsilon > 0$. Since f is uniformly continuous on [0,1], choose $\delta > 0$ such that $|f(t) - f(x)| \le \varepsilon$ for $|t - x| < \delta$. Choose k_0 so large that $2^{-k_0} < \delta$. Notice that if points t and x belong to the same $\Delta_j^{(k)}$ for some $k \ge k_0$ then $|t - x| < \delta$. Consequently, it follows from (2.3.2) that

$$|S_{2^k}(x,f) - f(x)| \le \varepsilon, \qquad k \ge k_0. \quad \blacksquare$$

We can strengthen this theorem by determining how the rate of the approximation to the function f by the partial sums $S_{2^k}(f)$ depends on the smoothness of the function f as measured by the *modulus of continuity* of the function f, i.e., the quantity

$$\omega(\delta,f) = \sup_{\substack{|t-x|\le\delta \\ t,x\in[0,1)}} |f(t) - f(x)|. \tag{2.3.3}$$

It is apparent from (2.3.2) that

$$|S_{2^k}(x,f) - f(x)| \le \omega\left(\frac{1}{2^k}, f\right) \tag{2.3.4}$$

for all $x \in [0,1)$. This formula allows us to obtain the necessary estimates for various classes of functions of a given smoothness, for example the Lipschitz classes. Recall that a function f belongs to the *Lipschitz class of order* α for some $\alpha > 0$, if $\omega(\delta,f) \le C\delta^\alpha$, where the constant C does not depend on δ. Clearly, (2.3.4) implies the following proposition:

2.3.2. *If f belongs to the Lipschitz class of order α, then*

$$|S_{2^k}(x,f) - f(x)| \leq C2^{-\alpha k}.$$

The definition (2.3.3) of the modulus of continuity uses $|t-x|$ to measure the distance between x and t. In view of the importance of the operation $\oplus$ for the theory of Walsh series, it is natural to look at another modulus of continuity. This one is defined using the metric $\rho^*(x,t)$ (see identity (1.2.19)) in the following way:

$$\overset{*}{\omega}(\delta, f) = \sup_{\substack{\rho^*(t,x)<\delta \\ t,x\in[0,1)}} |f(t) - f(x)|. \tag{2.3.5}$$

Since $|t-x| \leq \rho^*(t,x)$ (see (1.2.20)), the usual modulus of continuity and the one just defined are related by the relationship

$$\overset{*}{\omega}(\delta, f) \leq \omega(\delta, f). \tag{2.3.6}$$

As we mentioned in §1.2, the metric $\rho^*(x,t)$ gives rise to a new concept of continuity on [0,1). We shall define, for example, a generalization of uniform continuity. We shall say that a function f is *uniformly ρ^*-continuous* on [0,1) if for every $\varepsilon > 0$ there is a $\delta > 0$ such that $|f(t) - f(x)| < \varepsilon$ for all $\rho^*(t,x) < \delta$. We leave it to the reader to verify that although a Walsh function may be discontinuous in the usual sense, every function from the Walsh system is uniformly ρ^*-continuous on [0,1).

It is not difficult to verify that if x and $t \in \Delta_j^{(k)}$ then $\rho^*(t,x) < 1/2^k$. Hence it follows from (2.3.2) that estimate (2.3.4) can be strengthened as follows:

$$|S_{2^k}(x,f) - f(x)| \leq \overset{*}{\omega}\left(\frac{1}{2^k}, f\right), \qquad x \in [0,1). \tag{2.3.7}$$

In the same way we can strengthen Theorem **2.3.1** by replacing the condition about continuity of f on [0,1] by the condition that f is uniformly ρ^*-continuous.

The concept of a modulus of continuity and the estimates we obtained above are easily carried over to the group G using the system $\{\overset{*}{w}_\ell\}$ and the distance $\overset{*}{\rho}_G(\overset{*}{x}, \overset{*}{y})$ which was introduced in §1.2. We shall not dwell on this in any more detail.

The fact that the partial sums of the form $S_{2^k}(x,f)$ converge uniformly when f is continuous in some sense is closely connected to the fact that the Lebesgue constants of order 2^k are uniformly bounded (see (2.2.2)). This can be seen by a closer look at the proof of Theorem **2.3.1** in conjunction with (2.3.2).

We shall now obtain a general estimate which demonstrates the role that the Lebesgue constants play in questions concerning convergence of Walsh-Fourier series.

2.3.3. *If f is integrable on $[0,1)$ and n is any positive integer then*

$$|S_n(x,f)-f(x)| \le \overset{*}{\omega}\left(\frac{1}{2^k},f\right)\left(2+\frac{1}{2}L_n\right), \tag{2.3.8}$$

where k is determined by the relationship $n = 2^k + m$ for $m \le 2^k$.

PROOF. By (2.3.1) and (1.4.11) we have

$$\begin{aligned} S_n(x,f)-f(x) &= \int_0^1 (f(x\oplus t)-f(x))\, D_{2^k}(t)\,dt \\ &+ \int_0^1 (f(x\oplus t)-f(x))\, r_k(t) D_m(t)\,dt \\ &\equiv J_1 + J_2. \end{aligned} \tag{2.3.9}$$

To estimate the first integral repeat the steps which lead to the inequality (2.3.7). We obtain

$$|J_1| \le \overset{*}{\omega}(1/2^k,f). \tag{2.3.10}$$

The second integral can be written in the form

$$J_2 = \sum_{j=0}^{2^k-1} \int_{\Delta_j^{(k)}} (f(x\oplus t)-f(x))\, r_k(t) D_m(t)\,dt \equiv \sum_{j=0}^{2^k-1} J_2^{(j)}. \tag{2.3.11}$$

Recall that $D_m(t)$ is constant on each $\Delta_j^{(k)} = \Delta_{2j}^{(k+1)} \cup \Delta_{2j+1}^{(k+1)}$ and that $r_k(t) = 1$ for $t \in \Delta_{2j}^{(k+1)}$ and $r_k(t) = -1$ for $t \in \Delta_{2j+1}^{(k+1)}$. Since $t \in \Delta_{2j}^{(k+1)}$ implies

$$t \oplus 1/2^{k+1} \in \Delta_{2j+1}^{(k+1)},$$

it follows that

$$\begin{aligned} J_2^{(j)} &= \int_{\Delta_{2j}^{(k+1)}} (f(x\oplus t)-f(x))\, D_m(t)\,dt \\ &\quad - \int_{\Delta_{2j+1}^{(k+1)}} (f(x\oplus t)-f(x))\, D_m(t)\,dt \\ &= \int_{\Delta_{2j}^{(k+1)}} (f(x\oplus t)-f(x))\, D_m(t)\,dt \\ &\quad - \int_{\Delta_{2j}^{(k+1)}} \left(f(x\oplus t\oplus \frac{1}{2^{k+1}})-f(x)\right) D_m(t)\,dt \\ &= \int_{\Delta_{2j}^{(k+1)}} \left(f(x\oplus t)-f(x\oplus t\oplus \frac{1}{2^{k+1}})\right) D_m(t)\,dt \end{aligned} \tag{2.3.12}$$

But for each $t \in \Delta_{2j}^{(k+1)}$ the points $x \oplus t$ and $x \oplus t \oplus 1/2^{k+1}$ both belong to the same interval of rank k. Consequently,

$$\overset{*}{\rho}\left(x \oplus t, x \oplus t \oplus \frac{1}{2^{k+1}}\right) < \frac{1}{2^k}.$$

Therefore, we obtain from (2.3.12) that

$$|J_2^{(j)}| \le \overset{*}{\omega}(\frac{1}{2^k}, f) \int_{\Delta_{2j}^{(k+1)}} |D_m(t)|\, dt = \frac{1}{2}\overset{*}{\omega}(\frac{1}{2^k}, f) \int_{\Delta_j^{(k)}} |D_m(t)|\, dt.$$

Summing this estimate over j, we have by (2.3.9) through (2.3.11) that

$$\begin{aligned} |S_n(x,f) - f(x)| &\le \overset{*}{\omega}(\frac{1}{2^k}, f)\left(1 + \frac{1}{2}\int_0^1 |D_m(t)|\, dt\right) \\ &= \overset{*}{\omega}(\frac{1}{2^k}, f)\left(1 + \frac{1}{2} L_m\right). \end{aligned} \tag{2.3.13}$$

Since $L_m \le L_n + 1$ by (1.4.11) and (1.4.13), we continue (2.3.13) to arrive at (2.3.8). ∎

Theorem **2.3.3** implies the following result concerning uniform convergence of a given subsequence of partial sums of a Walsh- Fourier series:

2.3.4. *Let $\{n_i\} \equiv \{2^{k_i} + m_i\}$, $m_i \le 2^{k_i}$, be an increasing sequence of natural numbers. If for some function f the condition*

$$\lim_{i\to\infty} \overset{*}{\omega}\left(\frac{1}{2^{k_i}}, f\right) L_{n_i} = 0 \tag{2.3.14}$$

holds, then the subsequence of partial sums $\{S_{n_i}(x,f)\}$ converges to f uniformly on $[0,1)$.

PROOF. By (1.4.10) it is clear that $L_n = \int_0^1 |D_n(t)|\, dt \ge 1$ for $n \ge 1$. Thus (2.3.14) implies that $\overset{*}{\omega}(1/2^{k_i}, f) \to 0$ as $i \to \infty$. In particular,

$$\lim_{i\to\infty} \overset{*}{\omega}\left(\frac{1}{2^{k_i}}, f\right)\left(2 + \frac{L_{n_i}}{2}\right) = 0.$$

Consequently, the proof is completed by an application of Theorem **2.3.3**. ∎

Since estimate (2.2.3) holds for all natural numbers n, **2.3.4** contains the following test for uniform convergence of Walsh-Fourier series which is an analogue of the Dini-Lipschitz test from the classical theory of Fourier series.

2.3.5. *If the modulus of continuity of a function f satisfies*

$$\lim_{k\to\infty} k\overset{*}{\omega}\left(\frac{1}{2^k}, f\right) = 0,$$

then the Walsh-Fourier series of f converges to f uniformly on $[0,1)$.

PROOF. If $4 \le 2^k < n \le 2^{k+1}$, then it follows from (2.2.3) that

$$L_n \le \log_2 n \le k+1.$$

Hence the hypothesis of this theorem implies that (2.3.14) is satisfied by the sequence of natural numbers $n_i = i$. ∎

For continuous functions, it is more natural to formulate Theorem **2.3.5** in terms of the usual modulus of continuity.

2.3.6. *If the modulus of continuity of a function f satisfies*

$$\lim_{n\to\infty} \omega\left(\frac{1}{n}, f\right)\ln n = 0,$$

then the Walsh-Fourier series of f converges to f uniformly on $[0,1)$.

PROOF. This result follows directly from **2.3.5** if we notice that according to definition (2.3.3), the modulus of continuity is a monotone increasing function of δ and satisfies $\omega(2\delta, f) \le 2\omega(\delta, f)$. Consequently,

$$\frac{1}{2}\omega\left(\frac{1}{2^k}, f\right) \le \omega\left(\frac{1}{2^{k+1}}, f\right) \le \omega\left(\frac{1}{n}, f\right)$$

and we finish the proof by appealing to (2.3.6). ∎

As a special case of Theorem **2.3.4** we see that if a subsequence of the Lebesgue constants $\{L_{n_i}\}$ is bounded, then the corresponding subsequence of partial sums $\{S_{n_i}(x,f)\}$ converges to f uniformly if f is continuous on [0,1] or if f is uniformly ρ^*-continuous on [0,1). (These results contain Theorem **2.3.1** as a special case.)

For the case when the subsequence $\{L_{n_i}\}$ is unbounded, it is not difficult to verify using the Banach-Steinhaus Theorem (see **A5.3.3**) that given any point $x \in [0,1)$ there is a function, continuous on [0,1], such that the corresponding subsequence of partial sums $\{S_{n_i}(x,f)\}$ of its Walsh- Fourier series diverges at the given point x. We shall not give a more explicit proof of this fact, which is itself a useful exercise which involves application of the Banach-Steinhaus Theorem to the subsequence of functionals $\{F_i(f)\} \equiv \{S_{n_i}(x,f)\}$.

§2.4. Other tests for uniform convergence.

We shall now establish a test for uniform convergence of Walsh- Fourier series which is useful in the sense that from it one can deduce simpler and more readily applicable tests for uniform convergence.

2.4.1. *Let f be uniformly ρ^*-continuous on $[0,1)$ and consider the sum*

$$T_k(x) \equiv \sum_{j=1}^{2^k-1} \frac{1}{j} \left| f\left(x \oplus \frac{2j}{2^{k+1}}\right) - f\left(x \oplus \frac{2j+1}{2^{k+1}}\right) \right| . \tag{2.4.1}$$

If $T_k(x) \to 0$ uniformly on $[0,1)$, as $k \to \infty$, then the Walsh-Fourier series of f converges to f uniformly on $[0,1)$.

PROOF. To estimate the difference $S_n(x,f) - f(x)$, use identities (2.3.9), (2.3.11), (2.3.12) and inequality (2.3.10). For $n = 2^k + m$, $m \leq 2^k$ we obtain

$$|S_n(x,f) - f(x)| \leq \overset{*}{\omega}\left(\frac{1}{2^k}, f\right)$$
$$+ \left| \sum_{j=0}^{2^k-1} \int_{\Delta_{2j}^{(k+1)}} \left(f(x \oplus t) - f(x \oplus t \oplus \frac{1}{2^{k+1}})\right) D_m(t)\,dt \right| .$$

In this last, sum change variables in each integral from t to $t \oplus 2j/2^{k+1}$. If we denote by $D_{m,j}$ the constant value which the kernel $D_m(t)$ assumes on the interval $\Delta_j^{(k)}$, then we obtain

$$|S_n(x,f) - f(x)| \leq \overset{*}{\omega}\left(\frac{1}{2^k}, f\right) \tag{2.4.2}$$

$$+ \left| \sum_{j=0}^{2^k-1} D_{m,j} \int_{\Delta_0^{(k+1)}} \left(f(x \oplus t \oplus \frac{2j}{2^{k+1}}) - f(x \oplus t \oplus \frac{2j+1}{2^{k+1}})\right) D_m(t)\,dt \right| .$$

Observe that $t \geq j2^{-k}$ for $t \in \Delta_j^{(k)}$. Consequently we have by (1.4.16) that

$$|D_{m,j}| \leq \frac{2^k}{j}, \qquad j = 1, 2, \ldots, 2^k - 1. \tag{2.4.3}$$

Moreover, $D_{m,0} = m$ (see **1.4.3**). Hence we can estimate the first term of the sum in (2.4.2) by the quantity $m2^{-k}\overset{*}{\omega}(1/2^k, f)$, and the remaining terms by using (2.4.3). We obtain

$$|S_n(x,f) - f(x)| \leq 2\overset{*}{\omega}\left(\frac{1}{2^k}, f\right) \tag{2.4.4}$$

$$+2^k \int_{\Delta_0^{(k+1)}} \left(\sum_{j=1}^{2^k-1} \frac{1}{j} \left| f(x \oplus t \oplus \frac{2j}{2^{k+1}}) - f(x \oplus t \oplus \frac{2j+1}{2^{k+1}}) \right| \right) dt.$$

Of course we recognize the sum inside this last integral as $T_k(x \oplus t)$ (see (2.4.1)). Consequently, for any $\varepsilon > 0$ we can choose k sufficiently large so that $|T_k(x \oplus t)| < \varepsilon$ for all x and t. Hence the second part of (2.4.4) does not exceed the quantity $2^k \varepsilon |\Delta_0^{(k+1)}| = \varepsilon/2$. Since the first part of (2.4.4) can be made as small as one wishes by using uniform ρ^*-continuity and the modulus (2.3.5), we have completed the proof of this theorem. ∎

This theorem gives another proof of the Dini-Lipschitz test formulated in **2.3.5** (or in **2.3.6**). Indeed,

$$T_k(x) \le \overset{*}{\omega}\left(\frac{1}{2^k}, f\right) \sum_{j=1}^{2^k-1} \frac{1}{j} < \overset{*}{\omega}\left(\frac{1}{2^k}, f\right)(1 + \ln 2^k) \le 2k\omega\left(\frac{1}{2^k}, f\right)$$

holds for all natural numbers k. In particular, the hypotheses of Theorem **2.3.5** imply those of Theorem **2.4.1**.

Another corollary of Theorem **2.4.1** is the following result which is an analogue of the *Jordan test* for trigonometric series.

2.4.2. *If f is uniformly ρ^*-continuous on $[0, 1)$ and is of bounded variation on $[0, 1)$ then its Walsh- Fourier series converges to f uniformly on $[0, 1)$.*

PROOF. Recall that a function f defined on some interval (a, b) is said to be of *bounded variation* if the sums $\sum_{j=1}^{i} |f(b_j) - f(a_j)|$ are bounded by some absolute constant for any collection $\{(a_j, b_j)\}_{j=1}^{i}$ of non- overlapping intervals lying in (a, b). The supremum of such sums taken over all such collections of non-overlapping intervals lying in (a, b) is called the variation of f on (a, b) and will be denoted by $V_a^b[f]$.

It is not difficult to verify that as j ranges over the integers $0, 1, \ldots, 2^k - 1$, the points $x \oplus (2j/2^{k+1})$ and $x \oplus ((2j+1)/2^{k+1})$ belong to different intervals of rank k and thus generate non-overlapping intervals in [0,1). Consequently,

$$\sum_{j=m}^{2^k-1} \left| f\left(x \oplus \frac{2j}{2^{k+1}}\right) - f\left(x \oplus \frac{2j+1}{2^{k+1}}\right) \right| \le V_0^1[f] \tag{2.4.5}$$

for all integers $0 \le m < 2^k$.

Choose a non-decreasing sequence of natural numbers $\{m_k\}$ such that $m_k < 2^k - 1$ for each k and $m_k \to \infty$ as $k \to \infty$ but $\omega(1/2^k, f) \ln m_k \to 0$ as $k \to \infty$. This is always possible if we choose $\{m_k\}$ tending to ∞ sufficiently slowly, namely, if we set $m_k = e^{o((\omega(2^{-k}, f)^{-1})}$.

Divide the sum $T_k(x)$ (see (2.4.1)) into two pieces and apply (2.4.5). We obtain the estimate

$$
\begin{aligned}
T_k(x) &= \left(\sum_{j=1}^{m_k} + \sum_{j=m_k+1}^{2^k-1}\right)\frac{1}{j}\left| f\left(x\oplus\frac{2j}{2^{k+1}}\right) - f\left(x\oplus\frac{2j+1}{2^{k+1}}\right)\right| \\
&\le \overset{*}{\omega}\left(\frac{1}{2^k}, f\right)\sum_{j=1}^{m_k}\frac{1}{j} + \frac{1}{m_k+1}V_0^1[f] \\
&\le 2\overset{*}{\omega}\left(\frac{1}{2^k}, f\right)\ln m_k + \frac{1}{m_k+1}V_0^1[f].
\end{aligned}
$$

By the choice of the sequence m_k it is now apparent that $T_k(x)$ tends to zero uniformly as $k \to \infty$. It remains to apply **2.4.1.** ∎

We notice by (2.3.6) that Theorems **2.4.1** and **2.4.2** still hold if ρ^*-continuity is replaced by continuity (in the usual sense) on the interval [0,1]. Since every monotone bounded function is of bounded variation, we notice also that Theorem **2.4.2** is valid for such functions.

§2.5. The localization principle. Tests for convergence of a Walsh-Fourier series at a point.

A generalization of definition (2.3.5) is the concept of the *integral modulus of continuity*, namely, the expression

$$
\overset{*}{\omega}{}^{(1)}(\delta, f) = \sup_{h<\delta}\int_0^1 |f(x\oplus h) - f(x)|\,dx. \tag{2.5.1}
$$

Such a modulus of continuity satisfies two basic properties.

2.5.1. *If ϕ is continuous on $[0,1]$ and $\omega(\delta, f)$ is its usual modulus of continuity, then*

$$
\overset{*}{\omega}{}^{(1)}(\delta,\phi) \le \overset{*}{\omega}(\delta,\phi) \le \omega(\delta,\phi).
$$

PROOF. It is enough to notice that $\rho^*(x\oplus h, x) \le \delta$ for $h \le \delta$ and thus from (2.3.5) and (2.3.6) we obtain

$$
|\phi(x\oplus h) - \phi(x)| \le \overset{*}{\omega}(\delta,\phi) \le \omega(\delta,\phi).
$$

The fact that for each fixed h the sum $x\oplus h$ may not be defined for countably many values of x (see §1.2) does not play any part in the integration. ∎

2.5.2. *If f is Lebesgue integrable on $[0,1)$ then*

$$
\lim_{\delta\to 0}\overset{*}{\omega}{}^{(1)}(\delta, f) = 0.
$$

PROOF. By a property of the Lebesgue integral (see **A5.2.1**), for any $\varepsilon > 0$ we can choose a function ϕ, continuous on [0,1], such that

$$\int_0^1 |f(x) - \phi(x)|\, dx < \varepsilon.$$

Since the integral is translation invariant under $\oplus$, we obtain

$$\begin{aligned}\int_0^1 |f(x \oplus h) - f(x)|\, dx &\leq \int_0^1 |f(x \oplus h) - \phi(x \oplus h)|\, dx + \int_0^1 |f(x) - \phi(x)|\, dx \\ &\quad + \int_0^1 |\phi(x \oplus h) - \phi(x)|\, dx \\ &\leq 2\varepsilon + \int_0^1 |\phi(x \oplus h) - \phi(x)|\, dx.\end{aligned}$$

Thus by **2.5.1** we see that

$$\overset{*}{\omega}{}^{(1)}(\delta, f) \leq 2\varepsilon + \overset{*}{\omega}{}^{(1)}(\delta, \phi) \leq 2\varepsilon + \omega(\delta, \phi).$$

Since ϕ is continuous, the expression $\omega(\delta, \phi)$ can be made as small as one wishes for sufficiently small δ. This proves **2.5.2.** ∎

We shall now establish that the behavior on some interval of the partial sums $S_n(x)$ of a Fourier series of an integrable function is totally determined, to within a quantity which tends to zero as $n \to \infty$, by the values of that function on that interval.

2.5.3. *If f is integrable on $[0, 1)$ and s is a natural number then*

$$\left| S_n(x, f) - \int_{\Delta_0^{(s)}} f(x \oplus t) D_n(t)\, dt \right| \leq \frac{1}{2|\Delta_0^{(s)}|} \overset{*}{\omega}{}^{(1)}\left(\frac{1}{2^k}, f\right), \qquad n \geq 2^s,$$

where k is determined by $n = 2^k + m$, $1 \leq m \leq 2^k$.

PROOF. Use the integral representation of $S_n(x, f)$ (see (2.1.10)) to write

$$\left| S_n(x, f) - \int_{\Delta_0^{(s)}} f(x \oplus t) D_n(t)\, dt \right| = \left| \int_{[0,1)\setminus\Delta_0^{(s)}} f(x \oplus t) D_n(t)\, dt \right|.$$

Thus it suffices to show that

$$\left| \int_{[0,1)\setminus\Delta_0^{(s)}} f(x \oplus t) D_n(t)\, dt \right| \leq \frac{1}{2|\Delta_0^{(s)}|} \overset{*}{\omega}{}^{(1)}\left(\frac{1}{2^k}, f\right), \qquad n \geq 2^s. \tag{2.5.2}$$

Combine formula (1.4.11) for the Dirichlet kernel with (1.4.13) which shows that $D_{2^k}(t) = 0$ for $t \in [0,1) \setminus \Delta_0^{(s)}$ and $k \geq s$. Thus verify that $D_n(t) = r_k(t) D_m(t)$ for $t \in [0,1) \setminus \Delta_0^{(s)}$. Notice by (1.4.14) that $|D_m(t)| \leq 2^{s-1}$ for the same t and

$$\Delta_0^{(s)} = \bigcup_{j=0}^{2^s-1} \Delta_j^{(k)}.$$

Repeating the steps which lead to (2.3.12), we obtain

$$\begin{aligned}
\Big| \int_{[0,1)\setminus\Delta_0^{(s)}} f(x \oplus t) D_n(t)\, dt \Big| &\leq \Big| \sum_{j=2^s}^{2^k-1} \int_{\Delta_j^{(k)}} f(x \oplus t) D_n(t)\, dt \Big| \\
&\leq \sum_{j=2^s}^{2^k-1} 2^{s-1} \int_{\Delta_{2j}^{(k+1)}} |f(x \oplus t) - f(x \oplus t \oplus \frac{1}{2^{k+1}})|\, dt \\
&\leq 2^{s-1} \int_0^1 |f(x \oplus t) - f(x \oplus t \oplus \frac{1}{2^{k+1}})|\, dt \\
&= 2^{s-1} \int_0^1 |f(t) - f(t \oplus \frac{1}{2^{k+1}})|\, dt \\
&\leq 2^{s-1} \overset{*}{\omega}{}^{(1)} \left(\frac{1}{2^k}, f \right) \\
&= \frac{1}{2|\Delta_0^{(s)}|} \overset{*}{\omega}{}^{(1)} \left(\frac{1}{2^k}, f \right).
\end{aligned}$$

This verifies inequality (2.5.2) and completes the proof of this theorem. ∎

An application of **2.5.3** gives two results which make up the *localization principle* for Walsh-Fourier series.

2.5.4. *If a function f, integrable on $[0,1)$, vanishes identically on some dyadic interval $\Delta_j^{(s)}$ then the Walsh-Fourier series of this function converges to zero uniformly on $\Delta_j^{(s)}$. Moreover,*

$$|S_n(x,f)| \leq \frac{1}{2|\Delta_j^{(s)}|} \overset{*}{\omega}{}^{(1)} \left(\frac{1}{2^k}, f \right), \qquad x \in \Delta_j^{(s)}, \quad n \geq 2^s, \tag{2.5.3}$$

where $n = 2^k + m$, $1 \leq m \leq 2^k$.

PROOF. It is clear that $x \oplus t \in \Delta_j^{(s)}$ when $x \in \Delta_j^{(s)}$ and $t \in \Delta_0^{(s)}$. Consequently, inequality (2.5.3) follows immediately from Theorem **2.5.3** and the hypotheses of this theorem. Moreover, uniform convergence to zero of the sums $S_n(x,f)$ follows directly from the property **2.5.2** of the integral modulus of continuity. ∎

2.5.5. *If two integrable functions f and g coincide on some dyadic interval Δ then their Walsh-Fourier series are uniformly equiconvergent on Δ, i.e., the difference of these series converges uniformly to zero on the interval Δ.*

PROOF. It is enough to apply Theorem **2.5.4** to the difference $f - g$ and recall from **2.1.1** that the Fourier series of the difference $f - g$ is the difference of the Fourier series of the functions f and g. ∎

In particular, we see that if two integrable functions coincide on some open interval containing a point x, then the Walsh-Fourier series of these functions are equiconvergent at the point x, i.e., at the point x either both these series converge or both these series diverge.

It is easy to see that the localization principle allows us to obtain local versions of the tests for uniform convergence which were proved in §2.3 and §2.4. The following result is a local version of Theorem **2.4.2**.

2.5.6. *If f is integrable on $[0, 1)$, uniformly ρ^*-continuous on some dyadic interval Δ, and of bounded variation on Δ, then its Walsh-Fourier series converges to f uniformly on Δ.*

PROOF. Clearly, the function

$$f_1(x) = \begin{cases} f(x) & \text{for } x \in \Delta, \\ 0 & \text{for } x \in [0,1) \setminus \Delta \end{cases}$$

is uniformly ρ^*-continuous and of bounded variation on [0,1). Consequently, by Theorem **2.4.2** the Fourier series of f_1 converges to f_1 uniformly on [0,1). In particular, this series converges to f uniformly on Δ. It remains to apply Theorem **2.5.5** to the functions f and f_1. ∎

We come now to a condition sufficient for convergence of a Walsh- Fourier series at a point which is an analogue of the Dini test for Fourier series.

2.5.7. *The Walsh-Fourier series of an integrable function f converges at some point x to a value c if the function $(f(u) - c)/(u - x)$ is Lebesgue integrable near x, i.e., if*

$$\int_{x-\delta}^{x+\delta} \frac{|f(u) - c|}{|u - x|}\, du < \infty \tag{2.5.4}$$

for some $\delta > 0$

PROOF. Since the indefinite Lebesgue integral is absolutely continuous, given $\varepsilon > 0$ we can choose s sufficiently large so that $x \in \Delta_j^{(s)} \subset (x - \delta, x + \delta)$ and

$$\int_{\Delta_j^{(s)}} \frac{|f(u) - c|}{|u - x|}\, du < \frac{\varepsilon}{2}. \tag{2.5.5}$$

If $x \in \Delta_j^{(s)}$, $t \in \Delta_0^{(s)}$, and $x \oplus t$ is defined then $x \oplus t \in \Delta_j^{(s)}$ (see **1.2.1**). Thus by the same kind of proof we used to establish translation invariance of the integral (see (2.1.2)), we can prove that

$$\int_{\Delta_0^{(s)}} \frac{|f(x \oplus t) - c|}{|(x \oplus t) - x|}\,dt = \int_{\Delta_j^{(s)}} \frac{|f(u) - c|}{|u - x|}\,du. \tag{2.5.6}$$

But by (1.2.20), $|(x \oplus t) - x| \le \rho^*(x \oplus t, x) = t$. Thus it follows from (2.5.6) and (2.5.5) that

$$\int_{\Delta_0^{(s)}} \frac{|f(x \oplus t) - c|}{t}\,dt \le \int_{\Delta_0^{(s)}} \frac{|f(x \oplus t) - c|}{|(x \oplus t) - x|}\,dt < \frac{\varepsilon}{2}. \tag{2.5.7}$$

In particular, it follows from the estimate (1.4.16) for the Dirichlet kernel that

$$\int_{\Delta_0^{(s)}} |f(x \oplus t) - c|\,|D_n(t)|\,dt \le \int_{\Delta_0^{(s)}} \frac{|f(x \oplus t) - c|}{t}\,dt < \frac{\varepsilon}{2} \tag{2.5.8}$$

for any n.

Apply Theorem **2.5.3** to the function $f(x) - c$ and the interval $\Delta_0^{(s)}$. Fix s and choose $k_0 \ge s$ so large that

$$\frac{1}{2|\Delta_0^{(s)}|}\omega^{(1)}\left(\frac{1}{2^k}, f\right) \le \frac{\varepsilon}{2}$$

for all $k \ge k_0$. Thus by (2.5.8) and the estimate from Theorem **2.5.3** we have

$$|S_n(x, f) - c| = |S_n(x, f - c)| < \frac{\varepsilon}{2} + \frac{\varepsilon}{2} = \varepsilon, \qquad n > 2^{k_0}.$$

This completes the proof of **2.5.7**. ∎

In particular, we see that the Walsh-Fourier series of a function f converges to $f(x)$ at a point x if the inequality

$$|f(u) - f(x)| \le c|u - x|^\alpha \tag{2.5.9}$$

is satisfied for some $\alpha > 0$ and $\delta > 0$ and all u such that $|u - x| \le \delta$. Condition (2.5.9) is obviously satisfied at each point where the function f is differentiable. Thus we have proved the following result:

2.5.8. *If f is integrable on $[0, 1)$ then the Walsh-Fourier series of f converges to $f(x)$ at every point x where f has a finite derivative.*

In §2.3 we remarked that the subsequence of partial sums of order 2^k play a fundamental role in questions about uniform convergence. Criteria for convergence of this subsequence at a given point are also of considerable interest and distinguished by unusual simplicity.

2.5.9. *Let* $\{\Delta_x^{(k)}\} \equiv \{[\alpha_k, \beta_k)\}_{k=0}^{\infty}$ *be the sequence of dyadic intervals which satisfy* $x \in \Delta_x^{(k)}$ *for all* $k \geq 0$ *and let* f *be integrable on* $[0,1)$ *with indefinite integral*

$$F(x) = \int_0^x f(t)\,dt.$$

The subsequence of partial sums $\{S_{2^k}(x, f)\}$ *of the Walsh- Fourier series of* f *at a point* x *converges to a number* a *if and only if*

$$\lim_{k\to\infty} \frac{F(\beta_k) - F(\alpha_k)}{\beta_k - \alpha_k} = a. \tag{2.5.10}$$

PROOF. For the proof it is enough to notice by (2.1.11) that

$$S_{2^k}(x, f) = \frac{1}{|\Delta_x^{(k)}|} \int_{\Delta_x^{(k)}} f(t)\,dt = \frac{F(\beta_k) - F(\alpha_k)}{\beta_k - \alpha_k}. \quad \blacksquare$$

The limit in (2.5.10) is called the ***derivative with respect to binary nets*** $\{\mathcal{N}_k\}$ (see §1.1), or the $\{\mathcal{N}_k\}$-*derivative*, and will be denoted by $D_{\{\mathcal{N}_k\}}F(x)$. Hence

$$D_{\{\mathcal{N}_k\}}F(x) = \lim_{k\to\infty} \frac{F(\beta_k) - F(\alpha_k)}{\beta_k - \alpha_k}. \tag{2.5.11}$$

The usual concept of differentiation and differentiation with respect to binary nets are related in the following way:

2.5.10. *If* F *is differentiable at a point* x *with* $F'(x) = f(x)$ *then* F *has a derivative with respect to binary nets and* $D_{\{\mathcal{N}_k\}}F(x) = f(x)$.

PROOF. Since the derivative $F'(x) = f(x)$ exists, we know that

$$F(\beta_k) - F(x) = f(x)(\beta_k - x) + o(|\Delta^{(k)}|),$$

and

$$F(\alpha_k) - F(x) = f(x)(\alpha_k - x) + o(|\Delta^{(k)}|),$$

as $k \to \infty$. Subtracting the second expression from the first we obtain

$$F(\beta_k) - F(\alpha_k) = f(x)(\beta_k - \alpha_k) + o(|\Delta^{(k)}|).$$

This completes the proof of the theorem. $\blacksquare$

Combining **2.5.9** with **2.5.10**, we are lead to the following test:

2.5.11. *The subsequence of partial sums $\{S_{2^k}(x,f)\}$ of the Walsh-Fourier series of an integrable function f with indefinite integral F converges at each point where F is differentiable, in which case*

$$\lim_{k\to\infty} S_{2^k}(x,f) = F'(x).$$

In view of the theorem about almost everywhere differentiability of the indefinite Lebesgue integral (see **A4.4.5**), namely $F'(x) = f(x)$ almost everywhere, the following result is a corollary of the preceding test.

2.5.12. *If f is integrable on $[0,1)$ then the subsequence of partial sums $\{S_{2^k}(x,f)\}$ of its Walsh- Fourier series converges to f almost everywhere on $[0,1)$.*

§2.6 The Walsh system as a complete, closed system.

Beginning with this section, we will systematically use the following standard notation for classes of integrable functions, each of which is a normed linear space (see **A5.1** and **A5.2**).

For any $p \ge 1$ denote by $\mathbf{L}^p = \mathbf{L}^p[0,1)$ the set of measurable functions f for which the quantity

$$\|f\|_p = \left(\int_0^1 |f(x)|^p \, dx\right)^{1/p} \tag{2.6.1}$$

is finite. This quantity satisfies all the properties of a *norm* if in $\mathbf{L}^p$ we do not distinguish between functions which coincide almost everywhere (more details about this can be found in **A5.2**).

For the space $\mathbf{L}^1$ we shall frequently use the simpler notation $\mathbf{L}$ or $\mathbf{L}[0,1)$.

The space of functions continuous on [0,1] will be denoted by $\mathbf{C}[0,1]$ and its norm by

$$\|f\|_C = \sup_{0\le x\le 1} |f(x)|. \tag{2.6.2}$$

We shall also use the symbols $\mathbf{C}[0,1)$, respectively $\mathbf{C}_{\rho^*}[0,1)$, to denote the space of functions uniformly continuous, respectively uniformly ρ^*-continuous (see §2.3), on [0,1). These spaces use the same norm (2.6.2).

By **1.1.5** the Walsh system is orthonormal on [0,1). As is the case with any orthonormal system, it is very important to verify that a function is uniquely determined by its Walsh-Fourier coefficients. Since changing a function on a set of measure zero does not affect the value of the integral which defines its Fourier coefficients, we can only speak of uniqueness with precision up to values on a set of measure zero. In the case of continuous functions this restriction is not necessary. We shall see that the theorem of uniqueness does hold for the Walsh system, namely, we shall establish the following result:

2.6.1. *1) If $f \in \mathbf{L}[0,1)$ and $\widehat{f}(i) = 0$ for $i = 0,1,2,\dots,$ then $f(x) = 0$ almost everywhere ;*

2) if $f \in \mathbf{C}_{\rho^*}[0,1)$ and $\widehat{f}(i) = 0$ for $i = 0,1,2,\dots,$ then $f(x) = 0$ for $x \in [0,1)$.

The theorem of uniqueness for an orthonormal system is connected with the completeness of that system. Recall that an orthonormal system of functions is said to be *complete* in the space $\mathbf{C}_{\rho^*}[0,1)$ (or in $\mathbf{L}^p[0,1)$), if there does not exist a non-zero function (that is a function different from zero on some set of positive measure) which is orthogonal to each function in that system.

It is clear that the property of completeness is simply a reformulation of the theorem of uniqueness and thus Theorem **2.6.1** is equivalent to the following result:

2.6.2. *The Walsh system is complete in the spaces $\mathbf{L}^1[0,1)$ and $\mathbf{C}_{\rho^*}[0,1)$.*

We shall prove **2.6.1** and thus **2.6.2.**

PROOF. Since

$$\int_0^1 f(t)w_i(t)\,dt = 0, \qquad i = 0,1,2,\dots,$$

it is evident that

$$\int_0^1 f(t)T_n(t)\,dt = 0$$

for any Walsh polynomial $T_n(t)$. Notice by (1.2.15) that for each fixed x the translated Dirichlet kernel $D_n(x \oplus t)$ is a Walsh polynomial. Consequently, it follows from the previous identity that

$$\int_0^1 f(t)D_{2^k}(x \oplus t)\,dt = 0, \qquad k = 0,1,2,\dots.$$

But the left side of this expression is a partial sum of the Fourier series of the function f at the point x (see (2.1.10)). Hence

$$S_{2^k}(x,f) = 0, \qquad k = 0,1,2,\dots,$$

i.e.,

$$\lim_{k\to\infty} S_{2^k}(x,f) = 0$$

for every $x \in [0,1)$. But by Theorem **2.5.12**

$$\lim_{k\to\infty} S_{2^k}(x,f) = f(x)$$

almost everywhere so we have $f(x) = 0$ almost everywhere on [0,1). In the case when f is ρ^*-continuous, it follows that $f(x) = 0$ everywhere on [0,1). ∎

It is clear that **2.6.1** 1) holds if we substitute any space of integrable functions for the space $\mathbf{L}[0,1)$, in particular, it holds for the spaces $\mathbf{L}^p[0,1)$, $p \geq 1$. Thus,

2.6.3. *The Walsh system is complete in the spaces* $\mathbf{L}^p[0,1)$, $p \ge 1$.

In particular, the Walsh system is complete in the space $\mathbf{L}^2[0,1)$. For this space, a given system is complete if and only if it is closed (see **A5.4.1**). Never the less, it is easy to prove directly that the Walsh system is closed in $\mathbf{L}^2[0,1)$. Indeed, we will use Theorem **2.3.1** to prove that the Walsh system is closed in every $\mathbf{L}^p$, $p \ge 1$.

Recall that a system of functions is said to be ***closed*** in the space $\mathbf{L}^p$ if given $f \in \mathbf{L}^p$ and $\varepsilon > 0$ there is a polynomial $T(x)$ in this system which satisfies

$$\|f(x) - T(x)\|_p < \varepsilon. \tag{2.6.3}$$

It is well known (see **A5.2.1**) that the collection of continuous functions is dense in the space $\mathbf{L}^p[0,1)$ for each $p \ge 1$. But by Theorem **2.3.1** any continuous function ϕ can be approximated uniformly, hence in the $\mathbf{L}^p[0,1)$ norm, as closely as one wishes by the partial sums $S_{2^k}(x,\phi)$, i.e., by a polynomial in the Walsh system. Thus given $f \in \mathbf{L}^p[0,1)$ and $\varepsilon > 0$ choose first a continuous function ϕ such that $\|f - \phi\|_p < \varepsilon/2$, and then a Walsh polynomial T such that $\|\phi - T\|_p < \varepsilon/2$. By the triangle inequality, it follows that (2.6.3) holds. In particular, we have proved the following result:

2.6.4. *The Walsh system is closed in each space* $\mathbf{L}^p[0,1)$ $p \ge 1$ *and in* $\mathbf{C}[0,1)$.

Using other terminology, this result can be stated as follows.

2.6.4'. *The set of Walsh polynomials is dense in each space* $\mathbf{L}^p[0,1)$ $p \ge 1$ *and in* $\mathbf{C}[0,1)$.

Since the Walsh system is closed in the space $\mathbf{L}^2[0,1)$, the Parseval identity holds (see **A5.4.2**).

2.6.5. *If* $f \in \mathbf{L}^2[0,1)$ *then*

$$\sum_{i=0}^{\infty} |\hat{f}(i)|^2 = \int_0^1 |f(t)|^2 \, dt.$$

PROOF. For any polynomial $T_n = \sum_{i=0}^{n} \alpha_i w_i(t)$ and any function $f \in \mathbf{L}^2[0,1)$ we have

$$\begin{aligned} \|f - T_n\|_2^2 &= \int_0^1 \left(f(t) - \sum_{i=0}^{n} \alpha_i w_i(t) \right)^2 dt \\ &= \int_0^1 |f(t)|^2 \, dt - 2\sum_{i=0}^{n} \alpha_i \hat{f}(i) + \sum_{i=0}^{n} |\alpha_i|^2 \\ &= \int_0^1 |f(t)|^2 \, dt - \sum_{i=0}^{n} |\hat{f}(i)|^2 + \sum_{i=0}^{n} (\hat{f}(i) - \alpha_i)^2. \end{aligned} \tag{2.6.4}$$

In particular, if $\alpha_i = \widehat{f}(i)$, i.e., the polynomial T_n is taken to be the partial sum $S_n(f)$ then

$$\|f - S_n(f)\|_2^2 = \int_0^1 |f(t)|^2\, dt - \sum_{i=0}^{n} |\widehat{f}(i)|^2. \tag{2.6.5}$$

Therefore, $\sum_{i=0}^{n} |\widehat{f}(i)|^2 \le \int_0^1 |f(t)|^2\, dt$ and it follows that the series $\sum_{i=0}^{\infty} |\widehat{f}(i)|^2$ converges and satisfies

$$\sum_{i=0}^{\infty} |\widehat{f}(i)|^2 \le \int_0^1 |f(t)|^2\, dt. \tag{2.6.6}$$

On the other hand, let $\varepsilon > 0$. Since the Walsh system is closed we can choose a Walsh polynomial T such that $\|f - T\|_2 < \sqrt{\varepsilon}$. Therefore, by (2.6.4) we have

$$\int_0^1 |f(t)|^2\, dt - \sum_{i=0}^{n} |\widehat{f}(i)|^2 < \varepsilon$$

for any natural number n. We conclude by (2.6.6) that

$$\int_0^1 |f(t)|^2\, dt < \varepsilon + \sum_{i=0}^{n} |\widehat{f}(i)|^2 \le \varepsilon + \sum_{i=0}^{\infty} |\widehat{f}(i)|^2 \le \varepsilon + \int_0^1 |f(t)|^2\, dt.$$

Since ε was arbitrary, this completes the proof of the theorem. ∎

We notice that Parseval's identity and (2.6.5) imply that the Walsh-Fourier series of any function $f \in \mathbf{L}^2[0,1)$ converges to it in the $\mathbf{L}^2[0,1)$ norm. Of course the Walsh system is not special in this regard; according to the Riesz-Fischer Theorem (see **A5.4.2**), this holds for any complete orthonormal system.

We close this section with some properties of the system $\{w_{i2^m}(t)\}_{i=0}^{\infty}$ which will be used in Chapter 9. By (1.1.9), $w_{i2^m}(t) = w_i(2^m t)$. Thus the system $\{w_{i2^m}(t)\}_{i=0}^{\infty}$ can be viewed on each interval $\Delta_j^{(m)}$ as a contraction of the Walsh system in the variable t by a factor of 2^m. It is easy to verify that this contracted Walsh system inherits the properties of completeness and closure from the original system. Relationship (1.1.10) shows that this system is orthogonal on $\Delta_j^{(m)}$, and can be normalized by multiplying each function by $2^{m/2}$. Thus the following result is true:

2.6.6. *On each interval $\Delta_j^{(m)}$ of rank m, the system $\{2^{m/2} w_{i2^m}(t)\}_{i=0}^{\infty}$ is a complete orthonormal system in the space $\mathbf{L}^p(\Delta_j^{(m)})$.*

§2.7. Estimates of Walsh-Fourier coefficients. Absolute convergence of Walsh-Fourier series..

We shall obtain simple estimates of how rapidly Walsh-Fourier coefficients decay which will, in particular, show that Walsh- Fourier coefficients always tend to zero. These estimates are stated in terms of the modulii of continuity (2.3.3) and (2.3.5), and the integral modulus of continuity (2.5.1).

2.7.1. *Let $2^k \le n < 2^{k+1}$. Then the Walsh- Fourier coefficients of any $f \in \mathbf{L}[0,1)$ satisfies the inequality*

$$|\widehat{f}(n)| \le \frac{1}{2}\overset{*}{\omega}{}^{(1)}\left(\frac{1}{2^k}, f\right). \tag{2.7.1}$$

PROOF. From the definition of the Walsh functions and from identity (1.1.5) it is evident that $w_n(2^{-k-1}) = -1$ for $2^k \le n < 2^{k+1}$. Applying **2.1.2** for $a = 2^{-k-1}$ we obtain the following formula for the Walsh-Fourier coefficients of the function $f(t \oplus 2^{-k-1})$:

$$\int_0^1 f(t \oplus 2^{-k-1}) w_n(t)\, dt = \widehat{f}(n) w_n(2^{-k-1}) = -\widehat{f}(n).$$

Consequently,

$$2\widehat{f}(n) = \int_0^1 (f(t) - f(t \oplus 2^{-k-1}))\, w_n(t)\, dt, \tag{2.7.2}$$

and it follows that

$$|\widehat{f}(n)| \le \frac{1}{2}\int_0^1 |f(t) - f(t \oplus 2^{-k-1})|\, dt \le \frac{1}{2}\overset{*}{\omega}{}^{(1)}(2^{-k}, f). \qquad \blacksquare$$

In view of **2.5.1**, the following result for continuous functions is a corollary of **2.7.1**:

2.7.2. *The Walsh-Fourier coefficients of a function $f \in \mathbf{C}[0,1)$ satisfy*

$$|\widehat{f}(n)| \le \frac{1}{2}\omega(\frac{1}{2^k}, f),$$

and those of a function $f \in \mathbf{C}_{\rho^*}[0,1)$ satisfy

$$|\widehat{f}(n)| \le \frac{1}{2}\overset{*}{\omega}(\frac{1}{2^k}, f),$$

for $2^k \le n < 2^{k+1}$.

These estimates are in some sense exact (see Efimov [2]), and Rubinšteĭn [2], [3]). Notice also that Theorem **2.7.2** gives only an upper estimate for how rapidly the Walsh- Fourier coefficients of a function from $\mathbf{C}[0,1)$ and $\mathbf{C}_{\rho^*}[0,1)$ decay. One can ask what is the exact rate of decay for functions in these two classes. It turns out (see Bočkarev [2], [1]) that if $f \in \mathbf{C}[0,1)$ and $\widehat{f}(n) = O(d_n)$ for some $d_n \downarrow 0$ with $\sum_{n=1}^{\infty} d_n < \infty$, then $f(x)$ is identically constant. On the other hand, given $\sum_{n=1}^{\infty} d_n = \infty$, there exists a non- constant function $f_0 \in \mathbf{C}[0,1)$ such that $\widehat{f_0}(n) = O(d_n)$. For non-constant functions from $\mathbf{C}_{\rho^*}[0,1)$, the coefficients $\widehat{f}(n)$ can decay as slowly as one wishes.

Appealing to **2.5.2**, another corollary of **2.7.1** is that Walsh-Fourier coefficients converge to zero:

2.7.3. *If* $f \in \mathbf{L}[0,1)$ *then*

$$\lim_{n\to\infty} \widehat{f}(n) = 0.$$

For various classes of continuous functions one can obtain estimates sharper than Theorem **2.7.2**. Since any function from the Lipschitz class of order α satisfies $\omega(\delta, f) \leq C\delta^{\alpha}$, it is clear that the Walsh-Fourier coefficients of such functions must satisfy

$$\widehat{f}(n) = O(n^{-\alpha}), \qquad n \to \infty.$$

From the result of Bočkarev cited above, it follows that this estimate holds for $f \in \mathbf{C}[0,1)$ and $\alpha \leq 1$. For $f \in \mathbf{C}_{\rho^*}[0,1)$, α can take on any positive value.

Similar estimates can be obtained for classes of integrable functions, for example, for the integral Lipschitz classes defined with the integral modulii of continuity in place of the continuous one. We shall not give a more detailed statement of these results.

From inequality (2.7.1), one can also obtain estimates for Fourier coefficients of functions of bounded variation. We shall now prove the following result.

2.7.4. *If* f *is a function of bounded variation on* $[0,1)$ *with total variation* $V_0^1[f]$ *then*

$$\overset{*}{\omega}{}^{(1)}\left(\frac{1}{2^k}, f\right) \leq \frac{1}{2^k} V_0^1[f].$$

PROOF. If $h < 2^{-k}$ and $t \in \Delta_j^{(k)}$ then $t \oplus h \in \Delta_j^{(k)}$. Consequently, for such h and t we have

$$|f(t \oplus h) - f(t)| \leq \sup_{x\in\Delta_j^{(k)}} f(x) - \inf_{x\in\Delta_j^{(k)}} f(x).$$

Moreover, it is clear by the definition of bounded variation (see §2.4) that

$$\sum_{j=0}^{2^k-1} \left(\sup_{x\in\Delta_j^{(k)}} f(x) - \inf_{x\in\Delta_j^{(k)}} f(x) \right) \leq V_0^1[f].$$

It follows, therefore, that

$$\begin{aligned}
\overset{*}{\omega}{}^{(1)}\left(\frac{1}{2^k}, f\right) &= \sup_{h<1/2^k} \int_0^1 |f(t \oplus h) - f(t)|\, dt \\
&= \sup_{h<1/2^k} \sum_{j=0}^{2^k-1} \int_{\Delta_j^{(k)}} |f(t \oplus h) - f(t)|\, dt \\
&\leq \sum_{j=0}^{2^k-1} |\Delta_j^{(k)}| \left(\sup_{x\in\Delta_j^{(k)}} f(x) - \inf_{x\in\Delta_j^{(k)}} f(x) \right) \\
&\leq \frac{1}{2^k} V_0^1[f]. \qquad \blacksquare
\end{aligned}$$

2.7.5. *If f is a function of bounded variation on $[0,1)$ then*

$$|\hat{f}(n)| \leq \frac{1}{n} V_0^1[f], \qquad n = 1, 2, \ldots.$$

PROOF. Combining estimates **2.7.1** and **2.7.4** we see that

$$|\hat{f}(n)| \leq \frac{1}{2^{k+1}} V_0^1[f].$$

Since $n < 2^{k+1}$, the proof of this result is complete. ∎

We now consider absolute convergence of Walsh series. Suppose that a Walsh series $\sum_{i=0}^{\infty} a_i w_i(x)$ converges absolutely at some point x_0. Since $|a_i w_i(x_0)| = |a_i|$ for $i = 0, 1, \ldots$ it follows that $\sum_{i=0}^{\infty} a_i$ is absolutely convergent and thus the Walsh series itself converges uniformly and absolutely everywhere on [0,1). Consequently, in contrast to the trigonometric case it does not make sense to study absolute convergence of Walsh series on proper subsets of [0,1). Moreover, when we speak of an absolutely convergent Walsh series it is understood that the series of its coefficients $\sum_{i=0}^{\infty} a_i$ is absolutely convergent.

We shall identify several conditions on a function f sufficient to conclude that its Walsh-Fourier series is absolutely convergent, i.e.,

$$\sum_{n=0}^{\infty} |\hat{f}(n)| < \infty. \tag{2.7.3}$$

In order to state these conditions or prove the corresponding results, it is necessary to introduce a companion to the integral modulus of continuity defined in (2.5.1), namely, the $\mathbf{L}^2$- *modulus of continuity* of a function f which is defined by

$$\overset{*}{\omega}{}^{(2)}(\delta, f) = \sup_{h<\delta} \|f(x \oplus h) - f(x)\|_2. \tag{2.7.4}$$

Analogous to **2.5.1**, it is easy to see that

$$\overset{*}{\omega}{}^{(2)}(\delta, f) \leq \overset{*}{\omega}(\delta, f). \tag{2.7.5}$$

Moreover, it is also clear by the Cauchy-Schwarz inequality (see **A5.2.2**) that

$$\overset{*}{\omega}{}^{(1)}(\delta, f) \leq \overset{*}{\omega}{}^{(2)}(\delta, f). \tag{2.7.6}$$

The following gives a sufficient condition for absolute convergence.

2.7.6. *If $f \in \mathbf{L}^2[0,1)$ and*

$$\sum_{k=0}^{\infty} 2^{k/2} \overset{*}{\omega}{}^{(2)}(1/2^k, f) < \infty, \tag{2.7.7}$$

then (2.7.3) is satisfied.

PROOF. Let $2^k \le n < 2^{k+1}$. Observe by (2.7.2) that the n-th Walsh-Fourier coefficient of the function $f(t) - f(t \oplus (1/2^{k+1}))$ is precisely $2\widehat{f}(n)$. Hence by Parseval's identity **2.6.5**, applied to the function $f(t) - f(t \oplus (1/2^{k+1}))$, we obtain

$$4 \sum_{n=2^k}^{2^{k+1}-1} |\widehat{f}(n)|^2 \le \|f(t \oplus \frac{1}{2^{k+1}}) - f(t)\|_2^2.$$

Thus (2.7.4) implies

$$4 \sum_{n=2^k}^{2^{k+1}-1} |\widehat{f}(n)|^2 \le |\overset{*}{\omega}{}^{(2)}(1/2^k, f)|^2. \tag{2.7.8}$$

Use the Cauchy-Schwarz inequality (see **A5.2.2**) and then (2.7.8). We arrive at the estimate

$$\begin{aligned} \sum_{n=2^k}^{2^{k+1}-1} |\widehat{f}(n)| &\le \left(\sum_{n=2^k}^{2^{k+1}-1} |\widehat{f}(n)|^2 \right)^{1/2} \left(\sum_{n=2^k}^{2^{k+1}-1} 1^2 \right)^{1/2} \\ &\le 2^{k/2-1} \overset{*}{\omega}{}^{(2)}(\frac{1}{2^k}, f). \end{aligned}$$

Summing this estimate over k we obtain

$$\sum_{n=1}^{\infty} |\widehat{f}(n)| \le \frac{1}{2} \sum_{k=0}^{\infty} 2^{k/2} \overset{*}{\omega}{}^{(2)}(\frac{1}{2^k}, f).$$

Since the series on the right is finite by hypothesis, the proof of (2.7.3) is complete. ∎

2.7.7. *If f is a function which satisfies*

$$\sum_{k=0}^{\infty} 2^{k/2} \overset{*}{\omega}(1/2^k, f) < \infty, \tag{2.7.9}$$

then (2.7.3) is satisfied.

PROOF. By (2.7.5) and (2.7.9), the hypotheses of Theorem **2.7.6** are satisfied. Consequently, the Walsh-Fourier series if f converges absolutely. ∎

For continuous functions and the usual modulus of continuity, the condition analogous to (2.7.9) is written in the form

$$\sum_{n=1}^{\infty} \frac{\omega(1/n, f)}{n^{1/2}} < \infty. \tag{2.7.10}$$

2.7.8. *If a function $f \in \mathbf{C}[0,1)$ satisfies (2.7.10) then its Walsh-Fourier coefficients satisfy (2.7.3).*

PROOF. Since the terms of the series (2.7.10) are monotone decreasing, it is evident that

$$\sum_{n=1}^{\infty} \frac{\omega(1/n, f)}{n^{1/2}} = \sum_{k=1}^{\infty} \sum_{n=2^{k-1}}^{2^k-1} \frac{\omega(1/n, f)}{n^{1/2}} \geq \sum_{k=1}^{\infty} 2^{k-1} \frac{\omega(1/2^k, f)}{2^{k/2}}.$$

Consequently, hypothesis (2.7.10) implies that the series $\sum_{k=1}^{\infty} 2^{k/2}\omega(1/2^k, f)$ converges. In particular, it follows from (2.3.6) that (2.7.9) holds, and thus **2.7.7** applies. ∎

Theorems **2.7.6** through **2.7.8** allow us to identify conditions on the rate of decay of the modulii of continuity, as $\delta \to 0$, which are sufficient to conclude that a given function has an absolutely convergent Walsh-Fourier series. For example, the following result is true:

2.7.9. *The Walsh-Fourier series of a function f converges absolutely if for $\alpha > 1/2$ any one of the following conditions is satisfied:*

$$\overset{*}{\omega}{}^{(2)}(\delta, f) = O(\delta^\alpha), \; \overset{*}{\omega}(\delta, f) = O(\delta^\alpha), \quad \text{or} \quad \omega(\delta, f) = O(\delta^\alpha),$$

as $\delta \to 0$.

PROOF. It is enough to notice that each of these conditions forces convergence of the corresponding series in Theorems **2.7.6** through **2.7.8**. ∎

The following result is another consequence of Theorem **2.7.6**.

2.7.10. *If f is of bounded variation on $[0,1)$ and satisfies the condition*

$$\sum_{k=0}^{\infty} \sqrt{\overset{*}{\omega}(1/2^k, f)} < \infty, \tag{2.7.11}$$

then the Walsh-Fourier series of f converges absolutely.

PROOF. Clearly,

$$\left(\int_0^1 |f(t \oplus h) - f(t)|^2 \, dt\right)^{1/2}$$
$$\leq \left(\sup_{t\in[0,1)} |f(t \oplus h) - f(t)| \int_0^1 |f(t \oplus h) - f(t)| \, dt\right)^{1/2}.$$

Specializing to the case $h < 1/2^k$ we obtain

$$\overset{*}{\omega}^{(2)}\left(\frac{1}{2^k}, f\right) \leq \sqrt{\overset{*}{\omega}\left(\frac{1}{2^k}, f\right) \overset{*}{\omega}^{(1)}\left(\frac{1}{2^k}, f\right)}. \tag{2.7.12}$$

But **2.7.4** implies

$$\overset{*}{\omega}^{(1)}\left(\frac{1}{2^k}, f\right) \leq \frac{1}{2^k} V_0^1[f].$$

Substituting this into (2.7.12) we obtain

$$\overset{*}{\omega}^{(2)}\left(\frac{1}{2^k}, f\right) \leq \sqrt{\frac{1}{2^k}\overset{*}{\omega}\left(\frac{1}{2^k}, f\right) V_0^1[f]}.$$

Consequently,

$$\sum_{k=0}^{\infty} 2^{k/2}\overset{*}{\omega}^{(2)}\left(\frac{1}{2^k}, f\right) \leq \sqrt{V_0^1[f]}\sum_{k=0}^{\infty}\sqrt{\overset{*}{\omega}\left(\frac{1}{2^k}, f\right)}.$$

Since by (2.7.11) the series on the right side of this last inequality converges, it follows that the hypotheses of Theorem **2.7.6** hold. In particular, the function f satisfies (2.7.3). ∎

2.7.11. *If f is a function of bounded variation whose modulus of continuity satisfies*

$$\overset{*}{\omega}(\delta, f) = O\left((\ln\frac{1}{\delta})^{-\alpha}\right)$$

for some $\alpha > 2$ then its Walsh-Fourier series converges absolutely.

PROOF. The hypotheses imply that

$$\sum_{k=1}^{\infty}\sqrt{\overset{*}{\omega}\left(\frac{1}{2^k}, f\right)} \leq C\sum_{k=1}^{\infty} k^{-\alpha/2} < \infty.$$

Thus we can use Theorem **2.7.10.** ∎

We notice that the hypothesis of this theorem is satisfied when the function f is of bounded variation and belongs to the Lipschitz class of order $\alpha > 0$.

Here is a criterion for absolute convergence of a Walsh-Fourier series which uses the concept of convolution, which was introduced in §2.1.

2.7.12. *Let $f \in \mathbf{L}[0,1)$. Then the Walsh- Fourier series of f converges absolutely if and only if f can be written as the convolution of two functions g and h which belong to the space $\mathbf{L}^2[0,1)$.*

PROOF. Suppose first that $f = g * h$ for some $g, h \in \mathbf{L}^2[0,1)$. By Parseval's identity

$$\sum_{n=0}^{\infty} |\widehat{g}(n)|^2 = \|g\|_2^2 < \infty, \quad \sum_{n=0}^{\infty} |\widehat{h}(n)|^2 = \|h\|_2^2 < \infty.$$

Thus by (2.1.9) and the Cauchy-Schwarz inequality we have

$$\sum_{n=0}^{\infty} |\widehat{f}(n)| = \sum_{n=0}^{\infty} |\widehat{g}(n)\widehat{h}(n)| \le \|g\|_2\|h\|_2 < \infty.$$

Conversely, suppose that the Walsh-Fourier series of f converges absolutely, i.e., $\sum_{n=0}^{\infty} |\widehat{f}(n)| < \infty$. Choose by the Riesz-Fischer Theorem (see **A5.4.3**) functions $g \in \mathbf{L}^2[0,1)$ and $h \in \mathbf{L}^2[0,1)$ whose Walsh-Fourier coefficients satisfy

$$\widehat{g}(n) = \sqrt{|\widehat{f}(n)|}, \quad \widehat{h}(n) = \operatorname{sgn}\widehat{f}(n)\sqrt{|\widehat{f}(n)|}, \tag{2.7.13}$$

and notice that

$$\sum_{n=0}^{\infty} |\widehat{g}(n)|^2 = \sum_{n=0}^{\infty} |\widehat{h}(n)|^2 = \sum_{n=0}^{\infty} |\widehat{f}(n)| < \infty.$$

Consider the convolution $g * h$. By (2.1.9) and (2.7.13) we have

$$(g * h)^{\widehat{}}(n) = \widehat{f}(n), \qquad n = 0, 1, 2, \ldots. \tag{2.7.14}$$

Since by Theorem **2.6.1** functions in $\mathbf{L}[0,1)$ are uniquely determined by their Walsh-Fourier coefficients, it follows from (2.7.14) that the functions f and $g*h$ are identical in the space $\mathbf{L}[0,1)$. ∎

§2.8. Fourier series in multiplicative systems.

In §1.5 we defined a class of multiplicative systems which contained the Walsh system as a special case. All the results about Walsh-Fourier series which appear in the previous sections of this chapter have analogues for Fourier series with respect to these multiplicative systems. We shall explicitly mention here only the simplest of these results. Other results along these lines can be obtained as corollaries of the theorems in Chapter 6 about multiplicative transformations.

A feature which distinguishes these general multiplicative systems from the Walsh system, which is itself a special case, is that in general they are made up of complex valued functions.

The Fourier series of a function $f \in \mathbf{L}[0,1)$ with respect to the system $\{\chi_n(x)\}_{n=0}^{\infty}$ defined by (1.5.10) (or, more briefly, the $\{\chi_n(x)\}$-Fourier series of f) is the series

$$\sum_{n=0}^{\infty} \widehat{f}(n)\chi_n(x), \tag{2.8.1}$$

where

$$\widehat{f}(n) = \int_0^1 f(t)\overline{\chi_n(t)}\, dt. \tag{2.8.2}$$

According to the second identity of (1.5.7), the partial sums of the series (2.8.1) can be represented by using the Dirichlet kernels of this system (1.5.19) in the following way:

$$S_n(x) = \sum_{j=0}^{n-1} \int_0^1 f(t)\overline{\chi_n(t)}\chi_n(x)\, dt = \int_0^1 f(t)D_n(x \ominus t)\, dt. \tag{2.8.3}$$

Analogous to formula (2.1.11), it is easy to see by (1.5.21) that the partial sums of order m_k (see (1.5.4)) satisfy the identity

$$S_{m_k}(x,f) = \frac{1}{|\delta_j^{(k)}|} \int_{\delta_j^{(k)}} f(t)\, dt, \qquad x \in \delta_j^{(k)}, \tag{2.8.4}$$

where the intervals $\delta_j^{(k)}$ are defined by (1.5.13). As in the Walsh case, this identity can be used to establish the following analogue of Theorem **2.1.3**:

2.8.1. *A series in the system $\{\chi_n(x)\}$ is the Fourier series of some $f \in \mathbf{L}[0,1)$ if and only if the partial sums of this series of order m_k satisfy identity (2.8.4).*

Identity (2.8.4) implies

$$S_{m_k}(x,f) - f(x) = \frac{1}{|\delta_j^{(k)}|} \int_{\delta_j^{(k)}} (f(t) - f(x))\, dt, \qquad x \in \delta_j^{(k)}. \tag{2.8.5}$$

Consequently, we are lead to the following result:

2.8.2. *The sequence of partial sums $\{S_{m_k}(x,f)\}$ of the $\{\chi_n(x)\}$-Fourier series of a function f continuous on $[0,1]$ converges to f uniformly on $[0,1)$ and satisfies*

$$|S_{m_k}(x,f) - f(x)| \le \omega\left(\frac{1}{m_k}, f\right), \qquad x \in [0,1). \tag{2.8.6}$$

As we did in the Walsh case in §2.3, this result can be transferred to the class of discontinuous functions which are continuous with respect to a metric on [0,1) corresponding to the operation $\ominus$, analogous to the metric (1.2.19). We leave it to the reader to develop for the system $\{\chi_n(x)\}$ the corresponding concepts of the metric $\rho_{\mathbf{P}}^*$, uniform $\rho_{\mathbf{P}}^*$-continuity, the generalized modulus of continuity $\overset{*}{\omega}_{\mathbf{P}}(\delta, f)$, analogous to identity (2.3.5), and to formulate in these terms the generalized version of Theorem **2.8.2**.

We shall prove analogues for the system $\{\chi_n(x)\}$ of Theorems **2.5.9** and **2.5.12**.

2.8.3. *Let* $\{\delta_x^{(k)}\} \equiv \{[\alpha_k, \beta_k)\}_{k=1}^{\infty}$ *be the sequence of half open intervals of the form (1.5.13) which satisfy* $x \in \delta_x^{(k)}$ *for all* $k \geq 0$. *If* $F(x) = \int_0^x f(t)\,dt$ *for some* $f \in \mathbf{L}[0,1)$, *then the subsequence of partial sums* $\{S_{m_k}(x,f)\}$ *of the* $\{\chi_n(x)\}$*-Fourier series of* f *converges at a point* x *to a number* a *if and only if*

$$\lim_{k\to\infty} \frac{F(\beta_k) - F(\alpha_k)}{\beta_k - \alpha_k} = a. \tag{2.8.7}$$

PROOF. By formula (2.8.4)

$$S_{m_k}(x,f) = \frac{1}{|\delta_x^{(k)}|} \int_{\delta_x^{(k)}} f(t)\,dt = \frac{F(\beta_k) - F(\alpha_k)}{\beta_k - \alpha_k},$$

and the theorem follows immediately. ∎

The limit on the left side of identity (2.8.7) can be viewed as a derivative with respect to a net formed by the intervals $\delta_x^{(k)}$. In contrast to the nets $\{\mathcal{N}_k\}$ introduced in §1.1, these nets are not binary.

Analogous to Theorem **2.5.10** it is easy to show that at each point x where the derivative $F'(x)$ exists, the limit (2.8.7) exists and equals $F'(x)$. Thus since $F'(x) = f(x)$ the following result is a corollary of Theorem **2.8.3**.

2.8.4. *If* $f \in \mathbf{L}[0,1)$ *then the subsequence of partial sums* $\{S_{m_k}(x,f)\}$ *of the* $\{\chi_n(x)\}$*- Fourier series of* f *converges to* $f(x)$ *almost everywhere on* $[0,1)$.

This result shows us that the multiplicative systems are complete.

2.8.5. *The system* $\{\chi_n(x)\}$ *is complete in the spaces* $\mathbf{L}^p[0,1)$ *for every* $p \geq 1$ *and in* $\mathbf{C}[0,1)$.

PROOF. The proof proceeds along the same lines as that of Theorem **2.6.2**. Namely, the condition

$$\int_0^1 f(t)\overline{\chi_n(t)}\,dt = 0, \qquad n = 0,1,2,\ldots,$$

implies

$$\int_0^1 f(t) D_{m_k}(x \ominus t)\,dt = 0$$

for $x \in [0,1)$ and $k = 0,1,\ldots$, because for each fixed x the function $D_{m_k}(x \ominus t)$ can be considered as a polynomial in the system $\{\chi_n(x)\}$. By (2.8.3) this last identity means that

$$S_{m_k}(x) = 0 \qquad k = 0,1,2,\ldots.$$

Hence we conclude by Theorem **2.8.4** that

$$f(x) = \lim_{k\to\infty} S_{m_k}(x) = 0$$

for almost every $x \in [0,1)$. In the case that f is continuous, we also have that $f(x) = 0$ everywhere. ∎

Analogous to Theorem **2.6.4** we can prove the following result.

2.8.6. *The system* $\{\chi_n(x)\}$ *is closed in each of the spaces* $\mathbf{L}^p[0,1)$, $p \geq 1$, *and in* $\mathbf{C}[0,1)$.

In particular, Parseval's identity holds for multiplicative systems.

Chapter 3

GENERAL WALSH SERIES AND FOURIER-STIELTJES SERIES. QUESTIONS ON UNIQUENESS OF REPRESENTATIONS OF FUNCTIONS BY WALSH SERIES

In this chapter we shall consider general Walsh series, i.e., series whose coefficients are not necessarily Walsh-Fourier coefficients of some function. For the study of these series, the function which is the sum of series obtained by term by term integration of the given series plays an important role. This function allows us to formulate necessary and sufficient conditions for a given series to be a Walsh-Fourier series or a Walsh-Fourier-Stieltjes, i.e., a Walsh series with coefficients of the form

$$a_i = \int_0^1 w_i(x)\, d\psi(x).$$

(This integral is understood as a Stieltjes integral (see **A4.3**).) In the process we shall show that any Walsh series can be interpreted as a Walsh-Fourier-Stieltjes series if we generalize the Stieltjes integral suitably. This generalization of the Stieltjes integral will be based on binary nets which were introduced in §1.1.

In §2.5 (see **2.5.9**) we already noticed the connection between convergence of a Walsh series and questions of differentiability of functions with respect to binary nets. In this chapter this connection will be used frequently. Because of this, we shall mention below several results from the theory of differentiation with respect to binary nets. Our study of uniqueness of representation of functions by Walsh series will essentially be based on these results.

In §2.6 we established Theorem **2.6.1** which showed that a function is uniquely determined by its Walsh-Fourier coefficients. Here we examine uniqueness theorems of a different type. Namely, we shall be interested in questions concerning whether the coefficients of a given Walsh series which converges in some sense are uniquely determined by the sum of this series, and also in questions about reconstruction of the coefficients from this sum if such uniqueness holds. In particular, when the sum of this series is integrable we shall consider the question of whether these coefficients coincide with the Fourier coefficients of this sum.

Notice first of all that if the partial sums $S_n(x)$ of the series (1.4.1) converge in $\mathbf{L}^p[0,1)$ norm for some $p \geq 1$, or in $\mathbf{C}[0,1)$ norm, and $S(x)$ represents the limit of these partial sums, then (1.4.1) must be the Walsh-Fourier series of the function $S(x)$. Indeed, in this case for every $i \geq 0$ we can view a_i as a Fourier coefficient of each partial sum $S_n(x)$ for $n \geq i$. Thus

$$\begin{aligned} |a_i - \int_0^1 S(t)w_i(t)\,dt| &= |\int_0^1 (S_n(t) - S(t))\,w_i(t)\,dt| \\ &\leq \int_0^1 |S_n(t) - S(t)|\,dt \\ &= \|S_n - S\|_1 \to 0 \qquad \text{as } n \to \infty\,. \end{aligned}$$

Consequently, $a_i = \widehat{S}(i)$. By Hölder's inequality (see **A5.2.2**), these steps are still valid when the $\mathbf{L}^p$ norm replaces the $\mathbf{L}^1$ norm. The case $\mathbf{C}[0,1)$ is even simpler.

Therefor, uniqueness holds when the Walsh series converges in the norm of the spaces mentioned above. Moreover, this illustrates the leading role that Walsh-Fourier series play among the entire class of Walsh series.

The situation is more complicated when the Walsh series only converges pointwise or is summable in some sense. In §3.4 we shall show that convergence almost everywhere is not sufficient for a Walsh series to be uniquely determined by its sum. To show this we shall construct an example of a Walsh series which converges to zero almost everywhere whose coefficients are not identically zero, and consequently, cannot be the Fourier series of its sum.

If we consider only the Walsh series which converge everywhere, or everywhere except possibly on a countable set of points, then uniqueness holds. In fact, we shall show in §3.2 that if such a series converges to an integrable function then its coefficients can be reconstructed from its sum by means of the Fourier formula.

Similar uniqueness theorems can be established for certain methods of summability. We shall not discuss such questions here, except the case when the some subsequence of partial sums of the series converges, which in itself represents a kind of summability method.

§3.1 General Walsh series as a general Stieltjes series.

It is convenient to have a way to characterize a given Walsh series by means of a sequence piecewise constant functions.

3.1.1. *Let $\{k_j\}$ be an increasing sequence of natural numbers and let $\{\phi_j\}$ be a sequence of functions which are constant on each interval $\Delta_m^{(k_j)}$ of rank k_j. Suppose further that*

$$\int_{\Delta_m^{(k_j)}} \phi_j(t)\,dt = \int_{\Delta_m^{(k_j)}} \phi_{j+1}(t)\,dt \tag{3.1.1}$$

for each interval $\Delta_m^{(k_j)}$. Then there exists one and only one Walsh series S whose partial sums $S_{2^{k_j}}(x)$ coincide with $\phi_j(x)$[1].

[1] The sequence $\{\phi_j\}$ is a martingale (see for example [20]).

PROOF. By **1.3.2** each of the functions ϕ_j can be written uniquely as a polynomial of the form

$$\phi_j(x) = \sum_{i=0}^{2^{k_j}-1} a_i w_i(x). \tag{3.1.2}$$

It remains to see that for each $j_1 > j$ the coefficients of order $i \leq 2^{k_j} - 1$ of the polynomial representation of ϕ_{j_1} coincide with a_i given by (3.1.2), i.e., that the identity

$$\phi_{j_1}(x) = \phi_j(x) + \sum_{i=2^{k_j}}^{2^{k_{j_1}}-1} a_i w_i(x)$$

holds.

For this we need only notice that the coefficients of these polynomials are in fact Walsh-Fourier coefficients. Thus we need to verify that $\widehat{\phi}_{j_1}(i) = \widehat{\phi}_j(i)$ for $i < 2^{k_j}$. But for each $i < 2^{k_j}$ the function $w_i(t)$ is constant on the interval $\Delta_m^{(k_j)}$ (see **1.1.3**) with constant value $w_{i,m}$. Hence it follows from (3.1.1) that

$$\int_{\Delta_m^{(k_j)}} \phi_j(t)\, dt = \int_{\Delta_m^{(k_j)}} \phi_{j_1}(t)\, dt, \qquad j_1 > j.$$

In particular,

$$\begin{aligned}
\widehat{\phi}_{j_1}(i) &= \int_0^1 \phi_{j_1}(t) w_i(t)\, dt \\
&= \sum_{m=0}^{2^{k_j}-1} w_{i,m} \int_{\Delta_m^{(k_j)}} \phi_{j_1}(t)\, dt \\
&= \sum_{m=0}^{2^{k_j}-1} w_{i,m} \int_{\Delta_m^{(k_j)}} \phi_j(t)\, dt \\
&= \int_0^1 \phi_j(t) w_i(t)\, dt = \widehat{\phi}_j(i). \qquad \blacksquare
\end{aligned}$$

Given a Walsh series (1.4.1), we shall denote the indefinite integral of its partial sums $S_{2^k}(x)$ by

$$\psi_k(x) = \int_0^x S_{2^k}(t)\, dt. \tag{3.1.3}$$

It is clear by (1.4.7) that for points of the form $j/2^k$ we have

$$\begin{aligned}\psi_{k+1}(j/2^k) &= \int_0^{j/2^k} S_{2^{k+1}}(t)\,dt \\ &= \sum_{m=0}^{j-1} \int_{\Delta_m^{(k)}} S_{2^{k+1}}(t)\,dt \\ &= \sum_{m=0}^{j-1} \int_{\Delta_m^{(k)}} S_{2^k}(t)\,dt = \psi_k(j/2^k).\end{aligned}$$

Hence beginning with the index $i = k$, the sequence $\{\psi_i(x)\}_{i=0}^{\infty}$ determined by (3.1.3) is constant at each dyadic rational point of the form $j/2^k$, and thus converges at such points. Therefore, the limit $\psi(x) = \lim_{k\to\infty} \psi_k(x)$ exists for each dyadic rational point x. Moreover, $\psi(x) = \psi_k(x)$ for $x = j/2^k$, $j = 0, 1, \ldots, 2^k$, and consequently if $x \in [m/2^k, (m+1)/2^k)$ then

$$S_{2^k}(x) = \frac{1}{|\Delta_m^{(k)}|} \int_{\Delta_m^{(k)}} S_{2^k}(t)\,dt = \frac{\psi((m+1)/2^k) - \psi(m/2^k)}{|\Delta_m^{(k)}|}. \tag{3.1.4}$$

Notice that the function $\psi(x)$ just defined can be viewed as the term by term integral of the Walsh series (1.4.1). Moreover, $\psi(0) = 0$.

Consider now the reverse problem. Namely, let $\phi(x)$ be a function defined on the set of dyadic rational points. Thus for each interval $\Delta_m^{(k)} = \Delta_{2m}^{(k+1)} \cup \Delta_{2m+1}^{(k+1)}$ we have

(3.1.5)
$$\begin{aligned}\frac{\phi((m+1)/2^k) - \phi(m/2^k)}{|\Delta_m^{(k)}|} &= \frac{\phi((m+1)/2^k) - \phi((2m+1)/2^{k+1})}{2|\Delta_{2m}^{(k+1)}|} \\ &\quad + \frac{\phi((2m+1)/2^{k+1}) - \phi(m/2^k)}{2|\Delta_{2m+1}^{(k+1)}|}.\end{aligned}$$

If we set

$$\phi_k(x) = \frac{\phi((m+1)/2^k) - \phi(m/2^k)}{|\Delta_m^{(k)}|},$$

for $x \in \Delta_m^{(k)}$, $0 \le m \le 2^k - 1$, $k = 0, 1, 2, \ldots$, then we obtain

$$\int_{\Delta_m^{(k)}} \phi_k(x)\,dx = \int_{\Delta_{2m}^{(k+1)}} \phi_{k+1}(x)\,dx + \int_{\Delta_{2m+1}^{(k+1)}} \phi_{k+1}(x)\,dx = \int_{\Delta_m^{(k)}} \phi_{k+1}(x)\,dx.$$

Hence by **3.1.1** the sequence $\{\phi_k(x)\}$ uniquely determines a series whose partial sums satisfy $S_{2^k}(x) = \phi_k(x)$, i.e., whose partial sums satisfy the identity

$$S_{2^k}(x) = \frac{\phi((m+1)/2^k) - \phi(m/2^k)}{|\Delta_m^{(k)}|}$$

for $x \in \Delta_m^{(k)}$. At the same time these partial sums must satisfy (3.1.4), where $\psi(x)$ represents the term by term integral of the series S. Consequently,

$$\phi\left(\frac{m+1}{2^k}\right) - \phi\left(\frac{m}{2^k}\right) = \psi\left(\frac{m+1}{2^k}\right) - \psi\left(\frac{m}{2^k}\right).$$

Applying this equation to the dyadic interval $\Delta_0^{(k)} = [0, 1/2^k)$ and using the identity $\psi(0) = 0$, we see that $\phi(1/2^k) - \phi(0) = \psi(1/2^k)$. Continued applications of this equation to the intervals $\Delta_1^{(k)}, \Delta_2^{(k)}, \ldots, \Delta_{m-1}^{(k)}$ eventuates in

$$\phi(m/2^k) - \phi(0) = \psi(m/2^k).$$

Since k is arbitrary, we conclude that $\phi(x) - \phi(0) = \psi(x)$ for all dyadic rational points x.

Thus we have proved the following result:

3.1.2. *To each Walsh series* (1.4.1) *there corresponds a function $\psi(x)$, defined at each dyadic rational $x \in [0,1]$ ($\psi(0) = 0$) which represents the formal term by term integral of the series* (1.4.1) *and satisfies* (3.1.4). *Conversely, to each function $\psi(x)$ defined at all dyadic rationals $x \in [0,1]$ there corresponds a unique Walsh series whose partial sums are connected with the given function $\psi(x)$ by means of* (3.1.4) *and whose term by term integral is a series which coincides with $\psi(x) - \psi(0)$.*

Thus there is a 1-1 correspondence between Walsh series and functions $\psi(x)$ defined on the dyadic rationals. We shall call the function $\psi(x)$ which satisfies (3.1.4) the *function associated with* the Walsh series (1.4.1).

The limit, as $k \to \infty$, of the right side of (3.1.4) is precisely the definition of the derivative $D_{\{\mathcal{N}_k\}}\psi(x)$ of the function $\psi(x)$ with respect to the binary net $\{\mathcal{N}_k\}$. Thus equation (3.1.4) leads directly to a generalization of Theorem **2.5.9**.

3.1.3. *Suppose that a function $\psi(x)$ and a Walsh series (1.4.1) are related to one another in the sense of* **3.1.2**. *Then the subsequence $\{S_{2^k}\}$ of partial sums of this Walsh series converges at a point x if and only if the function $\psi(x)$ is differentiable with respect to the binary net $\{\mathcal{N}_k\}$ in which case*

$$\lim_{k\to\infty} S_{2^k}(x) = D_{\{\mathcal{N}_k\}}\psi(x). \tag{3.1.6}$$

Notice that in order for the derivative (2.5.11) to exist at a point x it is only necessary for the differentiable function to be defined at the endpoints of the intervals $\Delta_x^{(k)}$ which contain x. Thus for the situation described in **3.1.3** it is sufficient that the function $\psi(x)$ be defined at all dyadic rational points in the interval [0,1].

We generalize the concept of differentiation with respect to binary nets by introducing the *upper* and *lower derivative with respect to binary nets.* These are defined by

$$\overline{D}_{\{\mathcal{N}_k\}}\psi(x) = \limsup_{k\to\infty} \frac{\psi(\beta_x^k) - \psi(\alpha_x^k)}{|\Delta_x^{(k)}|}, \tag{3.1.7}$$

and

$$\underline{D}_{\{\mathcal{N}_k\}}\psi(x) = \liminf_{k\to\infty} \frac{\psi(\beta_x^k) - \psi(\alpha_x^k)}{|\Delta_x^{(k)}|},$$

where $\{\Delta_x^{(k)} \equiv [\alpha_x^k, \beta_x^k)\}$ is the sequence of all dyadic intervals which contain x. Recall that the limit supremum and limit infimum of a sequence of real numbers $\{a_k\}$ is defined by

$$\limsup_{k\to\infty} a_k = \lim_{k\to\infty}\left(\sup_{i\geq k} a_i\right), \quad \liminf_{k\to\infty} a_k = \lim_{k\to\infty}\left(\inf_{i\geq k} a_i\right).$$

By using (3.1.4), we can strengthen (3.1.6) in the following way:

(3.1.8) $$\limsup_{k\to\infty} S_{2^k}(x) = \overline{D}_{\{\mathcal{N}_k\}}\psi(x), \quad \liminf_{k\to\infty} S_{2^k}(x) = \underline{D}_{\{\mathcal{N}_k\}}\psi(x),$$

where the function $\psi(x)$ and the Walsh series S are related to each other in the sense of **3.1.2**.

We shall also need the concept of $\{\mathcal{N}_k\}$-continuity of a function, i.e., continuity with respect to binary nets. We shall say that a given function $\psi(x)$ defined on the set of dyadic rational points is $\{\mathcal{N}_k\}$-*continuous* at a point $x \in [0,1]$ if the endpoints of a sequence $\{\Delta_x^{(k)} = [\alpha_x^k, \beta_x^k)\}$ of dyadic intervals whose closures contain the point x satisfy

(3.1.9) $$\psi(\beta_x^k) - \psi(\alpha_x^k) \to 0, \qquad k\to\infty.$$

Notice that if x is itself a dyadic rational then there are two sequences of dyadic intervals whose closures contain x, a left one and a right one, and that condition (3.1.9) must be fulfilled by both these sequences in the case of $\{\mathcal{N}_k\}$-continuity at this point.

3.1.4. *Let $\psi(x)$ be a function which is defined at all dyadic rational points in the interval $[0,1]$. Then $\psi(x)$ is $\{\mathcal{N}_k\}$-continuous at a point $x \in [0,1]$ if and only if the corresponding Walsh series determined by (3.1.4) satisfies*

(3.1.10) $$S_{2^k}(x \pm 0) = o(2^k), \qquad k\to\infty$$

at the point x.

PROOF. If $x \in \Delta_x^{(k)} = [\alpha_x^k, \beta_x^k)$ then it is clear by (3.1.4) that

(3.1.11) $$S_{2^k}(x)2^{-k} = \psi(\beta_x^k) - \psi(\alpha_x^k).$$

On the other hand, if x belongs only to the closure of $\Delta_x^{(k)}$ but not to the dyadic interval itself, then x must be the right endpoint of $\Delta_x^{(k)}$. Since $S_{2^k}(x)$ is constant on this interval, it follows from (3.1.11) that

$$S_{2^k}(x-0)2^{-k} = \psi(\beta_x^k) - \psi(\alpha_x^k).$$

Therefore, this identity and (3.1.11) hold for all $k \geq 0$ and the proof of the theorem is complete. ∎

3.1.5. *If the coefficients a_i of a Walsh series $\sum a_i w_i(x)$ converge to zero as $i \to \infty$ then (3.1.10) holds uniformly for $x \in [0,1]$.*

PROOF. It is obvious that

$$|S_{2^k}(x)|2^{-k} \le 2^{-k} \sum_{i=0}^{2^k-1} |a_i|.$$

But the right side of this inequality is an arithmetic mean of the coefficients a_i. Since $a_i \to 0$ as $i \to \infty$ it follows that this arithmetic mean must also converge to zero as $k \to \infty$ (see Theorem **4.1.3** from the next chapter). ∎

The following is a corollary of **3.1.4** and **3.1.5**.

3.1.6. *If the coefficients of a Walsh series converge to zero then the associated function $\psi(x)$ is $\{\mathcal{N}_k\}$-continuous at each point in the interval $[0,1]$.*

Theorem **2.1.3** and formula (3.1.4) imply the following result:

3.1.7. *A Walsh series (1.4.1) is the Walsh-Fourier series of some function f integrable on $[0,1)$ if and only if the function $\psi(x)$, defined at dyadic rational points as the sum of the term by term integral of this series, coincides at these points with the indefinite integral $F(x) = \int_0^x f(t)\,dt$.*

In particular, it is clear that given any Walsh-Fourier series, the function ψ associated with it is everywhere $\{\mathcal{N}_k\}$- continuous on [0,1].

Theorem **3.1.7** can be extended to Fourier-Stieltjes series in the following way:

3.1.8. *A Walsh series (1.4.1) is the Fourier- Stieltjes series of some function ϕ, continuous from the left and of bounded variation on the interval $[0,1)$, if and only if the function $\psi(x)$, defined at dyadic rational points as the sum of the term by term integral of this series, satisfies*

$$\psi(x) = \phi(x) - \phi(0) \tag{3.1.12}$$

for all dyadic rational $x \in [0,1]$. In the case when $\phi(x)$ is continuous from the right, condition (3.1.12) is replaced by

$$\psi(x) = \phi(x-0) - \phi(0) \tag{3.1.13}$$

PROOF. We shall suppose that ϕ is continuous from the left. The case when ϕ is continuous from the right is handled similarly.

Write the function ϕ as the difference of two non-decreasing functions, say $\phi = \phi_1 - \phi_2$, where each ϕ_i is also continuous from the left (see **A4.3.4**). Let mes_{ϕ_1} and mes_{ϕ_2} be the corresponding Lebesgue-Stieltjes measures and recall (see **A3.3**) that

$$\text{mes}_{\phi_1}(\Delta_m^{(k)}) = \phi_1\left(\frac{m+1}{2^k}\right) - \phi_1\left(\frac{m}{2^k}\right),$$

and

$$\operatorname{mes}_{\phi_2}(\Delta_m^{(k)}) = \phi_2\left(\frac{m+1}{2^k}\right) - \phi_2\left(\frac{m}{2^k}\right)$$

for any dyadic interval $\Delta_m^{(k)} = [m/2^k, (m+1)/2^k)$. Since $\phi = \phi_1 - \phi_2$, it follows that

$$\phi\left(\frac{m+1}{2^k}\right) - \phi\left(\frac{m}{2^k}\right) = \operatorname{mes}_{\phi_1}(\Delta_m^{(k)}) - \operatorname{mes}_{\phi_2}(\Delta_m^{(k)}). \tag{3.1.14}$$

Suppose that the function ϕ satisfies (3.1.12). Then we can write (3.1.14) in the form

$$\psi\left(\frac{m+1}{2^k}\right) - \psi\left(\frac{m}{2^k}\right) = \operatorname{mes}_{\phi_1}(\Delta_m^{(k)}) - \operatorname{mes}_{\phi_2}(\Delta_m^{(k)}). \tag{3.1.15}$$

As we have already noticed, the coefficients a_i of a Walsh series can be viewed as the Fourier coefficients of its partial sums $S_n(x)$ when $i < n$. Thus

$$a_i = \int_0^1 S_{2^k}(x) w_i(x)\, dx, \qquad i < 2^k.$$

Since the function $w_i(x)$ has a constant value $w_{i,m}$ on each interval $\Delta_m^{(k)}$, it follows from (3.1.4) and (3.1.15) that

$$\begin{aligned} a_i &= \sum_{m=1}^{2^k-1} \int_{\Delta_m^{(k)}} S_{2^k}(x) w_i(x)\, dx \\ &= \sum_{m=1}^{2^k-1} w_{i,m} \left(\psi\left(\frac{m+1}{2^k}\right) - \psi\left(\frac{m}{2^k}\right)\right) \\ &= \sum_{m=1}^{2^k-1} w_{i,m} \operatorname{mes}_{\phi_1}(\Delta_m^{(k)}) - \sum_{m=1}^{2^k-1} w_{i,m} \operatorname{mes}_{\phi_2}(\Delta_m^{(k)}). \end{aligned}$$

But the expression on the right is precisely the Lebesgue-Stieltjes integral of the step function $w_i(x)$ with respect to the function ϕ (see **A4.3.8**). Consequently,

$$a_i = (LS) \int_0^1 w_i(x)\, d\phi(x). \tag{3.1.16}$$

Conversely, suppose that the given Walsh series is the Fourier- Stieltjes series of the function ϕ, i.e., that its coefficients are determined by formula (3.1.16).

Analogous to the derivation of formula (2.1.11), we can use (3.1.14) and **A4.3.8** to see that for any $x \in \Delta_m^{(k)}$,

$$\begin{aligned} S_{2^k}(x, d\phi) &= \int_0^1 D_{2^k}(t \oplus x)\, d\phi(t) \\ &= \frac{1}{|\Delta_m^{(k)}|} \int_{\Delta_m^{(k)}} d\phi(t) \\ &= \frac{1}{|\Delta_m^{(k)}|} \left(\text{mes}_{\phi_1}(\Delta_m^{(k)}) - \text{mes}_{\phi_2}(\Delta_m^{(k)}) \right) \\ &= \frac{1}{|\Delta_m^{(k)}|} \left(\phi\left(\frac{m+1}{2^k}\right) - \phi\left(\frac{m}{2^k}\right) \right). \end{aligned}$$

In particular, we conclude by Theorem **3.1.2** that $\psi(x) = \phi(x) - \phi(0)$, i.e., (3.1.12) holds as promised. ∎

By applying **3.1.8**, we can establish the following theorem:

3.1.9. *Suppose that the Walsh series* (1.4.1) *is the Fourier-Stieltjes series of some function* ϕ. *Then*

$$\lim_{k\to\infty} S_{2^k}(x, d\phi) = \phi'(x) \tag{3.1.17}$$

for almost every x *in* $[0,1)$ *and in fact at each point* x *where the derivative of* ϕ *exists.*

PROOF. Since ϕ is a function of bounded variation it is differentiable almost everywhere (see **A4.3.5**). But at each point x where ϕ is differentiable, the derivative $D_{\{\mathcal{N}_k\}}\phi(x)$ with respect to binary nets also exists (see Theorem **2.5.10**). On the other hand, **3.1.8** implies that at each dyadic rational point the function ϕ differs from ψ, the term by term integral of the given series, by at most a constant. Consequently, identity (3.1.17) follows directly from (3.1.6). ∎

The function ψ we considered above, which is the term by term integral of a Walsh series, allows one to view any Walsh series as a generalized Fourier-Stieltjes series.

We shall now consider the generalized Riemann-Stieltjes integral.

Let $\psi(x)$ be any function defined on the dyadic rationals in the interval [0,1) and let $f(x)$ be a function defined everywhere on [0,1). Consider the sequence of partitions of the interval [0,1) generated by the nets $\{\mathcal{N}_k\}$, namely, the k-th partition of [0,1) will be the partition consisting of the nodes of the net $\{\mathcal{N}_k\}$, i.e., consisting of the dyadic intervals $\Delta_m^{(k)}$ for $m = 0, 1, \ldots, 2^k - 1$. Corresponding to the k-th partition we shall define the *Riemann-Stieltjes sums*

$$I_k = \sum_{m=0}^{2^k-1} f(\xi_m) \left(\psi(\frac{m+1}{2^k}) - \psi(\frac{m}{2^k}) \right), \tag{3.1.18}$$

where ξ_m is any point belonging to the interval $\Delta_m^{(k)}$.

If for some constant I, $I_k \to I$ as $k \to \infty$ in the sense that given $\varepsilon > 0$ there is a natural number k_0 such that for all $k > k_0$ the sums I_k, *independent of the choice of* $\xi_m \in \Delta_m^{(k)}$, satisfy the inequality $|I_k - I| < \varepsilon$, then we shall say that $f(x)$ is *integrable on* [0,1) *with respect to the function* $\phi(x)$ through the sequence of nets $\{\mathcal{N}_k\}$, or more briefly, $\{\mathcal{N}_k\}$*-integrable with respect to* $\psi(x)$. We shall denote the corresponding integral by

$$I \equiv \{\mathcal{N}_k\} \int_0^1 f(x)\, d\psi.$$

The class of functions f integrable in this sense is obviously a vector space. We shall not require any deep properties of this integral I since it is only used to define Fourier-Stieltjes coefficients. For this it is sufficient to notice that any function which is constant on all intervals of the form $\Delta_m^{(k)}$ for some natural number k is integrable (in the sense above) with respect to any function $\psi(x)$. In particular, the integral

$$\{\mathcal{N}_k\} \int_0^1 w_i(x)\, d\psi \tag{3.1.19}$$

is defined for all $i \geq 0$.

The Walsh series whose coefficients have the form (3.1.19) will be called the *generalized Fourier-Stieltjes series of* $\psi(x)$.

We are now prepared to formulate the interesting result that any Walsh series can be viewed as a generalized Fourier-Stieltjes series.

3.1.10. *Suppose a Walsh series* (1.4.1) *and a function* $\psi(x)$ *are related to one another in the sense of* **3.1.2**. *Then*

$$a_i = \{\mathcal{N}_k\} \int_0^1 w_i(x)\, d\psi, \qquad i \geq 0.$$

PROOF. By the proof of Theorem **3.1.8** we have

$$a_i = \int_0^1 S_{2^k}(x)\, dx = \sum_{m=1}^{2^k-1} w_{i,m} \left(\psi(\frac{m+1}{2^k}) - \psi(\frac{m}{2^k})\right)$$

for any $i < 2^k$. But the expression on the right is a Riemann- Stieltjes sum of the type (3.1.18) for the function $w_i(x)$. Moreover, in this case the value I_k of this sum does not depend on the choice of the points ξ_m. Since I_k equals a_i for all $2^k > i$, it follows that a_i is the limit of these sums and thus coincides with the integral (3.1.19). ∎

The proof of this theorem shows that the partial sums $S_n(x)$ of any Walsh series (1.4.1) can be written in an integral form analogous to the equation (2.1.10) which was valid for Walsh- Fourier series, namely,

$$S_n(x) = \{\mathcal{N}_k\} \int_0^1 D_n(x \oplus t)\, d\psi(t).$$

§3.2. Uniqueness theorems for representation of functions by pointwise convergent Walsh series.

We shall establish several properties about functions which depend on the behavior of their $\{\mathcal{N}_k\}$-derivative and then use these properties to prove a uniqueness theorem about Walsh series which converge everywhere except perhaps on some countable subset of the interval [0,1).

We shall begin by proving the following auxiliary result related to the behavior of the partial sums of order 2^k of a Walsh series.

3.2.1. *Suppose that the partial sums of a Walsh series satisfies* (3.1.10) *for all points x in the closure of some dyadic interval $\Delta_m^{(k)}$ (at the endpoints one needs only assume the condition from within). Then for some $p > k$ there are two non-overlapping intervals $\Delta_{m_p}^{(p)}$ and $\Delta_{m_p'}^{(p)}$ of rank p which are subsets of $\Delta_m^{(k)}$ such that*

$$S_{2^i}(x) \le S_{2^k}(x)$$

for all $x \in \Delta_{m_p}^{(p)} \bigcup \Delta_{m_p'}^{(p)}$ and all i satisfying $k < i \le p$.

PROOF. Denote the constant value which the function $S_{2^i}(x)$ assumes on the interval $\Delta_m^{(i)}$ by $s_m^{(i)}$. By Theorem **1.4.2**,

$$s_m^{(i)} = \frac{s_{2m}^{(i+1)} + s_{2m+1}^{(i+1)}}{2}. \tag{3.2.1}$$

Consequently, $S_{2^{i+1}}(x) \le S_{2^i}(x)$ surely holds on at least one of the intervals Δ_{2m}^{i+1} or Δ_{2m+1}^{i+1}.

Let $\Delta_{m_1}^{(k+1)}$ denote one of the two intervals of rank $k+1$ whose union is $\Delta_m^{(k)}$ on which $s_{m_1}^{(k+1)} \le s_m^{(k)}$ (if this inequality is satisfied on both halves of $\Delta_m^{(k)}$ then choose the left-most one). Similarly, choose $\Delta_{m_2}^{(k+2)}$ from the two intervals whose union is $\Delta_{m_1}^{(k+1)}$ on which the corresponding inequality is satisfied for the partial sums of order 2^{k+2}. Continuing this argument we construct a sequence of nested intervals $\{\Delta_{m_j}^{(k+j)}\}_{j=1}^{\infty}$ for which

$$s_{m_j}^{(k+j)} \le s_{m_{j-1}}^{(k+j-1)}, \qquad j = 1, 2, \dots, m_0 \equiv m. \tag{3.2.2}$$

Each interval $\Delta_{m_j}^{(k+j)}$ has a neighbor $\Delta_{\tilde{m}_j}^{(k+j)}$ of the same rank which satisfies

$$\Delta_{m_j}^{(k+j)} \bigcup \Delta_{\tilde{m}_j}^{(k+j)} = \Delta_{m_{j-1}}^{(k+j-1)}.$$

However, the intervals $\{\Delta^{(k+j)}_{\tilde{m}_j}\}$ are not nested.

We shall show that there is a number $j_0 \geq 1$ such that

$$s^{(k+j_0)}_{\tilde{m}_{j_0}} \leq s^k_m. \tag{3.2.3}$$

Suppose to the contrary that

$$s^{(k+j)}_{\tilde{m}_j} > s^k_m, \qquad j = 1, 2, \ldots, \tag{3.2.4}$$

and thus in particular that

$$\delta \equiv s^{(k+1)}_{\tilde{m}_1} - s^k_m > 0. \tag{3.2.5}$$

Then by (3.2.1) we have

$$s^{(k+j)}_{m_j} + s^{(k+j)}_{\tilde{m}_j} = 2s^{(k+j-1)}_{m_{j-1}}, \qquad j = 1, 2, \ldots, m+0 \equiv m.$$

Successively applying this identity, (3.2.4) and (3.2.5) we obtain

$$s^{k+1}_{m_1} = 2s^k_m - s^{k+1}_{\tilde{m}_1} = s^k_m - (s^{k+1}_{\tilde{m}_1} - s^k_m) = s^k_m - \delta,$$

$$s^{k+2}_{m_2} = 2s^{k+1}_{m_1} - s^{k+2}_{\tilde{m}_2} < 2(s^k_m - \delta) - s^k_m = s^k_m - 2\delta,$$

$$s^{k+3}_{m_3} = 2s^{k+2}_{m_2} - s^{k+3}_{\tilde{m}_3} < 2(s^k_m - 2\delta) - s^k_m = s^k_m - 2^2\delta,$$

$$\ldots$$

$$s^{k+j}_{m_j} = 2s^{k+j-1}_{m_{j-1}} - s^{k+j}_{\tilde{m}_j} < 2(s^k_m - 2^{j-2}\delta) - s^k_m = s^k_m - 2^{j-1}\delta,$$

$$\ldots .$$

It follows that $s^{k+j}_{m_j} 2^{-(k+j)} < s^k_m 2^{-(k+j)} - \delta 2^{-(k+1)}$ and $s^{k+j}_{m_j} 2^{-(k+j)}$ does not tend to zero as $j \to \infty$. Let x be the point to which the intervals $\Delta^{(k+j)}_{m_j}$ shrink. (This point either belongs to the intersection of these intervals or is eventually an endpoint of these intervals from some point on.) Then contrary to hypothesis, condition (3.1.10) fails at the point x. Hence (3.2.4) cannot hold for all $j \geq 1$ and we have proved that there is a $j_0 \geq 1$ for which the inequality (3.2.3) holds.

Let $p = k + j_0$ and let $\Delta^{(p)}_{m_p}$ and $\Delta^{(p)}_{m'_p}$ represent the intervals $\Delta^{(k+j_0)}_{m_{j_0}}$ and $\Delta^{(k+j_0)}_{\tilde{m}_{j_0}}$. Using (3.2.2) it is easy to see that these intervals satisfy all the necessary properties. ∎

This result allows us to establish the following theorem:

3.2.2. *Suppose the partial sums of a Walsh series satisfy condition (3.1.10) at all points in the interval* $[0,1]$, *and satisfy*

$$\limsup_{k\to\infty} S_{2^k}(x) \geq 0 \tag{3.2.6}$$

for all points $x \in [0,1)$, *except perhaps for points in some countable set* E. *Then*

$$S_{2^k}(x) \geq 0, \qquad x \in [0,1),\ k \geq 0. \tag{3.2.7}$$

PROOF. Evidently, it suffices to show that for any dyadic interval $\Delta_m^{(k)}$ the following inequality holds:

$$S_{2^k}(x) \geq 0, \qquad x \in \Delta_m^{(k)}. \tag{3.2.8}$$

Suppose to the contrary that there exists an interval $\Delta_m^{(k)}$ which satisfies

$$S_{2^k}(x) < 0, \qquad x \in \Delta_m^{(k)}. \tag{3.2.9}$$

Apply **3.2.1** to this interval. Thus choose two intervals, which we shall denote by $\Delta_{(0)}$ and $\Delta_{(1)}$, on which $S_{2^i}(x) \leq S_{2^k}(x)$ for certain indices i. Apply **3.2.1** to each of the intervals $\Delta_{(0)}$ and $\Delta_{(1)}$, choosing intervals $\Delta_{(00)}, \Delta_{(01)}, \Delta_{(10)}\Delta_{(11)}$ such that $\Delta_{(00)} \cup \Delta_{(01)} \subset \Delta_{(0)}$ and $\Delta_{(10)} \cup \Delta_{(11)} \subset \Delta_{(1)}$. Continuing this process for s steps we obtain 2^s non-overlapping intervals $\Delta_{(\varepsilon_1\varepsilon_2\ldots\varepsilon_s)}$, where $\varepsilon_n = 0$ or 1, such that if p_s is the rank of these intervals then $S_{2^i}(x) \leq S_{2^k}(x)$ for $x \in \Delta_{(\varepsilon_1\varepsilon_2\ldots\varepsilon_s)}$ and for all indices i which satisfy $k < i \leq p_s$. There is one of these intervals $\Delta_{(\varepsilon_1\varepsilon_2\ldots\varepsilon_s)}$ which corresponds to each choice of a sequence $\{\varepsilon_n\}_{n=1}^{\infty}$ of zeroes and ones. Moreover, any such sequence can be interpreted as the dyadic expansion of some number a with $0 \leq a \leq 1$. Thus the set of such sequences, hence the collection of intervals $\{\Delta_{(\varepsilon_1\varepsilon_2\ldots\varepsilon_s)}\}_{s=1}^{\infty}$ has the power of the continuum. But each sequence $\{\Delta_{(\varepsilon_1\varepsilon_2\ldots\varepsilon_s)}\}_{s=1}^{\infty}$ is nested and shrinks to some point in the interval [0,1] which belongs to the closures of all these intervals. Moreover, this point cannot be related like this to more than two such sequences of nested dyadic intervals. Consequently, among the uncountable collection of such nested sequences there is a sequence which shrinks to a dyadic irrational point x_0 which does not belong to the countable set E. In particular, condition (3.2.6) holds for $x = x_0$. On the other hand, by construction

$$S_{2^i}(x_0) \leq S_{2^k}(x_0), \qquad i > k,$$

because the point x_0 is a dyadic irrational and therefore must lie in the interior of each interval from a nested sequence of the form $\{\Delta_{(\varepsilon_1\varepsilon_2\ldots\varepsilon_s)}\}_{s=1}^{\infty}$ which shrinks to x_0. Consequently, it follows from (3.2.9) that $\limsup_{i\to\infty} S_{2^i}(x_0) \leq S_{2^k}(x_0) < 0$, contradicting inequality (3.2.6) which, as we have already observed, holds for the point x_0. Consequently, (3.2.8) holds for any dyadic interval $\Delta_m^{(k)}$ and the proof of this theorem is complete. ∎

Theorem **3.2.2** is equivalent to the following:

3.2.3. *Let $\psi(x)$ be a function which is defined at all dyadic rational points in the interval $[0,1]$. Suppose further that ψ is $\{\mathcal{N}_k\}$-continuous everywhere on $[0,1]$ and satisfies the inequality*

$$\overline{D}_{\{\mathcal{N}_k\}}\psi(x) \geq 0 \tag{3.2.10}$$

for all points $x \in [0,1)$, except perhaps for points in some countable set E. Then $\psi(x)$ is non-decreasing on the set of dyadic rational points in $[0,1]$.

PROOF. By **3.1.2** construct a Walsh series corresponding to $\psi(x)$ whose partial sums satisfy (3.1.4). This Walsh series evidently satisfies (3.1.10) by Theorem **3.1.4** and the fact that $\psi(x)$ is $\{\mathcal{N}_k\}$-continuous everywhere on the interval [0,1]. This Walsh series satisfies (3.2.6) by the first identity in (3.1.8) and hypothesis (3.2.10). Hence we can apply Theorem **3.2.2** to this Walsh series verifying that its partial sums satisfy (3.2.7). It follows, therefore, from (3.1.4) that $\psi((m+1)/2^k) \geq \psi(m/2^k)$ for any m satisfying $0 \leq m \leq 2^k - 1$ for some $k \geq 0$. Since each pair a, b of dyadic rationals, the interval $[a, b)$ can be written as a finite union of non-overlapping, contiguous dyadic intervals, it follows that $\psi(b) \geq \psi(a)$. ∎

This test for monotonicity of the function $\psi(x)$ is used in the proof of the following theorem which is the main result of this section.

3.2.4. *Let $\psi(x)$ be a function which is defined at all dyadic rational points in the interval $[0,1]$. Suppose further that ψ is $\{\mathcal{N}_k\}$-continuous everywhere on $[0,1]$ and satisfies the inequality*

$$\underline{D}_{\{\mathcal{N}_k\}}\psi(x) \leq f(x) \leq \overline{D}_{\{\mathcal{N}_k\}}\psi(x), \tag{3.2.11}$$

for all points $x \in [0,1)$, except perhaps for points in some countable set E, where $f(x)$ is everywhere finite-valued and Lebesgue integrable on $[0,1)$. Then

$$\psi(x) = \psi(0) + \int_0^x f(t)\,dt \tag{3.2.12}$$

for all dyadic rationals x.

PROOF. Let $\varepsilon > 0$. Since $f \in \mathbf{L}[0,1)$ we can choose (see **A4.4.6**) a lower semicontinuous function $u(x)$ which satisfies the following properties:

a) $u(x) > -\infty$ everywhere on [0,1);
b) $u(x) \geq f(x)$ everywhere on [0,1);
c) $u(x) \in \mathbf{L}[0,1)$ and $\int_0^1 u(t)\,dt < \varepsilon + \int_0^1 f(t)\,dt$.

Set

$$U(x) = \int_0^x u(t)\,dt, \quad F(x) = \int_0^x f(t)\,dt.$$

By property c) we have

$$U(1) - F(1) < \varepsilon. \tag{3.2.13}$$

We shall show that since $u(x)$ is lower semicontinuous everywhere on the interval [0,1) we also have

$$\underline{D}_{\{\mathcal{N}_k\}}U(x) \geq u(x), \qquad x \in [0,1). \tag{3.2.14}$$

Fix a point x. By property a) we know that $u(x) > -\infty$. For each $A < u(x)$ choose by the definition of lower semicontinuity (see **A4.1**) $\delta > 0$ such that $u(t) > A$ for all $t \in [0,1)$ which satisfy $|t - x| < \delta$. Thus for all dyadic intervals $\Delta_x^{(k)}$ which satisfy $x \in \Delta_x^{(k)} \subset (x - \delta, x + \delta)$ we have

$$\int_{\Delta_x^{(k)}} u(t)\, dt > A|\Delta_x^{(k)}|.$$

Setting $\Delta_x^{(k)} \equiv [\alpha_x^k, \beta_x^k)$, it follows that

$$\frac{U(\alpha_x^k) - U(\beta_x^k)}{\alpha_x^k - \beta_x^k} > A.$$

This means that the lower $\{\mathcal{N}_k\}$-derivative of $U(x)$ at the point x is no less than A, and consequently, (3.2.14) holds. In particular, we see by properties a) and b) that

$$\underline{D}_{\{\mathcal{N}_k\}}U(x) \geq f(x), \qquad x \in [0,1), \tag{3.2.15}$$

and

$$\underline{D}_{\{\mathcal{N}_k\}}U(x) > -\infty, \qquad x \in [0,1) \tag{3.2.16}$$

Recall for any sequence of real numbers $\{a_k\}$ and $\{b_k\}$ that

$$\limsup_{k\to\infty}(a_k - b_k) \geq \liminf_{k\to\infty} a_k - \liminf_{k\to\infty} b_k.$$

Since the function f is finite-valued, it follows from inequalities (3.2.11) and (3.2.15) that

$$\overline{D}_{\{\mathcal{N}_k\}}(U(x) - \psi(x)) \geq \underline{D}_{\{\mathcal{N}_k\}}U(x) - \underline{D}_{\{\mathcal{N}_k\}}\psi(x) \geq f(x) - f(x) = 0$$

for every $x \in [0,1) \setminus E$. Since $U(x) - \psi(x)$ is obviously $\{\mathcal{N}_k\}$-continuous, it follows from Theorem **3.2.3** that the difference $U(x) - \psi(x)$ is non-decreasing on the set of dyadic rationals in [0,1).

Similarly, we can find a lower semicontinuous function $-v(x)$ corresponding to the function $-f(x)$ whose indefinite integral $-V(x)$ satisfies the inequalities

$$-V(1) + F(1) < \varepsilon, \tag{3.2.17}$$

and

$$\underline{D}_{\{\mathcal{N}_k\}}(-V(x)) \geq -f(x), \qquad x \in [0,1),$$

i.e.,

$$\overline{D}_{\{\mathcal{N}_k\}}V(x) \leq f(x), \qquad x \in [0,1).$$

As we did above for the difference $U(x) - \psi(x)$, we can show that

$$\overline{D}_{\{\mathcal{N}_k\}}(\psi(x) - V(x)) \geq \overline{D}_{\{\mathcal{N}_k\}}\psi(x) - \overline{D}_{\{\mathcal{N}_k\}}V(x) \geq 0$$

for all $x \in [0,1)\setminus E$. In particular, it follows from Theorem **3.2.3** that the difference $\psi(x) - V(x)$ is also non-decreasing on the dyadic rationals in the interval [0,1).

Since both the functions $U(x) - \psi(x)$ and $\psi(x) - V(x)$ are non-decreasing on the set of dyadic rationals it is easy to see that

$$V(x) \leq \psi(x) - \psi(0) \leq U(x) \tag{3.2.18}$$

for all dyadic rational points x. Moreover, the choice of the functions $U(x)$ and $V(x)$ directly imply

$$V(x) \leq F(x) \leq U(x), \qquad x \in [0,1],$$

and (3.2.13) and (3.2.17) clearly imply

$$0 \leq U(x) - V(x) \leq 2\varepsilon.$$

Consequently, it follows from inequality (3.2.18) that

$$|F(x) - (\psi(x) - \psi(0))| \leq 2\varepsilon$$

for all dyadic rationals x. Since $\varepsilon > 0$ was arbitrary, we conclude that (3.2.12) holds for all dyadic rational points x. ∎

We are now prepared to formulate a very general uniqueness theorem which contains as corollaries several useful results.

3.2.5. *Suppose the partial sums of a Walsh series satisfy condition (3.1.10) at every point in the interval* $[0,1]$ *and satisfy*

$$\liminf_{k\to\infty} S_{2^k}(x) \leq f(x) \leq \limsup_{k\to\infty} S_{2^k}(x) \tag{3.2.19}$$

for all points $x \in [0,1)$*, except perhaps for points in some countable set* E*, where* $f(x)$ *is everywhere finite-valued and Lebesgue integrable on* $[0,1)$*. Then this Walsh series is the Walsh-Fourier series of the function* f*.*

PROOF. This theorem is essentially a translation of Theorem **3.2.4** into the language of series. Indeed, by **3.1.2**, **3.1.4**, and equations (3.1.8), the function $\psi(x)$ associated with the given Walsh series is $\{\mathcal{N}_k\}$-continuous everywhere on [0,1] and satisfies inequality (3.2.11). Thus Theorem **3.2.4** applies to $\psi(x)$ and it follows that $\psi(x)$ satisfies (3.2.12) for each dyadic rational point x. Recall that the function associated with a Walsh series satisfies the condition $\psi(0) = 0$. Therefore, we conclude by Theorem **3.1.7** that the given Walsh series is the Walsh-Fourier series of the function f. ∎

The following theorem is a corollary of Theorem **3.2.5**.

3.2.6. *Suppose that a Walsh series (1.4.1) converges at all but countably many points in $[0, 1)$ to a finite-valued function $f \in \mathbf{L}[0, 1)$. Then this series is the Walsh-Fourier series of the function f.*

PROOF. Since $|a_i w_i(x)| = |a_i|$ for $i = 0, 1, 2, \ldots$, it is clear that if the Walsh series converges at even one point then $a_i \to 0$ as $i \to \infty$. Thus by Theorem **3.1.5** the partial sums of this Walsh series satisfy condition (3.1.10). Moreover, at any point where this series converges, condition (3.2.19) is satisfied. Therefore, the proof of this result is completed by an application of Theorem **3.2.5**. ∎

Theorem **3.2.6** contains the following result which is a uniqueness theorem in the fundamental sense of the word.

3.2.7. *If two Walsh series*

$$\sum_{i=0}^{\infty} a_i w_i(x), \quad \sum_{i=0}^{\infty} b_i w_i(x)$$

converge everywhere, except perhaps on some countable subset of $[0, 1)$, to a finite-valued function then these series are identical, i.e., $a_i = b_i$ for all $i = 0, 1, \ldots$.

PROOF. The difference of these two series is a Walsh series with coefficients $a_i - b_i$ which converges (except perhaps on some countable set) to the zero. Applying Theorem **3.2.6** to the function $f(x) = 0$ we see that the coefficients $a_i - b_i$ must be the Walsh-Fourier coefficients of the zero function, i.e., $a_i - b_i = 0$ for $i = 0, 1, \ldots$. ∎

§3.3. A localization theorem for general Walsh series.

In this section we shall use a technique called the *formal product* of a Walsh series

$$\sum_{m=0}^{\infty} a_m w_m(x) \tag{3.3.1}$$

with a Walsh polynomial

$$P(x) = \sum_{n=0}^{p} b_n w_n(x) \tag{3.3.2}$$

which is defined to be the series

$$\sum_{\ell=0}^{\infty} c_\ell w_\ell(x), \tag{3.3.3}$$

where each coefficient c_ℓ is the sum of the coefficients of the products $w_m(x)w_n(x)$ which equal $w_\ell(x)$ that result from taking the formal product of (3.3.1) with (3.3.2). Recall from (1.2.17) that $w_m(x)w_n(x) = w_{m\oplus n}(x)$. Since $n \oplus (\ell \oplus n) = \ell$ (see §1.2), it follows that the coefficients c_ℓ of the formal product (3.3.3) are determined by the equations

$$c_\ell = \sum_{n=0}^{p} b_n a_{\ell \oplus n}. \tag{3.3.4}$$

3.3.1. *Suppose the coefficients of (3.3.1) satisfy the condition*

$$\lim_{m\to\infty} a_m = 0 \tag{3.3.5}$$

and let (3.3.3) be the series obtained from a polynomial (3.3.2) by formula (3.3.4). Then (3.3.3) and the series

$$\sum_{m=0}^{\infty} (a_m w_m(x) P(x)) = P(x) \sum_{m=0}^{\infty} a_m w_m(x) \tag{3.3.6}$$

are uniformly equiconvergent, i.e., their difference converges uniformly to zero.

PROOF. Fix k so large that $p < 2^k$. Then the indices n from the sums (3.3.2) and (3.3.4) also satisfy the inequality $n < 2^k$. Notice for each fixed $n < 2^k$ that the transformation $\ell \to \ell \oplus n$ is a permutation of the collection of integers $\{\ell : 0 \le \ell \le s2^k - 1\}$ for any fixed natural number s, i.e., the given transformation is a 1-1 map from this collection onto itself. This last remark follows easily from the definition of the operation $\oplus$ (§1.2). Consequently,

$$\sum_{\ell=0}^{s2^k-1} a_{\ell \oplus n} w_{\ell \oplus n}(x) = \sum_{\ell=0}^{s2^k-1} a_\ell w_\ell(x). \tag{3.3.7}$$

It is also easy to verify that

$$\text{if } \ell \ge s2^k \text{ and } n < 2^k \text{ then } \ell \oplus n \ge s2^k . \tag{3.3.8}$$

Therefore, we obtain from (3.3.7) and (3.3.2) that

$$\begin{aligned} \sum_{\ell=0}^{s2^k-1} c_\ell w_\ell(x) &= \sum_{\ell=0}^{s2^k-1} \sum_{n=0}^{p} b_n a_{\ell\oplus n} w_n(x) w_{\ell \oplus n}(x) \\ &= \sum_{n=0}^{p} b_n w_n(x) \left(\sum_{\ell=0}^{s2^k-1} a_{\ell\oplus n} w_{\ell\oplus n}(x) \right) \\ &= P(x) \sum_{\ell=0}^{s2^k-1} a_\ell w_\ell(x). \end{aligned} \tag{3.3.9}$$

Let N be any natural number and choose s such that

$$s2^k \le N < (s+1)2^k. \tag{3.3.10}$$

Then (3.3.9) implies

$$\begin{aligned} &\Big|\sum_{\ell=0}^{N} c_\ell w_\ell(x) - P(x)\sum_{\ell=0}^{N} a_\ell w_\ell(x)\Big| \\ &= \Big|\sum_{\ell=s2^k}^{N} c_\ell w_\ell(x) - P(x)\sum_{\ell=s2^k}^{N} a_\ell w_\ell(x)\Big| \\ &\le \sum_{\ell=s2^k}^{N} |c_\ell| - \|P\|_C \sum_{\ell=s2^k}^{N} |a_\ell|. \end{aligned} \tag{3.3.11}$$

Let $\varepsilon > 0$. If N and $s2^k$ (see (3.3.10)) are sufficiently large then $|a_\ell| < \varepsilon$ for all $\ell \ge s2^k$. Thus by (3.3.4) and (3.3.8) the last line of display (3.3.11) does not exceed the value $\varepsilon\left(2^k\sum_{n=0}^{p}|b_n| + 2^k\|P\|_C\right)$, i.e., the difference in the first line of display (3.3.11) converges uniformly to zero as $N \to \infty$. ∎

An application of this theorem gives the following localization theorem for Walsh series.

3.3.2. *If (3.3.1) is a Walsh series whose coefficients satisfy condition (3.3.5) and if $[\alpha, \beta)$ is a half open interval with dyadic rational endpoints then there exists a Walsh series $\sum_{m=0}^{\infty} a_m^* w_m(x)$ whose coefficients satisfy $a_m^* \to 0$ as $m \to \infty$ which is uniformly equiconvergent with the series (3.3.1) on $[\alpha, \beta)$ and which converges uniformly to zero outside $[\alpha, \beta)$.*

PROOF. The interval $[\alpha, \beta)$ can be written as a finite union of dyadic intervals $\Delta_j^{(k)}$ for some k. Thus by **1.3.2** the characteristic function $\chi_{[\alpha,\beta)}(x)$ of this interval can be represented by a Walsh polynomial $P(x)$. Apply Theorem **3.3.1** to this series and this polynomial. Thus choose a series $\sum_{m=0}^{\infty} a_m^* w_m(x)$ which is uniformly equiconvergent with the series $\sum_{m=0}^{\infty} a_m w_m(x)\chi_{[\alpha,\beta)}(x)$. Evidently this last series is identically equal to the series (3.3.1) for $x \in [\alpha, \beta)$ and converges to zero outside $[\alpha, \beta)$.

Notice that the differences $a_m - a_m^*$ are the coefficients of a series which converges on $[\alpha, \beta)$. Consequently, $\lim_{m\to\infty}(a_m - a_m^*) = 0$. Thus it follows from (3.3.5) that $a_m^* \to 0$ as $m \to \infty$. Therefore, the series $\sum_{m=0}^{\infty} a_m^* w_m(x)$ satisfies the required conditions. ∎

3.3.3. *Let (3.3.1) be a Walsh series whose partial sums satisfy condition (3.3.5) and suppose that a subsequence $\{S_{2^{k_i}}\}$ of partial sums converges to zero on some interval $(a, b) \subset [0, 1)$. Then the full sequence of partial sums of the series (3.3.1)* *converges to zero on (a,b). In fact, this series converges to zero uniformly on each subinterval $[\alpha,\beta)$ of (a,b) with dyadic rational endpoints.*

PROOF. We need only verify the last part about uniform convergence on an interval $[\alpha,\beta) \subset (a,b)$ with dyadic rational endpoints. Apply Theorem **3.3.2** to the series (3.3.1) and the interval $[\alpha,\beta)$. Thus the partial sums $\{S^*_{2^{k_i}}(x)\}$ of the series $\sum_{m=0}^{\infty} a^*_m w_m(x)$ converges to zero on $[\alpha,\beta)$ since they are uniformly equiconvergent with the series (3.3.1). But outside $[\alpha,\beta)$ this same subsequence of partial sums also converges to zero since by Theorem **3.3.2**, the series $\sum_{m=0}^{\infty} a^*_m w_m(x)$ converges to zero for $x \notin [\alpha,\beta)$. Therefore, the partial sums of the series $\sum_{m=0}^{\infty} a^*_m w_m(x)$ satisfy the relationships

$$\liminf_{k\to\infty} S^*_{2^k}(x) \le \lim_{i\to\infty} S^*_{2^{k_i}}(x) = 0 \le \limsup_{k\to\infty} S^*_{2^k}(x)$$

for all $x \in [0,1)$. Applying Theorem **3.2.5** to the function $f(x) = 0$ we see that $a^*_m = 0$ for $m = 0, 1, \dots$. In particular, the series $\sum_{m=0}^{\infty} a^*_m w_m(x)$ is identically zero, whence it converges uniformly to zero. Since this series is uniformly equiconvergent with the series (3.3.1), we conclude that (3.3.1) converges uniformly to zero on $[\alpha,\beta)$. ∎

§3.4. Examples of null series in the Walsh system. The concept of U-sets and M-sets.

We shall construct a Walsh series which converges to zero almost everywhere on [0,1) but whose coefficients are not identically zero. In the theory of orthogonal series, such series are called null series.

We begin with the following lemma:

3.4.1. *Let E be a union of intervals $\Delta_j^{(n)}$ of rank $n \ge 0$. Then the function*

$$P_n(x;E) = \begin{cases} 0 & \text{for } x \in [0,1)\setminus E, \\ w_{2^{2n}+j2^n}(x) & \text{for } x \in \Delta_j^{(n)} \subset E \end{cases} \tag{3.4.1}$$

is a Walsh polynomial of the form

$$P_n(x;E) = \sum_{i=2^{2n}}^{2^{2n+1}-1} b_i^{(n)} w_i(x), \tag{3.4.2}$$

where

$$|b_i^{(n)}| \le 2^{-n}, \qquad 2^{2n} \le i < 2^{2n+1}. \tag{3.4.3}$$

Moreover, the set $E' \equiv \{x : P_n(x;E) = +1\}$ is a union of intervals of rank $2n+1$ and

$$\text{mes}E' = \frac{1}{2}\text{mes}E. \tag{3.4.4}$$

PROOF. The second half of this lemma concerning the set E' and especially (3.4.4) follows directly from the definition (3.4.1).

Since $j \leq 2^n - 1$, the index of a Walsh function which appears on the right side of (3.4.1) cannot exceed $2^{2n} + (2^n - 1)2^n < 2^{2n+1}$. Thus the function $P_n(x; E)$ is constant on each interval $\Delta^{(2n+1)}$ of rank $2n + 1$ (see **1.1.3**). Therefore, Theorem **1.3.2** implies that $P_n(x; E)$ is a Walsh polynomial of order no greater than $2^{2n+1} - 1$. In particular, the upper limit of the sum in (3.4.2) is correct.

On the other hand, recall from **1.1.4** that $\int_{\Delta^{(2n)}} w_i(x)\, dx = 0$ on each interval $\Delta^{(2n)}$ of rank $2n$ for $i \geq 2^{2n}$. Consequently, we have by (3.4.1) that

$$\int_{\Delta^{(2n)}} P_n(x; E)\, dx = 0. \tag{3.4.5}$$

Let $i < 2^{2n}$ and denote the constant value $w_i(x)$ assumes on each interval $\Delta_j^{(2n)}$ by $w_{i,j}$. Viewing the coefficients of the polynomial $P_n(x; E)$ as its Fourier coefficients, it follows from (3.4.5) that

$$\int_0^1 P_n(x; E) w_i(x)\, dx = \sum_{j=0}^{2^{2n}-1} w_{i,j} \int_{\Delta_j^{(2n)}} P_n(x; E)\, dx = 0, \qquad i < 2^{2n}.$$

This verifies that the lower limit of the sum in (3.4.2) is correct.

It remains to verify estimate (3.4.3). We have

$$\begin{aligned} b_i^{(n)} = \widehat{P}_n(i) &= \int_0^1 P_n(x; E) w_i(x)\, dx \\ &= \sum_{j=0}^{2^n-1} w_{i,j} \int_{\Delta_j^{(n)}} P_n(x; E) w_i(x)\, dx. \end{aligned} \tag{3.4.6}$$

It is clear by (3.4.1) that

$$\int_{\Delta_j^{(n)}} P_n(x; E) w_i(x)\, dx = 0, \qquad \Delta_j^{(n)} \subset [0, 1) \setminus E,\ i = 1, 2, \ldots. \tag{3.4.7}$$

If $\Delta_j^{(n)} \subset E$ and $2^{2n} \leq i < 2^{2n+1}$ then write i in the form $i = 2^{2n} + \ell 2^n + k$, for some $0 \leq \ell \leq 2^n - 1$, $0 \leq k \leq 2^n - 1$. Use relationship (1.2.17), equation (1.1.10) for the system $\{w_{j2^n}(x)\}_{j=0}^{\infty}$, and the fact that for $k < 2^n$ each Walsh function $w_k(x)$

is constant on the interval $\Delta_j^{(n)}$. We obtain

(3.4.8)

$$\begin{aligned}
\left| \int_{\Delta_j^{(n)}} P_n(x;E) w_i(x)\,dx \right| &= \left| \int_{\Delta_j^{(n)}} w_{2^{2n}+j2^n}(x) w_{2^{2n}+\ell 2^n+k}(x)\,dx \right| \\
&= \left| \int_{\Delta_j^{(n)}} w_{2^{2n}}^2(x) w_k(x) w_{j2^n}(x) w_{\ell 2^n}(x)\,dx \right| \\
&= \left| \int_{\Delta_j^{(n)}} w_{j2^n}(x) w_{\ell 2^n}(x)\,dx \right| \\
&= \begin{cases} 2^{-n} & \text{for } j=\ell\,, \\ 0 & \text{for } j\neq \ell, \end{cases}
\end{aligned}$$

for any $2^{2n} \le i < 2^{2n+1}$.

Substituting (3.4.7) and (3.4.8) into (3.4.6) we see that

$$|b_i^{(n)}| = \left| \int_{\Delta_\ell^{(n)}} P_n(x;E) w_i(x)\,dx \right| = \begin{cases} 2^{-n}, & \text{if } \Delta_\ell^{(n)} \subset E, \\ 0, & \text{if } \Delta_\ell^{(n)} \subset [0,1)\setminus E \end{cases}$$

for $i = 2^{2n} + \ell 2^n + k$, which establishes (3.4.3). ∎

We are now prepared to prove the fundamental result of this section.

3.4.2. *There exists a Walsh series whose coefficients are not all zero which converges to zero everywhere on $[0,1)$ except on some closed set of Lebesgue measure zero.*

PROOF. We shall use Lemma **3.4.1** to construct a sequence of sets $\{F_k\}_{k=0}^{\infty}$ and corresponding polynomials $\{P_{n_k}(x,F_k)\}_{k=0}^{\infty}$ in the following way.

Set $F_0 = [0,1)$ and $n_0 = 1$. Choose by Lemma **3.4.1** a polynomial

$$P_{n_0} \equiv P_1(x;[0,1)) = \sum_{i=2^2}^{2^3-1} b_i^{(1)} w_i(x)$$

and set

$$T_1(x) \equiv 1 + P_{n_0}(x;[0,1)).$$

By **3.4.1** it is clear that $P_{n_0}(x;[0,1)) = 1$ on some set which is a union of intervals of rank $n_1 = 3 = 2n_0 + 1$. Denote this set by F_1. Then we have by construction that $T_1(x) = 2$ for $x \in F_1$, mes$F_1 = 1/2$, mes$F_0 = 1$, $F_1 \subset F_0$ and $T_1(x) = 0$ on the set $[0,1)\setminus F_1$.

We shall now show how to execute the inductive step from k to $k+1$.

Suppose we have already chosen a sequence of sets

$$\{F_i\}_{i=0}^{\infty}, \quad F_k \subset F_{k-1} \subset \cdots \subset F_0, \quad \text{mes}F_i = 2^{-i} \tag{3.4.9}$$

(where each F_i is a union of intervals of rank $n_i = 2n_{i-1} + 1$) and a sequence of polynomials $\{P_{n_i}(x; F_i)\}_{i=0}^{k-1}$ of the form (3.4.2) such that if

$$T_k(x) \equiv 1 + \sum_{j=0}^{k-1} 2^j P_{n_i}(x; F_i)$$

then

$$F_k = \{x : T_k(x) = 2^k\}, \quad T_k(x) = 0 \quad \text{for } x \in [0,1) \setminus F_k. \tag{3.4.10}$$

Apply Lemma **3.4.1** to a function $P_{n_k}(x; F_k)$ of type (3.4.1) (i.e., with $E = F_k$) to verify that $P_{n_k}(x; F_k)$ can be written in the form (3.4.2) and moreover, that the polynomials $T_k(x)$ and $P_{n_k}(x; F_k)$ have no common terms with the same index. In fact, if we set

$$T_{k+1}(x) \equiv T_k(x) + 2^k P_{n_k}(x; F_k), \tag{3.4.11}$$

then it is clear that the polynomial $T_k(x)$ is a partial sum of the polynomial $T_{k+1}(x)$. From the construction of the polynomial $P_{n_k}(x; F_k)$ it is not difficult to verify that the set $F_{k+1} = \{x : T_{k+1}(x) = 2^{k+1}\}$ satisfies $F_{k+1} \subset F_k$, $T_{k+1}(x) = 0$ for all $x \in [0,1) \setminus F_{k+1}$, that $\text{mes} F_{k+1} = 1/2 \text{mes} F_k = 2^{-(k+1)}$ and F_{k+1} is a union of intervals of rank $n_{k+1} = 2n_k + 1$.

This completes the inductive step, and it follows that we can construct an infinite sequence of sets F_k and polynomials $T_k(x)$ which satisfy properties (3.4.9) and (3.4.10).

It is evident by (3.4.11) that the $T_{k+1}(x)$'s can be viewed as a partial sums of some Walsh series, and that these partial sums have the form

$$S_{2^{2n_k+1}}(x) = \sum_{i=0}^{2^{2n_k+1}-1} a_i w_i(x) = T_{k+1}(x).$$

By (3.4.10) we also have that

$$S_{2^{2n_k+1}}(x) = 0, \qquad x \in [0,1) \setminus F_{k+1}. \tag{3.4.12}$$

Set $F = \bigcap_{k=0}^{\infty} \overline{F}_k$ where $\overline{F}_k$ represents the closure of F_k. Notice that in this case, $\text{mes}\overline{F}_k = \text{mes} F_k$, so we have by (3.4.9) that $\text{mes} F = 0$. Moreover, by (3.4.12) we have

$$\lim_{k \to \infty} S_{2^{2n_k+1}}(x) = 0, \qquad x \in [0,1) \setminus F. \tag{3.4.13}$$

Notice that $n_k > 2n_{k-1} > 2^k$ and $a_i = 2^k b_i^{(n_k)}$. Consequently we have by (3.4.3) that $|a_i| = |2^k b_i^{(n_k)}| \le 2^k 2^{-n_k} \le 2^k 2^{-2^k}$ for $2^{2n_k} \le i < 2^{2n_k+1} = 2^{n_{k+1}}$ and $a_i = 0$

for $2^{n_k} \le i < 2^{2n_k}$. In particular, the coefficients a_i of the constructed Walsh series converge to zero as $i \to \infty$. Since this series also satisfies (3.4.13), we can apply Theorem **3.3.3** on each subinterval $[\alpha, \beta)$ of the open set $(0,1) \setminus F$, provided the endpoints α, β are dyadic rationals. It follows that the constructed Walsh series converges everywhere outside F to zero, i.e., it is the Walsh series we search for. ∎

In the theory of orthogonal series, the following concept is defined:

A set E is called a *set of uniqueness* or *U-set* for some system $\{\phi_n\}$ if the only series in this system which converges to zero outside E is identically zero, i.e., its coefficients are all equal to zero.

If there is a series in the system $\{\phi_n\}$ whose coefficients are not all zero which converges to zero outside some set E, then the set E is called an *M-set* for the system $\{\phi_n\}$.

In this terminology Theorem **3.4.2** shows us that *there is a closed non-empty M-set for the Walsh system which is of Lebesgue measure zero.*

It is not difficult to verify that any subset E of [0,1) of positive Lebesgue measure is an M-set for the Walsh system. For this it is enough to look at the Walsh-Fourier series of the characteristic function of a closed set F of positive measure which satisfies $F \subset E$. We leave it to the reader to verify the details.

On the other hand, Theorem **3.2.5** applied in the special case when $f(x) = 0$ shows us that any countable set is a U-set for the Walsh system.

Among the subsets of [0,1) which contain uncountably many points there are both M-sets (which we verified above) and U-sets. The first uncountable U-set for the Walsh system was constructed by A. A. Šneĭder (see the commentary for this chapter). We shall not give an account of his construction here. We only notice that for trigonometric series, there is a subtle theory of sets of uniqueness which contains, in particular, a profound characterization of certain perfect sets into the classes of M- sets and U-sets (see [2], Chapter 14). For the Walsh case, this theory is not yet sufficiently developed, although it is already clear that the solution to this problem will require just as complicated and delicate methods as the trigonometric case does (see the commentary).

Chapter 4

SUMMABILITY OF WALSH SERIES BY THE METHOD OF ARITHMETIC MEANS

As we saw in §2.3, there exist continuous functions whose Walsh- Fourier series diverge at a given point. We shall show in Chapter 9 that there exist integrable functions whose Walsh-Fourier series diverge everywhere on [0,1). Thus we see the necessity of identifying various methods of summability which allow a function to be recaptured from its Fourier series. In Chapter 2 we considered one such method of summability, convergence of Walsh- Fourier series through the sequence of partial sums of order 2^n. Here we shall examine another widely used method of summability which is called the method of arithmetic means, the first order method of Cesàro, or more briefly, the $(C, 1)$ method.

We shall show that the Walsh-Fourier series of every continuous functions is uniformly $(C, 1)$ summable and that any Fourier- Stieltjes series, in particular, any Walsh-Fourier series of an integrable function, is $(C, 1)$ summable almost everywhere.

§4.1. Linear methods of summability. Regularity of the arithmetic means.

This section is of an introductory nature. Recall that by a method of summability we mean a method which in general defines a sum of a divergent series.

The method of arithmetic means, or the $(C, 1)$ method is the following one. Let $\{S_n\}$ represent the partial sums of some series of numbers. Instead of insisting that the sequence S_n converges, we only look for convergence of the arithmetic means, or $(C, 1)$ *means*, of these sums. These means are defined by

$$\sigma_n \equiv \frac{S_1 + S_2 + \dots S_n}{n}.$$

If $\lim_{n\to\infty} \sigma_n$ exists and equals σ then we say that the sequence $\{S_n\}$, or the series itself, is $(C, 1)$ *summable* (or *summable by the method of arithmetic means*) to the number σ.

The $(C, 1)$ method is a special case of a class of methods called *linear*. These methods are defined in the following way.

Let

$$A = \begin{pmatrix} a_{11} & \dots & a_{1k} & \dots \\ a_{21} & \dots & a_{2k} & \dots \\ \dots & \dots & \dots & \dots \\ a_{n1} & \dots & a_{nk} & \dots \\ \dots & \dots & \dots & \dots \end{pmatrix}$$

be an infinite matrix whose rows, $a_{n1}, a_{n2}, \dots, a_{nk}, \dots$, are fixed sequences of real numbers for each $n = 1, 2, \dots$. (We shall denote both this matrix and the method of summation it defines by A.)

Given a sequence $\{S_n\}_{n=1}^{\infty}$ we define its *n-th order means by the method A* (or *A means*) to be the sum

$$\sigma_n^{(A)} \equiv \sum_{k=1}^{\infty} a_{nk} S_k,$$

when this sum converges. (In the event that this sum does not converge for some n, we say that the A means are not defined.)

If the A means $\sigma_n^{(A)}$ of some sequence $\{S_n\}_{n=1}^{\infty}$ are defined for $n = 1, 2, \dots$, and if $\lim_{n\to\infty} \sigma_n^{(A)} = \sigma$ then we shall say that the sequence $\{S_n\}$ (or the series whose partial sums are given by S_n) is *summable by the method* A to the number σ.

Such methods are called *linear* because, as can easily be seen, if two sequences $\{S_n'\}$ and $\{S_n''\}$ are summable by the method A to σ' and σ'', respectively, then any linear combination $\{\alpha S_n' + \beta S_n''\}$ of these sequences is also summable by the method A and is summable to $\alpha\sigma_n' + \beta\sigma_n''$.

The $(C, 1)$ method is a linear method of summation. Indeed, its means can be defined by using the matrix

$$\begin{pmatrix} 1 & 0 & 0 & \dots\dots\dots\dots & & \\ 1/2 & 1/2 & 0 & \dots\dots\dots\dots & & \\ \dots & \dots & \dots & \dots\dots\dots\dots & & \\ 1/n & 1/n & \dots & 1/n & 0 & \dots \\ \dots & \dots & \dots & \dots\dots\dots\dots & & \end{pmatrix}.$$

In each row of this matrix there are only finitely many non-zero entries. Thus the $(C, 1)$ method belongs to the class of *finite* linear methods of summation.

Another method of summation used in previous chapters is the method obtained by taking limits of the subsequence of partial sums $\{S_{2^n}\}$. This also is a finite linear method of summation. Indeed, the matrix which determines this method of summation is the one whose n-th row consists of a 1 at the 2^n-th place and zeroes elsewhere.

Of greatest practical interest among the linear methods of summation are the so-called *regular* methods of summation, i.e., those for which if the sequence $\{S_n\}$ converges to a number S in the conventional sense then the sequence is summable to the same number S.

We shall prove the following theorem:

4.1.1. *A linear method of summation generated by a matrix A is regular if the following conditions are satisfied:*

1) For each $k = 1, 2, \dots$,

$$\lim_{n\to\infty} a_{nk} = 0;$$

2) the series $\sum_{k=1}^{\infty} a_{nk}$ is absolutely convergent for each $n = 1, 2, \ldots$. In fact, if

$$B_n \equiv \sum_{k=1}^{\infty} |a_{nk}|,$$

then $B_n \leq C$ for $n = 1, 2, \ldots$, where C is an absolute constant which does not depend on n;

3) if $A_n \equiv \sum_{k=1}^{\infty} a_{nk}$ then $A_n \to 1$ as $n \to \infty$.

Conditions 1) through 3) were first identified by Toeplitz, who proved that these conditions are not only sufficient but also necessary for the method A to be regular. We prove here only that the Toeplitz conditions are sufficient.

PROOF. Suppose that a sequence $\{S_n\}$ converges to a number S. Then S_n can be written in the form $S_n = S + \varepsilon_n$, where $\varepsilon_n \to 0$ as $n \to \infty$. Thus

$$\sigma_n^{(A)} = \sum_{k=1}^{\infty} a_{nk} S_k = \sum_{k=1}^{\infty} a_{nk} S + \sum_{k=1}^{\infty} a_{nk}\varepsilon_k = SA_n + \sum_{k=1}^{\infty} a_{nk}\varepsilon_k.$$

By condition 3), SA_n converges to S. Hence it remains to prove that

$$\lim_{n\to\infty} \sum_{k=1}^{\infty} a_{nk}\varepsilon_k = 0.$$

Let $\varepsilon > 0$. Choose N so large that $|\varepsilon_k| < \varepsilon/(2C)$ for $k > N$, where C is the constant given by condition 2). Then

$$|\sum_{k=1}^{\infty} a_{nk}\varepsilon_k| \leq |\sum_{k=1}^{N} a_{nk}\varepsilon_k| + \frac{\varepsilon}{2C} \sum_{k=N+1}^{\infty} |a_{nk}|. \tag{4.1.1}$$

Since N is fixed, we see by property 1) that the first sum on the right side of (4.1.1) converges to zero as $n \to \infty$. Consequently, this sum is less than $\varepsilon/2$ for n sufficiently large. By condition 2), the second sum is no greater than $\varepsilon/2$. Therefore, $|\sum_{k=1}^{\infty} a_{nk}\varepsilon_k| < \varepsilon$ for n sufficiently large. We conclude that $\sum_{k=1}^{\infty} a_{nk}\varepsilon_k \to 0$ as $n \to \infty$. ∎

4.1.2. *Suppose A is a finite method of summation whose matrix contains only non-negative entries. Suppose further that A satisfies condition 1) in Theorem* **4.1.1** *and $A_n \equiv \sum_{k=1}^{\infty} a_{nk} = \sum_{k=1}^{n_k} a_{nk} = 1$. Then the method A is regular.*

PROOF. This result is an immediate corollary of Theorem **4.1.1**. Indeed, since the entries A are non-negative, condition 2) follows from condition 3). ∎

4.1.3. *The $(C, 1)$ method is regular.*

PROOF. Above, we identified the matrix which corresponds to this method of summation. It is clear from its form that condition 1) of **4.1.1** is satisfied. Moreover, $A_n = \sum_{k=1}^{n} 1/n = 1$. Thus **4.1.3** follows from **4.1.2**. ∎

§4.2. The kernel for the method of arithmetic means for Walsh-Fourier series.

We shall denote the arithmetic means (or $(C,1)$ means) of the partial sums of the Walsh-Fourier series of some integrable function f by

$$\sigma_n(x,f) = \frac{1}{n}\sum_{i=1}^{n} S_i(x,f). \tag{4.2.1}$$

Substituting the integral representation (2.1.10) of $S_i(x,f)$ into (4.2.1), we obtain

$$\sigma_n(x,f) = \int_0^1 f(x\oplus t)\frac{1}{n}\sum_{i=1}^{n} D_i(t)\,dt = \int_0^1 f(x\oplus t)K_n(t)\,dt, \tag{4.2.2}$$

where $K_n(t) \equiv \frac{1}{n}\sum_{i=1}^{n} D_i(t)$ is the kernel of the $(C,1)$ method.

We shall study the kernel $K_n(t)$.

By (1.4.11) it is clear that

$$\begin{aligned} nK_n(t) &= \sum_{i=1}^{2^k} D_i(t) + \sum_{i=2^k+1}^{2^k+m} D_i(t) \\ &= 2^k K_{2^k}(t) + \sum_{i=1}^{m}(D_{2^k}(t) + w_{2^k}(t)D_i(t)) \\ &= 2^k K_{2^k}(t) + mD_{2^k}(t) + mw_{2^k}(t)K_m(t) \end{aligned} \tag{4.2.3}$$

for $n = 2^k + m$, $1 \le m \le 2^k$. Specializing to the case $m = 2^k$ we find that

$$2^{k+1}K_{2^{k+1}}(t) = 2^k K_{2^k}(t) + 2^k D_{2^k}(t) + 2^k w_{2^k}(t)K_{2^k}(t) \tag{4.2.4}$$

or

$$2^{k+1}K_{2^{k+1}}(t) = 2^k(1 + w_{2^k}(t))K_{2^k}(t) + 2^k D_{2^k}(t). \tag{4.2.5}$$

From this formula we obtain a closed form for the kernel $K_{2^k}(t)$ and prove that it is non-negative.

4.2.1. *For each $k \ge 0$ the identity*

$$K_{2^k}(t) = \begin{cases} \dfrac{2^k+1}{2} & \text{for } t \in \Delta_0^{(k)}; \\ 2^{k-r-2} & \text{for } t \in \Delta_{2^r}^{(k)},\ r = 0,1,\ldots,k-1; \\ 0 & \text{otherwise} \end{cases} \tag{4.2.6}$$

holds. In particular, the kernel $K_{2^k}(t)$ is non-negative for all $t \in [0,1)$, all $k \geq 0$, and satisfies

$$\int_0^1 K_{2^k}(t)\,dt = \int_0^1 |K_{2^k}(t)|\,dt = 1. \tag{4.2.7}$$

PROOF. Identity (4.2.6) is verified by induction on k. It is obvious when $k = 0$, since in this case

$$K_{2^0}(t) = K_1(t) = D_1(t) = 1 = \frac{2^0+1}{2}, \qquad t \in \Delta_0^{(0)}.$$

Suppose (4.2.6) holds for some $k \geq 0$. We shall show it holds for $k+1$ in place of k by using formula (4.2.5).

Notice that $\Delta_0^{(k)} = \Delta_0^{(k+1)} \bigcup \Delta_1^{(k+1)}$ and $\Delta_{2^r}^{(k)} = \Delta_{2^{r+1}}^{(k+1)} \bigcup \Delta_{2^{r+1}+1}^{(k+1)}$. By combining (4.2.5) and (4.2.6), we obtain

$$K_{2^{k+1}}(t) = \frac{1}{2}(1 + w_{2^k}(t))\frac{2^k+1}{2} + \frac{1}{2}2^k = \begin{cases} \dfrac{2^{k+1}+1}{2} & \text{for } t \in \Delta_0^{(k+1)}, \\ 2^{k-1} & \text{for } t \in \Delta_1^{(k+1)} \end{cases}$$

for any $t \in \Delta_0^{(k)}$. To evaluate $K_{2^{k+1}}(t)$ for $t \in \Delta_{2^r}^{(k)}$ recall that $D_{2^k}(t) = 0$ for such t's. Thus on each interval $\Delta_{2^r}^{(k)}$, for $r = 0, 1, \ldots, k-1$, the value of $K_{2^{k+1}}(t)$ is completely determined by the first term on the right side of identity (4.2.5). Consequently,

$$K_{2^{k+1}}(t) = \frac{1}{2}(1 + w_{2^k}(t))2^{k-r-2} = \begin{cases} 2^{(k+1)-(r+1)-2} & \text{for } t \in \Delta_{2^{r+1}}^{(k+1)}, \\ 0 & \text{for } t \in \Delta_{2^{r+1}+1}^{(k+1)}. \end{cases}$$

It remains to observe by (4.2.5) that $K_{2^{k+1}}(t)$ takes on the value 0 everywhere that $K_{2^k}(t)$ does. This completes the inductive step from k to $k+1$ and verifies that (4.2.6) holds for all k. In particular, the kernel is non-negative and (4.2.7) holds as promised. ∎

Successive applications of (4.2.4) lead to the identity

$$2^k K_{2^k}(t) = 1 + \sum_{i=0}^{k-1} 2^i D_{2^i}(t) + \sum_{i=0}^{k-1} 2^i w_{2^i}(t) K_{2^i}(t). \tag{4.2.8}$$

We establish one more formula for the kernel $K_n(t)$ which generalizes (4.2.3). Let

$$n = \sum_{j=0}^{k} \varepsilon_j 2^j,\ \varepsilon_k = 1,\ n^{(i)} = \sum_{j=0}^{k-i} \varepsilon_j 2^j, \qquad i = 0, 1, \ldots, k. \tag{4.2.9}$$

Using this notation, (4.2.3) can be written in the following way:

$$nK_n(t) = 2^k K_{2^k}(t) + n^{(1)} D_{2^k}(t) + w_{2^k}(t) n^{(1)} K_{n^{(1)}}(t).$$

Applying this same formula to $n^{(1)}K_{n^{(1)}}(t)$, and continuing in this manner, we see by the choice of the coefficients ε_j that

(4.2.10)

$$\begin{aligned} nK_n(t) &= 2^k K_{2^k}(t) + n^{(1)} D_{2^k}(t) + w_{2^k}(t)(2^{k-1}\varepsilon_{k-1} K_{2^{k-1}}(t) \\ &\quad + n^{(2)}\varepsilon_{k-1} D_{2^{k-1}}(t) + (w_{2^{k-1}}(t))^{\varepsilon_{k-1}} n^{(2)} K_{n^{(2)}}(t)) \\ &= \dots \\ &= \sum_{j=0}^{k} \varepsilon_j 2^j \prod_{i=j+1}^{k} (w_{2^i}(t))^{\varepsilon_i} K_{2^j}(t) + \sum_{j=0}^{k} \varepsilon_j n^{(k-j+1)} \prod_{i=j+1}^{k} (w_{2^i}(t))^{\varepsilon_i} D_{2^j}(t) \\ &= \sum_{j=0}^{k} \varepsilon_j 2^j w_{n-n^{(j)}}(t) K_{2^j}(t) + \sum_{j=0}^{k} \varepsilon_j n^{(k-j+1)} w_{n-n^{(j)}}(t) D_{2^j}(t). \end{aligned}$$

This formula allows us to establish the following property for the kernel $K_n(t)$:

4.2.2. *For all* $n \geq 1$,

$$\int_0^1 |K_n(t)|\,dt < 2.$$

This property of the kernel $K_n(t)$ is called *quasi- positivity.*

PROOF. Combining (4.2.10) and (4.2.7), and bearing in mind that $n^{(i)} \leq 2^{k-i+1}$, we obtain

$$\int_0^1 |K_n(t)|\,dt \leq \frac{1}{n}\left(\sum_{j=0}^{k} \varepsilon_j 2^j \int_0^1 |K_{2^j}(t)|\,dt + \sum_{j=0}^{k} \varepsilon_j 2^j \int_0^1 |D_{2^j}(t)|\,dt\right) = 2. \quad \blacksquare$$

§4.3. Uniform $(C, 1)$ summability of Walsh-Fourier series of continuous functions.

In parallel with our study of uniform convergence in §2.3, we shall look not only at functions continuous in the usual sense, but also ρ^*-continuous functions for which the modulus of continuity is defined by identity (2.3.5). We shall obtain uniform $(C, 1)$ summability of Walsh-Fourier series of ρ^*-continuous functions as a special case of a result which estimates the rate of approximation to a function by the $(C, 1)$ means of its Walsh- Fourier series in terms of its modulus of continuity.

4.3.1. *Let* $\overset{*}{\omega}(\delta, f)$ *be the modulus of continuity of* f *defined by identity (2.3.5). Then the* $(C, 1)$ *means* $\sigma_n(x, f)$ *of the Walsh-Fourier series of the function* f *satisfies*

$$|\sigma_n(x, f) - f(x)| \leq C \sum_{i=0}^{k} 2^{i-k} \overset{*}{\omega}(2^{-i}, f) \tag{4.3.1}$$

for all $2^k < n \le 2^{k+1}$, where C is an absolute constant.

PROOF. Using (4.2.2) and the fact that $\int_0^1 K_n(t)\,dt = 1$, we obtain the following integral expression for the difference we must estimate:

$$\sigma_n(x,f) - f(x) = \int_0^1 (f(x\oplus t) - f(x))K_n(t)\,dt.$$

Let $n = 2^k + m$, $1 \le m \le 2^k$. Substitute the expression for $K_n(t)$ from (4.2.3) and use (4.2.8) to obtain

(4.3.2)

$$\begin{aligned}\sigma_n(x,f) - f(x) &= \frac{1}{n}\int_0^1 (f(x\oplus t) - f(x))\,dt \\ &+ \frac{1}{n}\sum_{i=0}^{k-1} 2^i \int_0^1 (f(x\oplus t) - f(x))D_{2^i}(t)\,dt \\ &+ \frac{1}{n}\sum_{i=0}^{k-1} 2^i \int_0^1 (f(x\oplus t) - f(x))w_{2^i}(t)K_{2^i}(t)\,dt \\ &+ \frac{m}{n}\int_0^1 (f(x\oplus t) - f(x))D_{2^k}(t)\,dt \\ &+ \frac{m}{n}\int_0^1 (f(x\oplus t) - f(x))w_{2^k}(t)K_m(t)\,dt.\end{aligned}$$

We estimate the last term first. Toward this, notice that for $m \le 2^k$ the kernel $K_m(t)$ is a sum of Dirichlet kernels each of which is constant on the intervals $\Delta_j^{(k)}$ (see **1.4.3**). Consequently, for each $j = 0, 1, \ldots, 2^{k-1}$, $K_m(t)$ is also constant on the interval $\Delta_j^{(k)}$. We shall denote this constant value by $K_{m,j}^{(k)}$.

Recall that $w_{2^k}(t) = r_k(t) = 1$ for $t \in \Delta_{2j}^{(k+1)}$ and $w_{2^k}(t) = -1$ for $t \in \Delta_{2j+1}^{(k+1)}$, where $\Delta_j^{(k)} = \Delta_{2j}^{(k+1)} \bigcup \Delta_{2j+1}^{(k+1)}$. Also observe for each $t \in \Delta_{2j}^{(k+1)}$ that $t \oplus 2^{-(k+1)}$ belongs to $\Delta_{2j+1}^{(k+1)}$ but the points $x \oplus t$ and $x \oplus t \oplus 2^{-k-1}$ belong to the same interval of rank k. Hence

$$|f(x\oplus t) - f(x \oplus t \oplus \frac{1}{2^{k+1}})| \le \overset{*}{\omega}(\frac{1}{2^k}, f).$$

Therefore, we can estimate the last term of (4.3.2) as follows:

$$
\begin{aligned}
&\left| \int_0^1 (f(x \oplus t) - f(x)) w_{2^k}(t) K_m(t)\, dt \right| \\
&= \left| \sum_{j=0}^{2^k-1} K_{m,j}^{(k)} \int_{\Delta_j^{(k)}} (f(x \oplus t) - f(x)) w_{2^k}(t)\, dt \right| \\
&= \left| \sum_{j=0}^{2^k-1} K_{m,j}^{(k)} \int_{\Delta_{2j}^{(k+1)}} (f(x \oplus t) - f(x)) - (f(x \oplus t \oplus \frac{1}{2^{k+1}}) - f(x))\, dt \right| \\
&= \left| \sum_{j=0}^{2^k-1} K_{m,j}^{(k)} \int_{\Delta_{2j}^{(k+1)}} (f(x \oplus t) - f(x \oplus t \oplus \frac{1}{2^{k+1}}))\, dt \right| \\
&\le \sum_{j=0}^{2^k-1} |K_{m,j}^{(k)}| \cdot |\Delta_{2j}^{(k+1)}| \cdot \overset{*}{\omega}(2^{-k}, f) \\
&= \overset{*}{\omega}(\frac{1}{2^k}, f) \frac{1}{2} \sum_{j=0}^{2^k-1} |K_{m,j}^{(k)}| \cdot |\Delta_j^{(k)}| \\
&= \overset{*}{\omega}(\frac{1}{2^k}, f) \frac{1}{2} \int_0^1 |K_m(t)|\, dt \le \overset{*}{\omega}(\frac{1}{2^k}, f).
\end{aligned}
$$

Similarly, we can estimate the third term on the right side of (4.3.2) by applying the same reasoning to $K_{2^i}(t)$ instead of $K_m(t)$. We obtain

$$
\int_0^1 (f(x \oplus t) - f(x)) w_{2^i}(t) K_{2^i}(t)\, dt \le \overset{*}{\omega}(\frac{1}{2^i}, f), \quad i = 0, 1, \ldots, k-1. \tag{4.3.4}
$$

The terms which have the kernels $D_{2^i}(t)$ as integrands can be estimated by using the facts that each such kernel is supported on the interval $\Delta_0^{(i)}$ and that for $t \in \Delta_0^{(i)}$ the points x and $x \oplus t$ always belong to the same interval of rank i. Thus

$$
\int_0^1 (f(x \oplus t) - f(x)) D_{2^i}(t)\, dt \le \overset{*}{\omega}(\frac{1}{2^i}, f), \quad i = 0, 1, \ldots, k. \tag{4.3.5}
$$

Finally, the integral in the first term on the right side of (4.3.2) can be estimated directly using the modulus of continuity $\overset{*}{\omega}(1, f)$. Substituting estimates (4.3.3) through (4.3.5) into (4.3.2) we conclude that

$$\begin{aligned}|\sigma_n(x,f)-f(x)| &\le \frac{1}{n}\overset{*}{\omega}(1,f)\\&+\frac{1}{n}\sum_{i=0}^{k-1}2^i\overset{*}{\omega}(\frac{1}{2^i},f)+\frac{1}{n}\sum_{i=0}^{k-1}2^i\overset{*}{\omega}(\frac{1}{2^i},f)+\frac{m}{n}\overset{*}{\omega}(\frac{1}{2^k},f)\\&\le \frac{1}{2^k}\overset{*}{\omega}(1,f)+2\sum_{i=0}^{k-1}2^{i-k}\overset{*}{\omega}(\frac{1}{2^i},f)+2\overset{*}{\omega}(\frac{1}{2^k},f)\\&\le 3\sum_{i=0}^{k}2^{i-k}\overset{*}{\omega}(\frac{1}{2^i},f).\end{aligned}$$

In particular, (4.3.1) holds with $C=3$. ∎

An immediate corollary of this result is the following theorem:

4.3.2. *If a function f is ρ^*-continuous on $[0,1)$ (in particular, if it is continuous in the classical sense on the interval $[0,1]$), then the Walsh-Fourier series of f is $(C,1)$ summable to f uniformly on $[0,1)$.*

PROOF. The hypotheses imply that $\overset{*}{\omega}(1/2^i,f)\to 0$ as $i\to\infty$. Consider the method of summability which is determined by the matrix

$$A=\begin{pmatrix}1 & 0 & 0 & 0 & \cdots\cdots & & \\ 1/2 & 1/2 & 0 & 0 & \cdots\cdots & & \\ 1/4 & 1/4 & 1/2 & 0 & \cdots\cdots & & \\ \cdots & \cdots & \cdots & \cdots & \cdots & \cdots & \cdots\\ 1/2^k & 1/2^k & 1/2^{k-1} & \ldots & 1/2 & 0 & \ldots\\ \cdots & \cdots & \cdots & \cdots & \cdots & \cdots & \cdots\end{pmatrix}.$$

It is evident that this method satisfies the hypotheses of Theorem **4.1.2**. Hence the summability method A is regular. Since the sequence $\{\overset{*}{\omega}(2^{-i},f)\}_{i=0}^{\infty}$ converges to 0, it follows that

$$\sigma_k^{(A)}=\frac{1}{2^k}\overset{*}{\omega}(1,f)+\sum_{i=1}^{k}\frac{2^{i-1}}{2^k}\overset{*}{\omega}(\frac{1}{2^i},f)$$

also converges to zero as $k\to\infty$. Since the sum on the right side of (4.3.1) is precisely $2\sigma_k^{(A)}-2^{-k}\overset{*}{\omega}(1,f)$, we conclude by Theorem **4.3.1** that $\sigma_n(x,f)$ converges to f uniformly on [0,1) as promised. ∎

4.3.3. *Suppose $0<\alpha\le 1$. If $\overset{*}{\omega}(2^{-i},f)=O(2^{-i\alpha})$ as $i\to\infty$ (in particular, if f belongs to the Lipschitz class on $[0,1)$ of order α) then*

$$\|\sigma_n(x,f)-f(x)\|=\begin{cases}O(1/n^\alpha) & \text{for } 0<\alpha<1,\\ O((\ln n)/n) & \text{for } \alpha=1.\end{cases}$$

PROOF. Substitute the estimate

$$\overset{*}{\omega}(\frac{1}{2^i}, f) \le C_1 \frac{1}{2^{i\alpha}}$$

into (4.3.1) to see that the difference $\|\sigma_n(x,f) - f(x)\|$ is dominated by

$$\frac{1}{2^k}\sum_{i=0}^{k} 2^i \cdot \frac{1}{2^{i\alpha}} \le C_2 \frac{1}{2^{(k+1)\alpha}} \qquad \text{for } 0 < \alpha < 1,$$

and

$$\frac{1}{2^k}\sum_{i=0}^{k} 2^i \cdot \frac{1}{2^i} = \frac{k+1}{2^k} \le C_3 \frac{k}{2^{k+1}} \qquad \text{for } \alpha = 1$$

where $2^k < n \le 2^{k+1}$. Since such integers n, k satisfy $k \le C_4 \ln n$, these inequalities establish the promised estimates. ∎

§4.4. $(C,1)$ summability of Fourier-Stieltjes series.

In this section we again use the technique of differentiation through binary nets, this time to establish that every Fourier- Stieltjes series in the Walsh system (in particular, every Walsh- Fourier series) is $(C,1)$ summable almost everywhere on [0,1).

We begin with two lemmas connected with the dyadic metric $\rho^*(x,t)$ defined in (1.2.19).

4.4.1. *Let $I = (a,b) \subset [0,1)$ be an interval and set $d(x) = \min\{\rho^*(x,a), \rho^*(x,b)\}$. Then $d(x) \le 2|I|$ for all $x \in I$.*

PROOF. Let $\Delta^{(k)}$ be the dyadic interval of minimal rank which is contained in the interval I. Thus for any $x \in I$, the interval $\Delta_x^{(k-1)}$ which contains x also contains either a or b. Hence $\min\{\rho^*(x,a), \rho^*(x,b)\} \le 1/2^{k-1}$. But $|I| \ge 1/2^k$ since $\Delta^{(k)}$ is a subset of I. Consequently, $d(x) \le 1/2^{k-1} \le 2|I|$. ∎

4.4.2. *Let E be a closed subset of $[0,1)$ and define a function p by*

$$p(x) \equiv \rho^*(x,E) \equiv \inf_{y\in E} \rho^*(x,y).$$

Then the series

$$\sum_{j=1}^{\infty} 2^j p(x \oplus \frac{1}{2^j})$$

converges for almost every x in E.

PROOF. Let $\{I_n\}$ be the collection of intervals contiguous to E, i.e.,

$$\bigcup_{n=1}^{\infty} I_n = (0,1) \setminus E.$$

It is clear that $\sum_{n=1}^{\infty} |I_n| \le 1$. Fix $0 < h < 1$ and consider the function $p(x \oplus h)$. If we denote the set $\{x = y \oplus h : y \in E\}$ by $E \oplus h$ then we have by translation invariance of Lebesgue integration that

$$\text{(4.4.1)} \qquad \begin{aligned} \int_E p(x \oplus h)\, dx &= \int_{E\oplus h} p(x)\, dx \\ &= \sum_{n=1}^{\infty} \int_{(E\oplus h)\cap I_n} p(x)\, dx \\ &= \sum_{n:|I_n|<h} + \sum_{n:|I_n|\ge h} \equiv S_1(h) + S_2(h). \end{aligned}$$

For each interval I_n define the function $d_n(x)$ as in **4.4.1** and notice that $p(x) \le d_n(x)$ for all $x \in I_n$. Hence by lemma **4.4.1** we have $p(x) \le 2|I_n|$ for every $x \in I_n$. Moreover, $\text{mes}((E \oplus h) \cup I_n) \le |I_n|$. Consequently,

$$\int_{(E\oplus h)\cap I_n} p(x)\, dx \le 2|I_n|^2.$$

Therefore,

$$S_1(h) \le \sum_{n:|I_n|<h} 2|I_n|^2.$$

Applying this estimate to $h = 2^{-j}$ for $j = 1, 2, \ldots$ results in

$$\text{(4.4.2)} \qquad \begin{aligned} \sum_{j=1}^{\infty} 2^j S_1(2^{-j}) &\le \sum_{j=1}^{\infty} 2^j \left(\sum_{n:|I_n|<2^{-j}} 2|I_n|^2 \right) \\ &= \sum_{n=1}^{\infty} 2|I_n|^2 \left(\sum_{j:2^j<|I_n|^{-1}} 2^j \right) \\ &< 2\sum_{n=1}^{\infty} |I_n|^2 |I_n|^{-1} = 2\sum_{n=1}^{\infty} |I_n| \le 2. \end{aligned}$$

To estimate the remaining sum $S_2(h)$ in (4.4.1), fix $|I_n| \ge h$ and observe by (1.2.20) that

$$|y \oplus h - y| \le \rho^*(y \oplus h, y) = h.$$

Thus it is clear that if $y \in E$ and $x = y \oplus h$ lies in I_n then x is no further from the endpoints of the interval I_n than h, i.e., $\text{mes}((E \oplus h) \cap I_n) \le 2|h|$. Moreover, it is obvious that $p(x) \le \rho^*(y \oplus h, y) = h$ for $x \in (E \oplus h) \cap I_n$, where $x = y \oplus h$ and $y \in E$. Consequently,

$$S_2(h) \le \sum_{n:|I_n|\ge h} h \cdot 2h = 2h^2 \left(\sum_{n:|I_n|\ge h} 1 \right).$$

Using this estimate for $h = 2^{-j}$ and summing over j we are lead to the following inequality:

(4.4.3)

$$\begin{aligned}\sum_{j=1}^{\infty} 2^j S_2(2^{-j}) &\le \sum_{j=1}^{\infty} 2^j \cdot 2 \cdot 2^{-2j} \left(\sum_{n:|I_n| \ge 2^{-j}} 1 \right) \\ &= \sum_{j=1}^{\infty} 2 \cdot 2^{-j} \left(\sum_{n:|I_n| \ge 2^{-j}} 1 \right) = \sum_{n=1}^{\infty} \left(\sum_{j:2^{-j} \le |I_n|} 2 \cdot 2^{-j} \right) \\ &\le \sum_{n=1}^{\infty} 4|I_n| \le 4.\end{aligned}$$

Therefore, it follows from (4.4.1) through (4.4.3) and a theorem of B. Levy (see **A4.4.2**) that

$$\begin{aligned}\int_E \sum_{j=1}^{\infty} 2^j p(x \oplus \frac{1}{2^j})\, dx &= \sum_{j=1}^{\infty} 2^j \int_E p(x \oplus \frac{1}{2^j})\, dx \\ &\le \sum_{j=1}^{\infty} 2^j S_1(2^{-j}) + \sum_{j=1}^{\infty} 2^j S_2(2^{-j}) \le 6.\end{aligned}$$

Thus the function $\sum_{j=1}^{\infty} 2^j p(x \oplus \frac{1}{2^j})$ is integrable on E, in particular, almost everywhere finite on E. ∎

The next result is the main technical lemma of this section. We shall use it to prove that each Fourier-Stieltjes series in the Walsh system is almost everywhere $(C, 1)$ summable.

4.4.3. *Let $\psi(x)$ be a function of bounded variation on $[0, 1)$ with $\psi'(x) = f(x)$ almost everywhere. For each fixed x, let $V_0^t(\psi(u) - f(x)u)$ represent the variation of the function $\psi(u) - f(x)u$ as u ranges over the interval $[0, t)$. Then*

$$\lim_{k\to\infty} \int_0^1 K_{2^k}(x \oplus t) dV_0^t(\psi(u) - f(x)u) = 0 \tag{4.4.4}$$

for almost everywhere x in $[0, 1)$, where $K_n(t)$ represents the kernel with respect to the Walsh system of the $(C, 1)$ method of summation.

PROOF. Let $\eta > 0$ and choose by Lusin's Theorem (see **A4.1.2**) a closed set $E_1 \subset [0, 1)$ such that $\text{mes} E_1 > 1 - \eta/2$, and such that the function $f(x)$ is continuous on the set E_1. We may also choose the set E_1 so that it contains no points from the set $P \bigcup \{\bigcup_{j=1}^{\infty} P \oplus 2^{-j}\}$, where P is some countable set which contains all the points of discontinuity of the function $V_0^t(\psi(u))$. We may also suppose that E_1 contains no dyadic rational points.

By Theorem A4.3.7 we know that

$$\lim_{h\to 0}\frac{1}{h}V_x^{x+h}(\psi(u)-f(x)u)=0$$

for almost every $x\in[0,1)$. Consequently, it follows from **2.5.10** that

$$\lim_{k\to 0}\frac{1}{|\Delta_x^{(k)}|}V_{\alpha_x^k}^{\beta_x^k}(\psi(u)-f(x)u)=0 \tag{4.4.5}$$

for almost every $x\in[0,1)$, where as usual $\{\Delta_x^{(k)}=[\alpha_x^k,\beta_x^k)\}_{k=0}^{\infty}$ represents the sequence of dyadic intervals which contain x. Apply Egoroff's Theorem (see **A4.1.1**) to this limit of functions. Thus choose a closed set E_2 such that $\text{mes}E_2>1-\eta/2$ and

$$\lim_{k\to\infty}\frac{1}{|\Delta_x^{(k)}|}V_{\alpha_x^k}^{\beta_x^k}(\psi(u)-f(x)u)=0 \text{ uniformly for } x\in E_2. \tag{4.4.6}$$

Let $E=E_1\cap E_2$. Then E is closed, $\text{mes}E>1-\eta$, and E contains no dyadic rational points and no points from the set $P\bigcup\{\bigcup_{j=1}^{\infty}P\oplus 2^{-j}\}$ mentioned above. The function $f(x)$ is continuous on E so there is a constant A_η such that

$$|f(x)|\le A_\eta,\qquad x\in E. \tag{4.4.7}$$

Moreover, if we set

$$M_k\equiv\sup_{s\ge k}\sup_{x\in E}\frac{1}{|\Delta_x^{(s)}|}V_{\alpha_x^s}^{\beta_x^s}(\psi(u)-f(x)u) \tag{4.4.8}$$

then we have by (4.4.6) that

$$\lim_{k\to\infty}M_k=0. \tag{4.4.9}$$

Since η was arbitrary, it suffices to prove (4.4.4) for almost every point x from the set E.

Use (4.2.6) to write

(4.4.10)

$$K_{2^k}(x\oplus t)dV_0^t(\psi(u)-f(x)u)=\frac{2^k-1}{2}\int_{x\oplus\Delta_0^{(k)}}dV_0^t(\psi(u)-f(x)u)$$
$$+\sum_{r=0}^{k-1}2^{k-r-2}\int_{x\oplus\Delta_{2^r}^{(k)}}dV_0^t(\psi(u)-f(x)u).$$

Fix $x \in E$ and let $x \in \Delta_x^{(k)} = [\alpha_x^k, \beta_x^k)$. Then with the exception of a countable set of points, $x \oplus \Delta_0^{(k)} = \Delta_x^{(k)}$ (see **1.2.1**). By (4.4.8), the first term on the right side of (4.4.10) is dominated by

$$\frac{2^k - 1}{2} V_{\alpha_x^k}^{\beta_x^k}(\psi(u) - f(x)u) < M_k,$$

hence must converge to zero by (4.4.6). Change the index in the sum on the right side of (4.4.10), setting $k - r = j$. Since with the exception of a countable set of points, $2^{-j} \oplus \Delta_0^{(k)} = \Delta_{2^{k-j}}^{(k)}$, for $j = 1, 2, \dots k$, we see that this sum can be written in the form

$$\frac{1}{4} \sum_{j=1}^{k} 2^j \int_{x \oplus 2^{-j} \oplus \Delta_0^{(k)}} dV_0^t(\psi(u) - f(x)u).$$

In particular, it remains to show that the sum

$$S \equiv \sum_{j=1}^{k} 2^j \int_{x \oplus 2^{-j} \oplus \Delta_0^{(k)}} dV_0^t(\psi(u) - f(x)u) \tag{4.4.11}$$

converges to zero almost everywhere on E.

Fix $\varepsilon > 0$ and choose by (4.4.9) and the continuity of f on E an index $q = q(\varepsilon)$ so large that

$$M_q < \varepsilon \tag{4.4.12}$$

and

$$|f(y) - f(x)| < \varepsilon, \ x, y \in E, \ |x - y| \le 2^{-q}. \tag{4.4.13}$$

Divide the sum (4.4.11) into two pieces:

$$S = \sum_{j=1}^{q} + \sum_{j=q+1}^{k} \equiv S_1 + S_2. \tag{4.4.14}$$

It is easy to see that $V_0^t(\psi(u) - f(x)u) \le V_0^t(\psi(u)) + |f(x)|t$. Therefore, we have by (4.4.7) that

$$\begin{aligned} S_1 &\le \sum_{j=1}^{q} 2^j \left(\int_{x \oplus 2^{-j} \oplus \Delta_0^{(k)}} dV_0^t(\psi(u)) + |\Delta^{(k)}|\,|f(x)| \right) \\ &\le \left(\max_{1 \le j \le q} \int_{x \oplus 2^{-j} \oplus \Delta_0^{(k)}} dV_0^t(\psi(u)) + A_\eta \cdot 2^{-k} \right) \sum_{j=1}^{q} 2^j. \end{aligned}$$

By the choice of E, the function $V_0^t(\psi(u))$ is continuous at each point of the form $t = x \oplus 2^{-j}$, for $x \in E$. Hence for each fixed q, the right part of this last inequality (and thus S_1 itself) converges to zero as $k \to \infty$. Consequently,

$$S_1 < \varepsilon \tag{4.4.15}$$

for k sufficiently large.

We now estimate the sum S_2. For $q < j \leq k$ and $x \in E$ let $y_j = y_j(x)$ represent a point from E for which the minimum $\min_{y \in E} \rho^*(x \oplus 2^{-j}, y)$ is attained. Thus y_j can be thought of as a point in E nearest to $x \oplus 2^{-j}$ as measured by the metric ρ^*, i.e., $\rho^*(x \oplus 2^{-j}, E) = \rho^*(x \oplus 2^{-j}, y_j)$. Notice that this minimum is attained since E is closed, contains no dyadic rational points, and thus by **1.2.5** the function $\rho^*(x \oplus 2^{-j}, y)$ is continuous in y on E. But $x \in E$. This means that

$$\rho^*(x \oplus 2^{-j}, y_j) \leq \rho^*(x \oplus 2^{-j}, x) = \frac{1}{2^j}. \tag{4.4.16}$$

Hence it follows from (1.2.20), (1.2.19), and the fact $j > q$ that

(4.4.17)
$$\begin{aligned} |x - y_j| &\leq \rho^*(x, y_j) = \lambda(g(x) \oplus g(y_j) \oplus g(2^{-j}) \oplus (2^{-j})) \\ &\leq \rho^*(x \oplus 2^{-j}, y_j) + 2^{-j} \leq 2 \cdot 2^{-j} \leq 2^{-q}. \end{aligned}$$

The integral in each term of the sum S_2 can be estimated by observing that the variation of a sum of functions does not exceed the sum of the variations of these functions:

$$\begin{aligned} \int_{x \oplus 2^{-j} \oplus \Delta_0^{(k)}} dV_0^t(\psi(u) - f(x)u) &\leq \int_{x \oplus 2^{-j} \oplus \Delta_0^{(k)}} dV_0^t(\psi(u) - f(y_j)u) \\ &\quad + \int_{x \oplus 2^{-j} \oplus \Delta_0^{(k)}} |f(y_j) - f(x)|\, dt \\ &\equiv I_1 + I_2. \end{aligned}$$

By (4.4.13) and (4.4.17), $I_2 \leq \varepsilon 2^{-k}$. To estimate I_1, let $j_1 \leq k$ be the maximal rank of a dyadic interval which contains both points $x \oplus 2^{-j}$ and y_j. By (4.4.16) and **1.2.4** it is clear that

$$j_1 \geq j - 1 \geq q. \tag{4.4.18}$$

Since the rank j_1 is maximal, the points $x \oplus 2^{-j}$ and y_j belong to different intervals of rank $j_1 + 1$. Consequently, we have by **1.2.4** in the notation of **4.4.2** that

$$p(x \oplus 2^{-j}) = \rho^*(x \oplus 2^{-j}, E) = \rho^*(x \oplus 2^{-j}, y_j) \geq 2^{-(j_1+1)}. \tag{4.4.19}$$

Since $j \le k$, the interval $\Delta_{y_j}^{(j_1)}$ which contains y_j also contains the interval

$$x \oplus 2^{-j} \oplus \Delta_0^{(k)}.$$

Since $y_j \in E$, we obtain from (4.4.8), (4.4.12), (4.4.18), and (4.4.19) that

$$\begin{aligned}
I_1 &\equiv \int_{x\oplus 2^{-j}\oplus\Delta_0^{(k)}} dV_0^t\,(\psi(u) - f(y_j)u) \\
&\le \int_{\Delta_{y_j}^{(j_1)}} dV_0^t\,(\psi(u) - f(y_j)u) \\
&= V_{\alpha_{y_j}^{j_1}}^{\beta_{y_j}^{j_1}}\,(\psi(u) - f(y_j)u) \\
&\le |\Delta_{y_j}^{(j_1)}|M_q \le \varepsilon 2^{-j_1} \le 2\varepsilon\, p(x\oplus 2^{-j}),
\end{aligned}$$

where $\Delta_{y_j}^{(j_1)} = [\alpha_{y_j}^{j_1}, \beta_{y_j}^{j_1})$ denotes the interval which contains y_j.

Combine these estimates of I_1 and I_2 to verify the inequality

$$\int_{x\oplus 2^{-j}\oplus\Delta_0^{(k)}} dV_0^t\,(\psi(u) - f(x)u) \le \varepsilon(2^{-k} + 2p(x\oplus 2^{-j})).$$

Applying these inequalities for all j which satisfy $q < j \le k$, we obtain

$$\begin{aligned}
S_2 &= \sum_{j=q+1}^{k} 2^j \cdot \int_{x\oplus 2^{-j}\oplus\Delta_0^{(k)}} dV_0^t\,(\psi(u) - f(x)u) \\
&\le \sum_{j=q+1}^{k} 2^j \varepsilon(2^{-k} + 2p(x\oplus 2^{-j})) \\
&\le 2\varepsilon + 2\varepsilon \sum_{j=1}^{\infty} 2^j p(x\oplus 2^{-j}).
\end{aligned}$$

In view of (4.4.15), we at last can estimate the sum (4.4.14):

$$S = S_1 + S_2 \le 3\varepsilon + 2\varepsilon \sum_{j=1}^{\infty} 2^j p(x\oplus 2^{-j}).$$

Since ε was arbitrary, it follows that the sum S converges to zero at every point x from the set E for which the series on the right side of this last inequality converges. By **4.4.2** this happens for almost every x in E. We conclude that (4.4.4) holds for almost every x. ∎

4.4.4. *Let $\psi(x)$ be a function of bounded variation on $[0,1)$ with $\psi'(x) = f(x)$ almost everywhere. Then the Fourier-Stieltjes series of the function $\psi(x)$ in the Walsh system is $(C,1)$ summable to $f(x)$ almost everywhere on $[0,1)$.*

PROOF. Analogous to formula (4.2.2) it is easy to see that the $(C,1)$ means of the given Fourier-Stieltjes series can be written in the form

$$\sigma_n(x, d\psi) = \int_0^1 K_n(x \oplus t)\, d\psi(t).$$

We need to show

$$\lim_{n\to\infty}\left(\int_0^1 K_n(x\oplus t)\, d\psi(t) - f(x)\right) = \lim_{n\to\infty}\int_0^1 K_n(x\oplus t)\, d\left(\psi(t) - f(x)t\right) = 0$$

almost everywhere on [0,1). We shall show that this expression holds at every point which satisfies (4.4.4) and (4.4.5).

Fix such a point $x \in [0,1)$. Use identity (A4.3.6) for Stieltjes integrals and apply formula (4.2.10) to the kernel $K_n(t)$, where n is written in the form (4.2.9). Since $n^{(i)} \leq 2^{k-i+1}$, we obtain the following estimate:

$$\begin{aligned}
\left|\int_0^1 K_n(x\oplus t)\, d(\psi(t) - f(x)t\right| &\leq \int_0^1 |K_n(x\oplus t)|\, dV_0^t\left(\psi(u) - f(x)u\right)\\
&\leq \frac{1}{n}\sum_{j=0}^{k}\varepsilon_j 2^j \int_0^1 K_{2^j}(x\oplus t)\, dV_0^t\left(\psi(u) - f(x)u\right)\\
&\quad + \frac{1}{n}\sum_{j=0}^{k}\varepsilon_j 2^j \int_0^1 D_{2^j}(x\oplus t)\, dV_0^t\left(\psi(u) - f(x)u\right)\\
&\equiv \Sigma_1^{(n)} + \Sigma_2^{(n)}.
\end{aligned}$$

Set

$$h_j(x) \equiv \int_0^1 K_{2^j}(x\oplus t)\, dV_0^t\left(\psi(u) - f(x)u\right)$$

and

$$g_j(x) = \int_0^1 D_{2^j}(x\oplus t)\, dV_0^t\left(\psi(u) - f(x)u\right) = 2^j V_{\alpha_x^k}^{\beta_x^k}\left(\psi(u) - f(x)u\right).$$

Notice as $j \to \infty$, that $h_j(x) \to 0$ by (4.4.4), and that $g_j(x) \to 0$ by (4.4.5). Moreover, the sums $\Sigma_1^{(n)}$ and $\Sigma_2^{(n)}$ can be viewed as n-th means of the sequences $\{h_j(x)\}$ and $\{g_j(x)\}$ for the finite linear method of summation induced by the matrix whose n-th row is given by $a_{jn} = \varepsilon_j 2^j/n$ for $j \leq k$ and 0 for $j > k$. Since

$A_n = \sum_{j=0}^{n} a_{jn} = 1$ for all n, it follows from **4.1.2** that the sums $\Sigma_1^{(n)}$ and $\Sigma_2^{(n)}$ also converge to zero. This completes the proof of **4.4.4.** ∎

It is clear that each Walsh-Fourier series of a function $f \in \mathbf{L}[0,1)$ can be viewed as a Fourier-Stieltjes series (in the Walsh system) of the absolutely continuous function $\psi(x) = \int_0^x f(t)\,dt$. Hence Theorem **4.4.4** contains the following result as a special case:

4.4.5. *The Walsh-Fourier series of a function* $f \in \mathbf{L}[0,1)$ *is* $(C,1)$ *summable to* $f(x)$ *almost everywhere on* $[0,1)$.

Chapter 5

OPERATORS IN THE THEORY OF WALSH-FOURIER SERIES

§5.1. Some information from the theory of operators on spaces of measurable functions.

In this chapter, and the next, we shall obtain several results about Walsh-Fourier series by using properties of operators which take one space of measurable functions to another. We begin with definitions and some simple properties of the class of operators we wish to use.

We shall define the *distribution function*[1] $\lambda_f(y)$ of a function f, measurable on $[0,1)$, by

$$\lambda_f(y) \equiv \text{mes}\{x : |f(x)| > y\}. \tag{5.1.1}$$

We shall establish a formula which allows us to express the $\mathbf{L}^p[0,1)$ norm of f in terms of the function $\lambda_f(y)$:

$$\int_0^1 |f(x)|^p\,dx = p\int_0^\infty y^{p-1}\lambda_f(y)\,dy, \qquad p \geq 1. \tag{5.1.2}$$

First, write the right side of (5.1.2) in the following form:

$$p\int_0^\infty y^{p-1}\lambda_f(y)\,dy = p\int_0^\infty y^{p-1}\left(\int_0^1 \chi_{\{x:|f(x)|>y\}}(x,y)\,dx\right)dy,$$

where, as usual, χ_E represents the characteristic function of the set E. Apply Fubini's Theorem (see **A4.4.4**) to this last integral and rewrite it using the fact that for a fixed x we have

$$\chi_{\{x:|f(x)|>y\}}(x,y) = \begin{cases} 1 & \text{for } y < |f(x)|, \\ 0 & \text{for } y \geq |f(x)|. \end{cases}$$

We obtain

$$\begin{aligned} p\int_0^\infty y^{p-1}\left(\int_0^1 \chi_{\{x:|f(x)|>y\}}(x,y)\,dx\right)dy &= \int_0^1\left(p\int_0^{|f(x)|} y^{p-1}\,dy\right)dx \\ &= \int_0^1 |f(x)|^p\,dx. \end{aligned}$$

[1] More precisely, this is the distribution function of $|f(x)|$.

This completes the proof of (5.1.2).

Besides linear operators in the following, we shall use the pointwise upper bound of a sequence of linear operators. Such an operator T turns out to be *sublinear* (i.e., convex downward). This means that if f and g belong to some linear space of functions which make up the domain of T then in the space of measurable functions which make up the range of T we have

$$|T(f+g)| \leq |Tf| + |Tg|, \qquad |T(cf)| = |c|\,|Tf| \tag{5.1.3}$$

almost everywhere.

It is easy to see that if T is a sublinear operator and f, g are functions in the domain of T then

$$\{x : |Tf| \leq y\} \bigcap \{x : |Tg| \leq y\} \subset \{x : |T(f+g)| \leq 2y\}.$$

By taking complements, we see that

$$\{x : |T(f+g)| > 2y\} \subset \{x : |Tf| > y\} \bigcup \{x : |Tg| > y\}.$$

In particular, using the notation of (5.1.1) we arrive at

$$\lambda_{T(f+g)}(2y) \leq \lambda_{Tf}(y) + \lambda_{Tg}(y). \tag{5.1.4}$$

If an operator T takes the space $\mathbf{L}^p[0,1)$ into itself for some $1 \leq p \leq \infty$ and satisfies the inequality

$$\|Tf\|_p \leq C\|f\|_p \tag{5.1.5}$$

(i.e., in the case that T is a linear operator it is a bounded operator in the usual sense), then we shall say that T is of *strong type* (p,p).

If

$$y^p \lambda_{Tf}(y) \leq C^p \|f\|_p^p \tag{5.1.6}$$

for $y > 0$ and $f \in \mathbf{L}^p[0,1)$, then we say that T is of *weak type* (p,p), and we shall refer to (5.1.6) as the *inequality of weak type* (p,p) for the operator T.

5.1.1. *If a sublinear operator T is of strong type (p,p) with constant C, then it is of weak type (p,p) with the same constant.*

PROOF. We need only observe that

$$y^p \lambda_{Tf}(y) = \int_{\{x:|Tf|>y\}} y^p\,dx \leq \int_0^1 |Tf(x)|^p\,dx. \quad \blacksquare$$

Notice that this last inequality contains the famous Chebyshev inequality:

$$\operatorname{mes}\{x : |f| > y\} \leq y^{-p} \int_0^1 |f(x)|^p\,dx. \tag{5.1.7}$$

We shall need the following special case of the Marcinkiewicz Interpolation Theorem:

5.1.2. *Let $1 \leq p_1 < p_2 < \infty$ be two numbers and suppose a sublinear operator T satisfies*

$$y^{p_j}\lambda_{Tf}(y) \leq C_j^{p_j}\int_0^1 |f(x)|^{p_j}\,dx$$

for $y > 0$, $f \in \mathbf{L}^{p_j}[0,1)$, and $j = 1,2$, where the C_j's are absolute constants independent of y and f. In the case when $p_2 = \infty$, suppose that T is of strong type, namely, $\|Tf\|_\infty \leq C_2\|f\|_\infty$. Then

$$\|Tf\|_p^p \leq p2^pC_1^{p_1\frac{p_2-p}{p_2-p_1}}C_2^{p_2\frac{p-p_1}{p_2-p_1}}\left(\frac{1}{p-p_1}+\frac{1}{p_2-p}\right)\|f\|_p^p$$

for all $f \in \mathbf{L}^p[0,1)$, $p_1 < p < p_2$.

PROOF. Let

$$A \equiv \begin{cases} C_1^{\frac{p_1}{p_2-p_1}}C_2^{\frac{-p_2}{p_2-p_1}} & \text{if } p_2 < \infty, \\ C_2^{-1} & \text{if } p_2 = \infty. \end{cases}$$

Fix $y > 0$, set

$$f^y(x) \equiv \begin{cases} f(x), & \text{if } |f(x)| \leq Ay, \\ 0, & \text{if } |f(x)| > Ay, \end{cases}$$

and $f_y(x) \equiv f(x) - f^y(x)$.

Since T is a sublinear operator and $f(x) = f_y(x) + f^y(x)$, we have by (5.1.4) that

$$\lambda_{Tf}(2y) \leq \lambda_{Tf_y}(y) + \lambda_{Tf^y}(y).$$

Thus in the case when $p_2 < \infty$ we have by hypotheses that

$$\begin{aligned}\lambda_{Tf}(2y) &\leq C_1^{p_1}y^{-p_1}\int_0^1 |f_y(x)|^{p_1}\,dx + C_2^{p_2}y^{-p_2}\int_0^1 |f^y(x)|^{p_2}\,dx \\ &= C_1^{p_1}y^{-p_1}\int_0^1 \chi_{\{x:|f(x)|>Ay\}}(x,y)|f(x)|^{p_1}\,dx \\ &\quad + C_2^{p_2}y^{-p_2}\int_0^1 \chi_{\{x:|f(x)|\leq Ay\}}(x,y)|f(x)|^{p_2}\,dx.\end{aligned}$$

Multiply this inequality by $p2^py^{p-1}$ and integrate the resulting product with respect to y from 0 to ∞. Apply Fubini's Theorem in the same way we did to prove (5.1.2).

Thus verify that

$$\begin{aligned}
\int_0^1 |Tf(x)|^p\,dx &= p\int_0^\infty (2y)^{p-1}\lambda_{Tf}(2y)\,d2y \\
&\le p2^pC_1^{p_1}\int_0^\infty y^{p-p_1-1}\int_0^1 \chi_{\{x:|f(x)|>Ay\}}(x,y)|f(x)|^{p_1}\,dx\,dy \\
&\quad + p2^pC_2^{p_2}\int_0^\infty y^{p-p_2-1}\int_0^1 \chi_{\{x:|f(x)|\le Ay\}}(x,y)|f(x)|^{p_2}\,dx\,dy \\
&= p2^pC_1^{p_1}\int_0^1 |f(x)|^{p_1}\int_0^{|f(x)|/A} y^{p-p_1-1}\,dy\,dx \\
&\quad + p2^pC_2^{p_2}\int_0^1 |f(x)|^{p_2}\int_{|f(x)|/A}^\infty y^{p-p_2-1}\,dy\,dx \\
&= p2^pC_1^{p_1}\int_0^1 |f(x)|^{p_1}\frac{1}{p-p_1}|f(x)|^{p-p_1}A^{p_1-p}\,dx \\
&\quad + p2^pC_2^{p_2}\int_0^1 |f(x)|^{p_2}\frac{1}{p_2-p}|f(x)|^{p-p_2}A^{p_2-p}\,dx \\
&= p2^pC_1^{p_1}A^{p_1-p}\frac{1}{p-p_1}\int_0^1 |f(x)|^p\,dx + p2^pC_2^{p_2}A^{p_2-p}\frac{1}{p_2-p}\int_0^1 |f(x)|^p\,dx.
\end{aligned}$$

Substituting the value for A chosen at the beginning of this proof establishes the promised inequality.

In the case $p_2=\infty$ we have $\|f^y\|_\infty \le Ay = C_2^{-1}y$. Hence $\|Tf^y\|_\infty \le C_2\|f^y\|_\infty \le y$ and we have by definition (5.1.1) that $\lambda_{Tf^y}(y)=0$. Consequently, in the string of inequalities above we can simply omit the terms involving p_2 when $p_2=\infty$. ∎

These observations lead to a useful result about sequences of operators.

5.1.3. *Let $\{T_n\}_{n=1}^\infty$ be a sequence of linear operators such that the sublinear operator*

$$f(x)\to Tf(x)=\sup_{n\ge1}|T_nf(x)|$$

is of weak type (p,p) for some $1\le p<\infty$. If $\lim_{n\to\infty}T_n\phi(x)=\phi(x)$ almost everywhere on $[0,1)$ for all functions ϕ in some dense subset of $\mathbf{L}^p[0,1)$, then

$$\lim_{n\to\infty}T_nf(x)=f(x)\qquad \textit{almost everywhere on } [0,1) \tag{5.1.8}$$

for all $f\in\mathbf{L}^p[0,1)$.

PROOF. Each point x for which (5.1.8) fails to hold satisfies the inequality

$$\limsup_{n\to\infty}|T_nf(x)-f(x)|>0.$$

Hence each such point belongs to the set

$$P_n \equiv \{x : \limsup_{k\to\infty} |T_k f(x) - f(x)| > \frac{1}{n}\}$$

for some natural number n. If P represents the set of all points x where (5.1.8) fails then $P = \bigcup_{n=1}^{\infty} P_n$. Since $P_{n+1} \supset P_n$ implies $\text{mes}P = \lim_{n\to\infty}\text{mes}P_n$, we will conclude that $\text{mes}P = 0$ if we show $\text{mes}P_n \le 1/n$ for $n = 1, 2, \ldots$. In particular, it suffices to show that

$$\text{mes}\{x : \limsup_{n\to\infty} |T_n f(x) - f(x)| > \varepsilon\} < \varepsilon. \tag{5.1.9}$$

Toward this, notice first that

$$\limsup_{n\to\infty} |T_n f(x) - f(x)| \le \limsup_{n\to\infty} |T_n f(x) - T_n\phi(x)| + \limsup_{n\to\infty} |T_n\phi(x) - \phi(x)| + |f(x) - \phi(x)|.$$

Repeating the argument which lead to (5.1.4), it is not difficult to see that

$$\begin{aligned} \text{mes}\{x : \limsup_{n\to\infty} |T_n f(x) - f(x)| > \varepsilon\} &\le \text{mes}\{x : \limsup_{n\to\infty} |T_n f(x) - T_n\phi(x)| > \varepsilon/3\} \\ &+ \text{mes}\{x : \limsup_{n\to\infty} |T_n\phi(x) - \phi(x)| > \varepsilon/3\} \\ &+ \text{mes}\{x : |f(x) - \phi(x)| > \varepsilon/3\}. \end{aligned} \tag{5.1.10}$$

Choose the function ϕ from the dense set which is mentioned in the hypotheses of this theorem. Thus the second term on the right side of (5.1.10) reduces to zero. Applying inequality (5.1.7) to the function $\phi(x) - f(x)$, we can estimate the third term by

$$\text{mes}\{x : |f(x) - \phi(x)| > \varepsilon/3\} \le \frac{3^p}{\varepsilon^p}\|f - \phi\|_p^p. \tag{5.1.11}$$

Finally, the set in the first term is evidently a subset of

$$\{x : \sup_{n\ge 1} |T_n(f(x) - \phi(x))| > \varepsilon/3\}.$$

Since the operator T is of weak type (p, p) (see (5.1.6)), it follows that the first term on the right side of (5.1.10) can be estimated by

$$\begin{aligned} &\text{mes}\{x : \limsup_{n\to\infty} |T_n(f(x) - \phi(x))| > \varepsilon/3\} \\ &\le \text{mes}\{x : |T(f(x) - \phi(x))| > \varepsilon/3\} \le C^p\frac{3^p}{\varepsilon^p}\|f - \phi\|_p^p. \end{aligned} \tag{5.1.12}$$

Substituting (5.1.11) and (5.1.12) into (5.1.10), we obtain

$$\operatorname{mes}\{x : \limsup_{n\to\infty} |T_n f(x) - f(x)| > \varepsilon\} \le (1 + C^p)\frac{3^p}{\varepsilon^p}\|f-\phi\|_p^p.$$

Thus complete the proof by choosing the function ϕ so near f, in the norm of $\mathbf{L}^p[0,1)$, that

$$\|f-\phi\|_p < \frac{\varepsilon^{1+1/p}}{3(1+C^p)^{1/p}}. \quad \blacksquare$$

§5.2. The Hardy-Littlewood maximal operator corresponding to sequences of dyadic nets.

In the theory of trigonometric series the *Hardy-Littlewood operator* plays a major role. This operator assigns to each integrable function f a "maximal function" defined by

$$Mf(x) = \sup_{h>0} \frac{1}{2h}\int_{x-h}^{x+h} |f(t)|\,dt.$$

In the theory of Walsh series a similar role is played by an operator based on the *dyadic maximal function*

$$\Lambda f(x) = \sup_{k\ge 0} \frac{1}{|\Delta_x^{(k)}|}\int_{\Delta_x^{(k)}} |f(t)|\,dt, \tag{5.2.1}$$

where $\{\Delta_x^{(k)}\}_{k=0}^{\infty}$ is the sequence of dyadic intervals which contain x, i.e., the sequence of nodes from the binary nets $\mathcal{N}_k$ (see §1.1) which shrink to x.

We shall establish a number of properties for the operator Λ.

5.2.1. *The operator* Λ *is of weak type* $(1,1)$, *namely, for all* $y>0$ *and* $f\in\mathbf{L}[0,1)$ *the inequality*

$$\operatorname{mes}\{x : \Lambda f(x) > y\} \le \frac{1}{y}\int_0^1 |f(t)|\,dt. \tag{5.2.2}$$

is satisfied.

PROOF. We introduce the notation

$$E_y = \{x : \Lambda f(x) > y\}. \tag{5.2.3}$$

Suppose first that $0 < y \le \int_0^1 |f(t)|\,dt$. Since by (5.2.1) it is clear that for all x

$$\Lambda f(x) \ge \int_0^1 |f(t)|\,dt,$$

our assumption implies that $E_y = [0,1)$ and thus (5.2.2) is obviously satisfied.

Suppose now that

$$y > \int_0^1 |f(t)|\,dt. \tag{5.2.4}$$

In this case we shall show that E_y can be written as a finite or countably infinite union of dyadic intervals Δ_j, perhaps of different ranks, such that the following conditions are satisfied:

$$\begin{cases} 1)\ \Delta_j \bigcap \Delta_i = \emptyset \text{ for } j \neq i, \\ 2)\ E_y = \bigcup_{j=1}^{\infty} \Delta_j, \\ 3)\ y < \dfrac{1}{|\Delta_j|} \displaystyle\int_{\Delta_j} |f(t)|\,dt \leq 2y \text{ for all } \Delta_j. \end{cases} \tag{5.2.5}$$

For each $x \in E_y$, consider the sequence of intervals $\{\Delta_x^{(k)}\}$ which contain x, and choose from among them the interval of smallest rank which satisfies

$$\frac{1}{|\Delta_x^{(k)}|} \int_{\Delta_x^{(k)}} |f(t)|\,dt > y.$$

Denote this interval by Δ_x. Notice by (5.2.4) that $\Delta_x \neq \Delta_x^{(0)} = [0,1)$ and by minimality that we can find a dyadic interval Δ_x^* from the sequence $\{\Delta_x^{(k)}\}$ of rank one less than that of Δ_x such that $\Delta_x \subset \Delta_x^*$, $|\Delta_x^*| = 2|\Delta_x|$, and

$$\frac{1}{|\Delta_x^*|} \int_{\Delta_x^*} |f(t)|\,dt \leq y.$$

Hence

$$\frac{1}{|\Delta_x|} \int_{\Delta_x} |f(t)|\,dt = \frac{2}{|\Delta_x^*|} \int_{\Delta_x} |f(t)|\,dt \leq \frac{2}{|\Delta_x^*|} \int_{\Delta_x^*} |f(t)|\,dt \leq 2y.$$

Moreover, by construction we see that Δ_x is a subset of E_y. Hence $E_y = \bigcup_{x \in E_y} \Delta_x$. In particular, it remains to choose from the family

$$\{\Delta_x : x \in E_y\} \tag{5.2.6}$$

a sequence of non-overlapping intervals whose union is E_y.

Toward this, recall that a pair of dyadic intervals are either non- overlapping or one is a subset of the other. Choose from (5.2.6) all intervals of rank 1, if they exist, then all intervals of rank 2, e.t.c., with the stipulation that at the n-th stage we choose only those intervals from (5.2.6) of rank n which are not contained in the intervals of lower rank already chosen. In this way we obtain a sequence $\{\Delta_j\}$ of non-overlapping intervals which satisfy all the conditions in (5.2.5). Using these conditions, we verify the inequality

$$\text{mes} E_y = \sum_{j=1}^{\infty} |\Delta_j| \leq \sum_{j=1}^{\infty} \frac{1}{y} \int_{\Delta_j} |f(t)|\,dt = \frac{1}{y} \int_{E_y} |f(t)|\,dt \leq \frac{1}{y} \int_0^1 |f(t)|\,dt.$$

Thus (5.2.2) is proved. ∎

5.2.2. *The operator Λ is of type (∞, ∞), namely,*

$$\|\Lambda f\|_\infty \le \|f\|_\infty.$$

PROOF. This result follows directly from definition (5.2.1) and the fact that

$$\frac{1}{|\Delta^{(k)}|}\int_{\Delta^{(k)}} |f(t)|\,dt \le \frac{1}{|\Delta^{(k)}|}\|f\|_\infty|\Delta^{(k)}| = \|f\|_\infty$$

for every $\Delta^{(k)}$. ∎

Combining **5.2.1** and **5.2.2** with the interpolation theorem **5.1.2** for $p_1 = 1$ and $p_2 = \infty$ leads to a proof of the following theorem:

5.2.3. *The operator Λ is of type (p, p) for all $1 < p \le \infty$, namely,*

$$\|\Lambda f\|_p \le C_p\|f\|_p.$$

Along with the operator Λ we shall also consider the operator H, which is defined for each function $f \in \mathbf{L}[0, 1)$ by

$$Hf = \sup_{k\ge 0} |S_{2^k}(x, f)|, \tag{5.2.7}$$

where $S_{2^k}(x, f)$ represents the partial sum of the Walsh- Fourier series of f of order 2^k.

By (2.1.11), we have

$$Hf(x) = \sup_{k\ge 0} \frac{1}{|\Delta_x^{(k)}|} \left| \int_{\Delta_x^{(k)}} f(t)\,dt \right| .$$

Thus it is clear that $0 \le Hf(x) \le \Lambda f(x)$ for all $x \in [0, 1)$. In particular, the following result is a corollary of **5.2.3**.

5.2.4. *The operator H is of type (p, p) for all $1 < p \le \infty$, namely,*

$$\|Hf\|_p \le C_p\|f\|_p.$$

§5.3. Partial sums of Walsh-Fourier series as operators.

Below we shall use the so-called "modified" Dirichlet kernel, which is defined by the identity

$$D_n^*(t) = w_n(t)D_n(t), \tag{5.3.1}$$

where $D_n(t)$ is the usual Dirichlet kernel in the Walsh system which we defined in (1.4.8). The kernel $D_n^*(t)$ can be used to generate the expression

$$S_n^*(x, f) = \int_0^1 f(t)D_n^*(x \oplus t)\,dt, \tag{5.3.2}$$

which is called the *modified partial sum*.

Using (5.3.1), (1.2.15), and (2.1.10), we see that

$$S_n^*(x,f) = w_n(x)S_n(x, w_n f). \tag{5.3.3}$$

If the dyadic expansion of an integer n has the form $n = \sum_{i=0}^{k} \varepsilon_i 2^i$, where $\varepsilon_k = 1$, and we denote the truncated expansion by $n_j = \sum_{i=k-j}^{k} \varepsilon_i 2^i$, then the definition of the Walsh functions (see §1.1) can be used to write the Dirichlet kernel in the following way:

$$\begin{aligned}
D_n(t) &= \sum_{i=0}^{n-1} w_i(t) \\
&= \sum_{i=0}^{2^k-1} w_i(t) + \varepsilon_{k-1}\left(\sum_{i=2^k}^{2^k+\varepsilon_{k-1}2^{k-1}-1} w_i(t)\right) + \dots \\
&\qquad + \varepsilon_{k-j} \sum_{i=n_{j-1}}^{n_j-1} w_i(t) + \dots + \varepsilon_0 \sum_{i=n_{k-1}}^{n-1} w_i(t) \\
&= D_{2^k}(t) + \varepsilon_{k-1} \sum_{i=0}^{2^{k-1}-1} w_{2^k+i}(t) + \dots \\
&\qquad + \varepsilon_{k-j} \sum_{i=0}^{2^{k-j}-1} w_{n_{j-1}+i}(t) + \dots + \varepsilon_0 w_{n_{k-1}}(t) \\
&= D_{2^k}(t) + \varepsilon_{k-1} w_{2^k}(t) \sum_{i=0}^{2^{k-1}-1} w_i(t) + \dots \\
&\qquad + \varepsilon_{k-j} \prod_{m=k-j+1}^{k} (w_{2^m}(t))^{\varepsilon_m} \sum_{i=0}^{2^{k-j}-1} w_i(t) + \dots + \varepsilon_0 \prod_{m=1}^{k} (w_{2^m}(t)))^{\varepsilon_m} w_0(t) \\
&= \sum_{j=0}^{k} \varepsilon_{k-j} \prod_{m=k-j+1}^{k} (w_{2^m}(t))^{\varepsilon_m} D_{2^{k-j}}(t) \\
&= \sum_{j=0}^{k} \varepsilon_j \prod_{m=j+1}^{k} (w_{2^m}(t))^{\varepsilon_m} D_{2^j}(t).
\end{aligned}$$

Multiply this expression by $w_n(t) = \prod_{m=0}^{k} (w_{2^m}(t))^{\varepsilon_m}$ and recall that $w_i^2(t) = 1$ for all i. We obtain the following formula for the kernel (5.3.1):

$$D_n^*(t) = \sum_{j=0}^{k} \varepsilon_j \prod_{m=0}^{j} (w_{2^m}(t))^{\varepsilon_m} D_{2^j}(t).$$

But each function $D_{2^j}(t)$ is different from zero only on the interval $\Delta_0^{(j)}$, and on this interval $w_{2^m}(t) = 1$ for $m < j$. Consequently, the modified Dirichlet kernel can be written in the form

$$D_n^*(t) = \sum_{j=0}^{k} \varepsilon_j w_{2^j}(t) D_{2^j}(t) = \sum_{j=0}^{k} \varepsilon_j D_{2^j}^*(t). \tag{5.3.4}$$

The following important property shows why the modified kernel $D_n^*(t)$ is more convenient to work with than the usual kernel $D_n(t)$.

5.3.1. *If $x \notin \Delta^{(s)}$ for some dyadic interval $\Delta^{(s)}$, $s \geq 0$, then the kernel $D_n^*(x \oplus t)$ takes on a constant value for all $t \in \Delta^{(s)}$.*

PROOF. By (5.3.4), it is enough to prove the theorem for the functions

$$D_{2^j}^*(x \oplus t) = w_{2^j}(x \oplus t) D_{2^j}(x \oplus t)$$

for all j. The function $D_{2^j}^*(t)$, and thus the function $D_{2^j}^*(x \oplus t)$, is constant on each interval $\Delta^{(j+1)}$ of rank $j+1$. Hence the theorem is obvious if the rank of the interval $\Delta^{(s)}$ satisfies $s \geq j+1$.

If $s < j+1$, i.e., $s \leq j$, then the conditions $t \in \Delta^{(s)}$ and $x \notin \Delta^{(s)}$ imply that $x \oplus t \notin \Delta_0^{(s)}$. Hence by (1.4.13) we have $D_{2^j}(x \oplus t) = 0$. In particular, the kernel $D_{2^j}^*(x \oplus t)$ is again constant with respect to t on $\Delta^{(s)}$. ∎

We shall now consider the maps which take an integrable function f to the partial sums $S_n(x, f)$ or to the modified partial sums $S_n^*(x, f)$. These maps are evidently linear operators, defined on the space of integrable functions, and we shall denote them by S_n and S_n^*. These operators enjoy the following properties:

5.3.2. *The operators S_n and S_n^* are of weak type $(1,1)$ and of strong type (p,p) for all $1 < p < \infty$ with constants which do not depend on n.*

PROOF. Notice first of all that if one of the operators S_n and S_n^* are of weak or strong type (p,p) then the other one is also of the same type. Indeed, if S_n is of strong type (p,p) with constant C_p then since $|w_n(t)| = 1$ it is obvious that

$$\|S_n^*(f)\|_p = \|S_n(w_n f)\|_p \leq C_p \|w_n f\|_p = C_p \|f\|_p.$$

Hence S_n^* is also of strong type (p,p) with the same constant C_p. On the other hand, if S_n^* is of strong type (p,p) then by (5.3.3), with the function f in place of $w_n f$, we have

$$\|S_n(f)\|_p = \|S_n^*(w_n f)\|_p \leq C_p \|w_n f\|_p = C_p \|f\|_p.$$

A similar argument shows that S_n is of weak type (p,p) if and only if the operator S_n^* is also. Consequently, it suffices to prove the theorem for one of the operators S_n or S_n^*. For the most part, it is more convenient to work the operator S_n^*.

We begin by noting that Parseval's formula (see **2.6.5**) for the Walsh system implies $\|S_n(f)\|_2 \le \|f\|_2$, i.e., S_n is of strong type (2,2) with constant 1. In view of what we have already proved, we see that S_n^* is also of type (2,2) with the same constant, i.e.,

$$\|S_n^* f\|_2 \le \|f\|_2. \tag{5.3.5}$$

We shall now prove that S_n^* is of weak type (1,1), i.e.,

$$\text{mes}\{x : |S_n^*(x,f)| > y\} \le \frac{C}{y}\int_0^1 |f(t)|\,dt, \tag{5.3.6}$$

where C does not depend on n. Fix $f \in \mathbf{L}[0,1)$. Since for $0 < y \le \int_0^1 |f(t)|\,dt$ the inequality (5.3.6) is evidently satisfied, with constant $C = 1$, we may suppose that $y > \int_0^1 |f(t)|\,dt$. Define the set E_y by (5.2.3) and decompose it into a union of dyadic intervals Δ_j which satisfy conditions (5.2.5). Using these intervals, define two auxiliary functions:

$$g(x) = \begin{cases} \dfrac{1}{|\Delta_j|}\displaystyle\int_{\Delta_j} f(t)\,dt & \text{for } x \in \Delta_j \subset E_y, \\ f(x) & \text{for } x \in [0,1) \setminus E_y \end{cases} \tag{5.3.7}$$

and $h(x) = f(x) - g(x)$. If $x \in E_y = \bigcup_{j=1}^{\infty} \Delta_j$, then it follows from (5.2.5) that

$$|g(x)| \le \frac{1}{|\Delta_j|}\int_{\Delta_j} |f(t)|\,dt \le 2y. \tag{5.3.8}$$

If $x \notin E_y$ then use the definitions of the set E_y and the maximal operator (5.2.1) to verify the inequality

$$\left|\frac{1}{|\Delta_x^{(k)}|}\int_{\Delta_x^{(k)}} f(t)\,dt\right| \le \frac{1}{|\Delta_x^{(k)}|}\int_{\Delta_x^{(k)}} |f(t)|\,dt \le y \tag{5.3.9}$$

for all dyadic intervals $\Delta_x^{(k)}$ containing x. But Theorem **2.5.10** implies

$$\lim_{k\to\infty} \frac{1}{|\Delta_x^{(k)}|}\int_{\Delta_x^{(k)}} f(t)\,dt = f(x)$$

for almost every $x \in [0,1)$. Consequently, we have by (5.3.9) that $|f(x)| \le y$ holds almost everywhere on $[0,1) \setminus E_y$. This inequality together with (5.3.8) and (5.3.7) shows that

$$|g(x)| \le 2y \qquad \text{almost everywhere on } [0,1).$$

Therefore,

$$\int_0^1 |g(t)|^2\,dt \le 2y \int_0^1 |g(t)|\,dt. \tag{5.3.10}$$

Moreover, (5.3.7) implies that

$$\int_{\Delta_j} |g(t)|\,dt = \Big| \int_{\Delta_j} f(t)\,dt \Big| \le \int_{\Delta_j} |f(t)|\,dt.$$

In particular, we have

$$\int_0^1 |g(t)|\,dt = \sum_j \int_{\Delta_j} |g(t)|\,dt + \int_{[0,1)\setminus E_y} |g(t)|\,dt \le \int_0^1 |f(t)|\,dt. \tag{5.3.11}$$

Combine the Chebyshev inequality (5.1.7) for $p = 2$, inequality (5.3.5), written for the function g, and also inequalities (5.3.10) and (5.3.11). We obtain

$$\begin{aligned}
\operatorname{mes}\{x : |S_n^*(x,g)| \ge \frac{y}{2}\} &\le \left(\frac{2}{y}\right)^2 \int_0^1 |S_n^*(t,g)|^2\,dt \\
&\le \frac{4}{y^2} \int_0^1 |g(t)|^2\,dt \\
&\le \frac{8}{y} \int_0^1 |g(t)|\,dt \le \frac{8}{y} \int_0^1 |f(t)|\,dt.
\end{aligned}$$

We shall now prove an analogous inequality for the function $h(x)$. Notice that $h(x) = 0$ for $x \in [0,1) \setminus E_y$ and $\int_{\Delta_j} h(t)\,dt = 0$ for any $\Delta_j \subset E_y$. Fix $x \notin E_y$ and notice by **5.3.1** that the kernel $D_n^*(x \oplus t)$ takes a constant value on Δ_j which we shall denote by $D_n^*(x \oplus \Delta_j)$. It follows that

$$\begin{aligned}
S_n^*(x,h) &= \int_0^1 h(t) D_n^*(x \oplus t)\,dt \\
&= \sum_j \int_{\Delta_j} h(t) D_n^*(x \oplus t)\,dt \\
&= \sum_j D_n^*(x \oplus \Delta_j) \int_{\Delta_j} h(t)\,dt = 0.
\end{aligned}$$

Consequently, $\{x : S_n^*(x,h) \ne 0\} \subset E_y$. In view of the third condition of (5.2.5), it follows that

$$\begin{aligned}
\operatorname{mes}\{x : S_n^*(x,h) > y/2\} &\le \operatorname{mes}\{x : S_n^*(x,h) > 0\} \le \sum_j |\Delta_j| \\
&\le \sum_j \frac{1}{y} \int_{\Delta_j} |f(t)|\,dt \le \frac{1}{y} \int_0^1 |f(t)|\,dt.
\end{aligned}$$

Use these inequalities together with (5.3.12) and (5.1.4) applied to the operator S_n^* and the functions g and h. Thus verify that

$$\text{mes}\{x : |S_n^*(x,f)| > y\} \le \frac{9}{y}\int_0^1 |f(t)|\,dt.$$

This proves that the operator S_n^* is of weak type (1,1) with constant 9. We have already noted (see (5.3.5)) that the operator S_n^* is of strong type (2,2), so by **5.1.1**, it is also of weak type (2,2) with constant 1. Hence by the interpolation theorem **5.1.2**, S_n^* is of strong type (p,p) for all $1 < p \le 2$, i.e.,

$$\|S_n^*(f)\|_p \le C_p\|f\|_p.$$

Furthermore, we see by **5.1.1** that the constant C_p depends only on the constants 1 and 9 and thus does not depend on n.

It remains to extend this last inequality to the case $2 < p < \infty$. To accomplish this we pass to the dual space (see **A5.2.2**). Let $f \in \mathbf{L}^p[0,1)$ for $2 < p < \infty$ and $\phi \in \mathbf{L}[0,1)^q$, where $1/p + 1/q = 1$. Thus $1 < q < 2$ and by the inequality already proved, we have

$$\|S_n^*(\phi)\|_q \le C_p\|\phi\|_q. \tag{5.3.13}$$

By Fubini's Theorem (see **A4.4.4**), Hölder's inequality (see **A5.2.2**), and (5.3.13) we obtain

(5.3.14)

$$\begin{aligned}
\left|\int_0^1 S_n^*(x,f)\phi(x)\,dx\right| &= \left|\int_0^1 \left(\int_0^1 f(t)D_n^*(x\oplus t)\,dt\right)\phi(x)\,dx\right| \\
&= \left|\int_0^1 f(t)\left(\int_0^1 D_n^*(x\oplus t)\phi(x)\,dx\right)dt\right| \\
&= \left|\int_0^1 f(t)S_n^*(t,\phi)\,dt\right| \\
&\le \|f\|_p\|S_n^*(\phi)\|_q \le C_q\|f\|_p\|\phi\|_q.
\end{aligned}$$

Since $F(\phi) = \int_0^1 S_n^*(x,f)\phi(x)\,dx$ is a functional on the space $\mathbf{L}^q[0,1)$ with norm precisely $\|S_n^*(x,f)\|_p$ (see **A5.3.2**), we see by (5.3.14) that this norm satisfies the inequality (see **A5.3.1**)

$$\|S_n^*(x,f)\|_p \le C_q\|f\|_p.$$

But this means that the operator S_n^* is of strong type (p,p) for $2 < p < \infty$. Thus the theorem is proved. ∎

§5.4. Convergence of Walsh-Fourier series in $\mathbf{L}^p[0,1)$.

Theorem **5.3.2** contains the following important result:

5.4.1. *If $f \in \mathbf{L}^p[0,1)$ for some $1 < p < \infty$ then*

$$\int_0^1 |f(x) - S_n(x,f)|^p\, dx \to 0, \qquad n \to \infty,$$

i.e., the partial sums of the Walsh-Fourier series of f converge to f in the $\mathbf{L}^p$ norm.

PROOF. We shall use the fact (see Theorem **2.6.4**) that the Walsh polynomials are a dense subset of $\mathbf{L}^p[0,1)$.

Notice for any polynomial T and any integer n which exceeds the degree of this polynomial, that $S_n(x,T) = T(x)$. For such an n we have by Theorem **5.3.2** that

$$\|T - S_n(f)\|_p = \|S_n(T - f)\|_p \le C_p\|T - f\|_p.$$

Consequently,

$$\|f - S_n(f)\|_p \le \|f - T\|_p + \|T - S_n(f)\|_p \le (1 + C_p)\|T - f\|_p.$$

Therefore, if we choose the polynomial T such that $\|T - f\|_p < \varepsilon/(1 + C_p)$ then $\|f - S_n(f)\|_p < \varepsilon$ for any integer n which exceeds the degree of the polynomial T. Since ε was arbitrary, this proves the theorem. ∎

We shall now show that the preceding theorem fails to hold for $p = 1$. Namely, the following result is true:

5.4.2. *There exists a Walsh-Fourier series which diverges in the norm of the space* $\mathbf{L}[0,1)$.

PROOF. We shall show that the function we search for can be defined by the series

$$f(x) = \sum_{k=1}^{\infty} \frac{1}{k^2} w_{2^{k^3}}(x) D_{2^{k^3}}(x). \tag{5.4.1}$$

First, the series (5.4.1) converges in the norm of $\mathbf{L}[0,1)$ and thus defines a function $f \in \mathbf{L}[0,1)$, since by (1.4.13)

$$\int_0^1 |w_{2^{k^3}}(x) D_{2^{k^3}}(x)|\, dx = \int_0^1 D_{2^{k^3}}(x)\, dx = 1.$$

Next, each term in (5.4.1) is by (1.2.17) a Walsh polynomial

$$\frac{1}{k^2} P_k(x) = \frac{1}{k^2} w_{2^{k^3}}(x) D_{2^{k^3}}(x) = \sum_{i=2^{k^3}}^{2^{k^3+1}-1} \frac{1}{k^3} w_i(x),$$

and for different k's the polynomials $P_k(x)$ do not contain any common terms. Hence it is not difficult to verify

$$\hat{f}(i) = \frac{1}{k^2}, \qquad 2^{k^3} \leq i \leq 2^{k^3+1} - 1.$$

Therefore,

$$|S_{2^{k^3}+m_k}(x,f) - S_{2^{k^3}}(x,f)| = \left| \frac{1}{k^2} \sum_{i=2^{k^3}}^{2^{k^3}+m_k-1} w_i(x) \right|$$

$$= \left| \frac{1}{k^2} w_{2^{k^3}}(x) \sum_{i=0}^{m_k-1} w_i(x) \right| = \frac{1}{k^2} |D_{m_k}(x)|$$

for any m_k which satisfies $2^{k^3} + m_k \leq 2^{k^3+1}$. Specializing this inequality to a number m_k given by Theorem **2.2.2** which also satisfies $2^{k^3-1} < m_k < 2^{k^3}$, we see that

$$\int_0^1 |S_{2^{k^3}+m_k}(x,f) - S_{2^{k^3}}(x,f)|\, dx \geq \frac{1}{k^2} \frac{\log_2 m_k}{4} > \frac{k^3-1}{4k^2} > \frac{k}{8}$$

for $k \geq 2$. In particular, the Walsh-Fourier series of the function f cannot converge in the $\mathbf{L}[0,1)$ norm. ∎

Chapter 6

GENERALIZED MULTIPLICATIVE TRANSFORMS

§6.1. Existence and properties of generalized multiplicative transforms.

Let $1 \leq p < \infty$. A complex valued function $f(x)$ is said to belong to $\mathbf{L}^p(0,\infty)$ if $\int_0^\infty |f(x)|^p \, dx < \infty$. The norm of $f(x)$ in the space $\mathbf{L}^p(0,\infty)$ will be denoted by $\|f\|_p$ and is defined by

$$\|f\|_p = \left(\int_0^\infty |f(x)|^p \, dx \right)^{1/p}.$$

The limit of a function $f(x,a)$ as $a \to \infty$ in the $\mathbf{L}^p(0,\infty)$ norm will be denoted by $\lim_{a\to\infty} {}_p f(x,a)$, i.e., the equation

$$f(x) = \lim_{a\to\infty} {}_p f(x,a)$$

means that

$$\|f(x) - f(x,a)\|_p \to 0, \qquad a \to \infty.$$

Let $f(x) \in \mathbf{L}^1(0,\infty)$. For the collection $\mathcal{P} = (\dots, p_{-j}, \dots, p_{-1}, p_1, \dots, p_j, \dots)$ given by (1.5.22), and the corresponding function $\chi(x,y)$ given by (1.5.33), define integral transforms of $f(x)$ by

$$\mathcal{F}[f](y) \equiv \widehat{f}(y) = \int_0^\infty f(x)\overline{\chi(x,y)}\, dx \tag{6.1.1}$$

and

$$\mathcal{F}^*[f](x) \equiv f^*(x) = \int_0^\infty f(t)\overline{\chi(x,t)}\, dt. \tag{6.1.2}$$

We shall call these *direct multiplicative transforms.*

Notice first of all that if the collection (1.5.22) is symmetric, i.e., $p_{-j} = p_j$ for all $j = 1, 2, \dots,$ then

$$\mathcal{F}[f](y) = \mathcal{F}^*[f](y). \tag{6.1.3}$$

The transforms (6.1.1) and (6.1.2) are analogues of the classical Fourier transform.

For a function $f \in \mathbf{L}^p(0,\infty)$ with $1 < p \leq 2$, the direct multiplicative $\mathbf{L}^p$-transforms will be defined by

$$\mathcal{F}[f](y) \equiv \widehat{f}(y) = \lim_{a\to\infty} {}_{p'} \int_0^a f(x)\overline{\chi(x,y)}\, dx \tag{6.1.4}$$

and

$$\mathcal{F}^*[f](x) \equiv f^*(x) = \lim_{a\to\infty} {}_{p'} \int_0^a f(t)\overline{\chi(x,t)}\, dt \tag{6.1.5}$$

where $1/p + 1/p' = 1$.

In view of the similarity of definitions (6.1.1) and (6.1.2), we need only look at the properties of the transform (6.1.1).

6.1.1. *The transform $\mathcal{F}[f]$ is additive and homogeneous, i.e., if $f = \sum_{j=1}^{k} \alpha_j f_j$ and $\widehat{f}_j$ represents $\mathcal{F}[f_j]$ for $j = 1, 2, \ldots, k$, then*

$$\mathcal{F}\left[\sum_{j=1}^{k} \alpha_j f_j\right] = \sum_{j=1}^{k} \alpha_j \mathcal{F}[f_j] = \sum_{j=1}^{k} \alpha_j \widehat{f}_j.$$

(See **2.1.1.**)

PROOF. This property follows immediately from the fact that the integral is both additive and homogeneous. ∎

6.1.2. *Translation of the argument of $f(x)$ by the quantity $\oplus h$ for $h > 0$ corresponds to multiplication of the transform $\mathcal{F}[f]$ by $\chi(h, y)$, i.e.,*

$$\int_0^\infty f(x \oplus h)\overline{\chi(x, y)}\, dx = \chi(h, y)\mathcal{F}[f].$$

Similarly, translation by $\ominus h$ corresponds to multiplication by $\overline{\chi(h, y)}$.

(See **2.1.2.**)

PROOF. Make the change of variables[1] $v = x \oplus h$ and use (1.5.34) together with the fact that $(x \oplus h) \ominus h = x = v \ominus h$. We obtain

$$\begin{aligned}\int_0^\infty f(x \oplus h)\overline{\chi(x, y)}\, dx &= \int_0^\infty f(v)\overline{\chi(v \ominus h, y)}\, dv \\ &= \int_0^\infty f(v)\overline{\chi(v, y)}\chi(h, y)\, dv = \chi(h, y)\mathcal{F}[f].\end{aligned}$$

Similarly, since $(x \ominus h) \oplus h = x = v \oplus h$ we have

$$\begin{aligned}\int_0^\infty f(x \ominus h)\overline{\chi(x, y)}\, dx &= \int_0^\infty f(v)\overline{\chi(v \oplus h, y)}\, dv \\ &= \int_0^\infty f(v)\overline{\chi(v, y)}\,\overline{\chi(h, y)}\, dv = \overline{\chi(h, y)}\mathcal{F}[f]. \quad ∎\end{aligned}$$

We shall say that a function $f(x)$ is the *$\mathcal{P}$-adic non- symmetric convolution* of a function $f_1(x)$ with another function $f_2(x)$ and will denote it by $f(x) = (f_1 \overset{\rightarrow}{*} f_2)(x)$ if

$$f(x) = \int_0^\infty f_1(v \oplus x)\overline{f_2(v)}\, dv.$$

---

[1] As we have seen, the transformation $x \to x \oplus h$ is defined on $[0, \infty)$ only up to a countable set but is measure preserving and 1-1 off this countable set.

The non-symmetric convolution of the function $f_2(x)$ with the function $f_1(x)$ will be denoted by $g(x)$:

$$g(x) = \int_0^\infty f_2(v \oplus x)\overline{f_1(v)}\,dv = (f_2 \overset{\rightarrow}{*} f_1)(x).$$

6.1.3. *If $f_j(x) \in \mathbf{L}^1(0,\infty)$ for $j = 1,2$ then the convolutions $f(x) = (f_1 \overset{\rightarrow}{*} f_2)(x)$ and $g(x) = (f_2 \overset{\rightarrow}{*} f_1)(x)$ exist, belong to $\mathbf{L}^1(0,\infty)$ and satisfy the inequalities*

$$\|f\|_1 \le \|f_1\|_1\|f_2\|_1 \quad \text{and} \quad \|g\|_1 \le \|f_1\|_1\|f_2\|_1.$$

PROOF. The proof is similar to that given in Chapter 2 for (2.1.7). Namely, notice that the functions $\omega_1(x,v) = f_1(v \oplus x)\overline{f_2(v)}$ and $\omega_2(x,v) = f_2(v \oplus x)\overline{f_1(v)}$ are measurable. Moreover, at each point where $f_2(v)$ (respectively, $f_1(v)$) is finite, i.e., almost everywhere, the identities

$$\int_0^\infty |\omega_1(x,v)|\,dx = |f_2(v)| \int_0^\infty |f_1(v \oplus x)|\,dx = |f_2(v)|\,\|f_1\|_1,$$

respectively,

$$\int_0^\infty |\omega_2(x,v)|\,dx = |f_1(v)|\,\|f_2\|_1$$

hold. Consequently,

$$\int_0^\infty dv \int_0^\infty |\omega_1(x,v)|\,dx = \int_0^\infty dv \int_0^\infty |\omega_2(x,v)|\,dx = \|f_1\|_1\|f_2\|_1.$$

Thus we conclude by Fubini's Theorem that

$$\begin{aligned}\int_0^\infty \Big| \int_0^\infty f_1(v \oplus x)\overline{f_2(v)}\,dv \Big|\,dx &\le \int_0^\infty \int_0^\infty |f_1(v \oplus x)\overline{f_2(v)}|\,dv\,dx \\ &= \int_0^\infty \int_0^\infty |f_1(v \oplus x)\overline{f_2(v)}|\,dx\,dv \\ &= \|f_1\|_1\|f_2\|_1.\end{aligned}$$

Similarly,

$$\int_0^\infty \Big| \int_0^\infty f_2(v \oplus x)\overline{f_1(v)}\,dv \Big|\,dx \le \|f_1\|_1\|f_2\|_1. \quad \blacksquare$$

6.1.4. *If $f_j(x) \in \mathbf{L}^1(0,\infty)$ for $j = 1,2$ then the convolutions defined above satisfy*

$$\mathcal{F}[f] = \mathcal{F}[f_1]\overline{\mathcal{F}[f_2]} \quad \text{and} \quad \mathcal{F}[g] = \overline{\mathcal{F}[f_1]}\mathcal{F}[f_2].$$

In particular, $\mathcal{F}[f] = \overline{\mathcal{F}[g]}$.

PROOF. By Fubini's Theorem and **6.1.2**, we have

$$\begin{aligned}\mathcal{F}[f] = \widehat{f}(y) &= \int_0^\infty \left(\int_0^\infty f_1(v \oplus x)\overline{f_2(v)}\, dv\right) \overline{\chi(x,y)}\, dx \\ &= \int_0^\infty \overline{f_2(v)}\, dv \int_0^\infty f_1(v \oplus x)\overline{\chi(x,y)}\, dx \\ &= \int_0^\infty \overline{f_2(v)}\chi(v,y)\, dv\, \mathcal{F}[f_1] = \mathcal{F}[f_1]\overline{\mathcal{F}[f_2]}.\end{aligned}$$

Similarly,

$$\begin{aligned}\mathcal{F}[g] = \widehat{g}(y) &= \int_0^\infty \left(\int_0^\infty f_2(v \oplus x)\overline{f_1(v)}\, dv\right) \overline{\chi(x,y)}\, dx \\ &= \int_0^\infty \overline{f_1(v)}\, dv \int_0^\infty f_2(v \oplus x)\overline{\chi(x,y)}\, dx \\ &= \int_0^\infty \overline{f_1(v)}\chi(v,y)\, dv\, \mathcal{F}[f_2] = \overline{\mathcal{F}[f_1]}\mathcal{F}[f_2].\end{aligned}$$

Consequently, $\mathcal{F}[f] = \overline{\mathcal{F}[g]}$. ∎

In this chapter it is convenient to modify somewhat the concepts of continuity and the moduli of continuity which were set forth in the first two chapters. We shall call a function $f(x)$ *$\mathcal{P}$- continuous at a point* $x \in [0,\infty)$ if

$$\sup_{0 \le h < 1/m_r} |f(x \oplus h) - f(x)| \to 0 \quad \text{as } r \to \infty. \tag{6.1.6}$$

It is not difficult to show that continuity at a point in the sense of (6.1.6) is equivalent (for the case $m_r = 2^r$) to continuity with respect to the metric ρ^* which was defined in (1.2.19).

The sequence defined by the relationship

$$\omega_r(f, \frac{1}{m_r}) = \omega_r(f) = \sup_{x \in [0,\infty)} \left(\max_{0 \le h < 1/m_r} |f(x \oplus h) - f(x)|\right) \tag{6.1.7}$$

will be called the *$\mathcal{P}$-modulus of continuity of the function* $f(x)$.

6.1.5. *If* $f(x) \in \mathbf{L}^1(0,\infty)$ *then the transform*

$$\mathcal{F}[f](y) = \widehat{f}(y)$$

is $\mathcal{P}'$-continuous on $[0,\infty)$, where $\mathcal{P}' = (\ldots, p_j, \ldots, p_1, p_{-1}, \ldots, p_{-j}, \ldots)$.

PROOF. By property (1.5.34) applied to the function $\chi(x,y)$ we have

$$\widehat{f}(y \oplus h) - \widehat{f}(y) = \int_0^\infty f(x)\left(\overline{\chi(x, y \oplus h)} - \overline{\chi(x,y)}\right) dx$$
$$= \int_0^\infty f(x)\overline{\chi(x,y)}\left(\overline{\chi(x,y)} - 1\right) dx.$$

Choose an integer N_0 so large that for a given $\varepsilon > 0$ and all $h > 0$ the following inequality holds:

$$\left| \int_{N_0}^\infty f(x)\overline{\chi(x,y)}\left(\overline{\chi(x,y)} - 1\right) dx \right| \le 2\int_{N_0}^\infty |f(x)|\, dx < \varepsilon.$$

Let $r > 0$ satisfy $m_{-r+1} > N_0$. Then for all h such that $0 \le h < 1/m_r$ and all $0 \le x \le N_0$, we have by (1.5.31) that $x_{-k} = 0$ for $k \ge r$, $h'_{-k} = 0$ for $k = 1, 2, \ldots$, and $h'_k = 0$ for $k \le r-1$. Consequently,

$$\chi(x,h) = \exp\left(2\pi i \sum_{k=1}^{\infty} \frac{x_{-k}h'_k + x_k h'_{-k}}{p_k}\right) = \exp\left(2\pi i \sum_{k=1}^{r-1} \frac{x_{-k}h'_k}{p_k}\right) = 1.$$

This means that

$$\int_0^{N_0} f(x)\overline{\chi(x,y)}\left(\overline{\chi(x,y)} - 1\right) dx = 0,$$

i.e., $|\widehat{f}(y \oplus h) - \widehat{f}(y)| < \varepsilon$ for $0 \le h < 1/m_r$. Thus the transform $\mathcal{F}[f]$ is $\mathcal{P}'$-continuous on $[0,\infty)$. ∎

Similarly, one can prove that the transform $\mathcal{F}^*[f]$ is $\mathcal{P}$-continuous on $[0,\infty)$.

6.1.6. *If a function $f(x) \in \mathbf{L}^1(0,\infty)$ is $\mathcal{P}$-continuous at the point $x = 0$ and the transform $\mathcal{F}[f] = \widehat{f}(y)$ is real and non-negative, i.e., $\widehat{f}(y) \ge 0$ for $y \in [0,\infty)$, then $\widehat{f}(y) \in \mathbf{L}^1(0,\infty)$ and*

$$\int_0^\infty \widehat{f}(y)\, dy = f(0).$$

PROOF. Since $\widehat{f}(y) \ge 0$, we can write

$$\int_0^A |\widehat{f}(y)|\, dy = \int_0^A \widehat{f}(y)\, dy.$$

To prove that the limit of this integral exists as $A \to \infty$ it is enough to show it converges through some subsequence of values A_r, for example, for $A_r = m_r$.

By (1.5.38) and (1.5.21), we have

$$\begin{aligned}\int_0^{A_r} \widehat{f}(y)\,dy &= \int_0^{m_r}\int_0^{\infty} f(x)\overline{\chi(x,y)}\,dx\,dy \\ &= \int_0^{\infty} f(x)\,dx \int_0^{m_r} \overline{\chi(x,y)}\,dy \\ &= \int_0^{\infty} f(x)\overline{D(x,\check{m}_r)}\,dx \\ &= \int_0^{1} f(x)\overline{D(x,\check{m}_r)}\,dx \\ &= m_r \int_0^{1/m_r} f(x)\,dx = m_r \int_0^{1/m_r} f(0\oplus x)\,dx.\end{aligned}$$

Consequently,

$$\int_0^{\infty} \widehat{f}(y)\,dy = \lim_{r\to\infty} \int_0^{m_r} \widehat{f}(y)\,dy = \lim_{r\to\infty} m_r \int_0^{1/m_r} f(0\oplus x)\,dx = f(0). \quad \blacksquare$$

We shall now prove that the $\mathbf{L}^p$-transform defined by (6.1.4) exists.

6.1.7. *Let $f \in \mathbf{L}^p$ for some $1 < p \le 2$. Then the $\mathbf{L}^p$-transform (6.1.4) exists and satisfies the inequality*

$$\|\widehat{f}(y)\|_{p'} \le \|f\|_p \tag{6.1.8}$$

where $1/p + 1/p' = 1$.

PROOF. Let $b > 0$, r be a natural number, $n = [bm_r]$, $\delta_\nu(r)$ be the half open interval $[\nu/m_r, (\nu+1)/m_r)$, and $\delta'_\nu(r) = [\nu/m_{-r}, (\nu+1)/m_{-r})$. Define numbers

$$a_\nu = \int_{\delta_\nu(r)} f(x)\,dx, \qquad \nu = 0, 1, \ldots,$$

and a function

$$\Phi_n(x) = \sum_{\nu=0}^{n-1} a_\nu \overline{\chi\left(x, \frac{\nu}{m_{-r}}\right)}.$$

In **1.5.6** we proved that the system

$$\{\phi_{\nu,r}(x)\} = \left\{\frac{1}{\sqrt{m_{-r}}}\chi\left(x, \frac{\nu}{m_{-r}}\right)\right\}_{\nu=0}^{\infty}$$

is uniformly bounded and orthonormal on the interval $[0, m_{-r})$ with

$$|\phi_{\nu,r}(x)| \le 1/\sqrt{m_{-r}} = M.$$

Set

$$\Phi(x) = \sum_{\nu=0}^{n-1} a_\nu \phi_{\nu,r}(x) = M\Phi_n(x)$$

and apply a theorem of Riesz ([17], p. 237). We obtain

$$\|\Phi\|_{p'} \le M^{\frac{2}{p}-1}\|a\|_p,$$

where

$$\|\Phi\|_{p'} = \left(\int_0^{m_r} |\Phi(x)|^{p'}\,dx\right)^{1/p'},$$

and

$$\|a\|_p = \left(\sum_{\nu=0}^{n-1} |a_\nu|^p\right)^{1/p}.$$

Thus $\Phi_n(x)$ can be estimated by

$$M\|\Phi_n\|_{p'} \le M^{\frac{2}{p}-1}\|a\|_p.$$

Hence it follows that

$$\int_0^{m_r} |\Phi(x)|^{p'}\,dx \le m_{-r}\left(\sum_{\nu=0}^{n-1} |a_\nu|^p\right)^{1/(p-1)}.$$

But Hölder's inequality implies

$$|a_\nu|^p \le \int_{\delta_\nu(r)} |f(x)|^p\,dx \left(\int_{\delta_\nu(r)} dx\right)^{p-1} = \frac{1}{m_{-r}^{p-1}} \int_{\delta_\nu(r)} |f(x)|^p\,dx.$$

Consequently,

$$\begin{aligned}
\text{(6.1.9)} \qquad \int_0^A |\Phi_n(x)|^{p'}\,dx &\le \int_0^{m_r} |\Phi_n(x)|^{p'}\,dx \\
&\le \left(\sum_{\nu=0}^{n-1} \int_{\delta_\nu(r)} |f(x)|^p\,dx\right)^{1/(p-1)} \\
&\le \left(\int_0^b |f(x)|^p\,dx\right)^{1/(p-1)}
\end{aligned}$$

for all $A \le m_{-r}$.

Now, we shall show that

$$\lim_{r\to\infty} \Phi_n(x) = \int_0^b f(y)\overline{\chi(y,x)}\,dy = \widehat{f}(x,b). \tag{6.1.10}$$

Indeed, by **1.5.4** we have $\overline{\chi(y,x)} = \overline{\chi(x,\nu/m_{-r})}$ for all $y \in \delta'_\nu(r)$ and $x < m_{-r}$. Thus

$$\begin{aligned} &| \Phi_n(x) - \int_0^b f(y)\overline{\chi(y,x)}\,dy \,| \\ &=| \sum_{\nu=0}^{n-1} \int_{\delta'_\nu(r)} f(y)\left(\chi(x,\frac{\nu}{m_{-r}}) - \overline{\chi(x,y)}\right)\,dy - \int_{n/m_{-r}}^b f(y)\overline{\chi(y,x)}\,dy \,| \\ &\le \int_{\delta'_n(r)} |f(y)|\,dy \to 0, \qquad r\to\infty. \end{aligned}$$

Therefore, if we let $r \to \infty$ while A remains fixed, we see by (6.1.9) and (6.1.10) that

$$\int_0^A \left| \int_0^b f(y)\overline{\chi(x,y)}\,dy \right|^{p'} dx \le \left(\int_0^b |f(x)|^p\,dx\right)^{1/(p-1)}.$$

In particular, if we let $A \to \infty$, we obtain

$$\int_0^\infty \left| \int_0^b f(y)\overline{\chi(x,y)}\,dy \right|^{p'} dx \le \left(\int_0^b |f(x)|^p\,dx\right)^{1/(p-1)}. \tag{6.1.11}$$

Apply (6.1.11) to the function

$$f_0(x) = \begin{cases} 0 & \text{for } 0 \le x \le a, \\ f(x) & \text{for } a < x < \infty. \end{cases}$$

We obtain

$$\begin{aligned} &\int_0^\infty \left| \int_0^b f_0(y)\,\overline{\chi(x,y)}\,dy \right|^{p'} dx \\ &\qquad = \int_0^\infty \left| \int_0^b f(y)\overline{\chi(x,y)}\,dy - \int_0^a f(y)\overline{\chi(x,y)}\,dy \right|^{p'} dx \\ &\qquad = \int_0^\infty |\widehat{f}(x,b) - \widehat{f}(x,a)|^{p'}\,dx \\ &\qquad \le \left(\int_0^b |f_0(x)|^p\,dx\right)^{1/(p-1)} \\ &\qquad = \left(\int_a^b |f(x)|^p\,dx\right)^{1/(p-1)}, \end{aligned}$$

$$\int_0^\infty |\widehat{f}(x,b)-\widehat{f}(x,a)|^{p'}\,dx \le \left(\int_a^b |f(x)|^p\,dx\right)^{1/(p-1)}. \tag{6.1.12}$$

Since $f(x)\in \mathbf{L}^p(0,\infty)$, the right side of (6.1.12) converges to zero as $a,b\to\infty$. Since $\mathbf{L}^{p'}(0,\infty)$ is complete, it follows that $\widehat{f}(x,a)$ converges in the $\mathbf{L}^{p'}(0,\infty)$ norm to some function $\widehat{f}(x)\in \mathbf{L}^{p'}(0,\infty)$. This means that the $\mathbf{L}^p$-transform of any $f(x)\in\mathbf{L}^p(0,\infty)$ exists for $1<p\le 2$. Moreover, if we let $b\to\infty$ in (6.1.11), we obtain

$$\left(\int_0^\infty |\widehat{f}(x)|^{p'}\,dx\right)^{1/p'} \le \left(\int_0^\infty |f(x)|^p\,dx\right)^{1/p}.$$

This verifies inequality (6.1.8). ∎

§6.2. Representation of functions in $\mathbf{L}^1(0,\infty)$ by their multiplicative transforms.

As we saw in §6.1, the direct multiplicative transform

$$\mathcal{F}[f](y)\equiv\widehat{f}(y)=\int_0^\infty f(x)\overline{\chi(x,y)}\,dx \tag{6.2.1}$$

exists for every function $f(x)\in\mathbf{L}^1(0,\infty)$. However, as in the case of the exponential Fourier transform, inversion

$$f(x)=\int_0^\infty \widehat{f}(y)\chi(x,y)\,dy \tag{6.2.2}$$

does not hold for every function $f(x)\in\mathbf{L}^1(0,\infty)$. Thus it is interesting to ascertain what kinds of conditions, on the functions $f(x)$ and $\widehat{f}(y)$, imply the inversion formula (6.2.2).

Before we begin identifying such conditions, we shall establish a result about a method of summability for the integral (6.2.2) which plays the same role that the methods of Fejér and Abel- Poisson do for the proof of inversion of the exponential Fourier transform.

6.2.1. *Let $f(x)\in\mathbf{L}^1(0,\infty)$ and $\widehat{f}(y)$ be its multiplicative transform (6.2.1). Then*

$$f(x)=\lim_{r\to\infty}\int_0^{m_r}\widehat{f}(y)\chi(x,y)\,dy,\qquad m_r=p_rm_{r-1},$$

at each point x of $\mathcal{P}$-continuity of the function $f(x)$.

PROOF. Since $\int_0^\infty |f(x)|\,dx<\infty$, we can apply Fubini's Theorem to the integral

$$I_{m_r}(f,x)=\int_0^{m_r}\left(\int_0^\infty f(t)\overline{\chi(t,y)}\,dt\right)\chi(x,y)\,dy$$

to change the order of integration, i.e.,

$$I_{m_r}(f,x) = \int_0^\infty f(t)\,dt \int_0^{m_r} \overline{\chi(t\ominus x, y)}\,dy = \int_0^\infty f(t)\overline{D(t\ominus x, \check{m}_r)}\,dt.$$

Recall from (1.5.38) and (1.5.21) that

$$D(t\ominus x, \check{m}_r) = \begin{cases} m_r, & \text{if } 0 \le t\ominus x < 1/m_r, \\ 0, & \text{if } 1/m_r \le t\ominus x < \infty. \end{cases}$$

Thus

$$I_{m_r}(f,x) = m_r\left(\int_{0\le t\ominus x<1/m_r} f(t)\,dt\right) = m_r \int_0^{1/m_r} f(x\oplus u)\,du.$$

Hence, it follows that if x is a point of $\mathcal{P}$-continuity for $f(x)$ then

$$\lim_{r\to\infty} I_{m_r}(f,x) = \lim_{r\to\infty} m_r \int_0^{1/m_r} f(x\oplus u)\,du = f(x). \quad \blacksquare$$

6.2.2. *Let $f(x) \in \mathbf{L}^1(0,\infty)$ and suppose that its multiplicative transform $\widehat{f}(y)$ belongs to $\mathbf{L}^1(0,\infty)$. Then equation (6.2.2) holds at each point x of $\mathcal{P}$-continuity of the function $f(x)$.*

PROOF. Since $\widehat{f}(y) \in \mathbf{L}^1(0,\infty)$, the integral in (6.2.2) converges absolutely and thus

$$\int_0^\infty \widehat{f}(y)\chi(x,y)\,dy = \lim_{r\to\infty} \int_0^{m_r} \widehat{f}(y)\chi(x,y)\,dy.$$

Applying **6.2.1** thus verifies **6.2.2.** $\blacksquare$

Theorems **6.1.6** and **6.2.2** contain the following corollary.

6.2.3. *If $x = 0$ is a point of $\mathcal{P}$-continuity of a function $f(x) \in \mathbf{L}^1(0,\infty)$ and if the transform $\widehat{f}(y)$ is real and non-negative, then equation (6.2.2) holds at each point of $\mathcal{P}$-continuity of the function $f(x)$.*

The Plancherel identity holds for multiplicative transforms:

6.2.4. *Suppose that $f(x) \in \mathbf{L}^1(0,\infty)$ is bounded and $\mathcal{P}$-continuous, and $\widehat{f}(y)$ is its multiplicative transform. Then*

$$\int_0^\infty |\widehat{f}(y)|^2\,dy = \int_0^\infty |f(x)|^2\,dx.$$

PROOF. Consider the $\mathcal{P}$-adic non-symmetric convolution of the function $f(x)$ with itself, i.e.,

$$\phi(x) = \int_0^\infty f(v\oplus x)\overline{f(v)}\,dv.$$

We know by **6.1.3** that $\|\phi\|_1 \le \|f\|_1^2$ and by **6.1.4** that

$$\mathcal{F}[\phi](y) = \mathcal{F}[f]\overline{\mathcal{F}[f](y)} = |\mathcal{F}[f](y)|^2 \ge 0, \qquad y \in [0,\infty).$$

Moreover, since $f(x) \in \mathbf{L}^1(0,\infty)$ is bounded, the value

$$\phi(0) = \int_0^\infty |f(v)|^2 dv$$

is finite, and since $f(x)$ is $\mathcal{P}$-continuous, the function $\phi(x)$ is $\mathcal{P}$-continuous at the point $x = 0$. Hence it follows from **6.1.6** that $\mathcal{F}[\phi](y) \in \mathbf{L}^1(0,\infty)$. Applying **6.2.3**, we obtain

$$\begin{aligned}\phi(x) &= \int_0^\infty f(v \oplus x)\overline{f(v)}\, dv \\ &= \int_0^\infty \mathcal{F}[\phi](y)\chi(x,y)\, dy \\ &= \int_0^\infty |\mathcal{F}[f](y)|^2 \chi(x,y)\, dy.\end{aligned}$$

Substituting $x = 0$ into this equation, we see that **6.2.4** holds. ∎

We shall presently establish (6.2.2) in the case when the $\mathcal{P}$- modulus of continuity (see (6.1.7)) of the function $f(x) \in \mathbf{L}^1(0,\infty)$ satisfies a particular growth condition. Before we do this we make some preliminary estimates.

6.2.5. *Let n be a natural number, $0 < \alpha < 1$, and $f(x) \in \mathbf{L}^1(0,\infty)$. Then*

$$\lim_{n\to\infty} \int_n^{n+\alpha} \hat{f}(y)\chi(x,y)\, dy = 0.$$

PROOF. In the following, we shall drop the index P on $D_n(x)_P$ and $\chi_n(x)_P$ but retain the index P' on $\chi_n(x)_{P'}$. By Fubini's Theorem,

$$\begin{aligned}I_n(f,\alpha) &= \int_n^{n+\alpha} \hat{f}(y)\chi(x,y)\, dy \\ &= \int_n^{n+\alpha} \left(\int_0^\infty f(t)\overline{\chi(t,y)}\, dt\right)\chi(x,y)\, dy \\ &= \int_0^\infty f(t) \int_n^{n+\alpha} \overline{\chi(t \ominus x, y)}\, dy\, dt \\ &= \int_0^\infty f(t)\left(\overline{D(t \ominus x, n \check{+} \alpha)} - \overline{D(t \ominus x, \check{n})}\right) dt.\end{aligned}$$

Apply (1.5.38). In the case $0 \le t \ominus x < 1$ we obtain

(6.2.3)
$$\begin{aligned} D(t \ominus x, n \check{+} \alpha) - D(t \ominus x, \check{n}) &= D_n(t \ominus x) + \alpha \chi_n(t \ominus x) - D_n(t \ominus x) \\ &= \alpha \chi_n(t \ominus x) \end{aligned}$$

and in the case $1 \le k \le t \ominus x < k+1$ we obtain

$$D(t \ominus x, n \check{+} \alpha) - D(t \ominus x, \check{n}) = \chi_n(\{t \ominus x\}) \int_0^\alpha \chi_k(v)_{P'}\, dv. \tag{6.2.4}$$

Make the change of variables $t = z \oplus x$ and use (6.2.3) and (6.2.4) to see that

$$\begin{aligned} I_n(f, \alpha) &= \int_0^\infty f(z \oplus x) \left(\overline{D(z, n \check{+} \alpha)} - \overline{D(z, \check{n})} \right) dz \\ &= \int_0^1 f(z \oplus x) \alpha \overline{\chi_n(z)}\, dz + \sum_{k=1}^\infty \int_k^{k+1} f(z \oplus x) \overline{\chi_n(\{z\})} \int_0^\alpha \overline{\chi_k(v)_{P'}}\, dv\, dz \\ &= \alpha \int_0^1 f(z \oplus x) \overline{\chi_n(z)}\, dz + \sum_{k=1}^\infty \int_0^1 f(k + z \oplus x) \overline{\chi_n(z)} \int_0^\alpha \overline{\chi_k(v)_{P'}}\, dv\, dz \\ &= \int_0^1 \left(\alpha f(z \oplus x) + \sum_{k=1}^\infty f(k + z \oplus x) \int_0^\alpha \overline{\chi_k(v)_{P'}}\, dv \right) \overline{\chi_n(z)}\, dz \\ &= \int_0^1 F(f, \alpha, z) \overline{\chi_n(z)}\, dz. \end{aligned}$$

Notice that

$$\begin{aligned} \int_0^1 |F(f, \alpha, z)|\, dz &\le \int_0^1 \left(\alpha |f(z \oplus x)| + \sum_{k=1}^\infty |f(k + z \oplus x)| \alpha \right) dz \\ &= \alpha \int_0^\infty |f(z \oplus x)|\, dz < \infty, \end{aligned}$$

i.e., $F(f, \alpha, z) \in \mathbf{L}^1(0,1)$. But by a theorem of Mercer (see [2], p. 66), the Fourier coefficients of an integrable function in any bounded orthonormal system $\{\chi_n(z)\}$ converge to zero. Consequently,

$$\lim_{n \to \infty} I_n(f, \alpha) = \lim_{n \to \infty} \int_0^1 \mathcal{F}(f, \alpha, z) \overline{\chi_n(z)}\, dz = 0. \quad \blacksquare$$

6.2.6. *Let r and n be whole numbers with $r \ge 1$ and $m_{r-1} < n < m_r$. If $\phi(x)$ is constant on the intervals $\delta_\nu(r-1) = [\nu/m_{r-1}, (\nu+1)/m_{r-1}]$ and $f(x)$ is $\mathcal{P}$-continuous, then*

$$\left| \int_0^{1/m_k} f(x) \phi(x) \overline{\chi_n(x)}\, dx \right| \le \frac{1}{m_k} \omega_{r-1}(f) M_k$$

for $k = 0, 1, \ldots, r-1$, *where* $M_k = \max_{0 \le x \le 1/m_k} |\phi(x)|$.

PROOF. Write the integral we wish to estimate in the form

$$(6.2.5) \qquad \int_0^{1/m_k} f(x)\phi(x)\overline{\chi_n(x)}\,dx = \sum_{\nu=0}^{(m_{r-1}/m_k)-1} \int_{\delta_\nu(r-1)} f(x)\phi(x)\overline{\chi_n(x)}\,dx.$$

Choose a point $x_\nu = \nu/m_{r-1} + 0$ from each $\delta_\nu(r-1)$ at which the function $f(x_\nu)$ is defined and set $M_\nu = |\phi(x_\nu)|$. Since

$$\int_{\delta_\nu(r-1)} \overline{\chi_n(x)}\,dx = 0, \qquad \nu = 0, 1, \ldots$$

(being a sum of p_r-th roots of unity) and since $\phi(x)$ is constant on $\delta_\nu(r-1)$ we have

$$\begin{aligned}
\left| \int_{\delta_\nu(r-1)} f(x)\phi(x)\overline{\chi_n(x)}\,dx \right| &= \left| \int_{\delta_\nu(r-1)} (f(x) - f(x_\nu))\phi(x)\overline{\chi_n(x)}\,dx \right| \\
&\le \int_{\delta_\nu(r-1)} |f(x) - f(x_\nu)| M_\nu\,dx \\
&= M_\nu \int_0^{1/m_{r-1}} |f(x_\nu \oplus u) - f(x_\nu)|\,du \\
&\le \frac{M_\nu}{m_{r-1}} \omega_{r-1}(f).
\end{aligned}$$

Substituting this estimate into (6.2.5) finishes the proof of this theorem. ∎

6.2.7. *Let* r *and* n *be whole numbers with* $r \ge 1$, *and* $m_{r-1} < n < m_r$. *Determine coefficients* n_{-j} *by* $n = n_{-1} + n_{-2}m_1 + \cdots + n_{-r}m_{r-1}$. *If* $f \in \mathbf{L}^1(0,\infty)$ *then*

$$\left| \int_{m_{r-1}}^n \widehat{f}(y)\chi(x,y)\,dy \right| \le (n_{-1} + n_{-2} + \cdots + n_{-r})\omega_{r-1}(f),$$

where $\omega_{r-1}(f)$ *is the* $\mathcal{P}$*-modulus of continuity of* $f(x)$.

PROOF. By Fubini's Theorem we can write

$$\begin{aligned}
J_{n,r}(f,x) &= \int_{m_{r-1}}^n \left(\int_0^\infty f(t)\overline{\chi(t,y)}\,dt \right) \chi(x,y)\,dy \\
&= \int_0^\infty f(t) \int_{m_{r-1}}^n \overline{\chi(t \ominus x, y)}\,dy\,dt \\
&= \int_0^\infty f(t)\,(D(t \ominus x, \check{n}) - D(t \ominus x, \check{m}_{r-1}))\,dt.
\end{aligned}$$

By (1.5.38), we have

$$J_{n,r}(f,x) = \int_{0\le t\ominus x<1} f(t)\left(\overline{D_n(t\ominus x)} - \overline{D_{m_{r-1}}(t\ominus x)}\right) dt.$$

Moreover, by (1.5.20) we can represent the Dirichlet kernel in the following form:

$$\begin{aligned} D_n(z) &= \sum_{j=0}^{m_{r-1}-1} \overline{\chi}_j(z) + \sum_{s=1}^{n_{-r}-1}\left(\sum_{j=sm_{r-1}}^{(s+1)m_{r-1}-1} \overline{\chi}_j(z)\right) \\ &\quad + \sum_{k=r-1}^{2}\left(\sum_{j=n_{-r}m_{r-1}+\cdots+n_{-k}m_{k-1}}^{n_{-r}m_{r-1}+\cdots+n_{-k+1}m_{k-2}-1} \overline{\chi}_j(z)\right) \\ &= D_{m_{r-1}}(z) + \sum_{s=1}^{n_{-r}-1} \overline{\chi_{sm_{r-1}}(z)} D_{m_{r-1}}(z) \\ &\quad + \overline{\chi_{n_{-r}m_{r-1}}(z)} \sum_{k=1}^{r-2} \overline{\chi_{n_{-r+1}m_{r-2}+\cdots+n_{-k-1}m_k}(z)} D_{m_{k-1}}(z) \sum_{s_k=0}^{n_{-k}-1} \overline{\chi_{s_k m_{k-1}}(z)}. \end{aligned}$$

Hence it follows from the equation

$$D_{m_\nu}(z) = \begin{cases} m_\nu & \text{for } 0\le z< 1/m_\nu, \\ 0 & \text{for } 1/m_\nu \le z < 1,\ \ \nu = 1,2,\ldots \end{cases}$$

that

$$\begin{aligned} J_{n,r}(f,x) &= m_{r-1} \sum_{s=1}^{n_{-r}-1}\left(\int_{0\le t\ominus x<1/m_r} f(t)\overline{\chi_{sm_{r-1}}(t\ominus x)}\, dt\right) \\ &\quad + \sum_{k=1}^{r-2} m_{k-1}\left(\int_{0\le t\ominus x<1/m_{k-1}} f(t)\phi_k(t\ominus x)\overline{\chi_{n_{-r}m_{r-1}}(t\ominus x)}\, dt\right), \end{aligned}$$

where

$$\phi_k(z) = \overline{\chi_{n_{-r+1}m_{r-2}+\cdots+n_{-k-1}m_k}(z)} \sum_{s_k=0}^{n_{-k}-1} \overline{\chi_{s_k m_{k-1}}(z)},$$

for $k = 1,2,\ldots,r-2$. But for each $1\le k\le r-2$ the function $\phi_k(z)$ is constant on the intervals $\delta_\nu(r-1)$, $\nu = 0,1,\ldots$, and satisfies $|\phi_k(z)| \le n_{-k}$. Consequently, we can use **6.2.6** to write

$$\left| \int_{0\le t\ominus x<1/m_{k-1}} f(t)\phi_k(t\ominus x)\overline{\chi_{n_{-r}m_{r-1}}(t\ominus x)}\, dt \right| \le \frac{1}{m_{k-1}}\omega_{r-1}(f) n_{-k}.$$

Moreover,

$$\left| \int_{0 \le t \ominus x < 1/m_r} f(t)\overline{\chi_{s m_{r-1}}(t \ominus x)}\, dt \right| \le \frac{1}{m_{r-1}} \omega_{r-1}(f).$$

Therefore,

$$\begin{aligned} |J_{n,r}(f,x)| &\le m_{r-1}(n_{-r}-1)\frac{1}{m_{r-1}}\omega_{r-1}(f) + \sum_{k=1}^{r-2} m_{k-1}\frac{1}{m_{k-1}} n_{-k}\omega_{r-1}(f) \\ &= (n_{-1}+n_{-2}+\cdots+n_{-r})\omega_{r-1}(f). \ \blacksquare \end{aligned}$$

We shall now prove a theorem about inversion of the multiplicative transform of a function $f(x) \in \mathbf{L}^1(0,\infty)$.

6.2.8. *Let $\mathcal{P} = \{\ldots, p_{-j}, \ldots, p_{-1}, p_1, \ldots, p_j, \ldots\}$. If $f(x) \in \mathbf{L}^1(0,\infty)$ and its $\mathcal{P}$-modulus of continuity satisfies*

$$\left(\sum_{k=1}^{r} p_k - r\right)\omega_{r-1}(f) = o(1), \tag{6.2.6}$$

then (6.2.2) holds at each point of $\mathcal{P}$-continuity of $f(x)$.

PROOF. Let x be a point of $\mathcal{P}$-continuity of $f(x)$. By **6.2.1** it suffices to show

$$\lim_{r\to\infty} \int_{m_{r-1}}^{a} \widehat{f}(y)\chi(x,y)\,dy = 0, \qquad a \in [m_{r-1}, m_r).$$

By **6.2.5** we may suppose that a is an integer, i.e., $a = n \in [m_{r-1}, m_r)$. But **6.2.7** allows us to write

$$\begin{aligned} \int_{m_{r-1}}^{n} \widehat{f}(y)\chi(x,y)\,dy &\le (n_{-1}+n_{-2}+\cdots+n_{-r})\omega_{r-1}(f) \\ &\le (p_1 - 1 + \cdots + p_r - 1)\omega_{r-1}(f) \\ &= \left(\sum_{k=1}^{r} p_k - r\right)\omega_{r-1}(f). \end{aligned}$$

Consequently, (6.2.6) implies

$$\lim_{r\to\infty} \int_{m_{r-1}}^{n} \widehat{f}(y)\chi(x,y)\,dy = 0. \ \blacksquare$$

The following result is a corollary of **6.2.8**.

6.2.9. *Let $\mathcal{P} = \{\dots, p_{-j}, \dots, p_{-1}, p_1, \dots, p_j, \dots\}$ and suppose that $\sum_{k=1}^{r} p_k \le cr$ for some $c > 0$. If $f(x) \in \mathbf{L}^1(0, \infty)$ satisfies*

$$\omega_{r-1}(f) = o(1/r), \tag{6.2.7}$$

then (6.2.2) holds at each point of $\mathcal{P}$-continuity of $f(x)$.

It is useful to notice that if $\sup_n p_n < \infty$ then hypothesis (6.2.7) is equivalent to the condition $\omega_r(f) = o(1/\ln m_r)$. This is clear from the relationship

$$\ln m_r = \sum_{k=1}^{r} \ln p_k \le r \ln(\sup_k p_k) = c_1 r,$$

and the fact that $p_k \ge 2$ implies

$$\ln m_r \ge r \ln 2 = c_2 r,$$

where $0 < c_2 < c_1$.

We shall establish a connection between finiteness of the transform $\widehat{f}(y)$ and the behavior of the function $f(x)$. For this we will use the following analogue of a theorem of Young (see [30], Vol. I, p. 160).

6.2.10. *Let $\{\omega_n(x)\}_{n=0}^{\infty}$ be a complete, orthonormal system in $\mathbf{L}^2[a, b]$ and $f(x)$, $g(x)$ be square integrable on $[a, b]$. If*

$$c_k = \int_a^b f(t)\omega_k(t)\, dt$$

then the series

$$\sum_{k=0}^{\infty} c_k \int_a^x g(t)\omega_k(t)\, dt$$

converges uniformly on $[a, b]$ and

$$\int_a^x f(t)g(t)\, dt = \sum_{k=0}^{\infty} c_k \int_a^x g(t)\omega_k(t)\, dt$$

for all $x \in [a, b]$.

PROOF. Let

$$s_n(t) = \sum_{k=0}^{n} c_k\omega_k(t).$$

We obtain from the Cauchy-Schwarz inequality that

$$| \int_a^x f(t)g(t)\,dt - \sum_{k=0}^n c_k \int_a^x g(t)\omega_k(t)\,dt | = | \int_a^x g(t)(f(t) - s_n(t))\,dt |$$
$$\leq \left(\int_a^b |g(t)|^2 dt \right)^{1/2} \left(\int_a^b |f(t) - s_n(t)|^2\,dt \right)^{1/2}.$$

By Parseval's identity, the right side of this inequality converges to zero as $n \to \infty$. Thus the series

$$\sum_{k=0}^{\infty} c_k \int_a^x g(t)\omega_k(t)\,dt$$

converges uniformly for $x \in [a, b]$ to the function $\int_a^x f(t)g(t)\,dt$. ∎

The following is an analogue of a theorem of Kotel'nikov [1] about representation of functions with finite Fourier spectrum by its values at equally spaced points.

6.2.11. *If $f(x) \in \mathbf{L}^1(0,\infty)$ is $\mathcal{P}$-continuous on the positive real axis $[0,\infty)$ and its multiplicative transform $\widehat{f}(y)$ satisfies $\widehat{f}(y) = 0$ for $y > \sigma$ and some $\sigma < m_r$, then the function $f(t)$ can be reconstructed from its values at the points $t_k = k/m_r$ for $k = 0, 1, \ldots$ by means of the formula*

$$f(t) = \frac{1}{m_r} \sum_{k=0}^{\infty} f(\frac{k}{m_r}) \overline{D(\frac{k}{m_r} \ominus t, \check{m}_r)},$$

where the kernel $D(u,\xi)$ is defined by (1.5.38).

PROOF. By **1.5.6** the system of functions $\psi_{\nu,r}(y) = 1/\sqrt{m_r}\chi(\nu/m_r, y)$ is orthonormal and bounded on the interval $[0, m_r)$. Thus by a result of Vilenkin [2] the system $\{\psi_{\nu,r}(y)\}_{\nu=0}^{\infty}$ is complete in $\mathbf{L}^2[0, m_r]$. Since $\widehat{f}(y) = 0$ for $y \geq m_r$ we can associate with $\widehat{f}(y)$ a series in the system $\{\overline{\psi_{\nu,r}(y)}\}_{\nu=0}^{\infty}$, i.e.,

$$\widehat{f}(y) \sim \sum_{\nu=0}^{\infty} b_\nu \overline{\psi_{\nu,r}(y)}, \qquad y \in [0, m_r],$$

where

$$b_\nu = \int_0^{m_r} \widehat{f}(y)\psi_{\nu,r}(y)\,dy = \frac{1}{\sqrt{m_r}} \int_0^{m_r} \widehat{f}(y)\chi(\frac{\nu}{m_r}, y)\,dy \tag{6.2.8}$$

for $\nu = 0, 1, \ldots$. On the other hand, $\widehat{f}(y)$ belongs to $\mathbf{L}^1(0,\infty)$ since it has compact support, and therefore by **6.2.1** we have

$$f(x) = \lim_{k\to\infty} \int_0^{m_k} \widehat{f}(y)\chi(x,y)\,dy = \int_0^{m_r} \widehat{f}(y)\chi(x,y)\,dy. \tag{6.2.9}$$

Now, (6.2.8) and (6.2.9) imply that

$$b_\nu = \frac{1}{\sqrt{m_r}} f(\frac{\nu}{m_r}), \qquad \nu = 0, 1, \ldots.$$

Consequently,

$$\widehat{f}(y) \sim \sum_{\nu=0}^{\infty} b_\nu \overline{\psi_{\nu,r}(y)} = \sum_{\nu=0}^{\infty} \frac{1}{m_r} f(\frac{\nu}{m_r}) \overline{\chi(\frac{\nu}{m_r}, y)}. \tag{6.2.10}$$

Since the system $\{\psi_{\nu,r}(x)\}$ is bounded and the function $g(y) = \chi(x,y)$ belongs to $\mathbf{L}^2(0, m_r)$, we have by **6.2.10** that the series for the product $\widehat{f}(y)\chi(x,y)$ can be obtained by integration. Therefore, it follows from (6.2.9) that

$$\begin{aligned} f(x) &= \int_0^{m_r} \chi(x,y) \sum_{\nu=0}^{\infty} \frac{1}{m_r} f(\frac{\nu}{m_r}) \overline{\chi(\frac{\nu}{m_r}, y)}\, dy \\ &= \frac{1}{m_r} \sum_{\nu=0}^{\infty} f(\frac{\nu}{m_r}) \int_0^{m_r} \chi(\frac{\nu}{m_r} \ominus x, y)\, dy, \end{aligned}$$

i.e.,

$$f(x) = \frac{1}{m_r} \sum_{\nu=0}^{\infty} f(\frac{\nu}{m_r}) \overline{D(\frac{\nu}{m_r} \ominus x, \check{m}_r)}. \quad \blacksquare$$

A consequence of **6.2.11** is the following.

6.2.12. *Let $f(x) \in \mathbf{L}^1(0,\infty)$ be $\mathcal{P}$-continuous on $[0,\infty)$ with finite spectrum, i.e, its multiplicative transform satisfies $\widehat{f}(y) = 0$ for $y \geq m_r$. Then $f(x)$ is a step function, constant on the intervals of rank r of the form $\delta_\nu(r) = [\nu/m_r, (\nu+1)/m_r)$, $\nu = 0, 1, \ldots$.*

PROOF. The proof is an easy consequence of **6.2.11** and the fact that

$$D(\frac{\nu}{m_r} \ominus x, \check{m}_r) = \begin{cases} m_r & \text{for } 0 \leq (\nu/m_r) \ominus x < 1/m_r, \\ 0 & \text{for } 1/m_r \leq (\nu/m_r) \ominus x < \infty. \end{cases} \quad \blacksquare$$

6.2.13. *If $f(x) \in \mathbf{L}^1(0,\infty)$ is a step function, constant on the intervals of rank r of the form $\delta'_\nu(r) = [\nu/m_{-r}, (\nu+1)/m_{-r})$, then it has finite spectrum, i.e., its multiplicative transform satisfies $\widehat{f}(y) = 0$ for $y \geq m_{-r}$.*

PROOF. Since the step function $f(x)$ is constant on the intervals $\delta'_\nu(r)$, we can write it in the form

$$f(x) = \sum_{\nu=0}^{\infty} c_\nu \frac{1}{m_{-r}} D(\check{m}_r, \frac{\nu}{m_{-r}} \ominus x),$$

for some constants c_ν. Thus

$$\begin{aligned}\widehat{f}(y) &= \frac{1}{m_{-r}} \int_0^\infty \sum_{\nu=0}^\infty c_\nu(\check{m}_r, \frac{\nu}{m_{-r}} \ominus x)\overline{\chi(t,y)}\,dt \\ &= \frac{1}{m_{-r}} \sum_{\nu=0}^\infty c_\nu \left(\int_{0 \le (\nu/m_{-r}) \ominus t < 1/m_{-r}} m_{-r}\overline{\chi(t,y)}\,dt \right).\end{aligned}$$

If we change variables by $v = (\nu/m_{-r}) \ominus t$, i.e., $\nu/m_{-r} = v \oplus t$ and $t = (\nu/m_{-r}) \ominus v$ we obtain

$$\begin{aligned}\widehat{f}(y) &= \sum_{\nu=0}^\infty c_\nu \int_0^{1/m_{-r}} \overline{\chi(\frac{\nu}{m_{-r}} \ominus v, y)}\,dv \\ &= \sum_{\nu=0}^\infty c_\nu \overline{\chi(\frac{\nu}{m_{-r}}, y)} \int_0^{1/m_{-r}} \chi(v,y)\,dv.\end{aligned}$$

But

$$\int_0^{1/m_{-r}} \chi(v,y)\,dv = \int_0^{1/m_{-r}} \chi_{[y]}(v)\,dv = 0$$

for $y \ge m_{-r}$. Thus $\widehat{f}(y) = 0$ for $y \ge m_{-r}$. ∎

Under the condition that the collection $\mathcal{P}$ is symmetric, i.e., $\mathcal{P}' = \mathcal{P}$, we shall establish theorems which are in a sense duals of **6.2.12** and **6.2.13**.

6.2.14. *Let $f(x) \in \mathbf{L}^1(0,\infty)$ and suppose that $f(x) = 0$ for $x \ge m_r$. Then the transform $\widehat{f}(y)$ is a step function constant on the intervals $\delta_\nu(r)$ of rank r.*

PROOF. Since $f(x) = 0$ for $x \ge m_r$, we can write the transform $\widehat{f}(y)$ in the form

$$\widehat{f}(y) = \sum_{k=0}^\infty \int_k^{k+1} f(x)\overline{\chi_{[y]}(\{x\})}\,\overline{\chi_k(\{y\})}\,dx = \sum_{k=0}^\infty c_{[y]}(f_k)\overline{\chi_k(\{y\})},$$

where $f_k(x) = f(k+x)$ and $c_{[y]}(f_k) = 0$ for $k \ge m_r$. Consequently,

$$\widehat{f}(y) = \sum_{k=0}^{m_r - 1} c_{[y]}(f_k)\overline{\chi_k(\{y\})}.$$

But for $k \le m_r - 1$, the functions $\chi_k(\{y\})$ are step functions and are constant on $\delta_\nu(r)$. ∎

6.2.15. *Let $f(x) \in \mathbf{L}^1(0,\infty)$ be $\mathcal{P}$-continuous on $[0,\infty)$. If its transform $\widehat{f}(y)$ is a step function, constant on the intervals $\delta_\nu(r)$ of rank r, then $f(x) = 0$ for $x \geq m_r$.*

PROOF. Suppose to the contrary that there exists a point $x_0 \geq m_r$ such that $f(x_0) \neq 0$. Since $f(x)$ is $\mathcal{P}$- continuous there is an interval $[x_0, a) \subset [\ell, \ell+1)$ for some natural number $\ell \geq m_r$ such that $f(x) \neq 0$ for all $x \in [x_0, a)$.

Consider the function $f_{[x_0]}(t) = f([x_0] + t)$ defined for $t \in [0,1)$. Choose a natural number ν such that $c_\nu(f_{[x_0]}) \neq 0$. For $y \in [\nu, \nu+1)$ the transform $\widehat{f}(y)$ can be written in the form

$$\widehat{f}(y) = \widehat{f}_1(y) + \widehat{f}_2(y) + \widehat{f}_3(y),$$

where

$$\begin{aligned}
\widehat{f}_1(y) &= \int_0^{m_r} f(t)\overline{\chi(t,y)}\,dt \\
&= \sum_{k=0}^{m_r-1} \int_k^{k+1} f(t)\overline{\chi(t,y)}\,dt \\
&= \sum_{k=0}^{m_r-1} \int_0^1 f_k(t)\overline{\chi_k(y)}\,\overline{\chi_{[y]}(t)}\,dt \\
&= \sum_{k=0}^{m_r-1} c_\nu(f_k)\overline{\chi_k(y)}
\end{aligned}$$

is a step function of rank r,

$$\widehat{f}_2(y) = \sum_{k=m_r}^{m_n-1} c_\nu(f_k)\overline{\chi_k(y)},$$

and

$$\widehat{f}_3(y) = \sum_{k=m_n}^{\infty} c_\nu(f_k)\overline{\chi_k(y)},$$

for n chosen so that $m_r \leq m_{n-1} \leq [x_0] < m_n$. Since the functions $\chi_k(y)$ are linearly independent and $c_\nu(f_{[x_0]}) \neq 0$, there exist neighboring intervals $\delta_\mu(n)$ and $\delta_{\mu+1}(n)$, belonging to the same interval $\delta_{\overline{\mu}}(r) \subset [\ell, \ell+1)$ of rank r, on which $\widehat{f}_2(y)$ takes two different values. Since $\widehat{f}_3(y)$ is periodic of period $1/m_n$, it follows that $\widehat{f}_2(y) + \widehat{f}_3(y)$ is not constant on $\delta_{\overline{\mu}}(r)$. This contradicts the hypothesis of this theorem. ∎

§6.3. Representation of functions in $\mathbf{L}^p(0,\infty)$, $1 < p \le 2$, by their multiplicative transforms.

We noticed in Theorem **6.1.7** that given any function $f \in \mathbf{L}^p(0,\infty)$, for $1 < p \le 2$, the multiplicative transform $\widehat{f}(y)$ exists and belongs to $\mathbf{L}^{p'}(0,\infty)$. In the case of zero-dimensional, locally compact groups, inversion of the $\mathbf{L}^2$-transform was proved in [1] (see p. 84). We shall prove that inversion holds for the $\mathbf{L}^p$- transform for any $1 < p \le 2$.

Throughout this section we shall suppose that the sequence $\mathcal{P}$ is symmetric, i.e., that $p_{-k} = p_k$ for $k = 1, 2, \ldots$. We first consider integrability of the kernel $D(u,\xi)$ defined in (1.5.37).

6.3.1. *For all $\xi \ge 0$, the kernel $D(u,\xi)$ belongs to $\mathbf{L}^p(0,\infty)$ for all $1 < p < \infty$.*

PROOF. We have

$$\int_0^\infty |D(u,\xi)|^p \, du = \int_0^1 |D(u,\xi)|^p \, du + \sum_{\ell=1}^\infty \int_\ell^{\ell+1} |D(u,\xi)|^p \, du.$$

The first term can be estimated by

$$\int_0^1 |D(u,\xi)|^p \, du = \int_0^1 \Big| \int_0^\xi \chi(u,x) \, dx \Big|^p \, du \le \xi^p.$$

Let $\ell \ge 1$. Use (1.5.38) to write

$$\begin{aligned}\int_\ell^{\ell+1} |D(u,\xi)|^p \, du &= \int_\ell^{\ell+1} \Big| \chi_{[\xi]}(u) \int_0^{\{\xi\}} \chi_\ell(x) \, dx \Big|^p \, du \\ &= \int_\ell^{\ell+1} |\chi_{[\xi]}(u)|^p \, du \, \Big| \int_0^{\{\xi\}} \chi_\ell(x) \, dx \Big|^p \\ &= \Big| \int_0^{\{\xi\}} \chi_\ell(x) \, dx \Big|^p .\end{aligned}$$

Consider the function

$$g(x) = \begin{cases} 1 & \text{for } 0 \le x < \{\xi\}, \\ 0 & \text{for } \{\xi\} \le x \le 1. \end{cases}$$

We have

$$\int_0^{\{\xi\}} \chi_\ell(x) \, dx = \int_0^1 g(x)\chi_\ell(x) \, dx = \overline{c_\ell(g)},$$

i.e., $\int_0^{\{\xi\}} \chi_\ell(x)\,dx$ is the ℓ-th Fourier coefficient of the function $g(x)$ with respect to the system $\{\chi_k(x)\}$. Consequently,

$$\int_0^\infty |D(u,\xi)|^p\,du \le \xi^p + \sum_{\ell=1}^\infty |c_\ell(g)|^p = \xi^p + \sum_{r=1}^\infty \sum_{\ell=m_{r-1}}^{m_r-1} |c_\ell(g)|^p = \xi^p + \sum_{r=1}^\infty \sum_{\ell_{-r}=1}^{p_r-1} \sum_{\ell=\ell_{-r}m_{r-1}}^{(\ell_{-r}+1)m_{r-1}-1} |c_\ell(g)|^p. \tag{6.3.1}$$

To estimate the coefficients $c_\ell(g)$, let $\ell \in [\ell_{-r}m_{r-1}, (\ell_{-r}+1)m_{r-1}-1]$, $\ell_{-r} \neq 0$ and introduce the notation

$$q = \exp\frac{2\pi i\ell_{-r}}{p_r} \quad \text{and} \quad \{\xi\}_\nu = \sum_{k=1}^{\nu} \frac{\xi_k}{m_k}.$$

Then

$$c_\ell(g) = \int_0^{\{\xi\}} \chi_\ell(x)\,dx = \int_0^{\{\xi\}_{r-1}} \chi_\ell(x)\,dx + \int_{\{\xi\}_{r-1}}^{\{\xi\}_r} \chi_\ell(x)\,dx + \int_{\{\xi\}_r}^{\{\xi\}} \chi_\ell(x)\,dx.$$

Since $m_{r-1} \le \ell < m_r$ implies

$$\int_{(\nu-1)/m_r}^{\nu/m_r} \chi_\ell(x)\,dx = 0, \qquad \nu = 1, 2, \ldots, r-1,$$

and the interval $[0, \{\xi\}_{r-1}]$ can be represented as aunion of intervals of the form $[(\nu-1)/m_r, \nu/m_r]$, we have $\int_0^{\{\xi\}_{r-1}} \chi_\ell(x)\,dx = 0$. It follows that

$$\begin{aligned}
\overline{c_\ell(g)} &= \int_{\{\xi\}_{r-1}}^{\{\xi\}_r} \chi_\ell(x)\,dx + \int_{\{\xi\}_r}^{\{\xi\}} \chi_\ell(x)\,dx \\
&= \sum_{\lambda=0}^{\xi_r-1} \left(\int_{\{\xi\}_{r-1}+\lambda/m_r}^{\{\xi\}_{r-1}+(\lambda+1)/m_r} \chi_\ell(x)\,dx \right) + \int_{\{\xi\}_r}^{\{\xi\}} \chi_\ell(x)\,dx \\
&= \frac{1}{m_r} \sum_{\lambda=0}^{\xi_r-1} q^\lambda + \int_{\{\xi\}_r}^{\{\xi\}} \chi_\ell(x)\,dx \\
&= \frac{1}{m_r} \frac{1-q^{\xi_r}}{1-q} + \int_{\{\xi\}_r}^{\{\xi\}} \chi_\ell(x)\,dx.
\end{aligned}$$

Therefore,

$$|c_\ell(g)| \le \frac{1}{m_r}\left|\frac{\sin((\pi\ell_{-r}\xi_r)/p_r)}{\sin((\pi\ell_{-r})/p_r)}\right| + \frac{1}{m_r}.$$

Substituting this estimate into (6.3.1) we obtain

$$\begin{aligned}
\int_0^\infty |D(u,\xi)|^p\,du &\le \xi^p + \sum_{r=1}^{\infty}\sum_{\ell_{-r}=1}^{p_r-1} m_{r-1}\frac{1}{m_r^p}\left(1+\frac{|\sin((\pi\ell_{-r}\xi_r)/p_r)|}{\sin((\pi\ell_{-r})/p_r)}\right)^p \\
&\le \xi^p + \sum_{r=1}^{\infty}\frac{m_{r-1}}{m_r^p}2\sum_{\ell_{-r}=1}^{[p_r/2]}\left(1+\frac{1}{\sin((\pi\ell_{-r})/p_r)}\right)^p \\
&\le \xi^p + 2^{p+1}\sum_{r=1}^{\infty}\frac{m_{r-1}}{m_r^p}\sum_{\ell_{-r}=1}^{[p_r/2]}\left(\frac{1}{(2/\pi)(\pi\ell_{-r})/p_r}\right)^p \\
&\le \xi^p + 2\sum_{r=1}^{\infty}\frac{m_{r-1}}{m_{r-1}^p p_r^p}p_r^p\sum_{\ell=1}^{\infty}\frac{1}{\ell^p} \\
&\le \xi^p + A\sum_{r=1}^{\infty}\frac{1}{m_{r-1}^{p-1}}.
\end{aligned}$$

This last series converges since $p > 1$. We conclude that $D(u,\xi) \in \mathbf{L}^p(0,\infty)$ for any $\xi \ge 0$. ∎

6.3.2. *Let $f \in \mathbf{L}^p(0,\infty)$ for some $1 < p \le 2$ and*

$$\widehat{f}(y) = \lim_{a\to\infty}{}_{p'} \int_0^a f(x)\overline{\chi(x,y)}\,dx,$$

i.e.,

$$\widehat{f}(y) \in \mathbf{L}^{p'}(0,\infty)$$

where $1/p + 1/p' = 1$. Then

$$\widehat{f}(y) = \frac{d}{dx}\int_0^\infty f(y)\int_0^x \overline{\chi(y,t)}\,dt\,dy, \tag{6.3.2}$$

and

$$f(x) = \frac{d}{dx}\int_0^\infty \widehat{f}(y)\int_0^x \chi(y,t)\,dt\,dy \tag{6.3.3}$$

both hold for almost every $x \in [0,\infty)$.

PROOF. We introduce the notation

$$\widehat{f}(y,a) = \int_0^a f(x)\overline{\chi(x,y)}\,dx.$$

For any $\xi \in [0,\infty)$ we have

$$\left| \int_0^\xi (\widehat{f}(y) - \widehat{f}(y,a))\, dy \right| \le \left(\int_0^\xi |\widehat{f}(y) - \widehat{f}(y,a)|^{p'}\, dy \right)^{1/p'} \left(\int_0^\xi 1^p\, dy \right)^{1/p}$$
$$= \xi^{1/p} \|\widehat{f}(y) - \widehat{f}(y,a)\|_{p'}.$$

By hypothesis it follows that

$$\lim_{a\to\infty} \|\widehat{f}(y) - \widehat{f}(y,a)\|_{p'} = 0. \tag{6.3.4}$$

Consequently, for all $\xi \in [0,\infty)$ we have

$$\int_0^\xi \widehat{f}(y)\, dy = \lim_{a\to\infty} \int_0^\xi \widehat{f}(y,a)\, dy,$$

i.e.,

$$\begin{aligned} \int_0^\xi \widehat{f}(y)\, dy &= \lim_{a\to\infty} \int_0^\xi \int_0^a f(t)\overline{\chi(t,y)}\, dt\, dy \\ &= \lim_{a\to\infty} \int_0^a f(t) \int_0^\xi \overline{\chi(t,y)}\, dy\, dt \\ &= \int_0^\infty f(t)\overline{D(t,\xi)}\, dt. \end{aligned}$$

Therefore,

$$\widehat{f}(\xi) = \frac{d}{d\xi} \int_0^\infty f(t)\overline{D(t,\xi)}\, dt$$

holds for almost every $\xi \in [0,\infty)$. This verifies (6.3.2).

To prove (6.3.3) it is enough to show

$$\int_0^\xi f(x)\, dx = \int_0^\infty \widehat{f}(y) \int_0^\xi \chi(x,y)\, dx\, dy. \tag{6.3.5}$$

Define a sequence of functions f_k on [0,1) by $f_k(x) = f(k+x)$. Consider the integral

$$\int_0^\infty \widehat{f}(u,a) D(u,\xi)\, du.$$

By property **1.5.3** for the function $\chi(x,y)$, we have

$$\begin{aligned} \widehat{f}(u,a) &= \int_0^a f(x)\overline{\chi_\ell(x)}\,\overline{\chi_{[x]}(u)}\, dx \\ &= \sum_{k=0}^{[a]-1} c_\ell(f_k)\overline{\chi_k(u)} + \int_{[a]}^a f(x)\overline{\chi_\ell(x)}\,\overline{\chi_{[a]}(u)}\, dx \end{aligned}$$

for $[u]=\ell$. For each $0<\xi\leq a$ we obtain by (1.5.38) that

(6.3.6)

$$I_0(a,\xi)=\int_0^1 \widehat{f}(u,a)D(u,\xi)\,du$$
$$=\int_0^1\left(\sum_{k=0}^{[a]-1} c_\ell(f_k)\overline{\chi_k(u)}+\int_{[a]}^a f(x)\overline{\chi_\ell(x)}\,dx\overline{\chi_{[a]}(u)}\right)(D_{[\xi]}(u)+\{\xi\}\chi_{[\xi]}(u))\,du.$$

Since the system $\{\chi_k(u)\}$ is orthogonal on [0,1], we have

$$I_0(a,\xi)=\sum_{k=0}^{[\xi]-1} c_0(f_k)+c_0(f_{[\xi]})\{\xi\}=\int_0^{[\xi]} f(y)\,dy+c_0(f_{[\xi]})\overline{c_0(g)} \tag{6.3.7}$$

where

$$g(x)=\begin{cases}1, & \text{if } 0\leq x<\{\xi\},\\ 0, & \text{if } \{\xi\}\leq x<1.\end{cases}$$

Similarly,

(6.3.8)

$$\int_1^\infty \widehat{f}(u,a)D(u,\xi)\,du$$
$$=\sum_{\ell=1}^\infty\int_\ell^{\ell+1}\left(\sum_{k=0}^{[a]-1} c_\ell(f_k)\overline{\chi_k(u)}+\int_{[a]}^a f(x)\overline{\chi_\ell(x)}\,dx\overline{\chi_{[a]}(u)}\right)\times$$
$$\times\left(\chi_{[\xi]}(u)\int_0^{\{\xi\}}\chi_\ell(t)\right)dt$$
$$=\sum_{\ell=1}^\infty c_\ell(f_{[\xi]})\overline{c_\ell(g)}.$$

Since $f_{[\xi]}(x)\in \mathbf{L}^p(0,1)$ and $g(x)\in\mathbf{L}^{p'}(0,1)$, it follows from Parseval's identity that

$$\sum_{\ell=0}^\infty c_\ell(f_{[\xi]})\overline{c_\ell(g)}=\int_0^1 f_{[\xi]}(x)g(x)\,dx=\int_{[\xi]}^\xi f(x)\,dx. \tag{6.3.9}$$

Identities (6.3.7) through (6.3.9) imply

$$\int_0^\infty \widehat{f}(u,a)D(u,\xi)\,du=\int_0^\xi f(x)\,dx. \tag{6.3.10}$$

Hence we have by Hölder's inequality that

$$\left| \int_0^\infty (\widehat{f}(u) - \widehat{f}(u,a)) D(u,\xi)\, du \right| \leq \|\widehat{f}(u) - \widehat{f}(u,a)\|_{p'} \|D(u,\xi)\|_p .$$

Therefore, we conclude from (6.3.1), (6.3.4), and (6.3.10) that

$$\int_0^\infty \widehat{f}(u) D(u,\xi)\, du = \int_0^\xi f(x)\, dx$$

i.e., (6.3.5) holds. ∎

Chapter 7

WALSH SERIES WITH MONOTONE DECREASING COEFFICIENTS

§7.1. Convergence and integrability.

Let $\{c_n\}$, $n = 0, 1, \ldots$, be a monotone decreasing sequence of real numbers, i.e., $\Delta c_n \equiv c_n - c_{n+1} \geq 0$ for $n = 0, 1, \ldots$. If the sequence also converges to zero, we shall write $c_n \downarrow 0$.

We shall frequently use the following easily verified identity, which is called Abel's transformation (see [15], p. 9):

$$\sum_{k=m}^{n-1} a_k b_k = \sum_{k=m}^{n-2} \Delta a_k B_{k+1} + a_{n-1} B_n - a_m B_m,$$

where $B_0 = 0$ and $B_k = \sum_{i=0}^{k-1} b_i$ for $k \geq 1$.

7.1.1. *If $c_n \downarrow 0$ then the series*

$$\sum_{n=0}^{\infty} c_n w_n(x) \tag{7.1.1}$$

converges on the interval $(0, 1)$ and is uniformly convergent on the interval $[\delta, 1)$ for each $0 < \delta < 1$.

PROOF. Let $S_n(x) = \sum_{k=0}^{n-1} c_k w_k(x)$ be the partial sums of (7.1.1). Let $1 \leq m < n$ and notice by Abel's transformation that

$$\begin{aligned} S_n(x) - S_m(x) &= \sum_{k=m}^{n-1} c_k w_k(x) \\ &= \sum_{k=m}^{n-2} \Delta c_k D_{k+1} + c_{n-1} D_n - c_m D_m, \end{aligned} \tag{7.1.2}$$

where $D_k(x) = \sum_{j=0}^{k-1} w_j(x)$ is the Dirichlet kernel. Recall from §1.4 that

$$|D_k(x)| < \frac{1}{x} \tag{7.1.3}$$

for $0 < x < 1$ and $k = 1, 2, \ldots$. Consequently, from (7.1.2) we obtain

$$|S_n(x) - S_m(x)| \le \frac{1}{x} \sum_{k=m}^{n-2} \Delta c_k + \frac{c_{n-1}}{x} + \frac{c_m}{x} = \frac{2c_m}{x}. \tag{7.1.4}$$

Since $c_m \to 0$ and $n > m$, (7.1.4) implies that the series (7.1.1) converges at each point $x \in (0, 1)$. Moreover, if $\delta \in (0, 1)$ is fixed, then

$$\sup_{\delta \le x < 1} |S_n(x) - S_m(x)| \le \frac{2c_m}{\delta}, \qquad n > m.$$

Hence the partial sums S_n are uniformly Cauchy and it follows that (7.1.1) converges uniformly on $(\delta, 1)$. ∎

We shall now prove that the sum of (7.1.1) need not be Lebesgue integrable on (0,1).

7.1.2. *There exists a sequence $\{c_n\}$ which converges monotonically to zero such that the function*

$$f(x) = \sum_{n=0}^{\infty} c_n w_n(x) \tag{7.1.5}$$

is not integrable on the interval $(0, 1)$.

PROOF. We proved in §2.2 that the Lebesgue constants

$$L_n = \int_0^1 |D_n(x)|\, dx, \qquad n = 1, 2, \ldots$$

satisfy the inequality

$$L_n \le \log_2 n \qquad \text{for } n \ge 4, \tag{7.1.6}$$

and that there is a subsequence of natural numbers $n_0 < n_1 < \ldots$ such that

$$L_{n_k} \ge \frac{1}{4} \log_2 n_k, \qquad k = 0, 1, \ldots. \tag{7.1.7}$$

Moreover, the sequence $\{n_k\}$ satisfies the condition $2^k \le n_k < 2^{k+1}$ for $k = 0, 1, \ldots$ (see (2.2.4)).

Choose a subsequence $\{m_k\}$ of $\{n_k\}$ which satisfies the inequalities

$$\begin{cases} \sum_{k=1}^{s-1} \sqrt{\log_2 m_k} < \frac{1}{s} \sqrt{\log_2 m_s}, \\ \sum_{k=s+1}^{\infty} \frac{\nu_s}{\sqrt{\log_2 m_k}} < \frac{1}{s} \sqrt{\log_2 m_s}, \end{cases} \tag{7.1.8}$$

where $2^{\nu_s} \le m_s < 2^{\nu_s+1}$. Define the sequence $\{c_n\}$ by

$$\tag{7.1.9} \begin{cases} c_0 = c_1 = \cdots = c_{m_1-1} = \displaystyle\sum_{k=0}^{\infty} \frac{1}{\sqrt{\log_2 m_k}}, \ldots, \\ c_{m_0} = c_0, \ c_{m_k} = c_{m_k+1} = \cdots = c_{m_{k+1}-1}, \ldots, \\ c_{m_k} - c_{m_{k+1}} = \dfrac{1}{\sqrt{\log_2 m_k}}, \ k = 0, 1, \ldots. \end{cases}$$

Obviously, $c_n \downarrow 0$. Hence by **7.1.1** the series on the right side of (7.1.5) converges uniformly on the interval $(\delta, 1)$ for each $0 < \delta < 1$. Hence the function $f(x)$ is integrable (even in the Riemann sense) on $[\delta, 1)$. We introduce the notation

$$I(\delta) = \int_\delta^1 |f(x)|\,dx, \qquad 0 < \delta < 1$$

and will show that $I(\delta) \to \infty$ as $\delta \to 0$. Surely, this will verify that the function $f(x)$ is not integrable on (0,1).

Apply Abel's transformation. By (7.1.9) we have

(7.1.10)

$$\begin{aligned} I(2^{-\nu_s}) &= \int_{2^{-\nu_s}}^1 \Big| \sum_{n=0}^{\infty} c_n w_n(x) \Big| \,dx \\ &= \int_{2^{-\nu_s}}^1 \Big| \sum_{n=0}^{\infty} \Delta c_n D_{n+1}(x) \Big| \,dx \\ &= \int_{2^{-\nu_s}}^1 \Big| \sum_{k=1}^{\infty} \frac{D_{m_k}(x)}{\sqrt{\log_2 m_k}} \Big| \,dx \\ &\ge \frac{1}{\sqrt{\log_2 m_s}} \left(\int_0^1 |D_{m_s}(x)|\,dx - \int_0^{2^{-\nu_s}} |D_{m_s}(x)|\,dx \right) \\ &\quad - \sum_{k=1}^{s-1} \frac{L_{m_k}}{\sqrt{\log_2 m_k}} - \sum_{k=s+1}^{\infty} \frac{1}{\sqrt{\log_2 m_k}} \int_{2^{-\nu_s}}^1 |D_{m_k}(x)|\,dx. \end{aligned}$$

Since the sequence $\{m_k\}$ is a subsequence of $\{n_k\}$ we have by (7.1.7) that

$$L_{m_k} \ge \frac{1}{4}\log_2 m_k, \qquad k = 0, 1, \ldots.$$

Using this observation together with inequalities (7.1.3), (7.1.7), (7.1.8) and the estimate $|D_{m_s}(x)| \le m_s$, we see by (7.1.10) that

$$\begin{aligned} I(2^{-\nu_s}) &\ge \frac{1}{4}\sqrt{\log_2 m_s} - \frac{2}{\sqrt{\log_2 m_s}} - \sum_{k=1}^{s-1} \sqrt{\log_2 m_s} - \sum_{k=s+1}^{\infty} \frac{\nu_s}{\sqrt{\log_2 m_k}} \\ &\ge \left(\frac{1}{4} - \frac{2}{s}\right)\sqrt{\log_2 m_s} - \frac{2}{\sqrt{\log_2 m_s}} \to \infty \end{aligned}$$

as $s \to \infty$. Since the function $I(\delta)$ is monotone, it follows that $I(\delta) \to \infty$ as $\delta \to 0$. ∎

Thus we see that in order to insure that the sum of a Walsh series with monotone decreasing coefficients is integrable, it is necessary to impose some further restriction on the coefficients. Such a restriction is contained in the following theorem:

7.1.3. *Let $\{c_n \geq 0\}$ be a monotone decreasing sequence which satisfies*

$$\sum_{n=1}^{\infty} \frac{c_n}{n} < \infty. \tag{7.1.11}$$

Then the series (7.1.1) converges on the interval $(0,1)$ to an integrable function and is its Walsh-Fourier series.

PROOF. First notice since $\{c_n\}$ is monotone decreasing that (7.1.11) implies

$$c_k \log_2 k \leq C \sum_{n=[\sqrt{k}]}^{k} \frac{c_n}{n} \to 0, \qquad k \to \infty, \tag{7.1.12}$$

so in particular, $c_k \to 0$. Consequently, we have by Theorem **7.1.1** that (7.1.1) converges on the interval (0,1) to some function $f(x)$. We shall prove that $f \in \mathbf{L}[0,1)$.

Let ν be any natural number and let $1/(\nu+1) < x \leq 1/\nu$. By (7.1.4) we have

$$\begin{aligned}
\sup_{n>m} \left| \sum_{k=m}^{n-1} c_k w_k(x) \right| &\leq \sum_{k=m}^{m+\nu-1} c_k + \sup_{n>m+\nu} \left| \sum_{k=m+\nu}^{n-1} c_k w_k(x) \right| \\
&\leq \sum_{k=m}^{m+\nu-1} c_k + \frac{2c_{m+\nu}}{x} \\
&\leq \sum_{k=m}^{m+\nu-1} c_k + 4\nu c_{m+\nu} \leq 5 \sum_{k=m}^{m+\nu-1} c_k.
\end{aligned}$$

Consequently,

$$\begin{aligned}
I_m &\equiv \int_0^1 \sup_{n>m} \left| \sum_{k=m}^{n-1} c_k w_k(x) \right| dx \\
&= \sum_{\nu=1}^{\infty} \int_{1/(\nu+1)}^{1/\nu} \sup_{n>m} \left| \sum_{k=m}^{n-1} c_k w_k(x) \right| dx \\
&\leq 5 \sum_{\nu=1}^{\infty} \frac{1}{\nu(\nu+1)} \sum_{k=m}^{m+\nu-1} c_k.
\end{aligned}$$

But by Abel's transformation,

$$\sum_{\nu=1}^{N} \frac{1}{\nu(\nu+1)} \sum_{k=m}^{m+\nu-1} c_k \leq \sum_{\nu=1}^{N-1} \frac{1}{\nu} c_{m+\nu-1}.$$

Therefore,

$$I_m \leq 5 \sum_{\nu=1}^{\infty} \frac{1}{\nu} c_{m+\nu-1} = 5 \sum_{n=m}^{\infty} \frac{c_n}{n-m+1} \tag{7.1.13}$$

$$= 5 \left(\sum_{n=m}^{2m-1} + \sum_{n=2m}^{\infty} \right) \frac{c_n}{n-m+1}$$

$$\leq 5 \left(c_m \sum_{\ell=1}^{m} \frac{1}{\ell} + 2 \sum_{n=2m}^{\infty} \frac{c_n}{n} \right)$$

$$\leq C(c_m \log_2 m + \sum_{n=2m}^{\infty} \frac{c_n}{n}).$$

Since (7.1.11) converges and satisfies condition (7.1.12), the right side of (7.1.13) converges to zero as $m \to \infty$. Thus the inequality

$$\int_0^1 |s_n(x) - s_m(x)| \, dx \leq I_m \qquad \text{for } n > m,$$

implies that the partial sums $s_n(x)$ of the series (7.1.1) are Cauchy in the $\mathbf{L}(0,1)$ norm. Since this space is complete (see **A2.1**, **A5.1**, and **A5.2**), it follows that (7.1.1) converges in the $\mathbf{L}(0,1)$ norm to some integrable function $s(x)$. But earlier we noted that (7.1.1) converges everywhere on (0,1) to the function $f(x)$. Consequently (see **A5.2.4**), $s(x) = f(x)$ almost everywhere. In particular, the function $f(x)$ must be integrable.

To see that (7.1.1) is the Walsh-Fourier series of $f(x)$, notice that

$$\lim_{n\to\infty} \int_0^1 |s_n(x) - f(x)| \, dx = 0$$

implies

$$\lim_{n\to\infty} \int_0^1 w_m(x)(s_n(x) - f(x)) \, dx = 0 \tag{7.1.14}$$

for $m = 0, 1, \ldots$. But for each $n \geq m$ we have by orthogonality that

$$\int_0^1 w_m(x) s_n(x) \, dx = c_m.$$

We conclude by (7.1.14) that

$$c_m = \int_0^1 f(x) w_m(x) \, dx, \qquad m = 0, 1, \ldots. \quad \blacksquare$$

An examination of the proof of Theorem **7.1.3** shows that we have in fact established the following theorem:

7.1.4. *Under the hypotheses of Theorem* **7.1.3**, *the sequence $s_n(x)$ of partial sums of the series (7.1.1) converges in* $\mathbf{L}(0,1)$ *and has integrable majorant, i.e.,*

$$\sup_{n\geq 1}|s_n(x)| \in \mathbf{L}(0,1).$$

If $c_n \downarrow 0$ then by Theorem **7.1.1** the series (7.1.1) converges uniformly on $(\delta, 1)$ for each $0 < \delta < 1$. Thus its sum $f(x)$ is Riemann integrable on $[\delta, 1)$ for each $\delta \in (0,1)$. It turns out that under these conditions the function $f(x)$ is improperly Riemann integrable on all of [0,1], i.e., that the following limit exists and is finite:

$$\lim_{\delta\to 0+}\int_\delta^1 f(x)\,dx.$$

In fact, the following theorem is true.

7.1.5. *If $c_n \downarrow 0$ then the function $f(x)$ defined by (7.1.5) is improperly Riemann integrable on $(0,1]$. Moreover, the series on the right side of (7.1.5) is its "improper" Walsh-Fourier series, i.e.,*

$$c_n = \lim_{\delta\to 0+}\int_\delta^1 f(x)w_n(x)\,dx, \qquad n = 0,1,\ldots. \tag{7.1.15}$$

PROOF. For the case $n = 0$, it is clear that (7.1.15) implies that $f(x)$ is improperly Riemann integrable. Hence it is enough to prove identity (7.1.15).

Since the hypotheses imply that the series on the right side of (7.1.5) converges uniformly on $[\delta, 1)$, $0 < \delta < 1$, we have for any $N = 0,1,\ldots$ that

$$\begin{aligned}\int_\delta^1 f(x)w_N(x)\,dx &= \sum_{n=0}^\infty c_n \int_\delta^1 w_n(x)w_N(x)\,dx \\ &= \sum_{n=0}^\infty c_n \left(\int_0^1 w_n(x)w_N(x)\,dx - \int_0^\delta w_n(x)w_N(x)\,dx\right) \\ &= c_N - \sum_{n=0}^\infty c_n \int_0^\delta w_n(x)w_N(x)\,dx.\end{aligned}$$

Choose $\delta > 0$ so small that $w_N(x) = 1$ on $[0, \delta]$. Then the previous identity can be written in the form

$$\int_\delta^1 f(x)w_N(x)\,dx = c_N - \sum_{n=0}^\infty c_n \int_0^\delta w_n(x)\,dx.$$

Thus the theorem will be proved if we show

$$I(\delta) \equiv \sum_{n=0}^\infty c_n \int_0^\delta w_n(x)\,dx \to 0 \tag{7.1.16}$$

as $\delta \to 0+$.

Write the number δ in the form $\delta = \sum_{k=1}^{\infty} 2^{-p_k}$ where $1 \le p_k < p_{k+1}$ and each p_k is a natural number. We shall use the inequality

(7.1.17)
$$|I(\delta)| \le \delta \sum_{n=0}^{2^{p_1}-1} c_n + \Big| \sum_{n=2^{p_1}}^{\infty} c_n \int_0^{\delta} w_n(x)\, dx \Big| .$$

By **1.1.4**

$$\int_0^{2^{-p_1}} w_n(x)\, dx = 0 \qquad \text{for } n \ge 2^{p_1}.$$

Thus

(7.1.18)
$$\begin{aligned} I_1(\delta) &\equiv \sum_{n=2^{p_1}}^{\infty} c_n \int_0^{\delta} w_n(x)\, dx \\ &= \sum_{n=2^{p_1}}^{\infty} c_n \int_{2^{-p_1}}^{\delta} w_n(x)\, dx \\ &= \sum_{q=1}^{\infty} \left(\sum_{n=2q\cdot 2^{p_1-1}}^{(2q+1)2^{p_1-1}-1} + \sum_{n=(2q+1)2^{p_1-1}}^{2(q+1)2^{p_1-1}-1} \right) c_n \int_{2^{-p_1}}^{\delta} w_n(x)\, dx. \end{aligned}$$

If $n = 2q2^{p_1-1} + s$ where $0 \le s < 2^{p_1-1}$ then

$$w_n(x) = w_{2q2^{p_1-1}}(x) w_s(x).$$

Since $w_s(x) = 1$ for $x \in [2^{-p_1}, 2^{-p_1+1}) \supset [2^{-p_1}, \delta)$, we have

(7.1.19)
$$\begin{aligned} &\sum_{n=2q\cdot 2^{p_1-1}}^{(2q+1)2^{p_1-1}-1} c_n \int_{2^{-p_1}}^{\delta} w_n(x)\, dx \\ &\qquad = \left(\sum_{n=2q\cdot 2^{p_1-1}}^{(2q+1)2^{p_1-1}-1} c_n \right) \int_{2^{-p_1}}^{\delta} w_{2q2^{p_1-1}}(x)\, dx. \end{aligned}$$

If $n = (2q+1)2^{p_1-1} + s$ where $0 \le s < 2^{p_1-1}$ then

$$w_n(x) = w_{2q2^{p_1-1}}(x) w_{2^{p_1-1}}(x) w_s(x).$$

Observe for each $x \in [2^{-p_1}, 2^{-p_1+1})$ that $w_s(x) = 1$ but $w_{2^{p_1-1}}(x) = r_{p_1-1}(x) = -1$. Consequently,

(7.1.20)
$$\sum_{n=(2q+1)2^{p_1-1}}^{2(q+1)2^{p_1-1}-1} c_n \int_{2^{-p_1}}^{\delta} w_n(x)\,dx$$
$$= -\left(\sum_{n=(2q+1)2^{p_1-1}}^{2(q+1)2^{p_1-1}-1} c_n\right) \int_{2^{-p_1}}^{\delta} w_{2q2^{p_1-1}}(x)\,dx.$$

Combine (7.1.18) through (7.1.20) to obtain

(7.1.21)
$$|I_1(\delta)| \le \delta \sum_{q=1}^{\infty} \Big| \sum_{n=2q\cdot 2^{p_1-1}}^{(2q+1)2^{p_1-1}-1} c_n - \sum_{n=(2q+1)2^{p_1-1}}^{2(q+1)2^{p_1-1}-1} c_n \Big|$$
$$\equiv \delta \sum_{q=1}^{\infty} |A_{2q,p_1} - A_{2q+1,p_1}|.$$

Since the sequence $\{c_n\}$ is monotone decreasing, so is the sequence $\{A_{s,p_1}\}_{s=2}^{\infty}$. Consequently,

$$\sum_{q=1}^{\infty} |A_{2q,p_1} - A_{2q+1,p_1}| = \sum_{s=2}^{\infty} (-1)^s A_{s,p_1} \le A_{2,p_1}$$
$$= \sum_{n=2^{p_1}}^{3\cdot 2^{p_1-1}-1} c_n \le 2^{p_1-1} c_{2^{p_1}}.$$

Thus it follows from (7.1.12) that

$$|I_1(\delta)| \le \delta 2^{p_1-1} c_{2^{p_1}}.$$

But $\delta \le 2\cdot 2^{-p_1}$ implies $I_1(\delta)| \le c_{2^{p_1}}$. Consequently, it follows from the definition of $I_1(\delta)$ (see (7.1.18)) and (7.1.17) that

$$|I(\delta)| \le 2^{-p_1+1} \sum_{n=0}^{2^{p_1}-1} c_n + c_{2^p} \equiv 2\sigma_{2^{p_1}} + c_{2^{p_1}}$$

where σ_n represents the $(C, 1)$ means of the sequence $\{c_n\}$. If $\delta \to 0$ then $p_1 \to \infty$. Since $c_n \to 0$ and the method $(C, 1)$ is a regular method of summability (see **4.1.3**), it follows that $\sigma_{2^{p_1}} \to 0$. Thus $I(\delta) \to 0$ as $\delta \to 0$ and the theorem is proved. ∎

When studying the problem of integrability of Walsh series with monotone decreasing coefficients, it is sometimes useful to replace the improper Riemann integral with a concept called the A- integral. This notion is defined in the following way.

A measurable function $f(x)$ defined on a measurable set E is called *A-integrable* if the following two conditions are satisfied:

1) $\lim_{N\to\infty} \operatorname{mes}\{x : x \in E,\ |f(x)| > N\} = 0$;

2) the limit

$$\lim_{N\to\infty} \int_E [f(x)]^N \, dx$$

exists and is finite, where

$$[f(x)]^N = \begin{cases} f(x) & \text{for } |f(x)| \le N, \\ 0 & \text{for } |f(x)| > N. \end{cases}$$

This limit is called the *A-integral of the function $f(x)$ on the set E* and will be denoted by

$$(A) \int_E f(x) \, dx.$$

If the function $f(x)$ is Lebesgue integrable on the set E, i.e., if $f(x)$ belongs to the class $\mathbf{L}(E)$, then it satisfies conditions 1) and 2). Hence any Lebesgue integrable function is A-integrable. However, the converse of this statement is not true. This can be verified by comparing **7.1.2** with the next theorem.

7.1.6. *If $c_n \downarrow 0$ then the function (7.1.5) is A-integrable on the interval $(0,1)$ and satisfies*

$$c_n = (A) \int_{(0,1)} f(x) w_n(x) \, dx, \qquad n = 0, 1, \dots, \tag{7.1.22}$$

i.e., (7.1.1) is the Walsh-Fourier series of the function f in the sense of the A-integral.

PROOF. We may suppose that $c_0 > 0$ since otherwise all the coefficients $c_n = 0$ and the result is trivial. Let $N > 0$ be given and set $n_0 = n_0(N) = [N/(2c_0)]$ where $[a]$ represents the greatest integer in the number a. Replace in (7.1.4) the index m by n_0 and let n tend to ∞. We obtain

$$\Big| \sum_{k=n_0}^{\infty} c_k w_k(x) \Big| \le \frac{2c_{n_0}}{x}, \qquad 0 < x < 1. \tag{7.1.23}$$

If $x \in (0,1)$ satisfies $|f(x)| > N$ then

$$\Big| \sum_{k=n_0}^{\infty} c_k w_k(x) \Big| > N/2$$

since $|\sum_{k=0}^{n_0-1} c_k w_k(x)| \le N/2$. Consequently, $N/2 < 2c_{n_0}/x$, i.e., $x < 4c_{n_0}/N$. Since $c_{n_0} \to 0$ as $N \to \infty$ it follows that

$$\text{mes}\{x : x \in (0,1),\ |f(x)| > N\} = o(\frac{1}{N}), \qquad N \to \infty.$$

Thus condition 1) (from the definition of A-integrability) is satisfied by the set $E = (0,1)$.

To verify condition 2) set $\varepsilon_N = \sup_{\{} x : |f(x)| > N\}$. Since $x < 4c_{n_0}/N$ holds for all x's which satisfy $|f(x)| > N$, it is evident that $\varepsilon_N \le 4c_{n_0}/N$. Consequently,

(7.1.24)

$$\begin{aligned} &| \int_0^1 [f(x)w_n(x)]^N\,dx - \int_{\varepsilon_N}^1 f(x)w_n(x)\,dx \,| \\ &\qquad = | \int_0^{\varepsilon_N} [f(x)w_n(x)]^N\,dx \,| \le N\varepsilon_N \le 4c_{n_0}. \end{aligned}$$

But by Theorem **7.1.5**,

$$\lim_{N\to\infty} \int_{\varepsilon_N}^1 f(x)w_n(x)\,dx = c_n$$

for $n = 0, 1, \ldots$. Thus it follows from (7.1.24) and the hypothesis $\lim_{N\to\infty} c_{n_0} = 0$ that

$$\lim_{N\to\infty} \int_0^1 [f(x)w_n(x)]^N\,dx = c_n, \qquad n = 0, 1, \ldots.$$

Hence by definition the function $f(x)$ is A-integrable and satisfies (7.1.22). ∎

We close this section with the following theorem:

7.1.7. *If $c_n \downarrow 0$ then the function $f(x)$ defined by identity (7.1.5) belongs to $\mathbf{L}^p(0,1)$ for every $0 < p < 1$ and*

$$\lim_{n\to\infty} \int_0^1 |f(x) - S_n(x)|^p\,dx = 0, \tag{7.1.25}$$

where $S_n(x) = \sum_{k=0}^{n-1} c_k w_k(x)$.

PROOF. Fix $0 < p < 1$. By (7.1.23) we have

$$|f(x) - S_n(x)|^p \le (2c_n/x)^p, \qquad 0 < x < 1,\ n = 1, 2, \ldots.$$

Integrate this inequality to obtain

$$\int_0^1 |f(x) - S_n(x)|^p\,dx \le (2c_n)^p \int_0^1 \frac{dx}{x^p}.$$

The right side of this last inequality tends to zero as $n \to \infty$. Hence (7.1.25) is true. But the conditions

$$\int_0^1 |S_n(x)|^p\,dx < \infty, \qquad \int_0^1 |f(x) - S_n(x)|^p\,dx < \infty$$

imply

$$\int_0^1 |f(x)|^p\,dx \le \int_0^1 |S_n(x)|^p\,dx + \int_0^1 |f(x) - S_n(x)|^p\,dx < \infty.$$

Thus the proof of **7.1.7** is complete. ∎

§7.2. Series with quasiconvex coefficients.

Let $\{c_n\}$, $n = 0, 1, \ldots$, be a sequence of real numbers. Set $\Delta c_n = c_n - c_{n+1}$, and $\Delta^2 c_n = \Delta(\Delta c_n) = c_n - 2c_{n+1} + c_{n+2}$ for $n = 0, 1, \ldots$. If $\Delta c_n \ge 0$ for $n = 0, 1, \ldots$, then the sequence $\{c_n\}$ is called monotone decreasing, and if $\Delta^2 c_n \ge 0$ for $n = 0, 1, \ldots$, then the sequence $\{c_n\}$ is called convex.

7.2.1. *1) If a sequence $\{c_n\}$ is convex and bounded above, then it is monotone decreasing.*

2) If a sequence $\{c_n\}$ is convex and bounded, then

$$n\Delta c_n \to 0, \qquad n \to \infty, \tag{7.2.1}$$

and

$$\sum_{n=0}^{\infty} (n+1)\Delta^2 c_n < \infty. \tag{7.2.2}$$

PROOF. 1) Let $\{c_n\}$ be convex and bounded above. We shall show that it is monotone decreasing, i.e., that $\Delta c_n \ge 0$ for $n = 0, 1, \ldots$. Suppose to the contrary that there is a non-negative integer m such that $\Delta c_m < 0$. Since $\Delta^2 c_n \ge 0$ for all integers $n \ge 0$, it follows that $\Delta c_n < 0$ and $|\Delta c_n| \ge |\Delta c_m|$ for all $n > m$. Consequently,

$$c_n - c_m = -\sum_{k=m}^{n-1} \Delta c_k = \sum_{k=m}^{n-1} |\Delta c_k| \ge (n-m)|\Delta c_m| \to \infty, \qquad n \to \infty,$$

which contradicts the fact that $\{c_n\}$ is bounded.

2) Let $\{c_n\}$ be convex and bounded. By 1), it is also monotone decreasing, i.e., $\Delta c_n \ge 0$ for $n = 0, 1, \ldots$. Since $\{c_n\}$ is bounded below it has a finite limit, say $c = \lim_{n\to\infty} c_n$. Consequently,

$$c_0 - c = \sum_{n=0}^{\infty} \Delta c_n < \infty. \tag{7.2.3}$$

But the terms of (7.2.3) are monotone decreasing because $\{c_n\}$ is convex. Therefore, convergence implies (7.2.1).

By Abel's transformation we have

$$\sum_{k=0}^{n} \Delta c_k = \sum_{k=0}^{n-1} (k+1)\Delta^2 c_k + (n+1)\Delta c_n. \tag{7.2.4}$$

But we already have verified that $(n+1)\Delta c_n \to 0$ as $n \to \infty$ and

$$\sum_{k=0}^{n} \Delta c_k = c_0 - c_{n+1} \to c_0 - c, \qquad n \to \infty.$$

Thus if we let $n \to \infty$ in (7.2.4), we obtain

$$\sum_{k=0}^{n-1} (k+1)\Delta^2 c_k = c_0 - c.$$

This proves (7.2.2). ∎

7.2.2. *If the sequence $\{c_n\}$ is convex and $c_n \to 0$ then $\{c_n\}$ is monotone decreasing and satisfies conditions (7.2.1) and (7.2.2).*

PROOF. This result follows directly from Proposition **7.2.1** since the condition $c_n \to 0$ implies that the sequence $\{c_n\}$ is bounded. ∎

If the sequence $\{c_n\}$ satisfies

$$\sum_{n=0}^{\infty} (n+1)\Delta^2 c_n < \infty \tag{7.2.5}$$

then it is called *quasiconvex*. Notice by Proposition **7.2.1** that every bounded convex sequence is quasiconvex.

7.2.3. *Any quasiconvex sequence $\{c_n\}$ satisfies (7.2.1).*

PROOF. Indeed,

$$\begin{aligned} |\Delta c_n| &= \left| \sum_{k=n}^{\infty} \Delta^2 c_k \right| \le \sum_{k=n}^{\infty} \frac{1}{k+1}(k+1)|\Delta^2 c_k| \\ &\le \frac{1}{n+1} \sum_{k=n}^{\infty} (k+1)|\Delta^2 c_k| = o\left(\frac{1}{n+1}\right) \qquad n \to \infty. \end{aligned}$$ ∎

The following theorem gives a sufficient condition for a series

$$\sum_{n=0}^{\infty} c_n w_n(x) \tag{7.2.6}$$

to be the Walsh-Fourier series of some function from the space $\mathbf{L}(0,1)$.

7.2.4. *If $c_n \to 0$ and the sequence $\{c_n\}$ is quasiconvex, then (7.2.6) is the Walsh-Fourier series of some function from the space* $\mathbf{L}(0,1)$.

The proof of this theorem requires the following lemma:

7.2.5. *If $c_n \downarrow 0$ and the sum $f(x)$ of the series (7.2.6) belongs to the space $\mathbf{L}(0,1)$, then (7.2.6) is the Walsh-Fourier series of f.*

PROOF. The estimate

$$(7.2.7) \qquad \left| \int_0^1 f(x)w_n(x)\,dx - \int_\delta^1 f(x)w_n(x)\,dx \right| \le \int_0^\delta |f(x)|\,dx$$

is obvious. Since f is integrable, we also have

$$(7.2.8) \qquad \lim_{\delta\to 0+} \int_0^\delta |f(x)|\,dx = 0.$$

Consequently, it follows from (7.1.15), (7.2.7), and (7.2.8) that

$$\int_0^1 f(x)w_n(x)\,dx = c_n, \qquad n = 0, 1, \ldots.$$

This completes the proof of **7.2.5.** ∎

PROOF OF **7.2.4**. Let

$$s_n(x) = \sum_{k=0}^{n-1} c_k w_k(x)$$

represent the partial sums of (7.2.6). Two applications of Abel's transformation yield

$$(7.2.9) \qquad s_n(x) = \sum_{k=0}^{n-3}(k+1)\Delta^2 c_k K_{k+1}(x) + (n-1)\Delta c_{n-2}K_{n-1}(x) + c_{n-1}D_n(x),$$

where $K_n(x) = (1/n)\sum_{k=1}^n D_k(x)$. Since $|D_n(x)| < 1/x$ and consequently $|K_n(x)| < 1/x$ for $0 < x < 1$, we have by Proposition **7.2.3** and the hypothesis $c_n \to 0$ that the last two terms on the right side of (7.2.9) converge to zero, as $n \to \infty$, for all $0 < x < 1$. Furthermore, the inequality $|K_n(x)| < 1/x$ implies

$$\sum_{k=0}^{\infty}(k+1)|\Delta^2 c_k K_{k+1}(x)| \le \frac{1}{x}\sum_{k=0}^{\infty}(k+1)|\Delta^2 c_k|.$$

But the series on the right side of this latest inequality must converge since the sequence $\{c_n\}$ is quasiconvex. Thus the series

$$\sum_{k=0}^{\infty}(k+1)\Delta^2 c_k K_{k+1}(x)$$

converges for all $0 < x < 1$. Consequently, the limit as $n \to \infty$ of the right side of inequality (7.2.9) exists and is finite for $0 < x < 1$. Passing to the limit in (7.2.9), as $n \to \infty$, we obtain

$$f(x) \equiv \lim_{n\to\infty} s_n(x) = \sum_{k=0}^{\infty}(k+1)\Delta^2 c_k K_{k+1}(x).$$

Thus

$$|f(x)| \le \sum_{k=0}^{\infty}(k+1)|\Delta^2 c_k|\,|K_{k+1}(x)|.$$

But **4.2.2** implies

$$\sum_{k=0}^{\infty}(k+1)|\Delta^2 c_k| \int_0^1 |K_{k+1}(x)|\,dx \le 2\sum_{k=0}^{\infty}(k+1)|\Delta^2 c_k| < \infty.$$

Therefore, $f(x) \in \mathbf{L}(0,1)$ with

$$\int_0^1 |f(x)|\,dx \le 2\sum_{k=0}^{\infty}(k+1)|\Delta^2 c_k|.$$

Since $f(x) \in \mathbf{L}(0,1)$, we conclude from Lemma **7.2.5** that (7.2.6) is the Walsh-Fourier series of f. ∎

§7.3. Fourier series of functions in $\mathbf{L}^p$.

In this section we shall identify necessary and sufficient conditions for the sum of a series

$$\sum_{n=0}^{\infty} c_n w_n(x) \tag{7.3.1}$$

with monotone coefficients to belong to $\mathbf{L}^p[0,1]$ for some $1 < p < \infty$.

First of all we establish a lemma.

7.3.1. *Let $1 < p < \infty$, $\phi(x) \ge 0$ be a function such that $\phi^p(x)x^{p-2}$ is integrable on $(1,\infty)$, and set*

$$\Phi(x) = \int_1^x \phi(t)\,dt. \tag{7.3.2}$$

Then $\Phi^p(x)x^{-2}$ is also integrable on $(1,\infty)$ and

$$\int_1^\infty \frac{\Phi^p(x)}{x^2}\,dx \le A_p \int_1^\infty \phi^p(x)x^{p-2}\,dx,$$

where A_p is an absolute constant which depends only on p.

PROOF. We begin by using Hölder's inequality to verify

(7.3.4)

$$\begin{aligned}\int_a^x \phi(t)\,dt &= \int_a^x \phi(t)t^{(p-2)/p}t^{-(p-2)/p}\,dt \\ &\le \left(\int_a^x \phi^p(t)t^{p-2}\,dt\right)^{1/p}\left(\int_a^x t^{-(p-2)/(p-1)}\,dt\right)^{(p-1)/p}\end{aligned}$$

for any real numbers $1 \le a < x$. We shall use this inequality to show that

$$\Phi(x) = \int_1^x \phi(t)\,dt = o(x^{1/p}), \qquad x \to \infty. \tag{7.3.5}$$

Indeed, the second factor on the right side of (7.3.4) is of order $x^{1/p}$ and by hypothesis, one can choose a sufficiently large so that the first factor is arbitrarily small. Consequently, given $\varepsilon > 0$ it is possible to choose an a such that

$$\int_a^x \phi(t)\,dt < \varepsilon x^{1/p}$$

holds for all $x > a$. Fix such an a. We obtain the estimate

$$\Phi(x) = \Phi(a) + \int_a^x \phi(t)\,dt < \Phi(a) + \varepsilon x^{1/p}, \qquad x > a.$$

But if $x > a$ is sufficiently large then $\Phi(a) < \varepsilon x^{1/p}$, and for such x's we have $\Phi(x) < 2\varepsilon x^{1/p}$. Since ε was arbitrary, (7.3.5) follows at once.

Integrate by parts to obtain

$$\int_1^x \frac{|\Phi(t)|^p}{t^2}\,dt = -\frac{|\Phi(t)|^p}{t}\Big|_1^x + p\int_1^x (\Phi(t))^{p-1}\frac{\phi(t)}{t}\,dt. \tag{7.3.6}$$

By (7.3.5), $(\Phi(x))^p/x \to 0$ as $x \to \infty$. Hence it remains to estimate the integral on the right side of (7.3.6). For this we shall use the identity

$$(\Phi(t))^{p-1}\frac{\phi(t)}{t} = \left((\Phi(t))^{p-1}t^{-1}t^{2/p-1}\right)\left(\phi(t)t^{(p-2)/p}\right).$$

Applying Hölder's inequality, we see that

$$\int_1^x (\Phi(t))^{p-1}\frac{\phi(t)}{t}\,dt \le \left(\int_1^x \frac{(\Phi(t))^p}{t^2}\,dt\right)^{(p-1)/p}\left(\int_1^x (\phi(t))^p t^{p-2}\,dt\right)^{1/p}.$$

Consequently, it follows from (7.3.6) that

$$\int_1^x \frac{|\Phi(t)|^p}{t^2}\,dt = o(1) + p\left(\int_1^x \frac{(\Phi(t))^p}{t^2}\,dt\right)^{(p-1)/p}\left(\int_1^x (\phi(t))^p t^{p-2}\,dt\right)^{1/p},$$

as $x \to \infty$. Dividing both sides of this inequality by the first factor of this last term, we obtain

$$\left(\int_1^x \frac{|\Phi(t)|^p}{t^2}\,dt\right)^{1/p} \le o(1) + p\left(\int_1^x (\phi(t))^p t^{p-2}\,dt\right)^{1/p}, \qquad x \to \infty.$$

Finally, (7.3.3) follows directly from this relationship. ∎

7.3.2. *If a sequence $\{c_n \ge 0\}$ is monotone decreasing and*

$$\sum_{n=1}^{\infty} c_n^p n^{p-2} < \infty, \qquad 1 < p < \infty, \tag{7.3.7}$$

then (7.3.1) is the Walsh-Fourier series of some function $f \in \mathbf{L}^p[0,1]$.

PROOF. For each natural number k set

$$I_k = \int_0^1 \left(\sup_{m\ge k} \Big|\sum_{n=k}^{m} c_n w_n(x)\Big|\right)^p dx. \tag{7.3.8}$$

We shall prove the inequalities

$$I_k \le \begin{cases} C_p\left(c_k^p k^{p-1} + \sum_{n=2k}^{\infty} \frac{c_n^p}{n^{2-p}}\right) & \text{for } 1 < p < 2, \\ C_p \sum_{n=k}^{\infty} c_n^p n^{p-2} & \text{for } p \ge 2 \end{cases} \tag{7.3.9}$$

where C_p is a constant which depends only on p.

By the proof of Theorem **7.1.3** in §7.1, if ν is any natural number which satisfies $1/(\nu+1) < x \le 1/\nu$, then

$$\sup_{n\ge k} \Big| \sum_{m=k}^{m} c_m w_m(x) \Big| \le 9 \sum_{m=k}^{k+\nu-1} c_m.$$

Thus

$$I_k \le 9^p \sum_{\nu=1}^{\infty} \frac{1}{\nu^2} \left(\sum_{m=k}^{k+\nu-1} c_m \right)^p.$$

Define a function $\phi(x)$ on $(1,\infty)$ by $\phi(x) = c_{n+k-1}$ for $n \le x < n+1$. It is easy to see by hypothesis (7.3.7) that the function $\phi^p(x)x^{p-2}$ is integrable on $(1,\infty)$. Thus **7.3.1** leads to the estimate

$$\sum_{\nu=1}^{\infty} \frac{1}{\nu^2} \left(\sum_{n=1}^{\nu} c_{n+k-1} \right)^p \le A_p \sum_{n=1}^{\infty} c_{n+k-1}^p n^{p-2}.$$

Consequently,

$$I_k \le 9^p A_p \sum_{n=k}^{\infty} c_n^p (n-k+1)^{p-2}. \tag{7.3.10}$$

For the case $p \ge 2$ we obtain the second estimate in (7.3.9) with $C_p = 9^p A_p$. For the case $1 < p < 2$, we have

$$\left(\sum_{n=k}^{2k-1} + \sum_{n=2k}^{\infty} \right) c_n^p (n-k+1)^{p-2} \le c_k^p \sum_{\ell=1}^{k} \ell^{p-2} + 2^{2-p} \sum_{n=2k}^{\infty} c_n^p n^{p-2}.$$

Hence the first estimate in (7.3.9) follows from this inequality and (7.3.10).

We have verified (7.3.9), and thus by (7.3.7) and (7.3.9), that $I_k \to 0$ as $k \to \infty$. Hence it follows from the definition of I_k (see (7.3.8)) that the sequence of partial sums of the series (7.3.1) is Cauchy in the space $\mathbf{L}^p[0,1]$. Since $\mathbf{L}^p[0,1]$ is complete, there exists a function $f(x) \in \mathbf{L}^p[0,1]$ such that

$$\lim_{n\to\infty} \left(\int_0^1 |s_n(x) - f(x)|^p \, dx \right)^{1/p} = 0.$$

In particular,

$$\lim_{n\to\infty} \int_0^1 |s_n(x) - f(x)| \, dx = 0.$$

From this, as we did in the proof of Theorem **7.1.3**, we conclude that (7.3.1) is the Walsh-Fourier series of the function f. ∎

In fact, we have proved the following somewhat stronger theorem:

7.3.3. *If the sequence $c_n \geq 0$ is monotone decreasing and satisfies condition (7.3.7) for some $p \in (1, \infty)$, then (7.3.1) converges in the $\mathbf{L}^p[0,1]$ norm to some function f and is its Walsh-Fourier series. Moreover, the partial sums $s_n(x)$ of the series (7.3.1) have a majorant $\sup_{n\geq 0} |s_n(x)|$ which belongs to the space $\mathbf{L}^p[0,1]$.*

The converse of Theorem **7.3.2** is also true. To prove this we need the following lemma:

7.3.4. *If $\phi(x) \in \mathbf{L}^p[0,1]$ for some $1 < p < \infty$ then*

$$\int_0^1 \left(\frac{\Phi(x)}{x}\right)^p dx \leq \left(\frac{p}{p-1}\right)^p \int_0^1 |\phi(x)|^p \, dx, \tag{7.3.11}$$

where

$$\Phi(x) = \int_0^x |\phi(t)| \, dt. \tag{7.3.12}$$

PROOF. We begin by fixing $\varepsilon > 0$ and integrating by parts:

$$\begin{aligned} I(\varepsilon) &\equiv \int_\varepsilon^1 \left(\frac{\Phi(x)}{x}\right)^p dx \\ &= -\frac{|\Phi(x)|^p}{(p-1)x^{p-1}} \Big|_\varepsilon^1 + \frac{1}{p-1} \int_\varepsilon^1 (\Phi^p(x))' x^{1-p} \, dx \\ &= -\frac{|\Phi(1)|^p}{(p-1)} + \frac{|\Phi(\varepsilon)|^p}{(p-1)\varepsilon^{p-1}} + \frac{p}{p-1} \int_\varepsilon^1 (\Phi(x))^{p-1} |\phi(x)| x^{1-p} \, dx, \end{aligned}$$

i.e.,

$$I(\varepsilon) \leq \frac{|\Phi(\varepsilon)|^p}{(p-1)\varepsilon^{p-1}} + \frac{p}{p-1} \int_\varepsilon^1 (\Phi(x))^{p-1} |\phi(x)| x^{1-p} \, dx. \tag{7.3.13}$$

Apply Hölder's inequality to (7.3.12) to obtain

$$\Phi(x) \leq \left(\int_0^x |\phi(t)|^p \, dt\right)^{1/p} \left(\int_0^x dt\right)^{(p-1)/p} = o(x^{(p-1)/p}), \qquad x \to 0+.$$

Consequently,

$$\frac{|\Phi(\varepsilon)|^p}{\varepsilon^{p-1}} \to 0, \qquad \varepsilon \to 0+.$$

By Hölder's inequality again, we also have

(7.3.14)

$$\begin{aligned} \int_\varepsilon^1 \left(\frac{\Phi(x)}{x}\right)^{p-1} |\phi(x)| \, dx &\leq \left(\int_\varepsilon^1 \left(\frac{\Phi(x)}{x}\right)^p dx\right)^{(p-1)/p} \left(\int_\varepsilon^1 |\phi(x)|^p \, dx\right)^{1/p} \\ &= (I(\varepsilon))^{(p-1)/p} \left(\int_\varepsilon^1 |\phi(x)|^p \, dx\right)^{1/p}. \end{aligned}$$

We find from (7.3.13) and (7.3.14) that

$$(I(\varepsilon))^{1/p} < o(1) + \frac{p-1}{p}\left(\int_\varepsilon^1 |\phi(x)|^p\, dx\right)^{1/p}, \qquad \varepsilon \to 0+.$$

Let $\varepsilon \to 0+$ to obtain

$$\left(\int_0^1 \left(\frac{\Phi(x)}{x}\right)^p dx\right)^{1/p} \le \frac{p-1}{p}\left(\int_0^1 |\phi(x)|^p\, dx\right)^{1/p}.$$

This verifies (7.3.11). ∎

7.3.5. *If $c_n \downarrow 0$ and the sum $f(x)$ of the series (7.3.1) is a function which belongs to $\mathbf{L}^p[0,1]$ for some $p > 1$ then condition (7.3.7) holds.*

PROOF. Set

$$F(x) = \int_0^x f(t)\, dt, \qquad F_1(x) = \int_0^x |f(t)|\, dt.$$

We shall use the identity

$$D_{2^n}(x) = \begin{cases} 2^n & \text{for } 0 \le x < 2^{-n}, \\ 0 & \text{for } 2^{-n} \le x < 1. \end{cases}$$

(see §1.4). Since the sequence $\{c_n\}$ is monotone, we have for any $n \ge 1$ that

$$F(2^{-n}) = 2^{-n}\int_0^1 f(x) D_{2^n}(x)\, dx = 2^{-n}\sum_{k=0}^{2^n-1} c_k \ge c_{2^n-1}.$$

A straightforward estimate establishes

$$\begin{aligned}\sum_{n=1}^{\infty} c_{2^n-1}^p 2^{n(p-1)} &\le \sum_{n=1}^{\infty} F^p(2^{-n}) 2^{n(p-1)} \\ &\le 2^p \sum_{n=1}^{\infty}\int_{2^{-n}}^{2^{-n+1}} \left(\frac{F_1(x)}{x}\right)^p dx \\ &= 2^p \int_0^1 \left(\frac{F_1(x)}{x}\right)^p dx.\end{aligned}$$

Hence it follows from **7.3.4** that

$$\sum_{n=1}^{\infty} c_{2^n-1}^p 2^{n(p-1)} \le \left(\frac{2p}{p-1}\right)^p \int_0^1 |f(x)|^p\, dx.$$

Bearing in mind that the sequence $\{c_n\}$ is monotone decreasing, we conclude

$$\begin{aligned}\sum_{n=1}^{\infty} c_n^p n^{p-2} &\le \sum_{k=1}^{\infty} c_{2^k-1}^p \sum_{n=2^k-1}^{2^k} n^{p-2} \\ &\le \max(2^{p-2}, 2^{2-p}) \sum_{k=1}^{\infty} c_{2^k-1}^p 2^{k(p-1)} \\ &\le C_p \int_0^1 |f(x)|^p \, dx.\end{aligned}$$

In particular, condition (7.3.7) follows from the fact that $f \in \mathbf{L}^p[0,1]$. ∎

Combining Theorems **7.3.2** and **7.3.5**, we see that the following is true:

7.3.6. *A series (7.3.1) with* $c_n \downarrow 0$ *is the Walsh-Fourier series of some function in* $\mathbf{L}^p[0,1]$, $1 < p < \infty$, *if and only if condition (7.3.7) is satisfied.*

Chapter 8

LACUNARY SUBSYSTEMS OF THE WALSH SYSTEM

§8.1. The Rademacher system.

The Rademacher system $\{r_k(x)\} = \{w_{2^k}(x)\}$, $k = 0, 1, \ldots$, which was used to define the Walsh system (see §1.1), is a typical example of what is called a *lacunary subsystem* of the Walsh system. We shall study these systems in the next several sections.

The Rademacher system and Rademacher series satisfy some interesting properties. The first such property we isolate is the multiplicative property which is expressed in the following proposition:

8.1.1. *If $n_1 \le n_2 \le \cdots \le n_k$, $k = 1, 2, \ldots$ is any increasing collection of non-negative integers, then the integral*

$$\int_0^1 r_{n_1}(x) r_{n_2}(x) \ldots r_{n_k}(x)\, dx$$

is either 0 or 1. In fact, this integral is always 0 unless the integrand consists only of products of identical pairs of factors.

PROOF. This result follows easily from **1.1.1** and orthogonality of the Walsh system. ▮

8.1.2. *If the series $\sum_{n=0}^{\infty} a_n^2$ converges then the Rademacher series*

$$\sum_{n=0}^{\infty} a_n r_n(x) \tag{8.1.1}$$

converges almost everywhere on $[0, 1)$.

PROOF. Since $r_n(x) = w_{2^n}(x)$, $n = 0, 1, \ldots$, (8.1.1) can be written in the form

$$\sum_{n=0}^{\infty} a_n w_{2^n}(x). \tag{8.1.2}$$

By the Riesz-Fischer Theorem (see **A5.4.3**), (8.1.2) is the Walsh-Fourier series of some function $f(x) \in \mathbf{L}^2[0, 1]$. Moreover, we also have

$$S_{2^k}(x, f) = \sum_{n=0}^{k-1} a_n w_{2^n}(x). \tag{8.1.3}$$

By Theorem **2.5.12** the subsequence of partial sums (8.1.3) of the Walsh-Fourier series of the function $f(x) \in \mathbf{L}^2[0, 1] \subset \mathbf{L}[0, 1]$ converges almost everywhere on [0,1). But $S_{2^k}(x, f)$ coincides with the partial sums of order k of the series (8.1.1). Thus (8.1.1) converges almost everywhere. ▮

8.1.3. *If $\sum_{n=0}^{\infty} a_n^2 = \infty$ then (8.1.1) diverges almost everywhere on $[0,1)$.*

PROOF. Set $s_n(x) = \sum_{k=0}^{n-1} a_k r_k(x)$. Suppose to the contrary that there exist coefficients which satisfy $\sum_{n=0}^{\infty} a_n^2 = \infty$ and a set $E_0 \subset [0,1)$ of positive Lebesgue measure such that the series (8.1.1) converges on E_0. By Egoroff's Theorem this series converges uniformly on some set $E \subset E_0$ also of positive Lebesgue measure. In particular, there is a constant $M > 0$ such that

$$|s_{n+p}(x) - s_n(x)| \le M, \qquad x \in E,$$

for any natural numbers n and p. Consequently,

$$\int_E (s_{n+p}(x) - s_n(x))^2 \, dx \le M^2 |E|,$$

i.e.,

(8.1.4)

$$\begin{aligned} M^2|E| &\ge \int_E \left(\sum_{k=n}^{n+p-1} a_k r_k(x) \right)^2 dx \\ &= |E| \sum_{k=n}^{n+p-1} a_k^2 + 2 \left(\sum_{n \le k < i \le n+p-1} a_k a_i \int_E r_k(x) r_i(x) \, dx \right). \end{aligned}$$

By **8.1.1** the system of functions $\{r_k(x) r_i(x)\}$, $0 \le k < i$, is orthonormal on [0,1). Hence it follows from Bessel's inequality that

$$\sum_{0 \le k < i < \infty} \left(\int_E r_k(x) r_i(x) \, dx \right)^2 \le \int_0^1 \chi_E^2(x) \, dx = |E|, \tag{8.1.5}$$

where $\chi_E(x)$ is the characteristic function of the set E, i.e., $\chi_E(x) = 1$ when $x \in E$ and $\chi_E(x) = 0$ when $x \notin E$. The expression on the left side of (8.1.5) is a series of squares of Fourier coefficients of the function $\chi_E(x)$ with respect to the system $\{r_k(x) r_i(x)\}$, $0 \le k < i$. Since $|E| \le 1$, this series must converge. Hence there is a natural number $N(E)$ such that

$$\sum_{n \le k < i \le n+p-1} \left(\int_E r_k(x) r_i(x) \, dx \right)^2 < \frac{|E|^2}{9}$$

for all $n \ge N(E)$ and $p = 1, 2, \ldots$. Continuing this inequality by the Cauchy-Schwarz inequality, we obtain

$$\begin{aligned} \left| \sum_{n \le k < i \le n+p-1} a_k a_i \int_E r_k(x) r_i(x) \, dx \right| &\le \frac{|E|}{3} \left(\sum_{n \le k < i \le n+p-1} a_k^2 a_i^2 \right)^{1/2} \\ &\le \frac{|E|}{3} \sum_{k=n}^{n+p-1} a_k^2. \end{aligned}$$

Consequently, by (8.1.4) we have

$$\sum_{k=n}^{n+p-1} a_k^2 \le 3M^2, \qquad n \ge N(E), \quad p = 1, 2, \dots.$$

In particular, the series $\sum_{n=0}^{\infty} a_n^2$ converges contrary to hypothesis. This contradiction proves **8.1.3.** ∎

8.1.4. *If $p > 0$ and n is any natural number then*

$$\left(\int_0^1 \Big| \sum_{k=0}^{n-1} a_k r_k(x) \Big|^p \, dx \right)^{1/p} \le \left(\frac{p}{2} + 1 \right)^{1/2} \left(\sum_{k=0}^{n-1} a_k^2 \right)^{1/2}. \tag{8.1.6}$$

PROOF. We first prove (8.1.6) in the special case when $p = 2r$ for some natural number r. It is evident that the identity

$$\left(\sum_{k=0}^{n-1} a_k r_k(x) \right)^{2r} = \sum \frac{(\alpha_1 + \dots \alpha_j)!}{\alpha_1! \dots \alpha_j!} a_{n_1}^{\alpha_1} \dots a_{n_j}^{\alpha_j} r_{n_1}^{\alpha_1}(x) \dots r_{n_j}^{\alpha_j}(x)$$

holds, where the second sum is taken over all non-negative integers $n_1, n_2, \dots, n_j, \alpha_1, \alpha_2, \dots, \alpha_j$ which satisfy $n_k < n$ and $\alpha_1 + \dots + \alpha_j = 2r$. Integrate this identity with respect to x over the unit interval [0,1] and use the multiplicative property **8.1.1** to obtain

$$\int_0^1 \left(\sum_{k=0}^{n-1} a_k r_k(x) \right)^{2r} dx = \sum \frac{(2\beta_1 + \dots 2\beta_j)!}{(2\beta_1)! \dots (2\beta_j)!} a_{n_1}^{2\beta_1} \dots a_{n_j}^{2\beta_j}, \tag{8.1.7}$$

where the second sum is taken over all non-negative integers $\beta_1, \beta_2, \dots, \beta_j$ which satisfy $\beta_1 + \dots + \beta_j = r$. We shall use the identity

$$\sum_{\beta_1 + \dots + \beta_j = r} \frac{(\beta_1 + \dots \beta_j)!}{\beta_1! \dots \beta_j!} a_{n_1}^{2\beta_1} \dots a_{n_j}^{2\beta_j} = \left(\sum_{k=0}^{n-1} a_k^2 \right)^r. \tag{8.1.8}$$

Indeed, (8.1.7) and (8.1.8) imply

$$\int_0^1 \left| \sum_{k=0}^{n-1} a_k r_k(x) \right|^{2r} dx \le C \left(\sum_{k=0}^{n-1} a_k^2 \right)^r,$$

where C is a constant which maximizes the ratios

$$\frac{(2\beta_1 + \dots 2\beta_j)!}{(2\beta_1)! \dots (2\beta_j)!} : \frac{(\beta_1 + \dots \beta_j)!}{\beta_1! \dots \beta_j!} = \frac{(2r)!\, \beta_1! \dots \beta_j!}{r!\, (2\beta_1)! \dots (2\beta_j)!} \le \frac{(2r)!}{r! 2^r} \le r^r.$$

This proves inequality (8.1.6) for the case $p = 2r$.

It is easy to see that the integral $\left(\int_0^1 |f(x)|^p\,dx\right)^{1/p}$ is an increasing function of the parameter $p \in (0, \infty)$. Indeed, if $0 < p < p' < \infty$ then by Hölder's inequality (see **A5.2.2**) we have

$$\int_0^1 |f(x)|^p\,dx \le \left(\int_0^1 |f(x)|^{p'}\,dx\right)^{p/p'} \left(\int_0^1 1^{p'/(p'-p)}\,dx\right)^{1-p/p'}$$
$$= \left(\int_0^1 |f(x)|^{p'}\,dx\right)^{p/p'}.$$

Consequently by the case already proved,

$$\left(\int_0^1 \left|\sum_{k=0}^{n-1} a_k r_k(x)\right|^p dx\right)^{1/p} \le \left(\int_0^1 \left|\sum_{k=0}^{n-1} a_k r_k(x)\right|^{2r} dx\right)^{1/(2r)}$$
$$\le r^{1/2}\left(\sum_{k=0}^{n-1} a_k^2\right)^{1/2}$$
$$\le \left(\frac{p}{2}+1\right)^{1/2}\left(\sum_{k=0}^{n-1} a_k^2\right)^{1/2}$$

for any p which satisfies $2r - 2 < p \le 2r$. ∎

One easy consequence of **8.1.4** is the following result.

8.1.5. *If $a_0, a_1, \ldots, a_{n-1}$ is any finite collection of real numbers then*

$$\text{(8.1.9)} \qquad \left(\sum_{k=0}^{n-1} a_k^2\right)^{1/2} \le 8\int_0^1 \left|\sum_{k=0}^{n-1} a_k r_k(x)\right| dx.$$

PROOF. Apply Hölder's inequality for $p = 3/2$ and $q = 3$. We obtain the inequality

$$\sum_{k=0}^{n-1} a_k^2 = \int_0^1 \left|\sum_{k=0}^{n-1} a_k r_k(x)\right|^2 dx \le \left(\int_0^1 |S_n(x)|\,dx\right)^{2/3}\left(\int_0^1 |S_n(x)|^4\,dx\right)^{1/3},$$

where

$$S_n(x) = \sum_{k=0}^{n-1} a_k r_k(x).$$

However, the following inequality was proved in **8.1.4**:

$$\int_0^1 |S_n(x)|^{2r}\,dx \le r^r \left(\sum_{k=0}^{n-1} a_k^2\right)^r.$$

Combining these two inequalities, we see that

$$\left(\sum_{k=0}^{n-1} a_k^2\right)^{1/3} \le 4\left(\int_0^1 |S_n(x)|\,dx\right)^{2/3},$$

which is equivalent to (8.1.9). ∎

§8.2. Other lacunary subsystems.

We shall call a sequence of natural numbers $\{n_k\}_{k=1}^{\infty}$ *lacunary in the Hadamard sense*, or simply *lacunary*, if there exists a number $q > 1$ such that $n_{k+1}/n_k > q$ for $k = 1, 2, \dots$. A system of functions $\{w_{n_k}\}$ will be called a *lacunary subsystem* of the Walsh system if $\{n_k\}$ is a lacunary sequence. Since $r_k(x) = w_{2^k}(x)$, $k = 0, 1, \dots$, the Rademacher system is a lacunary subsystem of the Walsh system.

A series will be called a *lacunary Walsh series* if it has the form

$$\sum_{k=1}^{\infty} a_k w_{n_k}(x), \tag{8.2.1}$$

where $\{n_k\}$ is a lacunary sequence. Many properties of Rademacher series actually hold for any lacunary Walsh series. For example, both **8.1.2** and **8.1.3** remain valid if "Rademacher series" is replaced by "lacunary Walsh series".

We shall prove the following theorem:

8.2.1. *If $s_m(x)$ represents the partial sum of order m of the lacunary Walsh series (8.2.1), and if*

$$\limsup_{m\to\infty} s_m(x) < \infty \tag{8.2.2}$$

holds at each point x in some interval $I \subset [0, 1)$, then

$$\sum_{k=1}^{\infty} |a_k| < \infty. \tag{8.2.3}$$

We shall need several lemmas for the proof of this theorem.

8.2.2. *Under the hypotheses of* **8.2.1** *there is an interval* $[a,b] \subset I$ *on which the partial sums* $s_m(x)$ *of the lacunary Walsh series (8.2.1) are uniformly bounded above, i.e..*

$$\sup_{a\le x\le b} s_m(x) \le M, \qquad m = 1,2,\dots. \tag{8.2.4}$$

PROOF. To prove this result, fix a dyadic interval $I' \subset I$. Let N be the subset of the group G which corresponds to I' under the transformation λ (see §1.2). We shall use the notation

$$E_{mk} = \{\overset{*}{x} \in N : |\overset{*}{S}_m(\overset{*}{x})| \le k\}, \qquad m,k = 1,2,\dots,$$

where

$$\overset{*}{S}_m(\overset{*}{x}) = \sum_{k=0}^{m-1} a_k \overset{*}{w}_k(\overset{*}{x}). \tag{8.2.5}$$

Since the partial sums (8.2.5) are all continuous on G, each set E_{mk} is closed in G. Thus the set $E_k = \bigcap_{m=1}^{\infty} E_{mk}$ is closed and the set $N' = \bigcup_{m=1}^{\infty} E_k$ is an $\mathcal{F}_\sigma$ set (recall that an $\mathcal{F}_\sigma$ set is any finite or countable union of closed sets). By construction, the set N' is simply the set of all sequences $\overset{*}{x} \in N$ which do not terminate in 1's. Hence for each such element $\overset{*}{x}$ there is a point $x \in I'$ such that $\lambda(x) = \overset{*}{x}$ and

$$\overset{*}{S}_m(\overset{*}{x}) = \overset{*}{S}_m(g(x)) = s_m(x).$$

In particular, since

$$\limsup_{m\to\infty} s_m(x) < \infty \tag{8.2.6}$$

we have that $\overset{*}{x} \in \bigcap_{m=1}^{\infty} E_{mk}$ for large k. Therefore, $N = Z \bigcup (\bigcup_{m=1}^{\infty} E_k)$, where Z is at most countable.

On the other hand, the set N is a neighborhood in the group G and is not of the first category, i.e., it cannot be expressed as a countable union of nowhere dense subsets of G. Hence one of the sets E_k is not nowhere dense. Let E_M be such a set. Thus it contains some neighborhood $\overline{N}$ of the group G. By the definition of E_M it follows that

$$\sup_{\overset{*}{x}\in\overline{N}} |\overset{*}{S}_m(\overset{*}{x})| \le M, \qquad m = 1,2,\dots. \tag{8.2.7}$$

Thus (8.2.4) is satisfied by any dyadic interval $[a,b)$ which is contained in I'. Since I' is contained in the image under the transformation λ of the neighborhood $\overline{N}$, such dyadic intervals exist. ∎

8.2.3. *Let $n_{k+1}/n_k > q > 3$ for $k = 1, 2, \dots,$ and suppose*

$$P_N(x) = \prod_{k=1}^{N}(1 + \varepsilon_k w_{n_k}(x)), \qquad \varepsilon_k = \pm 1. \tag{8.2.8}$$

Then for any interval $\Delta = \Delta_m^{(s)} = [m2^{-s}, (m+1)2^{-s})$, $m = 0, 1, \dots 2^s - 1$, $s = 0, 1, \dots$ the identity

$$\int_\Delta P_N(x)\,dx = |\Delta|. \tag{8.2.9}$$

holds for n_1 sufficiently large.

PROOF. Since $w_m(x)w_n(x) = w_{m\oplus n}(x)$ (see (1.2.17)), the product (8.2.8) can be written in the form

$$P_N(x) = 1 + \sum \alpha_\nu w_\nu(x), \tag{8.2.10}$$

where the sum on the right is a finite sum over indices ν. Indices ν over which the sum in (8.2.10) is taken satisfy the inequality $\nu \geq (q-2)n_1/(q-1)$. Consequently, if $(q-2)n_1/(q-1) \geq 2^s$ then for each such index ν we have by **1.1.4** that

$$\int_\Delta w_\nu(x)\,dx = 0.$$

Thus if we integrate (8.2.10) over Δ we obtain (8.2.9) for $(q-2)n_1/(q-1) \geq 2^s$. ∎

PROOF OF **8.2.1**. Let $\{n_k\}$ be a lacunary sequence with $n_{k+1}/n_k > q > 1$, $k = 1, 2, \dots$. Suppose $q = 1 + 2\varepsilon$ and choose $\varepsilon' > 0$ such that the following three conditions are satisfied:

$$\begin{cases} a) & (1+2\varepsilon)(1-\varepsilon') > \alpha > 1, \\ b) & \dfrac{1+2\varepsilon}{1+\varepsilon'} > \alpha > 1, \\ c) & 3\dfrac{1-\varepsilon'}{1+\varepsilon'} > \alpha > 1. \end{cases} \tag{8.2.11}$$

Find a number $Q = Q(\varepsilon') > 0$ such that $p > Q$ implies $(p-2)/(p-1) > 1 - \varepsilon'$ and $p/(p-1) < 1 + \varepsilon'$. Fix a natural number r such that $q^r > \max(3, Q)$ and divide the sequence $\{n_k\}$ into r subsequences $\{n_{kr+p}\}_{k=0}^{\infty}$, where $p = 1, 2, \dots, r$. Set

$$P_{N,p}(x) = \sum_{k=0}^{N-1}(1 + \varepsilon_{kr+p} w_{n_{kr+p}}(x)),$$

(8.2.12)
$$P_{N,p}^{(k)}(x) = \prod_{\substack{k'=0 \\ k'\neq p}}^{N-1} (1 + \varepsilon_{k'r+p} w_{n_{k'r+p}}(x)),$$

where $\varepsilon_{kr+p} = \pm 1$, $p = 1, 2, \ldots, r$, and let

$$s_{Nr}(x) = \sum_{k=1}^{Nr} a_k w_{n_k}(x)$$

represent the partial sums of the series (8.2.1). Then for any interval $\Delta \subset [0,1)$ we have

(8.2.13)
$$\begin{aligned}
\int_\Delta s_{N_r}(x) P_{N,p}(x)\,dx &= \sum_{k=0}^{N-1} a_{kr+p} \int_\Delta (1 + \varepsilon_{kr+p} w_{n_{kr+p}}(x)) w_{n_{kr+p}}(x) P_{N,p}^{(k)}(x)\,dx \\
&\quad + \sum_{\substack{p'=1 \\ p'\neq p}}^{r} \left(\sum_{k=0}^{N-1} a_{kr+p'} \int_\Delta w_{n_{kr+p'}}(x) P_{N,p}(x)\,dx \right) \\
&= \sum_{k=0}^{N-1} a_{kr+p} \int_\Delta P_{N,p}^{(k)}(x) w_{n_{kr+p}}(x)\,dx \\
&\quad + \sum_{k=0}^{N-1} a_{kr+p}\varepsilon_{kr+p} \int_\Delta P_{N,p}^{(k)}(x)\,dx \\
&\quad + \sum_{\substack{p'=1 \\ p'\neq p}}^{r} \sum_{k=0}^{N-1} a_{kr+p'} \int_\Delta P_{N,p}(x) w_{n_{kr+p'}}(x)\,dx \\
&\equiv I_1 + I_2 + I_3.
\end{aligned}$$

Let $\Delta = \Delta_m^{(s)} = [m2^{-s}, (m+1)2^{-s}) \subset [0,1)$, where $s \geq 0$ and $m \geq 0$ are integers. By **8.2.3** we have

$$\int_\Delta P_{N,p}^{(k)}(x)\,dx = |\Delta|$$

if n_1 is large enough. Consequently,

(8.2.14)
$$I_2 = |\Delta| \sum_{k=0}^{N-1} a_{kr+p}\varepsilon_{kr+p}$$

for n_1 sufficiently large.

To estimate the integrals I_{kN}^{rp} which appear in the terms of I_1, notice that

(8.2.15)
$$
\begin{aligned}
|I_{kN}^{rp}| &\equiv \Big| \int_\Delta P_{N,p}^{(k)}(x) w_{n_{kr+p}}(x)\,dx \Big| \\
&\le \Big| \int_\Delta w_{n_{kr+p}}(x)\,dx \Big| + \sum_{\nu=0}^{N-1} \Big| \int_\Delta w_{n_{kr+p}}(x) \sum_{j=0}^{\nu} G_{j\nu}(x)\,dx \Big| \\
&\equiv I_1^1 + I_1^2,
\end{aligned}
$$

where for each j the term $G_{j\nu}(x)$ is the sum of all products of $j+1$ Walsh functions whose indices n_{ir+p} do not exceed $n_{\nu r+p}$. Since $n_{kr+p} \ge n_1$, it is clear by **1.1.4** that

$$
I_1^1 = \Big| \int_\Delta w_{n_{kr+p}}(x)\,dx \Big| = 0 \tag{8.2.16}
$$

when $n_1 \ge 2^s$.

To estimate the sum I_1^2 notice that

$$
I_1^2 \le \sum_{\substack{\nu=0 \\ \nu\ne k}}^{\infty} \sum_{j=0}^{\nu} C_\nu^j \max_{w_\gamma \in G_{j\nu}} \Big| \int_\Delta w_{n_{kr+p}}(x) w_\gamma(x)\,dx \Big| \tag{8.2.17}
$$

where C_ν^j are the binomial coefficients. Since $n_{sr+p}/n_{(s-1)r+p} > q^r > \max(3, Q)$, each index γ which satisfies $w_\gamma(x) \in G_{j\nu}$ necessarily belongs to the interval $[(1-\varepsilon')n_{\nu r+p}, (1+\varepsilon')n_{\nu r+p}]$ where $\nu \ne k$. But if $s_{n_{\nu r+p}} = [\log_2 \gamma]$ then $s_{n_{\nu r+p}} \ne [\log_2 n_{kr+p}]$. Consequently, $\ell \equiv \gamma \oplus n_{kr+p} \ne 0$, and by **2.7.5** we have

(8.2.18)
$$
\begin{aligned}
\max_{w_\gamma \in G_{j\nu}} \Big| \int_\Delta w_{n_{kr+p}}(x) w_\gamma(x)\,dx \Big| &= \max_{w_\gamma \in G_{j\nu}} \Big| \int_\Delta X_\Delta(x) w_\ell(x)\,dx \Big| \\
&\le \frac{\mathrm{Var} X_\Delta}{\ell} \le \frac{2}{\ell},
\end{aligned}
$$

where $X_\Delta(x)$ is the characteristic function of the interval Δ.

If $k < \nu$ then

$$
\ell \ge \gamma - n_{kr+p} \ge (1-\varepsilon')n_{\nu r+p} - n_{kr+p} > (\frac{1}{2} - \frac{1}{3})n_{\nu r+p} = \frac{1}{6} n_{\nu r+p},
$$

since it follows from (8.2.11) that $\varepsilon' < 1/2$. On the other hand, if $\nu < k$ then

$$
\ell \ge n_{kr+p} - \gamma \ge n_{kr+p} - (1+\varepsilon')n_{\nu r+p} > (3 - \frac{3}{2})n_{\nu r+p} = \frac{3}{2} n_{\nu r+p}.
$$

Consequently, (8.2.18) implies

$$\max_{w_\gamma \in G_{j\nu}} \left| \int_\Delta w_{n_{kr+p}}(x) w_\gamma(x)\, dx \right| \leq \frac{12}{n_{\nu r+p}}.$$

Substituting this estimate into (8.2.17) we obtain

$$I_1^2 \leq \sum_{\nu=0}^{\infty} \frac{12}{n_{\nu r+p}} \sum_{j=0}^{\nu} C_\nu^j = 12 \sum_{\nu=0}^{\infty} \frac{2^\nu}{n_{\nu r+p}} \leq \frac{12 q^r}{(q^r-2)n_1}.$$

Combine this inequality with (8.2.15) and (8.2.16). We have

$$|I_{kN}^{rp}| \leq \frac{12 q^r}{(q^r-2)n_1}.$$

Let $\delta > 0$ and choose n_1 so large that $12q^r/((q^r-2)n_1) < \delta$. Then $|I_{kN}^{rp}| < \delta$. In particular, the sum I_1 in (8.2.13) can be estimated by

$$|I_1| < \delta \sum_{k=0}^{N-1} |a_{kr+p}|. \tag{8.2.19}$$

Finally, we estimate the integrals which appear in the terms of I_3:

(8.2.20)

$$\begin{aligned} |J_{kN}^{rp}| &\equiv \left| \int_\Delta w_{n_{kr+p'}}(x) P_{N,p}(x)\, dx \right| \\ &\leq \left| \int_\Delta w_{n_{kr+p'}}(x)\, dx \right| \\ &\quad + \sum_{\nu=0}^{N-1} \sum_{j=0}^{\nu} C_\nu^j \max_{w_\gamma \in G_{j\nu}} \left| \int_\Delta w_{n_{kr+p'}}(x) w_\gamma(x)\, dx \right| \\ &\equiv J_3^1 + J_3^2, \end{aligned}$$

where $p' \neq p$ but here ν may equal k. If n_1 is sufficiently large then

$$J_3^1 = 0. \tag{8.2.21}$$

This follows from **1.1.4**. As above, any index γ which satisfies $w_\gamma \in G_{j\nu}$ must belong to the interval $[(1-\varepsilon')n_{\nu r+p}, (1+\varepsilon')n_{\nu r+p}]$. This interval can always be located near the point $n_{kr+p'}$ on the real axis by the following choices of the pairs ν and p: a) $\nu = k$, $p = p' \pm 1$, b) $\nu = k-1$, $p' = 1$, $p = r$, c) $\nu = k+1$, $p' = r$, $p = 1$. In all these cases, $n_{\nu r+p}$ and $n_{kr+p'}$ are successive terms of the lacunary sequence

$\{n_k\}$. The expressions on the right side of (8.2.20) are greatest when γ and $n_{kr+p'}$ are nearest each other, i.e., in the following cases;

1) $\gamma \in [(1-\varepsilon')n_k, (1+\varepsilon')n_k]$ and $n_{kr+p'}$ plays the role of n_{k+1},

2) $\gamma \in [(1-\varepsilon')n_k, (1+\varepsilon')n_k]$ and $n_{kr+p'}$ plays the role of n_{k-1}.

In case 1), we have by (8.2.11) that

$$\frac{\gamma}{n_{k-1}} > \frac{(1-\varepsilon')n_k}{n_{k-1}} > q(1-\varepsilon') = (1+2\varepsilon)(1-\varepsilon') > \alpha > 1.$$

It is not difficult to verify that there exist constants A and B, depending only on α, such that $m > n > A$ and $m/n > \alpha > 1$ imply $m \oplus n \geq 2^{s_m - B}$, where $m = 2^{s_m} + m'$ and $0 \leq m' < 2^{s_m}$. Thus we have by estimate (8.2.18) that

$$\left| \int_\Delta w_{n_{kr+p'}}(x) w_\gamma(x)\, dx \right| = \left| \int_\Delta w_{\gamma \oplus n_{kr+p'}}(x)\, dx \right| \leq 2^{B+1-s_\gamma}.$$

Since $2^{s_\gamma} \leq \gamma < 2^{s_\gamma+1}$ and $\gamma \in [(1-\varepsilon')n_{\nu r+p}, (1+\varepsilon')n_{\nu r+p}]$ it follows that

$$\left| \int_\Delta w_{n_{kr+p'}}(x) w_\gamma(x)\, dx \right| \leq \frac{C}{n_{\nu r+p}}, \qquad C = \frac{2^{B+2}}{1-\varepsilon'}.$$

Combining this estimate with (8.2.20) and (8.2.21), we can dominate J_{kN}^{rp} as follows:

$$|J_{kN}^{rp}| \leq \sum_{\nu=0}^{N-1} \sum_{j=0}^{\nu} C_\nu^j \frac{C}{n_{\nu r+p}} = C \sum_{\nu=0}^{N-1} \frac{2^\nu}{n_{\nu r+p}} \leq C \sum_{\nu=0}^{\infty} \left(\frac{2}{3}\right)^\nu \frac{1}{n_p} \leq \frac{3C}{n_1}, \tag{8.2.22}$$

for n_1 sufficiently large. We shall suppose that n_1 is so large that the right side of (8.2.22) is less than the number $\delta > 0$ chosen earlier. Thus the sum I_3 in (8.2.13) satisfies the inequality

$$|I_3| \leq \delta \sum_{\substack{p'=1 \\ p' \neq p}}^{r} \sum_{k=0}^{N-1} |a_{n_{kr+p'}}|. \tag{8.2.23}$$

Hence it follows from (8.2.13), (8.2.14), (8.2.19), and (8.2.23) that

(8.2.24)
$$\int_\Delta s_{Nr}(x)\, P_{N,p}(x)\, dx \geq |I_2| - |I_1| - |I_3|$$
$$> \sum_{k=0}^{N-1} a_{kr+p} \varepsilon_{kr+p} |\Delta| - \delta \sum_{k=0}^{N-1} |a_{kr+p}| - \delta \sum_{\substack{p'=1 \\ p' \neq p}}^{r} \sum_{k=0}^{N-1} |a_{kr+p'}|.$$

This inequality holds for any sequence $\varepsilon_k = \pm 1$. Let $\varepsilon_k = \pm 1$ be the sequence which satisfies $\varepsilon_k a_k = |a_k|$, for $k = 1, 2, \ldots$. Then summing both sides of inequality (8.2.24) as p runs from 1 to r and using the fact that $\delta < |\Delta|/(2r)$ we arrive at

(8.2.25)

$$\sum_{p=1}^{r} \int_{\Delta} s_{Nr}(x) P_{N,p}(x)\,dx \geq (|\Delta| - \delta) \sum_{j=1}^{Nr} |a_j| - (r-1)\delta \sum_{j=1}^{Nr} |a_j|$$
$$\geq \frac{|\Delta|}{2} \sum_{j=1}^{Nr} |a_j|.$$

Suppose now that the hypotheses of Theorem **8.2.1** hold. Choose by **8.2.2** a dyadic interval $\Delta = [m2^{-s}, (m+1)2^{-s})$ and a constant $M > 0$ such that $s_k(x) \leq M$ for $k = 1, 2, \ldots$, and in particular, $s_{Nr}(x) \leq M$ for all N, r, and $x \in \Delta$. Using the last inequality and the fact that $P_{N,p}(x) \geq 0$ (see (8.2.12)), we obtain from (8.2.25) and (8.2.9) that

$$\frac{|\Delta|}{2} \sum_{j=1}^{Nr} |a_j| \leq M \sum_{p=1}^{r} \int_{\Delta} P_{N,p}(x)\,dx \leq Mr|\Delta|,$$

i.e., $\sum_{j=1}^{Nr} |a_j| \leq 2Mr$ for $N = 1, 2, \ldots$. By letting $N \to \infty$, we see that the series (8.2.3) converges.

Notice that the entire proof to this point is predicated on the fact that the first element n_1 of the lacunary sequence $\{n_k\}$ is sufficiently large. But by eliminating the first few terms of the sequence $\{n_k\}$ we may suppose this is true. Since the convergence of the series (8.2.3) does not depend on any finite number of terms, it follows that the proof of Theorem **8.2.1** is complete. ∎

Evidently, we can replace hypothesis (8.2.2) in Theorem **8.2.1** with the condition

$$\liminf_{m \to \infty} s_m(x) > -\infty,$$

since it can be obtained from (8.2.2) by multiplying each coefficient of the series (8.2.1) by -1.

If we combine this remark with Theorem **8.2.1**, it is not difficult to prove the following result:

8.2.4. *If a lacunary series (8.2.1) is the Walsh- Fourier series of some bounded function then it converges absolutely, i.e., it satisfies condition (8.2.3).*

PROOF. Since $\{n_k\}$ is lacunary we can choose a number $q > 1$ such that $n_{k+1}/n_k > q$, for $k = 1, 2, \ldots$. Suppose that (8.2.1) is the Walsh-Fourier series of some bounded, measurable function $f(x)$, i.e., $|f(x)| \leq M$, $x \in [0, 1]$, where

$M \geq 0$ is some fixed constant, and $\widehat{f}(n) = a_k$ for $n = n_k$ and $\widehat{f}(n) = 0$ for $n \neq n_k$, $k = 1, 2 \ldots$. Denote the arithmetic means of the series

$$\sum_{n=0}^{\infty} \widehat{f}(n) w_n(x) \tag{8.2.26}$$

by $\sigma_m(x)$. Then

$$\sigma_{n_N}(x) = \sum_{k=1}^{N-1} (1 - \frac{n_k}{n_N}) a_k w_{n_k}(x), \qquad N = 2, 3, \ldots .$$

Let $n_{N-1} < m \leq n_N$. Then the m-th partial sum $s_m(x, f)$ of the series (8.2.26) satisfies

$$\begin{aligned} |s_m(x,f) - \sigma_{n_N}(x)| &= \Big| \sum_{k=1}^{N-1} \frac{n_k}{n_N} a_k w_{n_k}(x) \Big| \\ &\leq \sum_{k=1}^{N-1} \frac{n_k}{n_N} |a_k| \leq \sum_{k=1}^{N-1} q^{-(N-k)} |a_k| \equiv d_N . \end{aligned} \tag{8.2.27}$$

Since Walsh-Fourier coefficients always converge to zero (see **2.7.3**), we know that $a_k \to 0$ as $k \to \infty$. Moreover, since $|f(x)| \leq M$ we have $|a_k| \leq M$ for $k = 1, 2, \ldots$. Let $\varepsilon > 0$ and choose N_1 so large that if $k > N_1$ then $M N_1 q^{-N_1} < \varepsilon/2$ and $|a_k| < (q-1)\varepsilon/2$. Then for $N > N_1$ we obtain

$$\begin{aligned} d_N &\leq \frac{q-1}{2} \varepsilon \sum_{j=1}^{N_1 - 1} q^{-j} + q^{-N_1} \sum_{j=1}^{N_1} |a_j| \\ &\leq \frac{q-1}{2} \varepsilon \frac{1}{q-1} + M N_q^{-N_1} < \varepsilon . \end{aligned}$$

We see then that $d_N \to 0$ as $N \to \infty$. But it is easy to see by (4.2.2) and **4.2.2** that $|\sigma_{n_N}(x)| \leq 2M$ for $N = 1, 2, \ldots$ and $x \in [0, 1)$ since $|f(x)| \leq M$. Consequently, we have by (8.2.27) that $|s_m(x, f)| \leq B$ for $m = 1, 2, \ldots$ and $x \in [0, 1)$, where $B = 2M + \max d_N$. Hence the partial sums of the lacunary series (8.2.26) are uniformly bounded and we conclude by Theorem **8.2.1** that this series converges absolutely. ∎

§8.3. The Central Limit Theorem for lacunary Walsh series.

Our main goal in this section is to prove the following theorem:

8.3.1. *Let $\{a_k\}$ be a sequence of real numbers which satisfies the conditions*

$$A_N \equiv (a_1^2 + a_2^2 + \cdots + a_N^2)^{1/2} \to \infty, \qquad N \to \infty, \tag{8.3.1}$$

and

$$\max_{1 \le k \le N} |a_k| = o(A_N), \qquad N \to \infty. \tag{8.3.2}$$

If $\{n_k\}$ is any lacunary sequence of natural numbers and $E \subset [0,1]$ is any Lebesgue measurable set of positive measure $|E|$, then

$$\lim_{N\to\infty} \frac{1}{|E|} \left| \left\{ x : x \in E, \ \frac{1}{A_N} \sum_{k=1}^{N} a_k w_{n_k}(x) \le y \right\} \right| = \frac{1}{\sqrt{2\pi}} \int_{-\infty}^{y} e^{-\lambda^2/2}\, d\lambda, \tag{8.3.3}$$

for all $-\infty < y < \infty$.

To prove this theorem we shall derive a preliminary result concerning lacunary sequences of natural numbers.

8.3.2. *Let $\{n_k\}$ be a lacunary sequence of natural numbers, i.e, there exists a $q > 1$ such that $n_{k+1}/n_k > q$ for $k = 1, 2, \ldots$. Then there is a number $T = T(q) > 0$ such that for any $r > T$ and any choice of $\gamma = 0, 1, \ldots$ the equation*

$$n_{k_1} \oplus n_{k_2} \oplus \cdots \oplus n_{k_p} = \gamma, \qquad n_{k_1} > n_{k_2} > \cdots > n_{k_p} \tag{8.3.4}$$

has only finitely many solutions when $n_{k_1}/n_{k_3} > q^r$. Moreover, given any γ there is a constant $C = C(\gamma, q) \ge 1$ such that equation (8.3.4) has no more that C^p solutions.

PROOF. We shall write γ in the form $\gamma = \sum_{i=0}^{s_\gamma} \varepsilon_i 2^i$ where $\varepsilon_i = 0$ or 1. Let $(n_{k_1}, n_{k_2}, \ldots, n_{k_p})$ denote a solution to equation (8.3.4) and write

$$n_{k_j} = \sum_{i=0}^{\ell_j} \varepsilon_{ji} 2^i, \qquad \varepsilon_{ji} = 0 \text{ or } 1, \ \varepsilon_{j\ell_j} = 1.$$

Let $g = \max_{1 \le j \le p} \ell_j$ and notice that $s_\gamma \le g$. By definition of the operation $\oplus$ (see §1.2), equation (8.3.4) can be interpreted in the following way:

$$\begin{array}{cccccccc}
 & & & B_1 & & & B_2 & \\
n_{k_1} & \varepsilon_{10} & \varepsilon_{11} & \cdots & \varepsilon_{1s_\gamma} & \varepsilon_{1s_\gamma+1} & \cdots & \varepsilon_{1g} \\
n_{k_2} & \varepsilon_{20} & \varepsilon_{21} & \cdots & \varepsilon_{2s_\gamma} & \varepsilon_{2s_\gamma+1} & \cdots & \varepsilon_{2g} \\
n_{k_3} & \varepsilon_{30} & \varepsilon_{31} & \cdots & \varepsilon_{3s_\gamma} & \varepsilon_{3s_\gamma+1} & \cdots & \varepsilon_{3g} \\
\vdots & \vdots & \vdots & & \vdots & \vdots & & \vdots \\
n_{k_p} & \varepsilon_{p0} & \varepsilon_{p1} & \cdots & \varepsilon_{ps_\gamma} & \varepsilon_{ps_\gamma+1} & \cdots & \varepsilon_{pg} \\
\gamma & \varepsilon_0 & \varepsilon_1 & \cdots & 1 & 0 & \cdots & 0
\end{array} \tag{8.3.5}$$

The elements of the matrix B_1 satisfy the equations

$$\sum_{j=1}^{p} \varepsilon_{ji} = \varepsilon_i \pmod 2, \qquad i = 0, \dots, s_\gamma, \tag{8.3.6}$$

and the elements of the matrix B_2 satisfy the equations

$$\sum_{j=1}^{p} \varepsilon_{ji} = 0 \pmod 2, \qquad i = s_\gamma + 1, \dots, g. \tag{8.3.7}$$

Since $q > 1$ we can choose a natural number j such that $1 + 2^{-j} \le q$. For each natural number a let s_a represent the largest integer n which satisfies the inequality $2^n \le a$, i.e., set $s_a = [\log_2 a]$. If for some γ the number of solutions of (8.3.4) is infinite then we can find solutions which satisfy

$$2^{s_{n_{k_1}}} = 2^g > 2^{s_\gamma}, \quad n_{k_1}/n_{k_3} \le 2^{g+1-\beta}, \quad \beta = s_{n_{k_3}}.$$

Hence it follows from (8.3.7) that n_{k_1} and n_{k_2} must have $g+1-\beta$ identical coefficients in their dyadic expansions, i.e.,

$$\varepsilon_{1i} = \varepsilon_{2i}, \qquad i = \beta, \beta + 1, \dots, g. \tag{8.3.8}$$

Suppose that $g - \beta + 1 \le g - s_\gamma$, i.e., $s_\gamma \le \beta - 1$. Then by (8.3.8) we have

(8.3.9)

$$\begin{aligned} \frac{n_{k_1}}{n_{k_2}} &\le \frac{2^g + \varepsilon_{g-1}2^{g-1} + \cdots + \varepsilon_\beta 2^\beta + (2^{\beta-1} + 2^{\beta-2} + \cdots + 2^0)}{2^g + \varepsilon_{g-1}2^{g-1} + \cdots + \varepsilon_\beta 2^\beta} \\ &\le 1 + \frac{2^\beta}{2^g} = 1 + 2^{g-\beta}. \end{aligned}$$

If $n_{k_1}/n_{k_3} > q^r$ for $r \ge T(q)$, then the inequality $n_{k_1}/n_{k_3} \le 2^{g+1-\beta}$ implies that $g - \beta \ge j$ for $T(q)$ chosen sufficiently large. But then we would have by (8.3.9) that $n_{k_1}/n_{k_2} \le 1 + 2^{-j} \le q$ which is impossible. Consequently, the assumption $s_\gamma \le \beta - 1$ leads to a contradiction. Thus $s_\gamma \ge \beta$ if $g - \beta \ge j$. We see then that all dyadic digits of rank greater than s_γ in the dyadic expansion of the number n_{k_3} must be 0, i.e., $\varepsilon_{3i} = 0$ for $i = s_\gamma + 1, \dots, g$. Hence

$$\begin{aligned} \frac{n_{k_1}}{n_{k_2}} &\le \frac{(2^g + \varepsilon_{g-1}2^{g-1} + \cdots + \varepsilon_{s_\gamma+1}2^{s_\gamma+1}) + (2^{s_\gamma} + \cdots + 2^0)}{2^g + \varepsilon_{g-1}2^{g-1} + \cdots + \varepsilon_{s_\gamma+1}2^{s_\gamma+1}} \\ &\le 1 + \frac{2^{s_\gamma+1}}{2^g} = 1 + 2^{-(g-(s_\gamma+1))}. \end{aligned}$$

If $g-(s_\gamma+1)\geq j$ then $n_{k_1}/n_{k_2}\leq 1+2^{-j}\leq q$ which again contradicts the fact that $\{n_k\}$ is lacunary. Hence $g-s_\gamma\leq j$. Therefore, all solutions of (8.3.4) necessarily satisfy $n_{k_1}\leq 2^{q+1}\leq 2^{s_\gamma+j+1}$ and the number of these solutions cannot exceed $2^{(s_\gamma+j+1)p}\equiv C^p(\gamma,p)$. ∎

To prove Theorem **8.3.1** we introduce the notation

$$F_N(y,E)=\frac{1}{|E|}\left|\left\{x: x\in E,\ \frac{1}{A_N}\sum_{k=1}^{N}a_kw_{n_k}(x)\leq y\right\}\right|. \tag{8.3.10}$$

Thus equation (8.3.3) takes on the form

$$\lim_{N\to\infty}F_N(y,E)=F(y)\equiv\frac{1}{\sqrt{2\pi}}\int_{-\infty}^{y}e^{-\lambda^2/2}\,d\lambda. \tag{8.3.11}$$

The function $F_N(y,E)$ is the distribution function of the sum $1/A_N\sum_{k=1}^N a_kw_{n_k}(x)$ on the set E. Consequently, according to the inverse limit theorem for characteristic functions (see, for example, [5], p. 241) formula (8.3.1) will be established if we prove that the characteristic functions

(8.3.12)
$$\Phi_N(\lambda,E)=|E|^{-1}\int_{-\infty}^{\infty}e^{i\lambda y}dF_N(y,E)=|E|^{-1}\int_E\exp\left(i\lambda\sum_{k=1}^{N}\frac{a_kw_{n_k}(x)}{A_N}\right)dx$$

of the distribution functions (8.3.10) converge uniformly to the function $e^{-\lambda^2/2}$, as $N\to\infty$, on any finite subinterval of the real axis.

We shall divide the sum $\sum_{k=1}^N a_kw_{n_k}(x)$ into blocks of r terms plus a certain remainder term in the following way. Suppose $N>r$. Set $N=Tr+N'$ where $0\leq N'<r$, and write

(8.3.13)
$$\begin{aligned}\frac{1}{A_N}i\lambda\sum_{k=1}^{N}a_kw_{n_k}(x)&=\frac{i\lambda}{A_N}\sum_{k=0}^{T-1}\sum_{j=1}^{r}a_{kr+j}w_{n_{kr+j}}(x)\\&\quad+\frac{i\lambda}{A_N}\sum_{j=1}^{N-Tr}a_{j+Tr}w_{n_{j+Tr}}(x)\\&\equiv\sum_{k=0}^{T-1}z_{N,k}(x)+z_{N,T}(x).\end{aligned}$$

Since

$$|z_{N,k}(x)|\leq\frac{|\lambda|}{A_N}\sum_{j=1}^{r}|a_{kr+j}|$$

and

$$|z_{N,T}(x)| \le \frac{|\lambda|}{A_N} \sum_{j=1}^{N-Tr} |a_{j+Tr}|,$$

we have by (8.3.2) that

$$\max_{0\le k\le T} |z_{N,k}(x)| \to 0, \qquad N \to \infty. \tag{8.3.14}$$

Since $e^z = (1+z)e^{z^2/2}e^{o(|z|^2)}$, as $z \to 0$, it follows from (8.3.12) and (8.3.13) that

$$\Phi_N(\lambda, E) = \frac{1}{|E|} \int_E \prod_{k=0}^{T} (1 + z_{N,k}(x)) \exp\left(\frac{1}{2}\sum_{k=0}^{T} z_{N,k}^2(x)\right) \exp \alpha_N(x)\, dx, \tag{8.3.15}$$

where

$$\alpha_N(x) = \exp\left(\sum_{k=0}^{T} o(|z_{N,k}(x)|^2)\right).$$

Let $\varepsilon > 0$. Choose by (8.3.14) a natural number $N_0 = N_0(\varepsilon)$ such that $N \ge N_0$ implies

(8.3.16)

$$\begin{aligned}
|\alpha_N(x)| &\le \varepsilon \sum_{k=0}^{T} |z_{N,k}(x)|^2 \\
&\le \lambda^2 \varepsilon A_N^{-2} \left(\sum_{k=0}^{T-1} \Big(\sum_{j=1}^{r} a_{kr+j}^2\Big) + \sum_{j=1}^{N-Tr} a_{j+Tr}^2 \right) \\
&\quad + 2\lambda^2 \varepsilon \mid A_N^{-2} \Big(\sum_{1\le i<j\le r} a_{kr+i} a_{kr+j} w_{n_{kr+i}\oplus n_{kr+j}}(x) \\
&\quad + \sum_{1\le i<j\le N'} a_{i+Tr} a_{j+Tr} w_{n_{i+Tr}\oplus n_{j+Tr}}(x) \Big) \mid \\
&\equiv \varepsilon\lambda^2 (1 + 2|P_N(x)|).
\end{aligned}$$

It is clear that $|P_N(x)| \le r(r-1)$ and thus that $\alpha_N(x)$ converges, as $N \to \infty$, uniformly to zero with respect to both $x \in [0,1)$ and $\lambda \in [a,b]$, for any interval $[a,b]$. Consequently,

$$\exp\left(\sum_{k=0}^{T} o(|z_{N,k}(x)|^2)\right) = 1 + \beta_N(x,\lambda), \tag{8.3.17}$$

where $\beta_N(x,\lambda) \to 0$, as $N \to \infty$, uniformly in x and λ.

We shall now estimate the middle factor of the integrand in (8.3.15). We obtain by (8.3.13) and (8.3.16) that

$$\sum_{k=0}^{T} z_{N,k}^2(x) = -\lambda^2(1+2P_N(x)). \tag{8.3.18}$$

Suppose for a moment that we have proved that $P_N(x)$ converges in measure to 0 (see **A4.1**), i.e., that

$$|\{x : x \in [0,1),\ |P_N(x)| \geq \delta\}| \to 0, \qquad N \to \infty,\ \delta > 0. \tag{8.3.19}$$

Then it would follow from (8.3.18) that

$$\exp\left(\frac{1}{2}\sum_{k=0}^{T} z_{N,k}^2(x)\right) \to \exp\left(-\frac{\lambda^2}{2}\right), \qquad N \to \infty \tag{8.3.20}$$

in measure, uniformly with respect to $\lambda \in [a,b]$, for any interval $[a,b]$.

Notice that

(8.3.21)
$$\begin{aligned} \left|\prod_{k=0}^{T}(1+z_{N,k}(x))\right| &\leq \left(\prod_{k=0}^{T}(1+|z_{N,k}(x)|^2)\right)^{1/2} \\ &\leq \exp\left(\frac{1}{2}\sum_{k=0}^{T}|z_{N,k}(x)|^2\right) \leq M(\lambda), \qquad x \in [0,1). \end{aligned}$$

In particular, it would follow from (8.3.15), (8.3.17), (8.3.20) and a version of the Lebesgue Dominated Convergence Theorem (see **A4.4.1**), that

$$\Phi_N(\lambda, E) - \frac{e^{-\lambda^2/2}}{|E|}\int_E \prod_{k=0}^{T}(1+z_{N,k}(x))\,dx \to 0, \qquad N \to \infty, \tag{8.3.22}$$

uniformly with respect to $\lambda \in [a,b]$, for any interval $[a,b]$.

We see, then, that it suffices to show (8.3.19) and the following proposition:

8.3.3. *If $-\infty < a < b < \infty$, and E is any measurable set then*

$$\int_E \prod_{k=0}^{T}(1+z_{N,k}(x))\,dx \to |E|, \qquad N \to \infty, \tag{8.3.23}$$

uniformly with respect to $\lambda \in [a,b]$.

Proof Of (8.3.19). We shall use Chebyshev's inequality (see (5.1.7)), namely, that

$$|\{x : x \in [0,1),\ |P_N(x)| \geq \delta\}| \leq \int_0^1 \delta^{-2}P_N^2(x)\,dx. \tag{8.3.24}$$

Hence it suffices to show that the right side of (8.3.24) converges to 0 as $N \to \infty$.

By construction (see (8.3.16)), we have

(8.3.25)

$$\begin{aligned}\int_0^1 P_N^2(x)\,dx &= \int_0^1 \frac{1}{A_N^4} \Big| \sum_{k=0}^{T-1} \Big(\sum_{1\le i<j\le r} a_{kr+i}a_{kr+j}w_{n_{kr+i}\oplus n_{kr+j}}(x)\Big) + \\ &\quad + \sum_{1\le i<j\le N'} a_{i+Tr}a_{j+Tr}w_{n_{i+Tr}\oplus n_{j+Tr}}(x)\Big|^2\,dx \\ &= \frac{1}{A_N^4}\left(\sum_{k=0}^{T-1}\Big(\sum_{1\le i<j\le r} a_{kr+i}^2 a_{kr+j}^2\Big) + \sum_{1\le i<j\le N'} a_{i+Tr}^2 a_{j+Tr}^2\right) \\ &\quad + \frac{2}{A_N^4}\int_0^1 \Sigma(x)\,dx \\ &\equiv S_N' + S_N'',\end{aligned}$$

where

$$\Sigma(x) = \sum a_{\alpha r+i}a_{\alpha r+j}a_{\beta r+u}a_{\beta r+v}w_{n_{\alpha r+i}\oplus n_{\alpha r+j}\oplus n_{\beta r+u}\oplus n_{\beta r+v}}(x)$$

and this sum is taken over indices $i,\ j,\ u,\ v$ ranging from 1 to r and over all indices $\alpha,\ \beta$ which satisfy $0 \le \alpha \le \beta \le T$ with the further restriction that $i \ne u,\ j \ne v$ when $\alpha = \beta$.

It is simple to estimate S_N':

(8.3.26)

$$\begin{aligned}S_N' &\le \frac{r(r-1)}{2A_N^4}\sum_{k=0}^{T-1}\max_{1\le i\le r}|a_{kr+i}|^4 + \frac{N'(N'-1)}{2A_N^4}\max_{1\le i\le N'}|a_{Tr+i}|^4 \\ &\le \frac{r(r-1)}{2A_N^4}\sum_{i=1}^{N}|a_i|^4 \le \frac{r(r-1)}{2A_N^2}\left(\max_{1\le i\le N}|a_i|\right)^2 \to 0, \qquad N\to\infty.\end{aligned}$$

To estimate S_N'' we shall break the sum $\Sigma(x)$ in (8.3.25) into two pieces $\overline{S}_N(x)$ and $\overline{\overline{S}}_N(x)$ in the following way. Let the sum $\overline{S}_N(x)$ represent all terms whose indices satisfy $\beta - \alpha \ge 2$, and let $\overline{\overline{S}}_N(x)$ represent the remaining terms, i.e., those terms for which $0 \le \beta - \alpha \le 1$.

We shall estimate the number of terms in the sum $\overline{\overline{S}}_N(x)$. If $0 \le \alpha \le T$ and $0 \le \beta \le T$ then there are precisely T pairs (α, β) which satisfy $\beta - \alpha = 1$ and precisely $T+1$ pairs (α,β) which satisfy $\beta = \alpha$. The number of terms of $\overline{\overline{S}}_N(x)$ which correspond to pairs (α, β) with $\beta = \alpha$ cannot exceed $(r(r-1)/2)^2$. Thus the total number of terms in $\overline{\overline{S}}_N(x)$ cannot exceed $2(T+1)(r(r-1)/2)^2$. Combining this observation with (8.3.2), we obtain

(8.3.27)
$$\begin{aligned}\frac{2}{A_N^4}\Big|\int_0^1 \overline{\overline{S}}_N(x)\,dx\Big| &\le \frac{4}{A_N^4}\left(\frac{r(r-1)}{2}\right)^2\sum_{\alpha=0}^{T}\max_{1\le i\le r}|a_{\alpha r+i}|^4\\ &\le r^2(r-1)^2A_N^{-4}\sum_{k=1}^{N}|a_k|^4\\ &\le \frac{r^2(r-1)^2}{A_N^2}\left(\max_{1\le k\le N}|a_k|\right)^2\to 0, \qquad N\to\infty.\end{aligned}$$

A typical term from the sum $\overline{S}_N(x)$ looks like

$$a_{\alpha r+i}a_{\alpha r+j}a_{\beta r+u}a_{\beta r+v}w_{n_{\alpha r+i}\oplus n_{\alpha r+j}\oplus n_{\beta r+u}\oplus n_{\beta r+v}}(x) \tag{8.3.28}$$

where $\beta-\alpha\ge 2$, $1\le i<j\le r$, $1\le u<v\le r$. Since $\alpha r+j\le(\beta r+u)-r$, there are at least r members of the given sequence $\{n_k\}$ which lie between $n_{\alpha r+j}$ and $n_{\beta r+u}$. Consequently, $n_{\beta r+u}/n_{\alpha r+j}>q^r$. But the integral over the interval $[0,1]$ of the expression (8.3.28) is different from zero only when the index of the Walsh function there equals zero. Hence to estimate the integral $\int_0^1 \overline{S}_N(x)\,dx$ we need only consider the terms in (8.3.28) for which

$$n_{\alpha r+i}\oplus n_{\alpha r+j}\oplus n_{\beta r+u}\oplus n_{\beta r+v}=0. \tag{8.3.29}$$

But by Lemma **8.3.2** (applied to the case $p=4$ and $\gamma=0$) we can choose an r such that the number of solutions of (8.3.29) does not exceed $C^4(0,q)$. Therefore,

$$\frac{2}{A_N^4}\Big|\int_0^1 \overline{S}_N(x)\,dx\Big|\le\frac{2C^4(0,q)}{A_N^4}\max|a_{n_{k_i}}|\to 0, \qquad N\to\infty, \tag{8.3.30}$$

where this maximum is taken over all n_{k_i} for which (8.3.29) has a solution.

Finally, observe that

$$S_N''=\frac{2}{A_N^4}\int_0^1\overline{S}_N(x)\,dx+\frac{2}{A_N^4}\int_0^1\overline{\overline{S}}_N(x)\,dx.$$

Therefore, we conclude from (8.3.25), (8.3.26), (8.3.27), and (8.3.30) that

$$\lim_{N\to\infty}\int_0^1 P_N^2(x)\,dx=0. \quad \blacksquare$$

PROOF OF **8.3.3**. By definition of the quantities $z_{N,k}(x)$ (see (8.3.13)) we have

$$\prod_{k=0}^{T}(1 + z_{N,k}(x)) = 1 + \sum_{\gamma \geq 0} \alpha_\gamma^{(N)} w_\gamma(x), \tag{8.3.32}$$

where

$$a_\gamma^{(N)} = \sum a_{n_{k_1}} a_{n_{k_2}} \dots a_{n_{k_p}} (i\lambda)^p, \tag{8.3.33}$$

and this sum is taken over all indices $(n_{k_1}, n_{k_2} \dots n_{k_p})$ which solve (8.3.4) for $p = 1, 2, \dots, N$. Notice that the factors $a_{n_{k_1}}, a_{n_{k_2}}, a_{n_{k_3}}$ in (8.3.33) are taken from three different sums $z_{N,k}(x)$, $k = 0, 1, \dots, N$. Consequently, there are at least r terms of the sequence $\{n_k\}$ which lie between n_{k_1} and n_{k_3}. Hence $n_{k_1}/n_{k_3} > q^r$ and we have by **8.3.2** that

$$|\alpha_\gamma^N| \leq \sum_{p=1}^{N} \left(|\lambda| C(\gamma, q) \max_{1 \leq i \leq N} \frac{|a_i|}{A_N} \right)^p, \qquad \gamma = 1, 2, \dots$$

But if $a \leq \lambda \leq b$ then given any $\varepsilon > 0$ there is an N_0 such that

$$|\lambda| C(\gamma, q) \max_{1 \leq i \leq N} \frac{|a_i|}{A_N} < \varepsilon$$

for $N \geq N_0$. Consequently, $\alpha_\gamma^N \to 0$ as $N \to \infty$ for any $\gamma \geq 1$. The same kind of estimate is also true for $\alpha_0^N - 1$, i.e., $\alpha_0^N \to 1$ as $N \to \infty$.

We have proved that

$$\alpha_\gamma^N \to 0, \quad \alpha_0^N \to 1, \qquad N \to \infty, \quad \gamma = 1, 2, \dots \tag{8.3.34}$$

uniformly for $\lambda \in [a, b]$. By integrating identity (8.3.32) over the set E, and applying (8.3.34), it is easy to verify that (8.3.23) holds uniformly for $\lambda \in [a, b]$, where $-\infty < a < b < \infty$. This completes the proof of **8.3.3** and thus the fundamental result **8.3.1** is established. ∎

Chapter 9

DIVERGENT WALSH-FOURIER SERIES. ALMOST EVERYWHERE CONVERGENCE OF WALSH-FOURIER SERIES OF L^2 FUNCTIONS

In Chapter 2 we saw that even for a continuous function it is necessary to impose additional conditions to insure that its Walsh- Fourier series converges at every point. Without such conditions, as we remarked in §2.3, the Fourier series of a continuous function may diverge at some points.

Given a particular continuous function, or a particular Lebesgue integrable function f, how large can the set of points be on which the Walsh-Fourier series of f diverges? We shall investigate this extremely complicated question in this chapter.

In §9.1 we shall establish that the Walsh-Fourier series of a Lebesgue integrable function need not converge at even one point.

It turns out that the Walsh-Fourier series of a continuous function can diverge only on a set of measure zero. This will follow as a special case of the general theorem proved in §9.2 which states that the Walsh-Fourier series of any $\mathbf{L}^2[0,1)$ function must converge almost everywhere.

§9.1. Everywhere divergent Walsh-Fourier series.

We shall prove an analogue of a theorem of Kolmogorov about the existence of functions whose Fourier series diverge everywhere. We notice that for the Walsh system, the construction is somewhat simpler in comparison to the classical case for the trigonometric system.

We shall prove the following lemma which plays a key role in this section:

9.1.1. *For any natural number n there is a polynomial of the form*

$$P_n(x) = 1 + \sum_{i=2^n}^{2^{n+2^n}-1} a_i w_i(x), \tag{9.1.1}$$

which satisfies the properties:

1) $P_n(x) \geq 0$ everywhere;

2) for each x there are integers ℓ_x and m_x with $2^n < \ell_x < m_x < 2^{n+2^n}$ such that the corresponding partial sums of the polynomial P_n satisfy

$$|S_{m_x}(x, P_n) - S_{\ell_x}(x, P_n)| \geq \frac{n}{4}.$$

PROOF. Fix n and choose a natural number m of the form (2.2.4) which satisfies $2^{n-1} < m < 2^n$. From inequality (2.2.5) and Theorem **2.2.2** we obtain

$$L_m \equiv \int_0^1 |D_m(t)|\, dt \geq \frac{n}{4}. \tag{9.1.2}$$

Since $m < 2^n$, the kernel $D_m(t)$ is constant on each of the intervals $\Delta_j^{(n)}$, for $j = 0, 1, \ldots, 2^n - 1$ (see **1.4.3**). By (1.2.15) and definition (1.4.8), it is evident that the kernel $D_m(x \oplus t)$ is constant for $x \in \Delta_i^{(n)}$ and $t \in \Delta_j^{(n)}$. Denote this constant value by $D_{j,i}$. Set

$$\gamma_{j,i} = \begin{cases} 0 & \text{for} \quad \operatorname{sgn} D_{j,i} = 1, \\ 1 & \text{for} \quad \operatorname{sgn} D_{j,i} = -1, \end{cases}$$

i.e.,

$$(-1)^{\gamma_{j,i}} = \operatorname{sgn} D_{j,i}. \tag{9.1.3}$$

From each $\Delta_j^{(n)}$ select the dyadic interval of rank $n + 2^n$ of the form

$$d_j \equiv \left[\frac{j}{2^n} + \sum_{i=0}^{2^n-1} \frac{\gamma_{j,i}}{2^{n+i+1}},\ \frac{j}{2^n} + \sum_{i=0}^{2^n-1} \frac{\gamma_{j,i}}{2^{n+i+1}} + \frac{1}{2^{n+2^n}}\right).$$

By construction it is clear that if $t \in d_j$ then the coefficients of $2^{-(n+i+1)}$, for $0 \leq i \leq 2^n - 1$, in the dyadic expansion of t are precisely $\gamma_{j,i}$. Thus by (1.2.12) we have

$$w_{2^{n+i}}(t) = (-1)^{\gamma_{j,i}}.$$

In particular, (9.1.3) implies that

$$w_{2^{n+i}}(t) D_m(x \oplus t) = |D_m(x \oplus t)|, \qquad x \in \Delta_i^{(n)}, \quad t \in d_j \subset \Delta_j^{(n)} \tag{9.1.4}$$

for each $0 \leq i \leq 2^n - 1$. We also notice that the interval d_j is of length

$$|d_j| = 2^{-(n+2^n)}. \tag{9.1.5}$$

We shall show that the polynomial

$$P_n(x) = \sum_{j=0}^{2^n-1} 2^{2^n} \chi_{d_j}(x), \tag{9.1.6}$$

satisfies the required conditions, where $\chi_{d_j}(x)$ is the characteristic function of the interval d_j.

Since the function $P_n(x)$ is constant on any dyadic interval $\Delta^{(n+2^n)}$, we have by Theorem **1.3.2** that it is a Walsh polynomial of order $2^{n+2^n}-1$, i.e., the upper limit on the sum (9.1.1) is correct. To show that the lower limit in (9.1.1) is correct, we shall examine the Walsh-Fourier series of $P_n(x)$ and show that all its Fourier coefficients $\widehat{P}_n(i)$ are zero for $1 \le i \le 2^n - 1$.

Notice first that

$$\int_{\Delta_{j_1}^{(n)}} P_n(t)\,dt = \int_{\Delta_{j_2}^{(n)}} P_n(t)\,dt$$

for all j_1, j_2, and thus that

$$\int_{\Delta_{j_1}^{(k)}} P_n(t)\,dt = \int_{\Delta_{j_2}^{(k)}} P_n(t)\,dt \tag{9.1.7}$$

for all $1 \le k \le n$ and all j_1, j_2.

Fix $1 \le i \le 2^n - 1$ and choose $k \le n$ with $2^{k-1} \le i < 2^k$ such that the function $w_i(t)$ is constant on each interval $\Delta_j^{(k)}$, and if $\Delta_j^{(k-1)} = \Delta_{2j}^{(k)} \bigcup \Delta_{2j+1}^{(k)}$ then $w_i(t)$ changes signs from $\Delta_{2j}^{(k)}$ to $\Delta_{2j+1}^{(k)}$. Using (9.1.7) we obtain

$$\int_{\Delta_j^{(k-1)}} P_n(t)w_i(t)\,dt = \left| \int_{\Delta_{2j}^{(k)}} P_n(t)\,dt - \int_{\Delta_{2j+1}^{(k)}} P_n(t)\,dt \right| = 0$$

for all $1 \le i \le 2^n - 1$ and $1 \le k \le n$. In particular,

$$\widehat{P}_n(i) = \int_0^1 P_n(t)w_i(t)\,dt = 0, \qquad 1 \le i \le 2^n - 1.$$

That $\widehat{P}_n(0) = 1$ is easy to verify by (9.1.5). Thus we have proved that (9.1.6) and (9.1.1) are two ways of representing the same function, namely, the polynomial $P_n(x)$.

Property 1) is obvious by (9.1.6). It remains to verify property 2).

Fix x and choose a natural number $i = i(x)$ such that $x \in \Delta_i^{(n)}$. Set $\ell_x = 2^{n+i}$ and $m_x = m + 2^{n+i}$, where m is chosen to satisfy (9.1.2). Since $0 \le i \le 2^n - 1$ and $2^{n-1} < m < 2^n$, it is clear that $2^n < \ell_x < m_x < 2^{n+2^n-1} + 2^n < 2^{n+2^n}$.

Use (2.1.10), (1.4.11) and apply (9.1.6), (9.1.4), and (9.1.2). We obtain for any

$x \in \Delta_i^{(n)}$ that

$$\begin{aligned}
|S_{m_x}(x,P_n) - S_{\ell_x}(x,P_n)| &= \Big| \int_0^1 P_n(t)(D_{m_x}(x\oplus t) - D_{2^{n+i}}(x\oplus t))\,dt \Big| \\
&= \Big| \int_0^1 P_n(t) w_{2^{n+i}}(x\oplus t) D_m(x\oplus t)\,dt \Big| \\
&= \Big| w_{2^{n+i}}(x) \int_0^1 P_n(t) w_{2^{n+i}}(t) D_m(x\oplus t)\,dt \Big| \\
&= \Big| \sum_{j=0}^{2^n} 2^{2^n} \int_{d_j} w_{2^{n+i}}(t) D_m(x\oplus t)\,dt \Big| \\
&= \sum_{j=0}^{2^n} 2^{2^n} \int_{d_j} |D_m(x\oplus t)|\,dt \\
&= \sum_{j=0}^{2^n} \int_{\Delta_j^{(n)}} |D_m(x\oplus t)|\,dt \\
&= \int_0^1 |D_m(t)|\,dt > \frac{n}{4}.
\end{aligned}$$

This verifies property 2) and the proof of the lemma is complete. ∎

We are now prepared to state and prove the fundamental result of this section.

9.1.2. *There is a function $f \in \mathbf{L}[0,1)$ whose Walsh-Fourier series diverges everywhere on $[0,1)$.*

PROOF. Let n_j be a sequence of natural numbers which satisfy

$$n_{j+1} > n_j + 2^{n_j} \tag{9.1.8}$$

and choose corresponding polynomials $P_{n_j}(x)$ by Lemma **9.1.1**. We notice by (9.1.8) and (9.1.1) that the polynomials $P_{n_j}(x)$ do not have common terms for different j's, except 1.

Consider the function defined by

$$f(x) = \sum_{j=1}^{\infty} \frac{1}{n_j} P_{n_j}(x). \tag{9.1.9}$$

Notice by (9.1.8) that $n_j > 2^{j-1}$ and thus that the series $\sum_{j=1}^{\infty} 1/n_j$ converges. Since the polynomials $P_{n_j}(x)$ are non-negative and satisfy $\int_0^1 P_{n_j}(t)\,dt = 1$, it follows from a theorem of Levy (see **A4.4.2**) that the series (9.1.9) converges to an $f \in \mathbf{L}[0,1)$. Moreover, since for each i the partial sums of the series $\sum_{j=1}^{\infty} \frac{1}{n_j} P_{n_j}(x) w_i(x)$ satisfy

$$| \sum_{j=1}^{k} \frac{1}{n_j} P_{n_j}(x) w_i(x) | \leq \sum_{j=1}^{k} \frac{1}{n_j} P_{n_j}(x) \leq f(x),$$

it follows from the Lebesgue Dominated Convergence Theorem (see **A4.4.1**) that this series can be integrated term by term. Hence we obtain from (9.1.8) that

$$\widehat{f}(i) = \frac{a_i^{(j)}}{n_j}$$

for $2^{n_j} \leq i < 2^{n_j + 2^{n_j}}$, where $a_i^{(j)}$ are the coefficients of the polynomial $P_{n_j}(x)$ defined by (9.1.1) for $n = n_j$. Consequently, if $2^{n_j} < \ell < m < 2^{n_j + 2^{n_j}}$ then the partial sums of f of order m and ℓ satisfy

$$S_m(x,f) - S_\ell(x,f) = \frac{1}{n_j}(S_m(x, P_{n_j}) - S_\ell(x, P_{n_j})).$$

In particular, this inequality holds for the numbers $m_x^{(j)}$ and $\ell_x^{(j)}$ (see **9.1.1**). Therefore, we obtain the following estimate for all $x \in [0,1)$:

$$| S_{m_x^{(j)}}(x,f) - S_{\ell_x^{(j)}}(x,f) | \geq \frac{1}{4}.$$

Since for each point x this inequality holds for all j, we conclude that the Walsh-Fourier series of the function $f(x)$ diverges everywhere on [0,1). ∎

§9.2. Almost everywhere convergence of Walsh- Fourier series of $\mathbf{L}^2[0,1)$ functions.

We shall prove the Walsh analogue of Carleson's Theorem which established that the trigonometric Fourier series of an $\mathbf{L}^2[0,2\pi]$ function converges to it almost everywhere.

Indeed, the following theorem is true:

9.2.1. *For any function* $f \in \mathbf{L}^2[0,1)$, *the partial sums* $S_n(x,f)$ *of its Walsh-Fourier series converges to* $f(x)$ *almost everywhere on* $[0,1)$.

This theorem, as we shall show, follows from the fact that the operator

$$f \to Mf(x) \equiv \sup_{n \geq 1} |S_n(x,f)|$$

is of weak type (2,2), i.e., from the following result:

9.2.2. *There is an absolute constant C such that*

$$\text{mes}\{x : \sup_{n\geq 1} |S_n(x,f)| > y\} \leq \frac{C}{y^2} \int_0^1 |f(t)|^2 \, dt$$

holds for all functions $f \in \mathbf{L}^2[0,1)$ and for all real numbers $y > 0$.

In turn, **9.2.2** follows from a lemma which is fundamental for this section:

9.2.3. *Given a function $f \in \mathbf{L}^2[0,1)$, a number $y > 0$, and a natural number N, there is a set $E = E(f,y,N)$ and some absolute constant C such that*

1) mes $E \leq \dfrac{C}{y^2} \displaystyle\int_0^1 |f(t)|^2 \, dt$;

2) $|S_n(x,f)| \leq Cy$ for $x \in [0,1) \setminus E$ and $1 \leq n < 2^N$.

We first show that **9.2.2** follows from **9.2.3**. Fix $y > 0$. By condition 2),

$$\mathcal{E}_N \equiv \left\{x : \sup_{1\leq n<2^N} |S_n(x,f)| > Cy\right\} \subset E,$$

and thus by condition 1) we have

$$\begin{aligned} \text{mes } \mathcal{E}_N &= \text{mes}\left\{x : \sup_{1\leq n<2^N} |S_n(x,f)| > Cy\right\} \\ &\leq \text{mes } E \leq \frac{C}{y^2} \int_0^1 |f(t)|^2 \, dt. \end{aligned}$$

Now, this inequality is true for any N, and the sets $\mathcal{E}_N$ satisfy $\mathcal{E}_{N+1} \supset \mathcal{E}_N$, and

$$\left\{x : \sup_{n\geq 1} |S_n(x,f)| > Cy\right\} = \bigcup_{N=1}^{\infty} \mathcal{E}_N.$$

Hence it follows that

$$\text{mes}\left\{x : \sup_{n\geq 1} |S_n(x,f)| > Cy\right\} = \lim_{N\to\infty} \text{mes}\mathcal{E}_N \leq \frac{C}{y^2} \int_0^1 |f(t)|^2 \, dt.$$

If we replace y by y/C in this last inequality, we obtain the inequality in **9.2.2** with absolute constant C^3.

Notice that in Lemma **9.2.3** we may exclude the case when the identity $S_n(x) = S_{2^N}(x)$ holds for some $n < 2^N$. Indeed, in this situation we can estimate the sums in **9.2.2** directly, using **5.2.4** and **5.1.1**, to see that

$$\text{mes}\{x : \sup_{N\geq 0} |S_{2^N}(x,f)| > y\} \leq \frac{C}{y^2} \int_0^1 |f(t)|^2 \, dt.$$

To prove that **9.2.1** follows directly from **9.2.2** we need only apply Theorems **5.1.3**, **5.4.1**, and the fact that for each Walsh polynomial Theorem **9.2.1** is trivial.

Thus to complete the proof of Theorem **9.2.1** it remains to establish Lemma **9.2.3**.

PROOF OF **9.2.3**. We shall need several new concepts and some additional notation.

For each dyadic interval Δ of rank m ($m > 0$), denote the unique dyadic interval of rank $m-1$ which contains Δ by Δ^*. Notice that $|\Delta| = 2^{-m}$ and $|\Delta^*| = 2|\Delta| = 2^{-m+1}$.

Corresponding to each number n with dyadic expansion $n = \sum_{j=0}^{N-1} \varepsilon_j 2^j$ and each interval Δ of rank m define a natural number by

$$n(\Delta) = \sum_{j=m}^{N-1} \varepsilon_j 2^j, \tag{9.2.1}$$

where the empty sum is defined to be identically zero. Then $n = n(\Delta) + s$ where

$$s = \sum_{j=0}^{m-1} \varepsilon_j 2^j < 2^m = |\Delta|^{-1}. \tag{9.2.2}$$

By definition of the operation $\oplus$ on the set of natural numbers (see §1.2), it is clear that $n = n(\Delta) \oplus s$ and thus by (1.2.17) that

$$w_n(t) = w_{n(\Delta)}(t) w_s(t). \tag{9.2.3}$$

Hence we see by (9.2.2) and **1.1.3** that the function $w_s(t)$ is constant on each interval Δ of rank m, taking on the value $+1$ or -1 on each of them.

We introduce "local" Fourier coefficients, which are defined by the formula

$$a_n(\Delta) \equiv a_n(\Delta, f) \equiv \frac{1}{|\Delta|} \int_\Delta f(t) w_n(t)\, dt. \tag{9.2.4}$$

Notice for $\Delta = [0,1)$ that these local coefficients are the usual Walsh-Fourier coefficients, namely, $a_n(f) = a_n([0,1), f)$. By (9.2.3), we also have

$$a_{n(\Delta)}(\Delta) = \pm a_n(\Delta). \tag{9.2.5}$$

If Δ is an interval of rank m and Δ' is another interval of the same rank such that $\Delta^* = \Delta \bigcup \Delta'$, then substituting $s = 2^{m-1} = 1/(2|\Delta|) = 1/|\Delta^*|$ into (9.2.3) and using the definition of $w_{2^{m-1}}(t)$, we see that the function $w_{n(\Delta)+|\Delta^*|^{-1}}$ takes on the value $w_{n(\Delta)}(t)$ on one of the intervals Δ, Δ' and takes on the value $-w_{n(\Delta)}(t)$ on the other interval. Consequently,

$$a_{n(\Delta)+|\Delta^*|^{-1}}(\Delta^*) = \pm\frac{1}{2}\left(a_{n(\Delta)}(\Delta) - a_{n(\Delta)}(\Delta')\right).$$

Since

$$a_{n(\Delta)}(\Delta^*) = \frac{1}{2}\left(a_{n(\Delta)}(\Delta) + a_{n(\Delta)}(\Delta')\right)$$

, we obtain

(9.2.6) $$a_{n(\Delta)}(\Delta) = a_{n(\Delta)}(\Delta^*) \pm a_{n(\Delta)+|\Delta^*|^{-1}}(\Delta^*).$$

Notice that 2^m divides $n(\Delta)$ and recall (see **2.6.6**) that $\{2^{m/2}w_{i2^m}(t)\}_{i=0}^{\infty}$ is a complete orthonormal system on each interval Δ of rank m. Moreover, notice that the Fourier coefficients of f with respect to this system are precisely the weighted local coefficients $2^{-m/2}a_{i2^m}(\Delta, f)$. Consequently, it follows from Parseval's identity (see **A5.4.2**) that

(9.2.7) $$\frac{1}{|\Delta|}\int_\Delta |f(t)|^2\,dt = \sum_{i=0}^{\infty} |a_{i2^m}(\Delta, f)|^2.$$

We introduce one more bit of notation:

(9.2.8) $$A_n(\Delta) = \max\{|a_n(\Delta')|, |a_n(\Delta'')|\}, \qquad \Delta = \Delta' \bigcup \Delta''.$$

Since $a_n(\Delta) = \frac{1}{2}\left(a_n(\Delta') + a_n(\Delta'')\right)$, we have

(9.2.9) $$|a_n(\Delta)| \le |A_n(\Delta)|.$$

The main technical step of this proof is the construction of a set Q^* which contains special pairs $(n(\Delta), \Delta)$. The set Q^* will be composed of subsets Q_k^* by means of $Q^* = \bigcup_{k=1}^{\infty} Q_k^*$. For each pair $(n(\Delta, \Delta) \in Q_k^*$ we shall define a partition of the interval Δ, which we shall denote by $\Omega = \Omega(n(\Delta), \Delta, k)$. By using the partition Ω we shall identify intervals $\overline{\Delta} \subset \Delta$ and corresponding sums

(9.2.10) $$S_{n(\Delta)}(x, f) - S_{n(\overline{\Delta})}(x, f)$$

which will be used to estimate the partial sum $S_n(x, f)$. The partition Ω will help estimate the sums (9.2.10). These estimates will be obtained for all x except those belonging to some sets of small measure. The choice of the pairs Q^*, the partition Ω, and the corresponding estimates will be carried out by using the numbers $A_n(\Delta)$ (see (9.2.8)).

We begin to carry out this plan.

The numbers $A_n(\Delta)$ will be controlled by the set

(9.2.11) $$S^* = \bigcup\left\{\Delta^* : \frac{1}{|\Delta|}\int_\Delta |f(t)|^2\,dt \ge y^2\right\},$$

where this union is taken over the collection of intervals Δ which satisfy the following property: if $\Delta = \Delta' \bigcup \Delta''$ then the corresponding inequality in (9.2.11) holds for at least one of the intervals Δ' or Δ''. Hence if Δ is not a subset of S^* then both of its halves Δ' and Δ'' satisfy the opposite inequality, so we have by Parseval's identity (9.2.7) that

$$|a_{n(\Delta')}(\Delta')|^2 \leq \frac{1}{|\Delta'|} \int_{\Delta'} |f(t)|^2 \, dt < y^2,$$

and

$$|a_{n(\Delta'')}(\Delta'')|^2 \leq \frac{1}{|\Delta''|} \int_{\Delta''} |f(t)|^2 \, dt < y^2,$$

for any n. By (9.2.5) the same kind of inequality holds for $|a_n(\Delta')|$ and $|a_n(\Delta'')|$. In particular (see **9.2.8**),

$$\text{(9.2.12)} \qquad \text{if } \Delta \text{ is not a subset of } S^* \text{ then } A_n(\Delta) < y \text{ for } n = 1, 2, \ldots.$$

It is clear by construction that the set S^* can be written as a union of non-overlapping intervals Δ_j^* such that for each j, one of the halves Δ_j of Δ_j^* satisfies the inequality

$$|\Delta_j| \leq \frac{1}{y^2} \int_{\Delta_j} |f(t)|^2 \, dt, \qquad j = 1, 2, \ldots.$$

Consequently,

$$\text{(9.2.13)} \qquad \operatorname{mes} S^* = \sum_j |\Delta_j^*| = 2 \sum_j |\Delta_j| \leq 2 \frac{1}{y^2} \int_0^1 |f(t)|^2 \, dt.$$

The set S^* is part of the set E for which we search.

Consider an interval which is not contained in S^*, i.e., an interval which satisfies (9.2.12). For each natural number k define a set Q_k of pairs $(n(\Delta), \Delta)$ which satisfy two conditions:

$$\text{(9.2.14)} \qquad |a_{n(\Delta)}(\Delta)| \geq \frac{y}{2^k} \qquad \text{for } (n(\Delta), \Delta) \in Q_k,$$

and

$$\text{(9.2.15)} \qquad \begin{cases} |a_{n(\Delta)}(\Delta)| \leq \frac{2y}{2^k} \text{ for } |\Delta| = \frac{1}{2}; \\ \text{if } 1/2^N \leq |\Delta| < 1/2 \text{ then } |a_{\overline{n}}(\overline{\Delta})| < y/2^k \text{ for all } \overline{n} \text{ and } \overline{\Delta} \\ \text{which satisfy } \overline{n}(\Delta) = n(\Delta) \text{ and } \overline{\Delta} \underset{\neq}{\supset} \Delta,\ |\overline{\Delta}| \leq 1/2. \end{cases}$$

By definition of the pairs in Q_k it is evident that

$$\text{(9.2.16)} \qquad \begin{cases} \text{if } (n(\Delta), \Delta) \in Q_k \text{ then } (\overline{n}(\overline{\Delta}), \overline{\Delta}) \notin Q_k \text{ for all} \\ \overline{\Delta} \underset{\neq}{\supset} \Delta \text{ and all } \overline{n} \text{ which satisfy } \overline{n}(\Delta) = n(\Delta). \end{cases}$$

Consider the polynomials

$$P_k(\Delta) \equiv P_k(x,\Delta) \equiv \sum_{\substack{(n(\overline{\Delta}),\overline{\Delta})\in Q_k \\ \overline{\Delta}\supset\Delta}} a_{n(\overline{\Delta})}(\overline{\Delta}) w_{n(\overline{\Delta})}(x). \tag{9.2.17}$$

Notice that

$$P_k(x,\Delta) = P_k(x,\Delta^*) + \sum_{(n(\Delta),\Delta)\in Q_k} a_{n(\Delta)}(\Delta) w_{n(\Delta)}(x). \tag{9.2.18}$$

If $(n(\Delta),\Delta) \in Q_k$ then by (9.2.16) the polynomial $P_k(x,\Delta^*)$ cannot contain terms $a_m w_m(x)$ whose indices satisfy $m(\Delta) = n(\Delta)$. Consequently, the polynomial $P_k(x,\Delta^*)$ is orthogonal to $w_{n(\Delta)}(x)$ on the interval Δ, i.e.,

$$\int_\Delta P_k(x,\Delta^*) w_{n(\Delta)}(x)\,dx = 0.$$

Thus for each $(n(\Delta),\Delta) \in Q_k$ we have

$$a_{n(\Delta)}(\Delta) = a_{n(\Delta)}(\Delta, f) = a_{n(\Delta)}(\Delta, f - P_k(\Delta^*)).$$

This formula, (9.2.18), and Parseval's identity imply

$$\begin{aligned} \int_\Delta |f(x) - P_k(x,\Delta)|^2\,dx &= \int_\Delta \Big| f(x) - P_k(x,\Delta^*) - \sum_{(n(\Delta),\Delta)\in Q_k} a_{n(\Delta)}(\Delta) w_{n(\Delta)}(x) \Big|^2\,dx \\ &= \int_\Delta |f(x) - P_k(x,\Delta^*)|^2\,dx - \sum_{(n(\Delta),\Delta)\in Q_k} |a_{n(\Delta)}(\Delta)|^2 |\Delta|. \end{aligned} \tag{9.2.19}$$

Notice that if $\Delta_1 \bigcup \Delta_2 = \Delta$, then

$$P_k(x,\Delta_1^*) = P_k(x,\Delta_2^*) = P_k(x,\Delta).$$

Consequently,

$$\sum_{|\Delta|=2^{-m}} \int_\Delta |f(x) - P_k(x,\Delta^*)|^2\,dx = \sum_{|\Delta|=2^{-m+1}} \int_\Delta |f(x) - P_k(x,\Delta)|^2\,dx. \tag{9.2.20}$$

Repeated applications of (9.2.19) and (9.2.20) eventuate in

$$
\begin{aligned}
0 &\le \sum_{|\Delta|=2^{-N}} \int_\Delta |f(x) - P_k(x,\Delta)|^2\,dx \\
&= \sum_{|\Delta|=2^{-N}} \int_\Delta |f(x) - P_k(x,\Delta^*)|^2\,dx - \sum_{\substack{(n(\Delta),\Delta)\in Q_k \\ |\Delta|=2^{-N}}} |a_{n(\Delta)}(\Delta)|^2|\Delta| \\
&= \sum_{|\Delta|=2^{-N+1}} \int_\Delta |f(x) - P_k(x,\Delta)|^2\,dx - \sum_{\substack{(n(\Delta),\Delta)\in Q_k \\ |\Delta|=2^{-N}}} |a_{n(\Delta)}(\Delta)|^2|\Delta| \\
&= \cdots = \int_0^1 |f(x)|^2\,dx - \sum_{(n(\Delta),\Delta)\in Q_k} |a_{n(\Delta)}(\Delta)|^2|\Delta|.
\end{aligned}
$$

Consequently, it follows from (9.2.14) that

$$
\int_0^1 |f(x)|^2\,dx \ge \sum_{(n(\Delta),\Delta)\in Q_k} |a_{n(\Delta)}(\Delta)|^2|\Delta| \ge \frac{y^2}{2^{2k}} \sum_{(n(\Delta),\Delta)\in Q_k} |\Delta|.
$$

In particular,

$$
\sum_{(n(\Delta),\Delta)\in Q_k} |\Delta| \le \frac{2^{2k}}{y^2} \int_0^1 |f(x)|^2\,dx. \tag{9.2.21}
$$

Let Q_k^* represent the set of pairs (n,Δ^*) such that $(n(\Delta),\Delta) \in Q_k$ and $n = n(\Delta^*)$. Thus each pair $(n(\Delta),\Delta) \in Q_k$ generates two pairs in Q_k^*: $(n(\Delta),\Delta^*)$ and $(n(\Delta) + \frac{1}{|\Delta^*|}, \Delta^*)$. Since $|\Delta^*| = 2|\Delta|$, we have by (9.2.21) that

$$
\sum_{(n(\Delta),\Delta)\in Q_k^*} |\Delta| \le 4\frac{2^{2k}}{y^2} \int_0^1 |f(x)|^2\,dx. \tag{9.2.22}
$$

Consider the set

$$
Q^* = \bigcup_{k=1}^{\infty} Q_k^*.
$$

Notice that

$$
\begin{cases} \text{if } \Delta = [0,1) \text{ and } \dfrac{y}{2^{\bar{k}}} \le A_{n(\Delta)}(\Delta) < \dfrac{y}{2^{\bar{k}-1}}, \\ \text{then } (n(\Delta),\Delta) \in Q_{\bar{k}}^*. \end{cases} \tag{9.2.23}
$$

This implication follows directly from definition (9.2.8), inequality (9.2.14) and the first inequality in (9.2.15).

Moreover, we shall prove that

$$(9.2.24)\quad \begin{cases} \text{if } \Delta \text{ is not a subset of } S^* \text{ and } \dfrac{y}{2^k} \le A_{n(\Delta)}(\Delta) \text{ then one can choose} \\ \text{a triple } (\tilde{n}, \tilde{\Delta}, \tilde{k}) \text{ such that } (\tilde{n}(\tilde{\Delta}), \tilde{\Delta}) \in Q^*_{\tilde{k}},\ \tilde{\Delta} \supset \Delta, \\ 1 \le \tilde{k} \le k, \text{ and such that if } \hat{\Delta} \text{ is any interval of rank one greater} \\ \text{than the rank of } \Delta\text{,i.e., } 2|\hat{\Delta}| = |\Delta|), \text{ then } \tilde{n}(\hat{\Delta}) = n(\hat{\Delta}) \,. \end{cases}$$

Indeed, if $\Delta = [0,1)$ then choose by (9.2.12) an integer $\tilde{k}$ which satisfies (9.2.23) and $1 \le \tilde{k} \le k$. Thus $(n(\Delta), \Delta) \in Q^*_{\tilde{k}}$ and the triple we want is given by $(n(\Delta), \Delta, \tilde{k})$.

If Δ is a proper subset of [0,1) and $(n(\Delta), \Delta) \notin Q^*_k$ then use definition (9.2.8) to see that at least one interval $\hat{\Delta} \subset \Delta$ of rank one greater than the rank of Δ must satisfy

$$|a_{n(\Delta)}(\hat{\Delta})| = |a_{n(\hat{\Delta})}(\hat{\Delta})| \ge \frac{y}{2^k},$$

i.e., the pair $(n(\hat{\Delta}), \hat{\Delta})$ satisfies (9.2.14). On the other hand, from the assumption $(n(\Delta), \Delta) \notin Q^*_k$ and the definition of Q^*_k it follows that $(n(\hat{\Delta}), \hat{\Delta}) \notin Q_k$. Moreover, we have $|\hat{\Delta}| < |\Delta| \le 1/2$. Thus $(n(\hat{\Delta}), \hat{\Delta})$ fails to satisfy (9.2.15) and there must be an $\overline{n}$ and a Δ' such that $\overline{n}(\hat{\Delta}) = n(\hat{\Delta})$, $\Delta' \supsetneq \hat{\Delta}$, $|\Delta'| \le 1/2$, and $|a_{\overline{n}}(\Delta')| \ge y/2^k$. In particular, $|A_{\overline{n}}(\overline{\Delta})| \ge y/2^k$ where $\overline{\Delta} = \Delta' \bigcup \Delta''$, $\overline{\Delta} \supsetneq \Delta$.

If $(\overline{n}(\overline{\Delta}), \overline{\Delta}) \in Q^*_k$ then the triple for which we search is given by $(\overline{n}(\overline{\Delta}), \overline{\Delta}, k)$. If $(\overline{n}(\overline{\Delta}), \overline{\Delta}) \notin Q^*_k$ then repeat the argument above, perhaps several times, until we either arrive at a pair $(n(\overline{\Delta}), \overline{\Delta}) \in Q^*_k$ or until $\overline{\Delta} = [0,1)$. But this last situation is considered above, because the fact that Δ is not a subset of S^* implies that $\overline{\Delta}$ is also not a subset of S^*. Therefore, the proof of (9.2.24) is complete.

We can stipulate that the $\tilde{k}$ chosen in (9.2.24) is minimal. Under this stipulation, it is easy to see that

$$(9.2.25)\quad \begin{cases} A_{\tilde{n}(\overline{\Delta})}(\overline{\Delta}) < \dfrac{y}{2^{\tilde{k}-1}} \text{ for all } \overline{\Delta} \text{ which satisfy} \\ \Delta \subset \overline{\Delta} \subset \tilde{\Delta} \text{ and all } \tilde{n}(\overline{\Delta}) \text{ such that } \tilde{n}(\tilde{\Delta}) = n(\tilde{\Delta}). \end{cases}$$

Indeed, suppose to the contrary that (9.2.25) is false, i.e., for some $\overline{\Delta}$ which satisfies $\Delta \subset \overline{\Delta} \subset \tilde{\Delta}$ we have $A_{\tilde{n}(\overline{\Delta})}(\overline{\Delta}) \ge y/2^{\tilde{k}-1}$. Then the triple $(\tilde{n}(\overline{\Delta}), \overline{\Delta}, \tilde{k}-1)$ satisfies (9.2.24), which contradicts the minimality of $\tilde{k}$.

We shall show that

$$(9.2.26)\quad \text{if } (n(\Delta), \Delta) \in Q^*_k \text{ then } A_{n(\Delta)}(\Delta) < \frac{y}{2^{k-1}}.$$

Suppose first that $|\Delta| < 1$. We have by the assumption $(n(\Delta), \Delta) \in Q_k^*$ that one of the halves Δ' of Δ must satisfy $(n(\Delta'), \Delta') \in Q_k$. Hence by (9.2.15)

$$|a_{n(\Delta')}(\Delta)| < \frac{y}{2^k}, \quad |a_{n(\Delta')+|\Delta|^{-1}}(\Delta)| < \frac{y}{2^k}.$$

Let Δ'' be the other half of Δ. Then it follows from $\Delta' \bigcup \Delta'' = \Delta$, (9.2.6), and $\Delta = (\Delta')^* = (\Delta'')^*$ that

$$|a_{n(\Delta')}(\Delta')| = |a_{n(\Delta')+|\Delta|^{-1}}(\Delta')| < \frac{y}{2^{k-1}}$$

and

$$|a_{n(\Delta')}(\Delta'')| = |a_{n(\Delta')+|\Delta|^{-1}}(\Delta'')| < \frac{y}{2^{k-1}}.$$

Since $n(\Delta)$ equals either $n(\Delta')$ or $n(\Delta') + |\Delta|^{-1}$, we conclude by definition (9.2.8) that (9.2.26) holds in the case $|\Delta| < 1$.

Notice by (9.2.26) and (9.2.14) that if $|\Delta| < 1$ then the pair $(n(\Delta), \Delta)$ can belong to only one Q_k^*. In sharp contrast, the pair $(n, [0, 1))$ can belong to two different sets $Q_{k_1}^*$ and $Q_{k_2}^*$, $k_1 < k_2$. Exclude $(n, [0, 1))$ from $Q_{k_2}^*$. Then we see by the definition of Q_k^* and by the first inequality in (9.2.15) that (9.2.26) holds in the case $\Delta = [0, 1)$.

We now show that for each interval Δ which appears in a pair $(n(\Delta), \Delta) \in Q_k^*$, it is possible to construct a partition $\Omega = \Omega((n(\Delta), \Delta, k)$ whose elements are dyadic intervals $\Delta' \in \Omega$ which satisfy the following three conditions:

(9.2.27)
$$\Delta' \subsetneq \Delta, \quad |\Delta'| \geq \frac{1}{2^N};$$

(9.2.28)
$$A_{n(\Delta)}(\overline{\Delta}) < \frac{y}{2^{k-1}}, \quad \Delta \supset \overline{\Delta} \supsetneq \Delta';$$

and

(9.2.29)
$$\text{either } |\Delta'| > \frac{1}{2^N} \text{ and } A_{n(\Delta)}(\Delta') \geq \frac{y}{2^{k-1}}, \quad \text{or } |\Delta'| = \frac{1}{2^N}.$$

We begin to construct the partition Ω by dividing each interval Δ into two halves (which are necessarily of rank one greater that the rank of Δ), and placing into Ω those halves $\hat{\Delta}$ which satisfy $A_{n(\Delta)}(\hat{\Delta}) \geq y/2^{k-1}$. (Notice by the definition of Q_k^* that $|\Delta| > 1/2^N$.) For those halves which did not get placed into Ω (it is possible to retain both halves) continue this process of dividing and placing. At each stage, the newly divided intervals $\overline{\Delta}$ are placed into Ω if $|\overline{\Delta}| = 1/2^N$ or if $A_{n(\Delta)}(\overline{\Delta}) \geq y/2^{k-1}$. On the other hand, those intervals $\overline{\Delta}$ which satisfy $A_{n(\Delta)}(\hat{\Delta}) < y/2^{k-1}$ and $|\overline{\Delta}| > 1/2^N$ are retained for further division. It is easy to verify that the partition of Δ obtained by this process generates non-overlapping dyadic intervals Δ' which satisfy conditions (9.2.27) through (9.2.29).

For each pair $(n(\Delta), \Delta) \in Q_k^*$ we shall call sums of the form

$$S_{n(\Delta)}(x, f) - S_{n(\overline{\Delta})}(x, f), \qquad x \in \overline{\Delta}, \quad \Delta' \subset \overline{\Delta} \subset \Delta, \quad \Delta' \in \Omega \tag{9.2.30}$$

distinguished.

Notice for each interval $\Delta'' \in \Omega$ that the intersection $\Delta'' \bigcap \overline{\Delta}$ is either empty or is a subset of Δ''. Let us examine one of the distinguished sums (9.2.30). Let

$$n(\Delta) = \sum_{j=m}^{N-1} \varepsilon_j 2^j, \quad n(\overline{\Delta}) = \sum_{j=\overline{m}}^{N-1} \varepsilon_j 2^j,$$

where m is the rank of the interval Δ and $\overline{m}$ is the rank of the interval $\overline{\Delta}$. Notice that $\overline{m} > m$ and thus $n(\overline{\Delta}) < n(\Delta)$. The sum (9.2.30) can be evaluated by means of the modified Dirichlet kernel using (5.3.1) in the form

$$D_n(t) = w_n(t) D_n^*(t).$$

Indeed, by (5.3.4) we have

(9.2.31)

$$\begin{aligned}
&S_{n(\Delta)}(x, f) - S_{n(\overline{\Delta})}(x, f) \\
&\quad = \int_0^1 f(t) w_{n(\Delta)}(x \oplus t) D_{n(\Delta)}^*(x \oplus t)\, dt - \int_0^1 f(t) w_{n(\overline{\Delta})}(x \oplus t) D_{n(\overline{\Delta})}^*(x \oplus t)\, dt \\
&\quad = \sum_{j=m}^{N-1} \varepsilon_j \int_0^1 f(t) w_{n(\Delta)}(x \oplus t) D_{2^j}^*(x \oplus t)\, dt \\
&\qquad - \sum_{j=\overline{m}}^{N-1} \varepsilon_j \int_0^1 f(t) w_{n(\overline{\Delta})}(x \oplus t) D_{2^j}^*(x \oplus t)\, dt
\end{aligned}$$

for all $x \in \overline{\Delta}$. Notice for all $j \geq \overline{m}$ that the function $D_{2^j}^*(t) = w_{2^j}(t) D_{2^j}(t)$ is different from zero only on the intervals $\Delta_0^j \subset \Delta_0^{\overline{m}}$ but for $j < \overline{m}$ the functions $w_{2^j}(t)$ are all constant on these same intervals, with value 1. Notice also that

$$w_{n(\Delta)}(t) = \prod_{j=m}^{\overline{m}-1} (w_{2^j}(t))^{\varepsilon_j} w_{n(\overline{\Delta})}(t) = w_{n(\overline{\Delta})}(t), \qquad t \in \Delta_0^{\overline{m}}.$$

Consequently,

$$w_{n(\overline{\Delta})}(t) D_{2^j}^*(t) = w_{n(\Delta)}(t) D_{2^j}^*(t), \qquad j \geq \overline{m}, \quad t \in [0, 1).$$

Combining this identity with (9.2.31), we see that the following expression is valid for all $x \in \overline{\Delta}$:

$$S_{n(\Delta)}(x,f) - S_{n(\overline{\Delta})}(x,f) = \sum_{j=m}^{\overline{m}-1} \varepsilon_j \int_0^1 f(t) w_{n(\Delta)}(x \oplus t) D^*_{2^j}(x \oplus t)\, dt. \tag{9.2.32}$$

We introduce an auxiliary function

$$g(t) \equiv g(t,\Omega) \equiv \begin{cases} a_{n(\Delta)}(\Delta') & \text{for } t \in \Delta' \in \Omega, \\ 0 & \text{for } t \notin \Delta'. \end{cases} \tag{9.2.33}$$

By (9.2.28) it is clear that

$$|g(t)| < \frac{y}{2^{k-1}}, \qquad t \in [0,1). \tag{9.2.34}$$

Fix $x \in \overline{\Delta}$. Choose an interval $\Delta' \in \Omega$ such that $x \in \Delta' \subset \overline{\Delta}$. The kernels $D^*_{2^j}(x \oplus t)$ are constant on the intervals of rank $j+1$ and thus constant for $j \le \overline{m}-1$ on all intervals of rank $\overline{m}$, in particular for $t \in \overline{\Delta}$, in fact for $t \in \Delta' \subset \overline{\Delta}$. These kernels $D^*_{2^j}(x \oplus t)$ are also constant in t on any interval $\Delta'' \in \Omega$, since in this case $x \notin \Delta''$ and we can apply **5.3.1**. Finally, for $j \ge m$ and $t \notin \Delta$ we have $D^*_{2^j}(x \oplus t) = 0$ since $x \in \Delta$ implies $x \oplus t \notin \Delta_0^{(m)}$, i.e., the point $x \oplus t$ does not belong to the support of the kernel $D^*_{2^j}(x \oplus t)$. If we combine these observations with (9.2.33) and (9.2.4), we see that the identity (9.2.32) can be continued in the following way:

(9.2.35)

$$\begin{aligned} S_{n(\Delta)}(x,f) - S_{n(\overline{\Delta})}(x,f) &= w_{n(\Delta)}(x) \sum_{j=m}^{\overline{m}-1} \varepsilon_j \sum_{\Delta' \in \Omega} \int_{\Delta'} f(t) w_{n(\Delta)}(t) D^*_{2^j}(x \oplus t)\, dt \\ &= w_{n(\Delta)}(x) \sum_{j=m}^{\overline{m}-1} \varepsilon_j \sum_{\Delta' \in \Omega} \int_{\Delta'} g(t) D^*_{2^j}(x \oplus t)\, dt \\ &= w_{n(\Delta)}(x) \sum_{j=m}^{\overline{m}-1} \varepsilon_j \int_0^1 g(t) D^*_{2^j}(x \oplus t)\, dt \end{aligned}$$

Use the notation (5.3.2) and formula (5.3.4). Since $n(\Delta) = \sum_{j=m}^{N-1} \varepsilon_j 2^j$, it follows that

$$\begin{aligned} S^*_{n(\Delta)}(x,g) &= \int_0^1 g(t) D^*_{n(\Delta)}(x \oplus t)\, dt \\ &= \sum_{j=m}^{N-1} \varepsilon_j \int_0^1 g(t) \left(w_{2^j}(x \oplus t) \sum_{i=0}^{2^j-1} w_i(x \oplus t) \right) dt \\ &= \sum_{j=m}^{N-1} \varepsilon_j \sum_{i=2^j}^{2^{j+1}-1} w_i(x) \int_0^1 g(t) w_i(t)\, dt. \end{aligned}$$

This last sum can be viewed as a partial sum of the Walsh-Fourier series of the function $S^*_{n(\Delta)}(x,g)$. Thus, we can interpret the sum on the right side of (9.2.35) as a partial sum of a Fourier series, i.e.,

$$\sum_{j=m}^{\overline{m}-1} \varepsilon_j \int_0^1 g(t) D^*_{2^j}(x \oplus t)\, dt = \sum_{j=m}^{\overline{m}-1} \varepsilon_j \sum_{i=2^j}^{2^{j+1}-1} w_i(x) \int_0^1 g(t) w_i(t)\, dt$$
$$= S_{2^{\overline{m}}}(x, S^*_{n(\Delta)}(g)).$$

Consequently, using the notation (5.2.7) and (9.2.35), we obtain

$$(9.2.36) \qquad |S_{n(\Delta)}(x,f) - S_{n(\overline{\Delta})}(x,f)| \le H(S^*_{n(\Delta)}(g))(x), \qquad x \in \overline{\Delta}.$$

We emphasize that the right side of this last estimate does not depend on $\overline{\Delta}$.

We now introduce a set on which the distinguished sums (9.2.30) can be relatively large. Set

$$(9.2.37) \qquad U^* \equiv U^*(n(\Delta), \Delta, k) \equiv \{x \in [0,1) : H(S^*_{n(\Delta)}(g))(x) > \frac{y}{2^{k/2}}\}.$$

Recall from **5.2.4** that the operator H is of strong type (p,p), hence (see **5.1.1**) of weak type (p,p) for all $1 < p < \infty$. Use this fact for $p = 6$. We obtain

$$\text{mes } U^* \le C \left(\frac{2^{k/2}}{y}\right)^6 \|S^*_{n(\Delta)}(g)\|_6^6.$$

Since the operator $S^*_{n(\Delta)}$ is also of type (6,6) (see **5.3.2**), we can continue this estimate using (9.2.34) and (9.2.33) as follows:

$$(9.2.38) \qquad \text{mes } U^* \le C' \frac{2^{3k}}{y^6} \|g\|_6^6 \le C'' \frac{2^{3k}}{y^6} \cdot \frac{y^6}{2^{6k}} |\Delta| = C'' \frac{|\Delta|}{2^{3k}}.$$

On the other hand, we have by (9.2.36) and (9.2.37) that

$$(9.2.39) \qquad |S_{n(\Delta)}(x,f) - S_{n(\overline{\Delta})}(x,f)| \le \frac{y}{2^{k/2}}, \qquad x \notin U^*, \quad x \in \overline{\Delta},$$

for any $\Delta' \subset \overline{\Delta} \subset \Delta$ and $\Delta' \in \Omega$.

We have completed construction of the set E, namely,

$$(9.2.40) \qquad E \equiv S^* \bigcup \bigcup_{k=1}^{\infty} \left(\bigcup_{(n(\Delta),\Delta) \in Q_k^*} U^*(n(\Delta), \Delta, k) \right).$$

By (9.2.13), (9.2.22), and (9.2.38), we have

$$\text{mes } E \le \left(2 + C''4\sum_{k=1}^{\infty}\frac{1}{2^k}\right)\frac{1}{y^2}\int_0^1 |f(t)|^2\,dt, \tag{9.2.41}$$

i.e., this set E satisfies condition 1) from the conclusion of Lemma **9.2.3** with $2{+}4C''$ as the absolute constant.

It remains to verify condition 2), i.e., to estimate $S_n(x, f)$ for any $x \notin E$ and $0 < n < 2^N$.

We may suppose that $A_n(\Delta_0) > 0$, where $\Delta_0 \equiv [0,1)$. Indeed, if this were not true then by (9.2.9) $a_n(\Delta_0) = a_n(f) = 0$ and either there is a natural number m, $n < m < 2^N$, such that $S_n(x, f) = S_m(x, f)$ for all $x \in [0,1)$ and $A_m(\Delta_0) > 0$, or $S_n(x, f) = S_{2^N}(x, f)$, and as we remarked at the beginning of this section (after the statement of Lemma **9.2.3**), we can exclude this case from consideration.

Fix $x \notin E$. Since $x \notin S^*$ we have $\Delta_0 \notin S^*$ and (9.2.12) holds for Δ_0. Thus we can choose $\tilde{k}_0 \ge 0$ such that

$$\frac{y}{2^{\tilde{k}_0}} \le A_n(\Delta_0) < \frac{y}{2^{\tilde{k}_0 - 1}},$$

and by (9.2.23) such that $(n, \Delta_0) \in Q^*_{\tilde{k}_0}$. Hence the partition $\Omega_0 = \Omega(n, \Delta_0, \tilde{k}_0)$ is defined.

Let $\Delta_1 \in \Omega_0$ satisfy $x \in \Delta_1$. By (9.2.39) we have

$$|S_n(x, f) - S_{n(\Delta_1)}(x, f)| \le \frac{y}{2^{\tilde{k}_0/2}}.$$

By (9.2.27), $\Delta_1 \subsetneq \Delta_0$. If $n(\Delta_1) = 0$ then estimate 2) is obtained for this point x. If $n(\Delta_1) \ne 0$ then we continue this process further.

Notice that if $n(\Delta_1) \ne 0$ then $|\Delta_1| > 1/2^N$ (see (9.2.1)). By (9.2.29)

$$A_{n(\Delta_1)}(\Delta_1) = A_{n(\Delta_0)}(\Delta_1) \ge \frac{y}{2^{\tilde{k}_0 - 1}}.$$

Since $x \notin S^*$ we see by (9.2.12) that $A_{n(\Delta_1)}(\Delta_1) < y$. Thus we can find a k_1, $1 \le k_1 < \tilde{k}_0$, such that

$$\frac{y}{2^{k_1}} \le A_{n(\Delta_1)}(\Delta_1) < \frac{y}{2^{k_1 - 1}} \tag{9.2.42}$$

The inequality on the left side of (9.2.42) allows us to apply (9.2.24) and (9.2.25). The result of this is we can choose a triple $(\tilde{n}_1, \tilde{\Delta}_1, \tilde{k}_1)$ such that the halves $\hat{\Delta}_1 \subset \Delta_1$, $2|\hat{\Delta}_1| = |\Delta_1|$, satisfy

$$\tilde{n}_1(\hat{\Delta}_1) = n(\hat{\Delta}_1), \tag{9.2.43}$$

such that $\tilde{\Delta}_1 \supset \Delta_1$, $\tilde{k}_1 \le k_1$, and $\tilde{k}_1$ is minimal such that $(\tilde{n}_1(\tilde{\Delta}_1), \tilde{\Delta}_1) \in Q^*_{\tilde{k}_1}$. Thus by (9.2.25) we see that all intervals $\overline{\Delta}$ which satisfy $\Delta_1 \subset \overline{\Delta} \subset \tilde{\Delta}_1$ also satisfy

$$A_{\tilde{n}_1(\tilde{\Delta}_1)}(\overline{\Delta}) = A_{\tilde{n}_1(\overline{\Delta})}(\overline{\Delta}) < \frac{y}{2^{\tilde{k}-1}}. \tag{9.2.44}$$

Thus the partition $\Omega_1 = \Omega(\tilde{n}_1(\tilde{\Delta}_1), \tilde{\Delta}_1, \tilde{k}_1)$ is defined. Choose $\Delta_2 \in \Omega_1$ such that $x \in \Delta_2$ and by (9.2.29) such that

$$A_{\tilde{n}_1(\tilde{\Delta}_1)}(\Delta_2) \ge \frac{y}{2^{\tilde{k}_1-1}}.$$

The interval Δ_2 cannot be one of the intervals $\overline{\Delta}$, $\Delta_1 \subset \overline{\Delta} \subset \tilde{\Delta}_1$, since such a $\overline{\Delta}$ satisfies the opposite inequality (9.2.44). Consequently, $\Delta_2 \subsetneq \Delta_1$ and $\Delta_2 \subset \hat{\Delta}_1$. Thus we see by (9.2.43) that $\tilde{n}_1(\Delta_2) = n(\Delta_2)$. We shall estimate

$$\begin{aligned}
&|S_{n(\Delta_1)}(x,f) - S_{n(\Delta_2)}(x,f)| \\
&\qquad \le |S_{n(\Delta_1)}(x,f) - S_{n(\hat{\Delta}_1)}(x,f)| + |S_{\tilde{n}_1(\tilde{\Delta}_1)}(x,f) - S_{n(\hat{\Delta}_1)}(x,f)| \\
&\qquad\qquad + |S_{\tilde{n}_1(\tilde{\Delta}_1)}(x,f) - S_{n(\Delta_2)}(x,f)| \\
&\qquad \equiv s_1 + s_2 + s_3.
\end{aligned}$$

To estimate s_1 notice that $n(\Delta_1)$ equals either $n(\hat{\Delta}_1)$ or $n(\hat{\Delta}_1) + |\Delta_1|^{-1}$ (see (9.2.1)). In the first case $s_1 = 0$. In the second case $|\Delta_1| = 2^{-m}$, $\Delta_1 = \hat{\Delta}_1 \bigcup \hat{\Delta}_1'$, and we obtain

$$\begin{aligned}
s_1 &= |S_{n(\hat{\Delta}_1)+2^m}(x,f) - S_{n(\hat{\Delta}_1)}(x,f)| \\
&= \left| \int_0^1 f(t) \left(\sum_{i=0}^{2^m-1} w_{n(\hat{\Delta}_1)+i}(x \oplus t) \right) dt \right| \\
&= \left| \int_0^1 f(t) w_{n(\hat{\Delta}_1)}(t) D_{2^m}(x \oplus t)\, dt \right| \\
&= \left| 2^m \left(\int_{\hat{\Delta}_1} f(t) w_{n(\hat{\Delta}_1)}(t)\, dt + \int_{\hat{\Delta}_1'} f(t) w_{n(\hat{\Delta}_1)}(t)\, dt \right) \right| \\
&\le \frac{1}{2}(|a_{n(\hat{\Delta}_1)}(\hat{\Delta}_1)| + |a_{n(\hat{\Delta}_1)}(\hat{\Delta}_1')|) \\
&= \frac{1}{2}(|a_{n(\Delta_1)}(\hat{\Delta}_1)| + |a_{n(\Delta_1)}(\hat{\Delta}_1')|) \le A_{n(\Delta_1)}(\Delta_1).
\end{aligned}$$

We used (9.2.5) and (9.2.8) to obtain these last estimates.

The terms s_2 and s_3 are distinguished sums and can be estimated by using (9.2.39) for $k = \tilde{k}_1$, where $\tilde{\Delta}_1$ plays the role of Δ and $\hat{\Delta}_1$ and Δ_2 play the role of $\overline{\Delta}$.

Combining these estimates for s_1, s_2, and s_3, using (9.2.42) and bearing in mind that $\tilde{k}_1 \leq k_1$, we obtain

$$|S_{n(\Delta_1)}(x,f) - S_{n(\Delta_2)}(x,f)| \leq A_{n(\Delta_1)}(\Delta_1) + 2\frac{y}{2^{\tilde{k}_1/2}}$$
$$< \left(\frac{1}{2^{k_1-1}} + \frac{2}{2^{\tilde{k}_1/2}}\right) y < 4 \cdot 2^{-\tilde{k}_1/2} y.$$

This change from Δ_1 to Δ_2 and the corresponding estimate obtained above are the first step in an inductive argument which we repeat if $n(\Delta_2) \neq 0$. This argument can be repeated only finitely many times. The result is that we obtain nested intervals

$$[0,1) = \Delta_0 \underset{\neq}{\supset} \Delta_1 \underset{\neq}{\supset} \cdots \underset{\neq}{\supset} \Delta_{s+1}$$

and integers $\tilde{k}_0 > \tilde{k}_1 > \cdots > \tilde{k}_s \geq 1$ such that $n(\Delta_j) \neq 0$ for $j \leq s$, $n(\Delta_{s+1}) = 0$ and

$$|S_{n(\Delta_j)}(x,f) - S_{n(\Delta_{j+1})}(x,f)| \leq 4 \cdot 2^{-\tilde{k}_j/2} y.$$

Summing these estimates over j we have

$$|S_n(x,f)| = \left| \sum_{j=0}^{s} S_{n(\Delta_j)}(x,f) - S_{n(\Delta_{j+1})}(x,f) \right| \leq 4y \sum_{k=1}^{\infty} 2^{-k/2}.$$

Adjusting the constants in this inequality and in (9.2.41) we see that condition 2) of Lemma **9.2.3** holds. This completes the proof of the lemma, and thus of Theorems **9.2.1** and **9.2.2**. ∎

Chapter 10

APPROXIMATIONS BY WALSH AND HAAR POLYNOMIALS

§10.1. Approximation in uniform norm.

Let $f(t)$ be a function continuous on the interval [0,1]. Consider the quantity

$$E_n(f) = \inf_{\{a_k\}} \|f(t) - \sum_{k=0}^{n-1} a_k w_k(t)\|, \qquad n = 1, 2, \ldots, \tag{10.1.1}$$

where the infimum is taken over all real coefficients $\{a_k\}$ and the norm is the uniform one, i.e.,

$$\|f\| = \sup_{0 \le t < 1} |f(t)|.$$

We shall call the quantity $E_n(f)$ the *best uniform approximation of the function f by Walsh polynomials of order n*. By definition it is clear that

$$E_n(f) \le \|f(t) - s_n(t, f)\|,$$

where $s_n(t, f) = \sum_{k=0}^{n-1} \hat{f}(k) w_k(t)$ is the partial sum of order n of the Walsh-Fourier series of f. As we showed in **2.3.1**,

$$\|f(t) - s_{2^n}(t, f)\| \to 0, \qquad n \to \infty$$

for all continuous functions f. Consequently, $E_{2^n}(f) \to 0$ as $n \to \infty$. Moreover, the sequence $\{E_n(f)\}_{n=0}^{\infty}$ is obviously non-increasing, i.e., $E_n(f) \ge E_{n+1}(f)$. Therefore, $E_n(f) \to 0$ as $n \to \infty$ for every function continuous on the interval [0,1].

Below, we shall obtain two-sided estimates for the best approximation $E_n(f)$ to a function f in terms of its modulus of continuity. Moreover, we shall establish estimates for the best approximation of a continuous function by Haar polynomials as well.

We begin by recalling the definition of the Haar system and its connection with the Walsh system (see §1.3). The Haar functions $h_n(x)$ are defined as follows: $h_0(x) = 1$ for $x \in [0, 1)$; if $n = 2^k + m$, $k = 0, 1, \ldots,$ $m = 0, 1, \ldots 2^k - 1$, then

$$h_n(x) = \begin{cases} 2^{k/2} & \text{for} \quad x \in \Delta_{2m}^{(k+1)}, \\ -2^{k/2} & \text{for} \quad x \in \Delta_{2m+1}^{(k+1)}, \\ 0 & \text{for} \quad x \in [0, 1) \setminus \Delta_m^{(k)}. \end{cases} \tag{10.1.2}$$

Here $\Delta_m^{(k)} = [m/2^k, (m+1)/2^k)$, $k = 0, 1, \ldots$, $m = 0, 1, \ldots 2^k - 1$. As we showed in §1.3, the identities

$$w_{2^k+m}(x) = 2^{-k/2} \sum_{n=0}^{2^k-1} w_{m,n}^{(k)} h_{2^k+n}(x), \qquad 0 \le m < 2^k, \tag{10.1.3}$$

and

$$h_{2^k+n}(x) = 2^{k/2} \sum_{m=0}^{2^k-1} w_{n,m}^{(k)} w_{2^k+m}(x), \qquad 0 \le n < 2^k, \tag{10.1.4}$$

hold for all $x \in [0, 1)$, where $(w_{n,m}^{(k)})$ is a symmetric orthogonal matrix of order 2^k.

Since $w_0(x) = h_0(x)$, we have by (10.1.3) and (10.1.4) that every Haar polynomial $\sum_{n=0}^{2^k-1} a_n h_n(x)$ of order 2^k is also a Walsh polynomial $\sum_{n=0}^{2^k-1} b_n w_n(x)$ of order 2^k and conversely. Consequently, we have by (10.1.1) that

$$E_{2^k}(f) = E_{2^k}(f)_h, \qquad k = 0, 1, \ldots,$$

where

$$E_{2^k}(f)_h = \inf_{\{a_k\}} \|f(t) - \sum_{k=0}^{n-1} a_k h_k(t)\| \tag{10.1.5}$$

is the best uniform approximation to the function f by Haar polynomials of order n. (Here, the infimum is taken over all choices of n real numbers $a_0, a_1, \ldots, a_{n-1}$.)

Notice that each Haar polynomial $\sum_{k=0}^{n-1} a_k h_k(t)$ is a step function of a special type. Namely, if $n = 2^m + \ell + 1$, $\ell = 0, 1, \ldots, 2^m - 1$, and $\Delta_i^{(m)} = [i/2^m, (i+1)/2^m)$, $i = 0, 1, \ldots 2^m - 1$, $m = 0, 1, \ldots$, then this Haar polynomial is a step function which is constant on the intervals $\Delta_i^{(m+1)}$, $0 \le i \le 2\ell + 1$, and $\Delta_j^{(m)}$, $\ell < j \le 2^m - 1$. This follows immediately from the definition of the Haar functions (see (10.1.2)). The converse of this statement is also true:

10.1.1. *If $n = 2^m + \ell + 1$, $\ell = 0, 1, \ldots, 2^m - 1$, $m = 0, 1, \ldots$, then any step function $\mathcal{P}_n(x)$ which is constant on the intervals $\Delta_i^{(m+1)}$, $0 \le i \le 2\ell+1$, and $\Delta_j^{(m)}$, $\ell < j \le 2^m - 1$ can be represented as a Haar polynomial $\sum_{k=0}^{n-1} a_k h_k(t)$ of order n.*

PROOF. Let $\mathcal{P}_n(x)$ be a step function which satisfies the hypotheses of Theorem **10.1.1**. We need to show that the equation

$$\sum_{k=0}^{n-1} a_k h_k(x) = \mathcal{P}_n(x), \qquad 0 \le x < 1, \tag{10.1.6}$$

can be solved for the coefficients a_k. We suppose at first that $n-1=2^m$ for some $m=0,1,\ldots$. In this case we need to prove that if $\mathcal{P}_{2^m+1}(x)$ a step function which takes the value $c_i^{(m)}$ on the interval $\Delta_i^{(m)}$ for $i=0,1,\ldots,2^m-1$, then there is a Haar polynomial of the form $\sum_{k=0}^{2^m-1} a_k h_k(x)$ which coincides with this function on the interval [0,1). But we have already noticed that a Haar polynomial of the form $\sum_{k=0}^{2^m-1} a_k h_k(x)$ is also a Walsh polynomial of the form $\sum_{k=0}^{2^m-1} b_k w_k(x)$, and conversely. Consequently, that equation (10.1.6) can be solved in the case $n=2^m+1$ follows immediately from **1.3.2**.

Suppose now that $n=2^m+\ell+1$, where $0\le \ell<2^m-1$, for some $m=1,2,\ldots$. Then we can write the left side of equation (10.1.6) in the form

$$\sum_{k=0}^{n-1} a_k h_k(x)=\sum_{k=0}^{2^m-1} a_k h_k(x)+\sum_{k=2^m}^{n-1} a_k h_k(x). \tag{10.1.7}$$

By what we just proved, we can choose coefficients for the first polynomial on the right side of this identity so that it coincides with any step function which is constant on intervals of the form $\Delta_s^{(m)}$, $s=0,1,\ldots 2^m-1$. To deal with the second polynomial on the right side of (10.1.7), notice by the definition of the Haar functions (10.1.2) that this second polynomial is a step function which vanishes on the intervals $\Delta_j^{(m)}$, $\ell<j\le 2^m-1$, and takes on constant values $b_i^{(m+1)}$ on the intervals $\Delta_i^{(m+1)}$ which satisfy the additional condition $b_{2s}^{(m+1)}=-b_{2s+1}^{(m+1)}$, $s=0,1,\ldots,\ell$. Consequently, if the step function $\mathcal{P}_n(x)$ on the right side of equation (10.1.6) takes the values $c_j^{(m+1)}$ on the intervals $\Delta_j^{(m+1)}$, $j=0,1,\ldots,2\ell+1$ and the values $c_j^{(m)}$ on the intervals $\Delta_j^{(m)}$, $j=\ell+1,\ldots,2^m-1$, then equation (10.1.6) is equivalent to the following system of linear equations:

$$\begin{gathered}
b_s^{(m)}+b_{2s}^{(m+1)}=c_{2s}^{(m+1)},\\
b_s^{(m)}-b_{2s}^{(m+1)}=c_{2s+1}^{(m+1)},\\
b_j^{(m)}=c_j^{(m)},\\
s=0,\ldots,\ell,\quad j=\ell+1,\ldots,2^m-1.
\end{gathered}$$

This system has a unique solution for any choice of the numbers on the right side. Therefore, (10.1.6) can be solved and the proof of Proposition **10.1.1** is complete. ∎

Given any function f defined on the interval [0,1), we shall use the notation

$$\alpha_s^{(m)}(f)=\sup_{x\in\Delta_s^{(m)}} f(x),\quad \beta_s^{(m)}(f)=\inf_{x\in\Delta_s^{(m)}} f(x), \tag{10.1.8}$$

$$\omega_s^{(m)}(f)=\alpha_s^{(m)}(f)-\beta_s^{(m)}(f), \tag{10.1.9}$$

for $s=0,1,\ldots,2^m-1$, $m=0,1,\ldots$.

10.1.2. *If a function $f(x)$ is continuous on $[0,1]$ and $\omega(\delta, f)$ is its modulus of continuity, then*

$$E_n(f)_h \le \omega(\frac{1}{n}, f) \le 6E_n(f)_h, \qquad n = 1, 2, \dots, \tag{10.1.10}$$

where $E_n(f)_h$ is defined by (10.1.5).

PROOF. The inequality is obvious for $n = 1$ since

$$\omega(1, f) = \alpha_0^{(0)}(f) - \beta_0^{(0)}(f), \quad \text{and} \quad E_1(f)_h = \frac{1}{2}\left(\alpha_0^{(0)}(f) - \beta_0^{(0)}(f)\right).$$

Suppose now that $n \ge 2$. Write n in the form $n = 2^m + \ell + 1$, $\ell = 0, 1, \dots, 2^m - 1$, $m = 0, 1, \dots$. We shall prove that

$$E_n(f)_h = \frac{1}{2}\max\left\{\max_{0 \le i \le 2\ell+1} \omega_i^{(m+1)}(f), \quad \max_{\ell < i \le 2^m} \omega_i^{(m)}(f)\right\}. \tag{10.1.11}$$

Toward this, let $\mathcal{P}_n(x) = \sum_{k=0}^{n-1} a_k h_k(x)$ be a Haar polynomial of order n which takes on the value $\frac{1}{2}(\alpha_s^{(m+1)}(f) + \beta_s^{(m+1)}(f))$ on each interval $\Delta_s^{(m+1)}$, $s = 0, 1, \dots, 2\ell+1$, and the value $\frac{1}{2}(\alpha_j^{(m)}(f) + \beta_j^{(m)}(f))$ on each interval $\Delta_j^{(m)}$, $j = \ell + 1, \dots, 2^m - 1$. Such a polynomial exists by **10.1.1**. This polynomial gives the best uniform approximation by constants to the function $f(x)$ on each of the intervals $\Delta_s^{(m+1)}$, $s = 0, 1, \dots 2\ell+1$, and on each of the intervals $\Delta_j^{(m)}$, $j = \ell+1, \dots, 2^m - 1$. Consequently, on the entire interval [0,1), this polynomial gives the best uniform approximation to the function $f(x)$ by Haar polynomials of order n. By the construction of this polynomial, we see that (10.1.11) holds. Using this identity together with the notation (10.1.8) and (10.1.9), we see that

$$\begin{aligned} E_n(f)_h &\le \frac{1}{2}\max\{\omega(2^{-m-1}, f), \omega(2^{-m}, f)\} \\ &= \frac{1}{2}\omega(2^{-m}, f) \le \omega(2^{-m-1}, f) \le \omega(\frac{1}{n}, f). \end{aligned}$$

In particular, the left hand inequality in (10.1.10) is proved. But, the definition of the modulus of continuity together with (10.1.9) and (10.1.11) imply

$$\begin{aligned} \omega(\frac{1}{n}, f) &\le \omega(2^{-m}, f) \\ &\le \max\{2\max_{0 \le s \le 2\ell+1} \omega_s^{(m+1)}(f), \quad 2\max_{\ell < j \le 2^m} \omega_j^{(m)}(f), \\ &\qquad \omega_{\ell+1}^{(m)}(f) + \omega_{2\ell}^{(m+1)}(f) + \omega_{2\ell+1}^{(m+1)}(f)\} \\ &\le \max\{3\max_{0 \le s \le 2\ell+1} \omega_s^{(m+1)}(f), \quad 3\max_{\ell < j \le 2^m} \omega_j^{(m)}(f)\} = 6E_n(f)_h. \end{aligned}$$

This proves the right hand inequality in (10.1.10). ∎

10.1.3. *If f is continuous on the interval $[0,1]$ then*

$$E_n(f) \le 2\omega(\frac{1}{n}, f) \le 24E_n(f), \qquad n = 1, 2, \dots. \tag{10.1.12}$$

PROOF. Let $n = 2^m$, $m = 0, 1, \dots$, and notice by (10.1.3) and (10.1.4) that

$$E_{2^m}(f) = E_{2^m}(f)_h. \tag{10.1.13}$$

Thus in this case (10.1.12) follows immediately from **10.1.2**.

Suppose now that $2^m < n < 2^{m+1}$ and write n in the form $n = 2^m + \ell$ where $0 < \ell < 2^m$. Since the sequence $\{E_n(f)\}$ is monotone, non-increasing, we see by identity (10.1.13) and **10.1.2** that

$$E_n(f) \le E_{2^m}(f) = E_{2^m}(f)_h \le \omega(2^{-m}, f) \le 2\omega(2^{-m-1}, f) \le 2\omega(1/n, f),$$

i.e., the left hand inequality in (10.1.12) is true.

Similarly,

$$\begin{aligned} \omega(1/n, f) &\le \omega(2^{-m}, f) \le 2\omega(2^{-m-1}, f) \\ &\le 12E_{2^{m+1}}(f)_h = 12E_{2^{m+1}}(f) \le 12E_n(f). \end{aligned}$$

This proves the right hand inequality in (10.1.12). ∎

As a corollary of **10.1.2** and **10.1.2** we obtain

10.1.4. *A function f, continuous on the interval $[0,1]$ belongs to the class Lip α for some $\alpha \in (0,1]$ if and only if either of the following two conditions are met:*

$$E_n(f)_h = O(n^{-\alpha}), \quad E_n(f) = O(n^{-\alpha})$$

In the theory of approximation, an upper estimate of the best approximation to a given function (in a fixed norm) by polynomials of a certain order in some system of functions, by means of the modulus of continuity of this function (in the same norm) is called a *direct theorem*. Results in which the best approximations are estimated below by the modulus of continuity are called *inverse theorems*. Direct theorems are represented by Jackson's Theorem, which uses the modulus of continuity on the interval $[0, 2\pi]$ of a continuous 2π-periodic function f to estimate the best uniform approximation $E_n(f)_T$ by trigonometric polynomials

$$t_n(x) = a_0 + \sum_{k=1}^{n-1}(a_k \cos kx + b_k \sin kx)$$

of order n as follows

$$E_n(f)_T \le 12\omega(\frac{1}{n}, f).$$

(More recently, the exact inequality

$$E_n(f)_T < \omega(\frac{\pi}{n}, f), \qquad f \neq \text{const}, \quad n = 1, 2, \dots,$$

was obtained by N. P. Korneĭčukom, who proved that the constant which precedes the modulus of continuity on the right side of this inequality cannot be replaced by any constant smaller than 1.)

The left hand inequalities in (10.1.12) and (10.1.10) can be viewed as analogues of Jackson's Theorem for the best approximation by Walsh and Haar polynomials. However, the right hand inequalities of (10.1.12) (respectively, (10.1.10)) are peculiar to the Walsh case (respectively, the Haar case) in the sense that there does not exist a constant C such that

$$\omega(1/n) \leq CE_n(f)_T, \qquad n = 1, 2, \dots$$

holds for all 2π-periodic, continuous functions for any constant C.

It is easy to use **10.1.2** and **10.1.3** to obtain the following result:

10.1.5. *If f is continuous on the interval $[0,1]$ and either $E_n(f)_h = o(1/n)$ or $E_n(f) = o(1/n)$, as $n \to \infty$ then $f \equiv const$.*

PROOF. If a continuous function f satisfies one of these conditions then we have by **10.1.2** or **10.1.3** that $\omega(1/n, f) \to 0$ as $n \to \infty$. Since the modulus of continuity is monotone, it follows that

$$\omega(\delta, f) = o(\delta), \qquad \delta \to 0+. \tag{10.1.14}$$

But $|f(x \pm \delta) - f(x)| \leq \omega(\delta, f)$ for any $x \in (0,1)$ and $x \pm \delta \in (0,1)$. Thus we obtain from (10.1.14) that

$$\limsup_{\delta \to 0+} \frac{|f(x \pm \delta) - f(x)|}{\delta} \leq \limsup_{\delta \to 0+} \frac{\omega(\delta, f)}{\delta} = \lim_{\delta \to 0+} \frac{\omega(\delta, f)}{\delta} = 0.$$

Since the left side of this display is always non-negative, it must be identically zero, i.e., $f'(x) = 0$. Since $x \in (0,1)$ was arbitrary, we see that the function f is constant on (0,1). Since it is continuous on the entire interval [0,1], it follows that f is constant on [0,1]. ∎

Thus we see that a non-constant, continuous function cannot be uniformly approximated by a Haar or Walsh polynomial of order n with a rate as fast as $o(1/n)$, as $n \to \infty$. This fact can be explained by the discontinuous nature of the Haar and Walsh functions. This is not a serious drawback. For a continuous but non-smooth function, i.e., a continuous function which does not have a continuous derivative, the Walsh and Haar polynomials provide sufficiently good uniform approximations on the interval [0,1]. Moreover, we should keep in mind the fact that for the best uniform approximation by Haar polynomials there is the simple formula (10.1.11) which in the case $n = 2^m$, $m = 0, 1, \ldots$, remains true for uniform approximation by Walsh polynomials as well (see (10.1.13). We notice also that the proof of Proposition **10.1.2** contains a simple and effective means to construct a Haar polynomial which attains the best uniform approximation to a given continuous function be Haar polynomials of order n. (Effective means for constructing trigonometric polynomials which are best uniform approximations to a given continuous, 2π-periodic functions are not yet known.)

§10.2. Approximation in the $\mathbf{L}^p$ norm.

We shall look at direct and inverse theorems for approximation by Walsh and Haar polynomials of functions in $\mathbf{L}^p[0,1)$ norm, $1 \le p < \infty$.

Let f be a function which belongs to $\mathbf{L}^p[0,1)$, for some $1 \le p < \infty$, i.e., f is Lebesgue measurable on [0,1) and the norm

(10.2.1)
$$\|f\|_p = \left(\int_0^1 |f(x)|^p \, dx \right)^{1/p}$$

is finite. The quantity

(10.2.2)
$$E_n^{(p)}(f) = \inf_{\{a_k\}} \|f(t) - \sum_{k=0}^{n-1} a_k w_k(t)\|_p,$$

where the infimum is taken over all choices of the n real numbers $a_0, a_1, \ldots, a_{n-1}$, will be called the *best approximation to the function* f *in the norm of* $\mathbf{L}^p[0,1)$ *by Walsh polynomials of order* n. It is clear by definition that

(10.2.3)
$$E_n^{(p)}(f) \le \|f(x) - s_n(x,f)\|_p,$$

where

(10.2.4)
$$s_n(x,f) = \sum_{k=0}^{n-1} \widehat{f}(k) w_k(t)$$

represents the partial sums of order n of the Walsh-Fourier series of the function f.

10.2.1. *If* $f \in \mathbf{L}^p[0,1)$, $1 \le p < \infty$, *then*

(10.2.5)
$$\|s_{2^n}(f)\|_p \le \|f\|_p, \qquad n = 0, 1, \ldots.$$

PROOF. Recall from §2.1 that these partial sums satisfy

(10.2.6)
$$s_{2^n}(f) = 2^n \int_{\Delta_i^{(n)}} f(t) \, dt, \qquad \text{for} \quad x \in \Delta_i^{(n)}, \quad i = 0, \ldots, 2^n - 1,$$

where $\Delta_i^{(n)} = [i/2^n, (i+1)/2^n)$. Let $\dfrac{1}{p} + \dfrac{1}{q} = 1$. We shall use Hölder's inequality

$$(10.2.7) \qquad \left| \int_a^b \phi(t)\psi(t)\, dt \right| \le \left(\int_0^1 |\phi(x)|^p\, dx \right)^{1/p} \left(\int_0^1 |\psi(x)|^q\, dx \right)^{1/q}$$

for the case $[a,b] = \Delta_i^{(n)}$, $\phi(t) = f(t)$, and $\psi(t) = 1$. Thus it follows from (10.2.6) that

$$\|s_{2^n}(f)\|_p \le \left(\sum_{i=0}^{2^n-1} \left| 2^n \int_{\Delta_i^{(n)}} f(x)\, dx \right|^p 2^{-n} \right)^{1/p}$$
$$\le 2^{n/q} \left(\sum_{i=0}^{2^n-1} \int_{\Delta_i^{(n)}} |f(x)|^p\, dx \right)^{1/p} 2^{-n/q} = \|f\|_p. \blacksquare$$

For the next several pages we shall assume that the function f and the Walsh functions are periodic of period 1. The quantity

$$(10.2.8) \qquad \omega_p(\delta, f) \equiv \sup_{0 \le h \le \delta} \left(\int_0^1 |f(x+h) - f(x)|^p\, dx \right)^{1/p}, \qquad \delta \ge 0,$$

is called the *modulus of continuity of the function* f *in the* $\mathbf{L}^p$ *norm*, or simply, the $\mathbf{L}^p$*-modulus of continuity.*

10.2.2. *If* $f \in \mathbf{L}^p[0,1)$, $1 \le p < \infty$, *then*

$$(10.2.9) \qquad \|f - s_{2^n}(x,f)\|_p \le 2^{1/p}\omega_p(2^{-n}, f), \qquad n = 0, 1, \ldots.$$

PROOF. By identity (10.2.6), we see by the definition (10.2.1) of the $\mathbf{L}^p$ norm that

$$\|f - s_{2^n}(f)\|_p = \left(\sum_{i=0}^{2^n-1} \int_{\Delta_i^{(n)}} \left| f(x) - 2^n \int_{\Delta_i^{(n)}} f(t)\, dt \right|^p dx \right)^{1/p}$$
$$= \left(\sum_{i=0}^{2^n-1} \int_{\Delta_i^{(n)}} \left| 2^n \int_{\Delta_i^{(n)}} (f(x) - f(t))\, dt \right|^p dx \right)^{1/p}.$$

Hence using Hölder's inequality (10.2.7) to estimate the inner integral, we obtain

$$\|f - s_{2^n}(f)\|_p \le \left(\sum_{i=0}^{2^n-1} 2^n \int_{\Delta_i^{(n)}} \left(\int_{\Delta_i^{(n)}} |f(x) - f(t))|^p\, dt \right) dx \right)^{1/p}$$
$$= \left(\sum_{i=0}^{2^n-1} 2^n \int_{\Delta_i^{(n)}} \left(\int_{i/2^n - x}^{(i+1)/2^n - x} |f(x) - f(x+t))|^p\, dt \right) dx \right)^{1/p}$$
$$\le \left(\sum_{i=0}^{2^n-1} 2^n \int_{\Delta_i^{(n)}} \left(\int_{-2^{-n}}^{2^{-n}} |f(x) - f(x+t))|^p\, dt \right) dx \right)^{1/p}.$$

By Fubini's Theorem (see **A4.4.4**), change the order of integration in this last expression and use the definition of the $\mathbf{L}^p$-modulus of continuity. Thus obtain

$$\begin{aligned}\|f - s_{2^n}(f)\|_p &\leq \left(2^n \int_{-2^{-n}}^{2^{-n}} \left(\sum_{i=0}^{2^n-1} \int_{\Delta_i^{(n)}} |f(x) - f(x+t))|^p \, dx\right) dt\right)^{1/p} \\ &= \left(2^n \int_{-2^{-n}}^{2^{-n}} \left(\int_0^1 |f(x) - f(x+t))|^p \, dx\right) dt\right)^{1/p} \\ &\leq 2^{1/p}\omega_p(2^{-n}, f). \blacksquare\end{aligned}$$

10.2.3. *If $f \in \mathbf{L}^p[0,1)$ for some $1 \leq p < \infty$ then*

$$E_n^{(p)}(f) \leq 2^{1+1/p}\omega_p(1/n, f), \qquad n = 1, 2, \ldots \tag{10.2.10}$$

PROOF. Write n in the form $n = 2^m + \ell$ where $\ell = 0, \ldots, 2^m - 1$, $m = 0, 1, \ldots$. By (10.2.3) and (10.2.9) we have

$$\begin{aligned}E_n^{(p)}(f) \leq E_{2^m}^{(p)}(f) &\leq 2^{1/p}\omega_p(2^{-m}, f) \\ &\leq 2^{1+1/p}\omega_p(2^{-m-1}, f) \\ &\leq 2^{1+1/p}\omega_p(1/n, f). \blacksquare\end{aligned}$$

Inequality (10.2.10) can be viewed as an analogue of the well-known estimate

$$E_n^{(p)}(f)_T \leq C_p\omega_p(1/n, f),$$

where $E_n^{(p)}(f)_T = \inf_{\{t_n\}} \|f - t_n\|_p$ is the best approximation to a function $f \in \mathbf{L}^p[0, 2\pi)$ in the $\mathbf{L}^p$ norm by trigonometric polynomials $t_n(x)$ of order n.

For $0 < \alpha \leq 1$, and $1 \leq p < \infty$, denote the class of functions $f \in \mathbf{L}^p[0,1)$ which satisfy $\omega_p(\delta, f) = O(\delta^\alpha)$ by Lip (α, p). The following result is a corollary of **10.2.3**:

10.2.4. *If $f \in Lip\ (\alpha, p)$, for some $0 < \alpha \leq 1$, $1 \leq p < \infty$, then*

$$E_n^{(p)} = O(n^{-\alpha}).$$

In order to prove a reverse theorem, that is one which estimates the $\mathbf{L}^p$-modulus of continuity from above by the best approximation in $L^p[0,1)$ norm, we need three preliminary results.

10.2.5. *If $f \in \mathbf{L}^p[0,1)$ for some $1 \leq p < \infty$ then*

$$E_{2^n}^{(p)}(f) \leq \|f - s_{2^n}(f)\|_p \leq 2E_{2^n}^{(p)}(f), \qquad n = 0, 1, \ldots.$$

PROOF. The left hand inequality is trivial by the definition of $E_{2^n}^{(p)}(f)$. To prove the right hand inequality let $\mathcal{P}_{2^n}(x)$ denote a Walsh polynomial of order 2^n which gives the best approximation to the function f in the $\mathbf{L}^p[0,1)$ norm, i.e., $E_{2^n}^{(p)}(f) = \|f - \mathcal{P}_{2^n}\|_p$. This polynomial exists since for $f \in \mathbf{L}^p[0,1)$ the quantity $\|f - \sum_{k=0}^{n-1} a_k w_k(x)\|_p$ is a continuous function of the coefficients $a_0, \ldots, a_{n-1}$. Apply the triangle inequality for the $\mathbf{L}^p$ norm

$$\|\phi + \psi\|_p \leq \|\phi\|_p + \|\psi\|_p, \tag{10.2.11}$$

which is frequently called *Minkowski's inequality*, and Proposition **10.2.1**. We obtain

$$\begin{aligned} \|f - s_{2^n}(f)\|_p &= \|f - \mathcal{P}_{2^n} - s_{2^n}(f - \mathcal{P}_{2^n})\|_p \\ &\leq \|f - \mathcal{P}_{2^n}\|_p + \|s_{2^n}(f - \mathcal{P}_{2^n})\|_p \\ &\leq 2\|f - \mathcal{P}_{2^n}\|_p = 2E_{2^n}^{(p)}(f). \blacksquare \end{aligned}$$

Let $h_n(x)$ be the Haar functions as defined in equation (10.1.2).

10.2.6. *For any $p > 0$ and any real numbers a_k, $2^n \leq k < 2^{m+1}$, $m = 0, 1, \ldots$, the following identity holds:*

$$\Big\| \sum_{k=2^m}^{2^{m+1}-1} a_k h_k \Big\|_p = 2^{m(\frac{1}{2} - \frac{1}{p})} \left(\sum_{k=2^m}^{2^{m+1}-1} |a_k|^p \right)^{1/p}. \tag{10.2.12}$$

PROOF. If $k = 2^m + i$, $0 \leq i < 2^m$, then $|h_k(x)| = 2^{m/2}$ for $x \in \Delta_i^{(m)} = [i/2^m, (i+1)/2^m)$ and $h_k(x) = 0$ for $x \in [0,1) \setminus \Delta_i^{(m)}$. From this it is easy to see that (10.2.12) holds. ∎

Let $f \in \mathbf{L}[0,1)$ be fixed. Let $\mathcal{P}_n(x) \equiv \mathcal{P}_n(x, f)$, $n = 1, 2, \ldots$ denote the partial sums of order n of the Haar-Fourier series of f, i.e., let

$$\mathcal{P}_n(x) = \sum_{k=0}^{n-1} a_k h_k(x), \tag{10.2.13}$$

where

$$a_k = a_k(f) = \int_0^1 f(x) h_k(x)\, dx \tag{10.2.14}$$

are the Haar-Fourier coefficients of the function f.

10.2.7. *If $n-1=2^m+\ell$ for some $\ell=0,1,\dots,2^m-1$, $m=0,1,\dots$, then*

$$\mathcal{P}_n(x)=\begin{cases} 2^{m+1}\int_{\Delta_i^{(m+1)}} f(y)\,dy, & \text{for } x\in\Delta_i^{(m+1)},\ 0\le i\le 2\ell+1,\\ 2^m\int_{\Delta_j^{(m)}} f(y)\,dy, & \text{for } x\in\Delta_j^{(m)},\ \ell<j<2^m,\end{cases}$$

where $\Delta_i^{(s)}=[i/2^s,(i+1)/2^s)$, $i=0,1,\dots,2^s-1$.

PROOF. Substitute the expression (10.2.14) for the Haar- Fourier coefficients of f into the right side of (10.2.13) to obtain

$$\mathcal{P}_n(x)=\int_0^1 f(y)K_n(x,y)\,dy, \tag{10.2.15}$$

where

$$K_n(x,y)=\sum_{i=0}^{n-1} h_k(x)h_k(y) \tag{10.2.16}$$

is the Haar kernel of order n. We shall isolate several properties of this kernel.

Recall that $h_0(x)=1$ for $0\le x<1$. Thus $K_1(x,y)=1$ for $0\le x,y<1$.

Notice by (10.1.2) that if $k\ge 1$ with $k=2^m+\ell$ for some $0\le\ell\le 2^m-1$ and some $m=0,1,\dots$, then the function $h_\ell^{(m)}(x)=h_k(x)$ satisfies

$$h_\ell^{(m)}(x)=\begin{cases} 2^{m/2} & \text{for } x\in\Delta_{2\ell}^{(m+1)},\\ -2^{m/2} & \text{for } x\in\Delta_{2\ell+1}^{(m+1)},\\ 0 & \text{for } x\in[0,1)\setminus\Delta_\ell^{(m+1)}.\end{cases}$$

Thus using the notation $Q_{ij}^{(m)}=\Delta_i^{(m)}\times\Delta_j^{(m)}$, we see that the function $h_1(x)h_1(y)$ equals 1 on the squares $Q_{00}^{(1)}$ and $Q_{11}^{(1)}$, and equals -1 on the squares $Q_{01}^{(1)}$ and $Q_{10}^{(1)}$. Thus the kernel $K_2(x,y)=K_1(x,y)+h_1(x)h_1(y)$ equals 2 for $(x,y)\in Q_{00}^{(1)}\bigcup Q_{11}^{(1)}$, and equals 0 for $(x,y)\in Q_{01}^{(1)}\bigcup Q_{10}^{(1)}$. In fact, it is easy to prove by induction that $K_{2^m}(x,y)=2^m$ on the squares $Q_{ii}^{(m)}$ for $i=0,1,\dots,2^m-1$ and $K_{2^m}(x,y)=0$ on the squares $Q_{ij}^{(m)}$ for $i\ne j$, $0\le i,j\le 2^m-1$, $m=0,1,\dots$.

Suppose now that $n=2^m+\ell+1$ for some $0\le\ell\le 2^m-1$ and $m=1,2,\dots$. Then by (10.2.16) we have

$$K_n(x,y)=K_{2^m}(x,y)+\sum_{k=0}^{\ell} h_k^{(m)}(x)h_k^{(m)}(y).$$

The function $h_k^{(m)}(x)h_k^{(m)}(y)$ equals 2^m on the squares $Q_{2k,2k}^{(m+1)}$, and $Q_{2k+1,2k+1}^{(m+1)}$, equals -2^m on the squares $Q_{2k,2k+1}^{(m+1)}$, and $Q_{2k+1,2k}^{(m+1)}$, and equals 0 on the remaining parts of the square $Q^0_{00} = [0,1)^2$. Since the four squares $Q^{(m+1)}_{2k,2k}$, $Q^{(m+1)}_{2k+1,2k+1}$, $Q^{(m+1)}_{2k,2k+1}$, and $Q^{(m+1)}_{2k+1,2k}$ divide the square $Q^{(m)}_{k,k}$ into four equal pieces, and since $K_{2^m}(x,y) = 2^m$ for $(x,y) \in Q^{(m)}_{k,k}$, it follows that $K_n(x,y) = 2^{m+1}$ on the squares $Q^{(m+1)}_{2k,2k}$ and $Q^{(m+1)}_{2k+1,2k+1}$, and $K_n(x,y) = 0$ on the squares $Q^{(m+1)}_{2k,2k+1}$ and $Q^{(m+1)}_{2k+1,2k}$. Consequently,

$$K_n(x,y) = \begin{cases} 2^{m+1} & \text{for } (x,y) \in Q^{(m+1)}_{2k,2k} \bigcup Q^{(m+1)}_{2k+1,2k+1}, \\ 0 & \text{for } (x,y) \in Q^{(m+1)}_{ij} \end{cases}$$

for $i \neq j$, $0 \leq i,j \leq 2\ell + 1$, and $k = 0, 1, \ldots \ell$.

Thus the kernel $K_n(x,y)$ is determined by its values on a union $\bigcup_{i,j=0}^{2\ell+1} Q^{(m+1)}_{ij}$ of squares of rank $m+1$. Moreover, since

$$\sum_{k=0}^{\ell} h^{(m)}_k(x) h^{(m)}_k(y) = 0, \qquad (x,y) \in Q^{(m)}_{jr}$$

for $\ell < j < 2^m - 1$ or $\ell < r < 2^m - 1$, we see that

$$K_n(x,y) = K_{2^m}(x,y) = 2^m, \qquad (x,y) \in Q^{(m)}_{jr}.$$

It is now easy to prove **10.2.7** using (10.2.15), (10.2.16) and the properties of the kernel $K_n(x,y)$ just established. ∎

10.2.8. *The inequality*

$$\omega_p(1/n, f) \leq \frac{C_p}{n^{1/p}} \sum_{k=1}^{n} k^{(1/p)-1} E^{(p)}_k(f), \qquad n = 1, 2, \ldots, \tag{10.2.17}$$

holds for any function $f \in \mathbf{L}^p[0,1)$, $1 \leq p < \infty$. *Moreover,*

$$0 < C_p = 8 \cdot 2^{1/p}(2^{1/p} + 4) \leq 96.$$

PROOF. As we noted in §10.1, every Walsh polynomial of order 2^m is also a Haar polynomial of order 2^m, and conversely. Let $s_{2^m}(x,f)$ denote the partial sums of order 2^m of the Walsh-Fourier series of the function f. From the definition (10.2.8) of the $\mathbf{L}^p$-modulus of continuity and Minkowski's inequality (10.2.11), we obtain

$$\begin{aligned} \omega_p(2^{-m}, f) &\leq \omega_p(2^{-m}, f - s_{2^m}) + \omega_p(2^{-m}, s_{2^m}) \\ &\leq 2\|f - s_{2^m}\|_p + \omega_p(2^{-m}, s_{2^m}) \\ &\leq 4E^{(p)}_{2^m}(f) + \omega_p(2^{-m}, s_{2^m}), \end{aligned} \tag{10.2.18}$$

where $s_{2^m} = s_{2^m}(f)$.

Notice by **10.2.7** and (10.2.6) that the partial sums of order 2^m of a Walsh-Fourier series are identical with the partial sums of order 2^m of the corresponding Haar-Fourier series. Hence

$$s_{2^m}(x) = \sum_{k=0}^{2^m-1} a_k h_k(x),$$

where $a_k = a_k(f)$ are the Haar-Fourier coefficients (10.2.14) of the function f. Let $0 < h \le 2^{-m}$, $m = 1, 2, \ldots$. Since $h_0(x) = 1$ it follows that

$$\|s_{2^m}(x+h) - s_{2^m}(x)\|_p \le \sum_{s=0}^{m-1} \| \sum_{k=2^s}^{2^{s+1}-1} a_k(h_k(x+h) - h_k(x))\|_p. \tag{10.2.19}$$

If $k = 2^s + \ell$ for some $\ell = 0, 1, \ldots, 2^s - 1$, and $s = 0, 1, \ldots$, then the term $h_k(x+h) - h_k(x)$ differs from zero on the set

$$A_k(h) = \left(\frac{\ell}{2^s} - h, \frac{\ell}{2^s}\right) \bigcup \left(\frac{2\ell+1}{2^{s+1}} - h, \frac{2\ell+1}{2^{s+1}}\right) \bigcup \left(\frac{\ell+1}{2^s} - h, \frac{\ell+1}{2^s}\right),$$

and is identically zero on the complement of the closure of the set $A_k(h)$, where this complement is taken with respect to any interval of length 1 which contains the set $A_k(h)$. Notice that the intervals which were used to define the set $A_k(h)$ are non-overlapping since $0 < h \le 2^{-m}$, $0 \le s < m$, $m = 1, 2, \ldots$. Moreover, it is obvious that if $A_k(h) \equiv A_\ell^{(s)}(h)$ then $A_\ell^{(s)}(h) \bigcap A_{\ell+2}^{(s)}(h) = \emptyset$ for $0 \le \ell < 2^m - 2$ and $|h_k(x+h) - h_k(x)| \le 2\sqrt{2^s}$ on the middle interval which defines the set $A_k(h)$ but $|h_k(x+h) - h_k(x)| \le \sqrt{2^s}$ on the other two intervals which define the set $A_k(h)$. Using the inequality

$$|a+b|^p \le 2^{p-1}(|a|^p + |b|^p), \qquad 1 \le p < \infty,$$

we obtain

$$\begin{aligned}
&\| \sum_{k=2^s}^{2^{s+1}-1} a_k(h_k(x+h) - h_k(x))\|_p \\
&\quad = \left(\int_0^1 | \sum_{\substack{k=2^s \\ k \text{ is even}}}^{2^{s+1}-1} + \sum_{\substack{k=2^s \\ k \text{ is odd}}}^{2^{s+1}-1} |^p \, dx \right)^{1/p} \\
&\quad \le 2^{1-1/p} \left(\int_0^1 \left\{ | \sum_{\substack{k=2^s \\ k \text{ is even}}}^{2^{s+1}-1} |^p + | \sum_{\substack{k=2^s \\ k \text{ is odd}}}^{2^{s+1}-1} |^p \right\} dx \right)^{1/p} \\
&\quad = 2^{1-1/p} \left(\int_0^1 \sum_{\ell=0}^{2^s-1} |a_{2^s+\ell}|^p \, |h_{2^s+\ell}(x+h) - h_{2^s+\ell}(x)|^p \, dx \right)^{1/p} \\
&\quad \le 2^{1-1/p} \left(\sum_{\ell=0}^{2^s-1} |a_{2^s+\ell}|^p (\sqrt{2^{sp}} + 2^p\sqrt{2^{sp}} + \sqrt{2^{sp}})h \right)^{1/p}.
\end{aligned}$$

Consequently, it follows from **10.2.6** and **10.2.5** that

$$\begin{aligned}
\| \sum_{k=2^s}^{2^{s+1}-1} a_k(h_k(x+h) - h_k(x))\|_p &\le 2(1+2^{p-1})^{1/p} 2^{s/p} \|s_{2^{s+1}}(f) - s_{2^s}(f)\|_p h^{1/p} \\
&\le 4 \cdot 2^{s/p} (\|s_{2^{s+1}}(f) - f\|_p + \|s_{2^s}(f) - f\|_p) h^{1/p} \\
&\le 16 \cdot 2^{s/p} h^{1/p} E_{2^s}^{(p)}(f).
\end{aligned}$$

Using this inequality in conjunction with (10.2.19) we see that

$$\|s_{2^m}(x+h) - s_{2^m}(x)\|_p \le 16 h^{1/p} \sum_{s=0}^{m-1} 2^{s/p} E_{2^s}^{(p)}(f).$$

Since $h \in (0, 2^{-m}]$ was arbitrary, we see by the definition of the $\mathbf{L}^p$-modulus of continuity (10.2.8) that

$$\omega_p(2^{-m}, s_{2^m}) \le 16 \cdot 2^{-m/p} \sum_{s=0}^{m-1} 2^{s/p} E_{2^s}^{(p)}(f).$$

Combining this estimate with (10.2.18) and the inequality $E_{2^m}^{(p)}(f) \le E_{2^m-1}^{(p)}(f)$, we obtain

$$\omega_p(2^{-m}, f) \le 4(2^{1/p} + 4) 2^{-m/p} \sum_{s=0}^{m-1} 2^{s/p} E_{2^s}^{(p)}(f).$$

Suppose that $n \geq 2$ is an integer. Write n in the form $n = 2^m + \ell$, for some $\ell = 0, 1, \ldots 2^m - 1$, $m = 1, 2, \ldots$. Then since the sequence $\{E_k^{(p)}(f)\}_{k=0}^{\infty}$ is monotone decreasing, we see that

$$\begin{aligned}\omega_p(1/n, f) &\leq \omega_p(2^{-m}, f)\\ &\leq C_p' 2^{-m/p} \sum_{s=0}^{m-1} 2^{s/p} E_{2^s}^{(p)}(f)\\ &\leq 2C_p' 2^{-m/p} \sum_{k=1}^{n} k^{(1/p)-1} E_k^{(p)}(f)\\ &\leq 2^{1+1/p} C_p' n^{-1/p} \sum_{k=1}^{n} k^{(1/p)-1} E_k^{(p)}(f),\end{aligned}$$

where $C_p' = 4(2^{1/p} + 4)$, i.e., we have proved (10.2.17) for the case $n \geq 2$.

If $n = 1$ then by definition of the $\mathbf{L}^p$-modulus of continuity and by (10.2.20) for $m = 1$, we have

$$\omega_p(1, f) \leq 2\omega_p(1/2, f) \leq 8(2^{1/p} + 4) 2^{-1/p} E_1^{(p)}(f). \quad \blacksquare$$

We notice since these norms are continuous as $p \to \infty$ that inequality (10.2.17) also holds for $p = \infty$ with a constant of 96.

10.2.9. *1) If $0 < \alpha < 1/p \leq 1$ then the conditions $E_n^{(p)}(f) = O(n^{-\alpha})$, as $n \to \infty$ and $f \in Lip\ (\alpha, p)$ are equivalent;*

2) If $0 < 1/p < \alpha \leq 1$ then the condition $E_n^{(p)}(f) = O(n^{-\alpha})$, as $n \to \infty$, implies that $f \in Lip\ (1/p, p)$;

3) If $0 < \alpha = 1/p \leq 1$ then the condition $E_n^{(p)}(f) = O(n^{-1/p})$, as $n \to \infty$, implies that $\omega_p(\delta, f) = O(\delta^{1/p} |\ln \delta|)$;

4) If the series

$$\sum_{k=1}^{\infty} k^{1/p-1} E_k^{(p)}(f) \tag{10.2.21}$$

converges for some $1 \leq p < \infty$, then $f \in Lip\ (1/p, p)$.

PROOF. Part 1) follows from **10.2.4** and **10.2.8**. Indeed, if $E_n^{(p)}(f) = O(n^{-\alpha})$, $n \to \infty$, where $0 < \alpha \leq 1$, $1 \leq p < \infty$, then by **10.2.8** we have

$$\omega_p(1/n, f) = O\left(n^{-1/p} \sum_{k=1}^{n} k^{(1/p)-\alpha-1}\right). \tag{10.2.22}$$

If $0 < \alpha < 1/p$ then $\sum_{k=1}^{n} k^{(1/p)-\alpha-1} = O(n^{(1/p)-\alpha})$, $n \to \infty$. Consequently, $\omega_p(1/n, f) = O(n^{-\alpha})$, $n \to \infty$, i.e., $\omega_p(1/n, f) \le Cn^{-\alpha}$, $n = 1, 2, \ldots$, where $C > 0$ is some constant. Let δ be any number which belongs to the interval $(0, 1]$. Choose a natural number $n = n(\delta)$ which satisfies $1/(n+1) < \delta \le 1/n$. Then

$$\omega_p(\delta, f) \le \omega_p(1/n, f) \le Cn^{-\alpha} \le C(n+1)^{-\alpha} \le 2C\delta^{\alpha}.$$

Thus $f \in \mathrm{Lip}\,(\alpha, p)$.

The fact that any $f \in \mathrm{Lip}\,(\alpha, p)$ must satisfy $E_n^{(p)}(f) = O(n^{-\alpha})$, as $n \to \infty$, follows directly from **10.2.4**. Thus the proof of 1) is complete.

If $E_n^{(p)}(f) = O(n^{-\alpha})$, as $n \to \infty$, for some $0 < 1/p < \alpha \le 1$ then as we showed above, (10.2.22) holds. Since $\alpha > 1/p$ we have $\sum_{k=1}^{n} k^{(1/p)-\alpha-1} = O(1)$, so it follows from (10.2.22) that $\omega_p(1/n, f) = O(n^{-1/p})$, $n \to \infty$. As above, this estimate implies that $f \in \mathrm{Lip}\,(1/p, p)$ and part 2) is established.

Suppose now that $E_n^{(p)}(f) = O(n^{-1/p})$, as $n \to \infty$. Use (10.2.22) for $\alpha = 1/p$. Since $\sum_{k=1}^{n} 1/k = O(\ln n)$ we see that

$$\omega_p(1/n, f) = O\left(n^{-1/p} \sum_{k=1}^{n} \frac{1}{k}\right) = O(n^{-1/p} \ln n), \qquad n \to \infty.$$

As we did in the proof of 1), we can use this relationship to verify that $\omega_p(\delta, f) = O(\delta^{1/p} |\ln \delta|)$, as $\delta \to 0+$. This completes the proof of 3).

Finally, if the series (10.2.21) converges for some $1 \le p < \infty$ then **10.2.8** implies that $\omega_p(1/n, f) = O(n^{-1/p})$, $n \to \infty$. From this it follows easily that $\omega_p(\delta, f) = O(\delta^{1/p})$, i.e., $f \in \mathrm{Lip}\,(1/p, p)$. ∎

A consequence of 4) is that if the series $\sum_{n=1}^{\infty} E_n^{(1)}(f)$ converges for some f in $\mathbf{L}[0, 1)$ then f satisfies the condition $\omega_1(\delta, f) = O(\delta)$. Thus the function f is equivalent to a function of bounded variation on [0,1), i.e., it is possible to change the values of f on a set of Lebesgue measure zero to obtain a function of bounded variation on [0,1).

Implications 2) through 4) of **10.2.9** are exact in the following sense:

10.2.10. *1) If $0 < 1/p < \alpha \le 1$ then the condition $E_n^{(p)}(f) = O(n^{-\alpha})$, as $n \to \infty$, does not necessarily imply that $\omega_p(\delta, f) = o(\delta^{1/p})$ as $\delta \to 0+$;*

2) The condition $E_n^{(p)}(f) = O(n^{-1/p})$, as $n \to \infty$, does not necessarily imply that $\omega_p(\delta, f) = o(\delta^{1/p} |\ln \delta|)$ as $\delta \to 0+$;

4) Convergence of the series (10.2.21) does not necessarily imply that $\omega_p(\delta, f) = o(\delta^{1/p})$ as $\delta \to 0+$.

PROOF. Consider the function $f_0(x) = w_1(x)$. Clearly, $E_n^{(p)}(f_0) = 0$ for $n \ge 2$, $1 \le p < \infty$, and thus $E_n^{(p)}(f_0) = O(n^{-\alpha})$, $n \to \infty$, for any $\alpha > 0$, in particular, for

$\alpha > 1/p$. But since $w_1(x) = 1$ for $0 \le x < 1/2$, $w_1(x) = -1$ for $1/2 \le x < 1$, and $w_1(x+1) = w_1(x)$, we have for any $0 < h < 1/2$ that

$$w_1(x+h) - w_1(x) = \begin{cases} -2 & \text{for} \quad x \in [1/2-h, 1/2), \\ 0 & \text{for} \quad x \notin [1/2-h, 1/2] \bigcap [0,1]. \end{cases}$$

Thus

$$\left(\int_0^1 |w_1(x+h) - w_1(x)|^p \, dx \right)^{1/p} = 2h^{1/p}.$$

Consequently, $\omega_p(\delta, f_0) = 2\delta^{1/p}$ for $0 < \delta < 1/2$, i.e., f_0 satisfies the condition $\omega_p(\delta, f_0) = O(\delta^{1/p})$ but fails to satisfy the condition $\omega_p(\delta, f_0) = o(\delta^{1/p})$, as $\delta \to 0+$. This completes the proof of 1).

Since $E_n^{(p)}(f_0) = 0$ for $n \ge 2$, it is obvious that the series $\sum_{n=1}^{\infty} n^{(1/p)-1} E_n^{(p)}(f_0)$ converges. Thus the function $f_0(x) = w_1(x)$ provides an example which verifies 3).

To prove 2), consider the function

$$f_1(x) = \sum_{n=0}^{\infty} 2^{-n/2} h_{2^n}(x).$$

This series obviously converges pointwise on the interval (0,1). We shall show that it converges in $\mathbf{L}^p[0,1)$ norm for all $p \ge 1$. Indeed, since

$$\sum_{n=0}^{\infty} 2^{-n/2} \|h_{2^n}\|_p = \sum_{n=0}^{\infty} 2^{-n/p} < \infty, \qquad 1 \le p < \infty,$$

we see that $f_1 \in \mathbf{L}^p[0,1)$ for $p \in [1, \infty)$. Moreover,

$$\begin{aligned} E_{2^m}^{(p)}(f_1) &\le \|f_1 - \sum_{n=0}^{m-1} 2^{-n/2} h_{2^n}\|_p \\ &= \|\sum_{n=m}^{\infty} 2^{-n/2} h_{2^n}\|_p \\ &\le \sum_{n=m}^{\infty} 2^{-n/2} \|h_{2^n}\|_p \\ &= \sum_{n=m}^{\infty} 2^{-n/p} = O(2^{-m/p}), \qquad m \to \infty. \end{aligned}$$

If $n = 2^m + \ell$, $\ell = 0, 1, \ldots 2^m - 1$, $m = 0, 1, \ldots$, then

$$E_n^{(p)}(f_1) \le E_{2^m}^{(p)}(f_1) = O(2^{-m/p}) = O(n^{-1/p}), \qquad m \to \infty.$$

Therefore, we have by **10.2.5** and the definition of the $\mathbf{L}^p$-modulus of continuity that

$$\begin{aligned}\omega_p(2^{-m}, f_1) &\geq \omega_p(2^{-m}, s_{2^m}(f_1)) - \omega_p(2^{-m}, f_1 - s_{2^m}(f_1))\\ &\geq \omega_p(2^{-m}, s_{2^m}(f_1)) - 2\|f_1 - s_{2^m}(f_1)\|_p\\ &\geq \omega_p(2^{-m}, s_{2^m}(f_1)) + O(2^{-m/p}), \qquad m \to \infty.\end{aligned}$$

It remains to prove that

$$\omega_p(2^{-m}, S_{2^m}(f_1)) \neq o(m2^{-m/p}), \qquad m \to \infty.$$

But this follows from the inequality

$$\begin{aligned}\omega_p(2^{-m}, S_{2^m}(f_1)) &\geq \left(\int_{1-2^{-m}}^{1} \left| \sum_{n=0}^{m-1} 2^{-n/2}(h_{2^n}(x) - h_{2^n}(x+2^{-m}))\right|^p dx\right)^{1/p}\\ &= \left(\int_{1-2^{-m}}^{1} \left| \sum_{n=0}^{m-1} 2^{-n/2} h_{2^n}(x+2^{-m})\right|^p dx\right)^{1/p} = m2^{-m/p}. \blacksquare\end{aligned}$$

§10.3. Connections between best approximations and integrability conditions.

We search for conditions on the sequence $\{E_n^{(p)}(f)\}$ of best approximations to a given function $f \in \mathbf{L}^p[0,1)$, $1 \leq p < \infty$, which will guarantee that the function f belongs to the space $\mathbf{L}^q[0,1)$ for some $q > p$. We shall also derive estimates of best approximations in the $\mathbf{L}^q[0,1)$ norm in this situation. Recall that the space $\mathbf{L}^\infty[0,1)$ consists of measurable functions which are essentially bounded on [0,1). A norm for this space is given by the *essential supremum of the function f on the interval* [0,1),

$$\|f\|_\infty = \operatorname*{ess\,sup}_{0 \leq x < 1} |f(x)| < \infty,$$

which equals the smallest constant $C \geq 0$ for which the inequality $|f(x)| \leq C$ holds almost everywhere on [0,1).

For each function $f \in \mathbf{L}[0,1)$, let $\mathcal{P}_n(x,f) \equiv \mathcal{P}_n(x)$, $n = 1, 2, \ldots$, represent the partial sum of order n of the Haar-Fourier series of f.

10.3.1. *If $n = 2^m + \ell$ for some $\ell = 0, 1, \ldots 2^m - 1$, and some $m = 0, 1, \ldots$, then*

$$\|f - \mathcal{P}_n(f)\|_p \leq 2^{1/p}\omega_p(2^{-m}, f)$$

holds for all functions $f \in \mathbf{L}^p[0,1)$, and all $1 \leq p < \infty$.

PROOF. This result has already been proved in the case $n = 2^m$ (see **10.2.2**). For arbitrary n, the proof can be based on **10.2.7** exactly as **10.2.2** was based on (10.2.6). The details are left to the reader. $\blacksquare$

10.3.2. *If $f \in \mathbf{L}^p[0,1)$ for some $1 \le p < \infty$ then*

$$\|\mathcal{P}_n(f)\|_p \le \|f\|_p.$$

PROOF. In the case $n = 2^m$ this result reduces to **10.2.1**, since $\mathcal{P}_{2^m}(f) = s_{2^m}(f)$, where $s_{2^m}(f)$ represents the partial sum of order 2^m of the Walsh-Fourier series of the function f. For arbitrary n, the proof can be based on **10.2.7** exactly as **10.2.1** was based on (10.2.6). The details are left to the reader. ∎

Proposition **10.3.2** (and **10.3.4** below) can also be obtained from the following facts. The sequence $\{\mathcal{P}_n(f)\}$ is a martingale (see [20]), $\{\|\mathcal{P}_n(f)\|_p\}$ is increasing for any $p \ge 1$, and $\lim_{n\to\infty} \|\mathcal{P}_n(f)\|_p = \|f\|_p$ since the Haar system is a basis in the space $\mathbf{L}^p[0,1)$.

10.3.3. *If $f \in \mathbf{L}^p[0,1)$ for some $1 \le p < \infty$ then*

$$E_n^{(p)}(f)_h \le \|f - \mathcal{P}_n(f)\|_p \le 2E_n^{(p)}(f)_h, \qquad n = 1, 2, \ldots.$$

PROOF. In the case $n = 2^m$ this result reduces to **10.2.5**. For arbitrary n, the proof can be based on **10.3.2** exactly as **10.2.5** was based on **10.2.1**. ∎

10.3.4. *The inequality*

$$\|\sum_{k=n+1}^{n+m} a_k h_k\|_p \le 2\|\sum_{k=0}^{n+m} a_k h_k\|_p \tag{10.3.1}$$

holds for any real numbers a_k and any $1 \le p \le \infty$.

PROOF. Set $f(x) = \sum_{k=0}^{n+m} a_k h_k(x)$ and notice that the numbers a_k are the Haar-Fourier coefficients of the function f. Thus **10.3.2** implies

$$\|\sum_{k=0}^{n} a_k h_k\|_p \le \|\sum_{k=0}^{n+m} a_k h_k\|_p, \qquad 1 \le p < \infty.$$

We continue this estimate by Minkowski's inequality (10.2.11) and obtain

$$\|\sum_{k=0}^{n+m} a_k h_k\|_p \ge \|\sum_{k=n+1}^{n+m} a_k h_k\|_p - \|\sum_{k=0}^{n} a_k h_k\|_p \ge \|\sum_{k=n+1}^{n+m} a_k h_k\|_p - \|\sum_{k=0}^{n+m} a_k h_k\|_p.$$

This proves inequality (10.3.1) in the case when $p \ne \infty$. But $\|f\|_\infty = \lim_{p\to\infty} \|f\|_p$, so the case $p = \infty$ can be obtained from (10.3.1) by taking the limit as $p \to \infty$. ∎

10.3.5. *If $1 \le p < \infty$ and $m = 0, 1, \dots,$ then*

(10.3.2)

$$(4\sqrt{2})^{-1}\, 2^{m(\frac{1}{2}-\frac{1}{p})} \left(\sum_{n=2^m}^{2^{m+2}-1} |a_n|^p \right)^{1/p}$$

$$\le \| \sum_{n=2^m}^{2^{m+2}-1} a_n h_n \|_p \le 2\sqrt{2} \cdot 2^{m(\frac{1}{2}-\frac{1}{p})} \left(\sum_{n=2^m}^{2^{m+2}-1} |a_n|^p \right)^{1/p}$$

holds for any choice of real numbers a_n.

PROOF. From

$$|a+b|^p \le 2^{p-1}(|a|^p + |b|^p), \qquad 1 \le p < \infty,$$

and **10.2.6** we obtain

$$\| \sum_{n=2^m}^{2^{m+2}-1} a_n h_n \|_p$$

$$\le 2^{1-\frac{1}{p}} \left(\int_0^1 | \sum_{n=2^m}^{2^{m+1}-1} a_n h_n(x) |^p \, dx + \int_0^1 | \sum_{n=2^{m+1}}^{2^{m+2}-1} a_n h_n(x) |^p \, dx \right)^{1/p}$$

$$= 2^{1-\frac{1}{p}} \left(2^{m(\frac{p}{2}-1)} \sum_{n=2^m}^{2^{m+1}-1} |a_n|^p + 2^{(m+1)(\frac{p}{2}-1)} \sum_{n=2^{m+1}}^{2^{m+2}-1} |a_n|^p \right)^{1/p}$$

$$\le 2^{1-\frac{1}{p}} \sqrt{2} \cdot 2^{m(\frac{1}{2}-\frac{1}{p})} \left(\sum_{n=2^m}^{2^{m+1}-1} |a_n|^p \right)^{1/p}.$$

Since $2^{-1/p} < 1$ for any $1 \le p < \infty$, the proof of the right hand inequality in (10.3.2) is established.

To prove the right hand inequality in (10.3.2), let

$$\Sigma_m(x) = \sum_{n=2^m}^{2^{m+1}-1} a_n h_n.$$

By **10.3.4** we have

$$\max\{\|\Sigma_m(x)\|_p, \|\Sigma_{m+1}(x)\|_p\} \le 2 \| \sum_{n=2^m}^{2^{m+2}-1} a_n h_n \|_p. \tag{10.3.3}$$

Use Proposition **10.2.6** and then inequality (10.3.2). Thus verify

$$\left(\sum_{n=2^m}^{2^{m+2}-1} |a_n|^p\right)^{1/p} =$$

$$= \left(2^{m(1-\frac{1}{p})}\int_0^1 \mid \Sigma_m(x)\mid^p dx + 2^{(m+1)(1-\frac{p}{2})}\int_0^1 \mid \Sigma_{m+1}(x)\mid^p dx\right)^{1/p}$$

$$\leq \max\{1, 2^{\frac{1}{p}-\frac{1}{2}}\}\cdot 2^{m(\frac{1}{p}-\frac{1}{2})}\cdot 2^{1+\frac{1}{p}}\|\sum_{n=2^m}^{2^{m+2}-1} a_n h_n\|_p.$$

Since $1 \leq p < \infty$, this completes the proof of the left hand inequality of (10.3.2). ∎

10.3.6. *Suppose $f \in \mathbf{L}^p[0,1)$ for some $1 \leq p < \infty$ and $n = 2^m + k$ for some $k = 0, 1, \ldots, 2^m - 1$, $m = 0, 1, \ldots$. Then*

$$|a_n(f)| \leq 4\cdot 2^{m(\frac{1}{p}-\frac{1}{2})}E_n^{(p)}(f)_h,$$

where $a_n(f)$ are the Haar-Fourier coefficients of the function f.

PROOF. Indeed, by **10.3.4** and **10.3.3** we have

$$|a_n(f)|\,\|h_n\|_p = \|a_n h_n\|_p = \|\mathcal{P}_{n+1}(f) - \mathcal{P}_n(f)\|_p$$
$$\leq \|\mathcal{P}_n(f) - f\|_p + \|\mathcal{P}_{n+1}(f) - f\|_p \leq 4E_n^{(p)}(f)_h.$$

Since $\|h_n\|_p = 2^{m(\frac{1}{p}-\frac{1}{2})}$, the promised inequality is established. ∎

We come now the main result of this section:

10.3.7. *Let $f \in \mathbf{L}^p[0,1)$, $1 \leq p < q \leq \infty$ and suppose the series*

$$\sum_{n=1}^{\infty} E_n^{(p)}(f)_h n^{\frac{1}{p}-\frac{1}{q}-1} \tag{10.3.4}$$

converges. Then $f \in \mathbf{L}^q[0,1)$ and

$$E_n^{(q)}(f)_h \leq 128\left(E_n^{(p)}(f)_h n^{\frac{1}{p}-\frac{1}{q}} + \sum_{k=n+1}^{\infty} E_k^{(p)}(f)_h k^{\frac{1}{p}-\frac{1}{q}-1}\right) \tag{10.3.5}$$

for $n = 1, 2, \ldots$, where $E_n^{(p)}(f)_h$ is the best approximation in the $\mathbf{L}^p$ norm to the function f by Haar polynomials of order no greater than n.

PROOF. Since $f \in \mathbf{L}^p[0,1)$, we have by **10.3.1** that the Haar-Fourier series of the function f converges in $\mathbf{L}^p$ norm, in fact, $\|f - \mathcal{P}_n(f)\|_p \to 0$ as $n \to \infty$. Consequently, for all n the inequality

$$f \overset{\mathbf{L}^p}{=} \mathcal{P}_n(f) + \sum_{k=0}^{\infty}(\mathcal{P}_{n2^{k+1}}(f) - \mathcal{P}_{n2^k}(f))$$

holds, where equality is understood in the sense that the series on the right side converges to f in the $\mathbf{L}^p[0,1)$ norm.

Let $n = 2^m + r$, $r = 0, 1, \ldots, 2^m - 1$, $m = 0, 1, \ldots$. Then $2^{k+m} \le n2^k$, and $n2^{k+1} < 2^{k+m+2}$. Since $1 \le p < q < \infty$, it follows from **10.3.5**, **10.3.6**, and **10.3.3** that

(10.3.7)

$$\|\mathcal{P}_{n2^{k+1}}(f) - \mathcal{P}_{n2^k}(f)\|_p \le 2^{(k+m)(\frac{1}{2}-\frac{1}{q})} \cdot 2\sqrt{2} \left(\sum_{s=n2^k}^{n2^{k+1}-1} |a_s|^q \right)^{1/q}$$

$$\le 2\sqrt{2} \cdot 2^{(k+m)(\frac{1}{2}-\frac{1}{q})} \left[\left(\sum_{s=n2^k}^{n2^{k+1}-1} |a_s|^p \right)^{1/p} \right]^{p/q} \left(\max_{n2^k \le s < n2^{k+1}} |a_s| \right)^{1-\frac{p}{q}}$$

$$\le 2\sqrt{2} \cdot 2^{(k+m)(\frac{1}{2}-\frac{1}{q})} \left[4\sqrt{2} \cdot 2^{(k+m)(\frac{1}{p}-\frac{1}{2})} \|\mathcal{P}_{n2^{k+1}}(f) - \mathcal{P}_{n2^k}(f)\|_p \right]^{p/q} \times$$

$$\times \left[\max\left(1, 2^{\frac{1}{p}-\frac{1}{2}}\right) \cdot 2^{(k+m)(\frac{1}{p}-\frac{1}{2})} \cdot 4E^{(p)}_{n2^k}(f)_h \right]^{1-\frac{p}{q}}$$

$$\le 2\sqrt{2} \cdot 2^{(k+m)(\frac{1}{p}-\frac{1}{q})} \left[16\sqrt{2} E^{(p)}_{n2^k}(f)_h \right]^{p/q} \left[4\sqrt{2} E^{(p)}_{n2^k}(f)_h \right]^{1-\frac{p}{q}}$$

$$\le 64 (n2^k)^{\frac{1}{p}-\frac{1}{q}} E^{(p)}_{n2^k}(f)_h.$$

Let $q \to \infty$ and verify that this estimate also holds for $1 \le p < q = \infty$. Thus we obtain by (10.3.6) and (10.3.7) that

$$E^{(q)}_n(f)_h \le \|f - \mathcal{P}_n(f)\|_q$$
$$\le \sum_{k=0}^{\infty} \|\mathcal{P}_{n2^{k+1}}(f) - \mathcal{P}_{n2^k}(f)\|_q$$
$$\le 64 \sum_{k=0}^{\infty} (n2^k)^{\frac{1}{p}-\frac{1}{q}} E^{(p)}_{n2^k}(f)_h.$$

holds for any $1 \le p < q \le \infty$. Since

$$\sum_{k=0}^{\infty} (n2^k)^{\frac{1}{p}-\frac{1}{q}} E^{(p)}_{n2^k}(f)_h \le n^{\frac{1}{p}-\frac{1}{q}} E^{(p)}_n(f)_h + 2 \sum_{k=n+1}^{\infty} E^{(p)}_k(f)_h k^{\frac{1}{p}-\frac{1}{q}-1},$$

we conclude by the previous inequality that (10.3.5) holds as promised. ∎

The following result[1] is a corollary of the previous one.

[1]This result is an analogue of a result of A.A. Konjuškov and S.B. Stečkin for approximation by trigonometric polynomials.

10.3.8. *Let $f \in \mathbf{L}^p[0,1)$, $1 \le p < q \le \infty$ and suppose the series*

$$\sum_{k=1}^{\infty} E_k^{(p)}(f) k^{\frac{1}{p}-\frac{1}{q}-1} \tag{10.3.8}$$

converges. Then $f \in \mathbf{L}^q[0,1)$ and

$$E_{2^n}^{(q)}(f) \le 128 \left(E_{2^n}^{(p)}(f) 2^{n(\frac{1}{p}-\frac{1}{q})} + \sum_{k=2^n+1}^{\infty} E_k^{(p)}(f) k^{\frac{1}{p}-\frac{1}{q}-1} \right) \tag{10.3.9}$$

for $n = 1, 2, \dots$, where $E_n^{(p)}(f)$ is the best approximation in the $\mathbf{L}^p$ norm to the function f by Walsh polynomials of order no greater than n.

PROOF. It is easy to see that the series (10.3.8) is equiconvergent with the series

$$\sum_{n=0}^{\infty} E_{2^n}^{(p)}(f) 2^{n(\frac{1}{p}-\frac{1}{q})}$$

and thus with the series

$$\sum_{n=0}^{\infty} E_{2^n}^{(p)}(f)_h 2^{n(\frac{1}{p}-\frac{1}{q})},$$

since $E_{2^n}^{(p)}(f) = E_{2^n}^{(p)}(f)_h$ for $n = 0, 1, \dots$. But this last series is equiconvergent with the series (10.3.4). Consequently, if (10.3.8) converges then so does (10.3.4). Thus it follows from **10.3.7** that $f \in \mathbf{L}^q[0,1)$ and (10.3.5) holds. If we let $n = 2^m$ and use the fact that the sequence $E_k^{(p)}(f)_h$ is monotone, we obtain

$$E_{2^m}^{(q)}(f)_h \le 128 \left(E_{2^m}^{(p)}(f)_h 2^{m(\frac{1}{p}-\frac{1}{q})} + \sum_{s=m}^{\infty} 2^{s(\frac{1}{p}-\frac{1}{q})} E_{2^s}^{(p)}(f)_h \right).$$

Since $E_{2^m}^{(p)}(f)_h = E_{2^m}^{(p)}(f)$, the proof of (10.3.9) is complete. ∎

We notice for $f \in \mathbf{L}^p[0,1)$, $1 \le p < q < \infty$, that the inequalities

$$E_n^{(q)}(f)_h \le C_1(p,q) \left(E_n^{(p)}(f)_h n^{\frac{1}{p}-\frac{1}{q}} + \sum_{k=n+1}^{\infty} E_k^{(p)}(f)_h k^{\frac{1}{p}-\frac{1}{q}-1} \right)$$

and

$$E_n^{(q)}(f) \le C_1(p,q) \left(E_n^{(p)}(f) n^{\frac{1}{p}-\frac{1}{q}} + \sum_{k=n+1}^{\infty} E_k^{(p)}(f) k^{\frac{1}{p}-\frac{1}{q}-1} \right)$$

where the constants $C_i(p,q)$ depend on p and q, will be derived by another method in the following section as corollaries of a more precise theorem.

A corollary of **10.3.7** deserves special mention:

10.3.9. *If $f \in \mathbf{L}^p[0,1)$, $1 \le p < \infty$ and the series*

$$\sum_{n=1}^{\infty} n^{\frac{1}{p}-1} E_n^{(p)}(f)_h$$

converges, then $f \in \mathbf{L}^\infty[0,1)$ and

$$E_n^{(\infty)}(f)_h \le 128 \left(n^{1/p} E_n^{(p)}(f)_h + \sum_{k=n+1}^{\infty} k^{\frac{1}{p}-1} E_k^{(p)}(f)_h \right)$$

This result together with **10.1.2** imply

10.3.10. *If a function f is continuous on $[0,1]$ then*

$$\omega(1/n, f) \le 768 \left(n^{1/p} E_n^{(p)}(f)_h + \sum_{k=n+1}^{\infty} k^{\frac{1}{p}-1} E_k^{(p)}(f)_h \right)$$

for all $1 \le p < \infty$ and $n = 1, 2, \ldots$.

The following result is a special case of **10.3.8**:

10.3.11. *If a function f is continuous on $[0,1]$ then*

$$E_{2^m}^{(\infty)}(f) \le 128 \left(2^{m/p} E_{2^m}^{(p)}(f) + \sum_{s=m}^{\infty} 2^{s/p} E_{2^s}^{(p)}(f) \right)$$

for all $1 \le p < \infty$ and $m = 0, 1, \ldots$.

§10.4. Connections between best approximations and integrability conditions (continued).

In this section we shall prove more precise results of the type illustrated in **10.3.7** and **10.3.8**. Under the hypotheses $1 \le p < q < \infty$, these results will contain Theorems **10.3.7** and **10.3.8** as corollaries, but the constants on the right sides of the corresponding inequalities will depend on p and q. For the proofs of these results we shall need several preliminary observations.

The modulus of continuity of a function $f \in \mathbf{L}^p[0,1)$, $1 \le p < \infty$ will be defined somewhat differently than in the previous sections, namely, we shall set

$$\omega_p^*(\delta, f) = \sup_{0 \le h \le \delta} \left(\int_0^{1-h} |f(x+h) - f(x)|^p \, dx \right)^{1/p}, \qquad 0 \le \delta \le 1. \tag{10.4.1}$$

Notice that this $\mathbf{L}^p$-modulus of continuity depends only on the values f takes on the interval [0,1), in contrast to the $\mathbf{L}^p$-modulus of continuity given by (10.2.8) which was defined for periodic functions of period 1.

10.4.1. *If $f \in \mathbf{L}^p[0,1)$, $1 \le p < \infty$, then*

$$I_p \equiv \int_a^b \int_a^b |f(x) - f(y)|^p \, dx \, dy \le 2 \int_0^{b-a} |\omega_p^*(t,f)|^p \, dt. \tag{10.4.2}$$

for any $0 \le a < b \le 1$.

PROOF. By changing variables, reversing the order of integration using Fubini's Theorem, and applying definition (10.4.1), we obtain

$$\begin{aligned}
I_p &= \int_a^b \left(\int_a^b |f(x) - f(y)|^p \, dx \right) dy \\
&= \int_a^b \left(\int_{a-y}^{b-y} |f(u+y) - f(y)|^p \, du \right) dy \\
&= \int_0^{b-a} \left(\int_a^{b-u} |f(u+y) - f(y)|^p \, dy \right) du \\
&\quad + \int_{a-b}^0 \left(\int_{a-u}^b |f(u+y) - f(y)|^p \, dy \right) du \\
&= \int_0^{b-a} \left(\int_a^{b-u} |f(u+y) - f(y)|^p \, dy \right) du \\
&\quad + \int_0^{b-a} \left(\int_{a+v}^b |f(y-v) - f(y)|^p \, dy \right) dv \\
&= \int_0^{b-a} \left(\int_a^{b-u} |f(u+y) - f(y)|^p \, dy \right) du \\
&\quad + \int_0^{b-a} \left(\int_a^{b-v} |f(t) - f(t+v)|^p \, dt \right) dv \\
&= 2 \int_0^{b-a} \left(\int_a^{b-t} |f(t+y) - f(y)|^p \, dy \right) dt \\
&\le 2 \int_0^{b-a} |\omega_p^*(t,f)|^p \, dt. \quad \blacksquare
\end{aligned}$$

10.4.2. *If $f \in \mathbf{L}^p[0,1)$, $1 \le p < \infty$, and*

$$\phi_h(t) = \int_t^{t+h} f(x) \, dx \qquad 0 \le t \le 1-h, \quad 0 \le h < 1,$$

then

$$\int_0^{1-h} |f(t) - \phi_h(t)|^p \, dt \le (\omega_p^*(h,f))^p. \tag{10.4.3}$$

PROOF. First suppose that $1 < p < \infty$. By Hölder's inequality (10.2.7) we have

$$\begin{aligned}
\int_0^{1-h} |f(t) - \phi_h(t)|^p \, dt &\le \int_0^{1-h} \Big| \frac{1}{h} \int_0^h [f(t) - f(t+u)] \, du \Big|^p \, dt \\
&\le h^{-p} \int_0^{1-h} \left(\int_0^h |f(t) - f(t+u)|^p \, du \right) dt \cdot h^{(1-\frac{1}{p})p} \\
&= \frac{1}{h} \int_0^h \left(\int_0^{1-h} |f(t) - f(t+u)|^p \, dt \right) du \\
&\le \frac{1}{h} \int_0^h \left(\int_0^{1-u} |f(t) - f(t+u)|^p \, dt \right) du \\
&\le \frac{1}{h} \int_0^h (\omega_p^*(u,f))^p \, du \le (\omega_p^*(h,f))^p.
\end{aligned}$$

This verifies inequality (10.4.3) for $1 < p < \infty$. If $p = 1$ then the proof is similar, but Hölder's inequality is not needed. ∎

10.4.3. *If* $f \in \mathbf{L}^p[0,1)$, $1 \le p < \infty$, *and*

$$\psi_h(t) = n \int_{k/n}^{(k+1)/n} f(u) \, du, \qquad k/n \le t \le (k+1)/n, \quad k = 0, \ldots, n-1,$$

then

$$I_n^{(p)} \equiv \int_0^{1-1/n} |\phi_{1/n}(t) - \psi_n(t)|^p \, dt \le (\omega_p^*(1/n, f))^p \tag{10.4.4}$$

for $n = 1, 2, \ldots$.

PROOF. For the case $n = 1$, (10.4.4) is obvious since in this case the left side reduces to zero and the right side is evidently non-negative. Suppose now that

$n \geq 2$ and $1 < p < \infty$. By Hölder's inequality we obtain

$$
\begin{aligned}
I_n^{(p)} &= n^p \sum_{k=0}^{n-2} \int_{k/n}^{(k+1)/n} \left| \int_t^{t+1/n} f(u)\,du - \int_{k/n}^{(k+1)/n} f(u)\,du \right|^p dt \\
&= n^p \sum_{k=0}^{n-2} \int_{k/n}^{(k+1)/n} \left| \int_{(k+1)/n}^{t+1/n} f(u)\,du - \int_{k/n}^{t} f(u)\,du \right|^p dt \\
&\leq n^p \sum_{k=0}^{n-2} \int_{k/n}^{(k+1)/n} \left(\int_{k/n}^{t} |f(u+\frac{1}{n}) - f(u)|\,du \right)^p dt \\
&\leq n^p \sum_{k=0}^{n-2} \int_{k/n}^{(k+1)/n} \left(\int_{k/n}^{(k+1)/n} |f(u+\frac{1}{n}) - f(u)|\,du \right)^p dt \cdot n^{(\frac{1}{p}-1)p} \\
&= \int_0^{1-\frac{1}{n}} |f(u+\frac{1}{n}) - f(u)|^p\,du \leq (\omega_p^*(1/n, f))^p,
\end{aligned}
$$

i.e., (10.4.4) holds for $1 < p < \infty$. If $p = 1$ then the proof is similar, but Hölder's inequality is not needed. ∎

10.4.4. *If $f \in \mathbf{L}^p[0,1)$, $1 \leq p < \infty$, then*

$$
J_n^{(p)} \equiv \left(\int_0^1 |f(t) - \psi_n(t)|^p\,dt \right)^{1/p} \leq 4\omega_p^*(\frac{1}{n}, f), \qquad n = 1, 2, \ldots. \tag{10.4.5}
$$

PROOF. Using the inequality $|a+b|^{1/p} \leq |a|^{1/p} + |b|^{1/p}$, $1 \leq p < \infty$, we see that

$$
J_n^{(p)} \leq \left(\int_0^{1-\frac{1}{n}} |f(t) - \psi_n(t)|^p\,dt \right)^{1/p} + \left(\int_{1-\frac{1}{n}}^1 |f(t) - \psi_n(t)|^p\,dt \right)^{1/p}.
$$

Applying the triangle inequality (10.2.11) to the first term on the right side of this last inequality, we obtain

$$
\begin{aligned}
J_n^{(p)} \leq &\left(\int_0^{1-\frac{1}{n}} |f(t) - \phi_{1/n}(t)|^p\,dt \right)^{1/p} + \left(\int_0^{1-\frac{1}{n}} |\phi_{1/n}(t) - \psi_n(t)|^p\,dt \right)^{1/p} \\
&+ \left(\int_{1-\frac{1}{n}}^1 |f(t) - \psi_n(t)|^p\,dt \right)^{1/p}.
\end{aligned}
$$

Estimate the first two terms on the right side by using (10.4.3) and (10.4.4):

$$
\begin{aligned}
J_n^{(p)} &\leq 2\omega_p^*(\frac{1}{n}, f) + n \left(\int_{1-\frac{1}{n}}^1 \left| \int_{1-\frac{1}{n}}^1 (f(t) - f(x))\,dx \right|^p dt \right)^{1/p} \\
&\equiv 2\omega_p^*(\frac{1}{n}, f) + I_n^{(p)}.
\end{aligned}
$$

But by Hölder's inequality (10.2.7) and estimate (10.4.2), we obtain

$$I_n^{(p)} \le n\left(\int_{1-\frac{1}{n}}^{1} n^{1-p}\left[\int_{1-\frac{1}{n}}^{1} |f(t)-f(x)|^p\,dx\right]dt\right)^{1/p}$$
$$\le 2^{1/p}n^{1/p}\left(\int_0^{1/n}[\omega_p^*(\delta,f)]^p\,d\delta\right)^{1/p} \le 2\omega_p^*(\frac{1}{n},f).$$

This together with the preceding inequality imply (10.4.5). ∎

10.4.5. *If $f \in \mathbf{L}^p[0,1)$, $1 \le p < \infty$, then*

$$\omega_p^*(\delta,|f|) \le \omega_p^*(\delta,f), \qquad 0 \le \delta \le 1. \tag{10.4.6}$$

PROOF. This inequality follows directly from the definition (10.4.1) of the modulus of continuity and the inequality $||a|-|b|| \le |a-b|$. ∎

10.4.6. *Let $f \in \mathbf{L}^p[0,1)$ and $1 \le p < \infty$.*
1) If $p=1$ then for $n=1,2,\ldots$ we have

$$\omega_1^*(\frac{1}{n},f) \ge \frac{1}{17}\left(\sup_{\substack{E\subset[0,1]\\|E|\le 1/n}}\int_E |f(t)|\,dt - \inf_{\substack{E\subset[0,1]\\|E|\ge 1/n}}\int_E |f(t)|\,dt\right); \tag{10.4.7}$$

2) If $1<p<\infty$ then for $n=2,3,\ldots$ we have

(10.4.8)
$$\omega_p^*(\frac{1}{n},f) \ge \frac{n^{1-1/p}}{17}\sup_{\substack{E\subset[0,1]\\|E|=1/n}}\left(\int_E |f(t)|\,dt - \sup_{\substack{E_1\subset[0,1]\setminus E\\|E_1|=1/n}}\int_{E_1}|f(t)|\,dt\right)$$
$$\equiv \frac{n^{1-1/p}}{17}F_n.$$

PROOF. Suppose first that $p=1$. Inequality (10.4.7) is obvious for $n=1$ since the right side is identically zero and the left side is non-negative. For $n \ge 2$ set

$$\psi(t) \equiv \psi_n(t) = n\int_{k/n}^{(k+1)/n} |f(u)|\,du = c_k, \tag{10.4.9}$$

$$k/n \le t < (k+1)/n, \quad k=0,1,\ldots n-1.$$

It is clear that

$$\begin{aligned}\omega_1^*(1/n,\psi) &\geq \int_0^{1-\frac{1}{n}} \left| \psi(t+\frac{1}{n}) - \psi(t) \right| dt \\ &= \sum_{k=0}^{n-2} \int_{k/n}^{(k+1)/n} \left| \psi(t+\frac{1}{n}) - \psi(t) \right| dt \\ &= \frac{1}{n} \sum_{k=0}^{n-2} |c_{k+1} - c_k| \geq \frac{1}{n} \left(\max_{0\leq i\leq n-1} c_i - \min_{0\leq i\leq n-1} c_i \right).\end{aligned}$$

But since

$$\max_{0\leq i\leq n-1} c_i = n \sup_{\substack{E\subset[0,1]\\ |E|\leq 1/n}} \int_E |\psi(t)|\,dt,$$

and

$$\min_{0\leq i\leq n-1} c_i = n \inf_{\substack{E\subset[0,1]\\ |E|\geq 1/n}} \int_E |\psi(t)|\,dt,$$

it follows that

$$\omega_1^*(\frac{1}{n},\psi) \geq \sup_{|E|\leq 1/n} \int_E \psi(t)\,dt - \inf_{|E|\geq 1/n} \int_E \psi(t)\,dt. \tag{10.4.10}$$

Consequently, we have by **10.4.4**, **10.4.5**, and (10.4.10) that

$$\begin{aligned}&\omega_1^*(\frac{1}{n},f) \geq \omega_1^*(\frac{1}{n},|f|) \\ &= \sup_{0\leq h\leq 1/n} \int_0^{1-h} \big|\,|f(t+h)| - |f(t)|\,\big|\,dt \\ &= \sup_{0\leq h\leq 1/n} \int_0^{1-h} \big|\,|f(t+h)| - \psi(t+h) + \psi(t+h) - \psi(t) + \psi(t) - |f(t)|\,\big|\,dt \\ &\geq \sup_{0\leq h\leq 1/n} \int_0^{1-h} |\psi(t+h) - \psi(t)|\,dt - 2\int_0^1 \big|\,|f(t)| - \psi(t)\big|\,dt \\ &\geq \sup_{|E|\leq 1/n} \int_E \psi(t)\,dt - \inf_{|E|\geq 1/n} \int_E \psi(t)\,dt - 8\omega_1^*(\frac{1}{n},|f|)\end{aligned}$$

$$= \sup_{|E|\le 1/n} \int_E [\psi(t) - |f(t)| + |f(t)|]\, dt$$
$$- \inf_{|E|\ge 1/n} \int_E [\psi(t) - |f(t)| + |f(t)|]\, dt - 8\omega_1^*(\frac{1}{n}, |f|)$$
$$\ge \sup_{|E|\le 1/n} \int_E |f(t)|\, dt - \inf_{|E|\ge 1/n} \int_E |f(t)|\, dt$$
$$- 2\int_0^1 ||f(t)| - \psi(t)|\, dt - 8\omega_1^*(\frac{1}{n}, |f|)$$
$$\ge \sup_{|E|\le 1/n} \int_E |f(t)|\, dt - \inf_{|E|\ge 1/n} \int_E |f(t)|\, dt - 16\omega_1^*(\frac{1}{n}, |f|).$$

This proves (10.4.7).

Suppose now that $1 < p < \infty$ and ψ is defined by (10.4.9). Let $0 \le d_0 \le d_1 \le \cdots \le d_{n-1}$ represent the sequence $\{c_k\}$, $k = 0, 1, \ldots, n-1$ arranged in increasing order. Then

$$\omega_p^*(1/n, \psi) \ge \left(\int_0^{1-\frac{1}{n}} \left| \psi(t + \frac{1}{n}) - \psi(t) \right|^p dt\right)^{1/p}$$
$$\ge \left(\sum_{k=0}^{n-2} |c_{k+1} - c_k|^p \frac{1}{n}\right)^{1/p}$$
$$\ge n^{-1/p}(d_{n-1} - d_{n-2}).$$

Two cases are possible: a) $d_{n-1} = d_{n-2}$; b) $d_{n-1} > d_{n-2}$. In case a), the function ψ achieves its maximum value on a set of points $A \subset [0,1]$ of measure $|A| \ge 2/n$, and thus the right side of (10.4.8) is identically zero, i.e., we have derived the inequality:

$$\omega_p^*(1/n, \psi) \ge n^{1-1/p} \sup_{\substack{E\subset[0,1] \\ |E|=1/n}} \left(\int_E \psi(t)\, dt - \sup_{\substack{E_1\subset[0,1]\setminus E \\ |E_1|=1/n}} \int_{E_1} \psi(t)\, dt \right). \tag{10.4.11}$$

In case b), set $B = \{t : t \in [0,1],\ \psi(t) = d_{n-1}\}$. Evidently,

(10.4.13)
$$\omega_p^*(1/n, \psi) \ge n^{-1/p}(d_{n-1} - d_{n-2})$$
$$= n^{1-1/p}\left(\int_B \psi(t)\, dt - \frac{1}{n} d_{n-2}\right) \equiv n^{1-1/p} D,$$

where $D \geq 0$. But

$$D \geq \int_E \psi(t)\,dt - \inf_{\substack{E_1 \subset [0,1]\setminus E \\ |E_1|=1/n}} \int_{E_1} \psi(t)\,dt \equiv D_1 \tag{10.4.13}$$

holds for any set $E \subset [0,1]$ of Lebesgue measure $|E| = 1/n$. Indeed, if $|E \bigcap B| = 0$ then $D_1 < 0$ and thus $D > D_1$. If $|E \setminus B| + |B \setminus E| = 0$ then $D = D_1$. And if $|E \bigcap B| > 0$, $|E \setminus B| + |B \setminus E| > 0$ then $D > D_1$. Thus (10.4.13) holds as promised. In particular, we see by (10.4.12) that (10.4.11) also holds in case b).

We have shown that given any function ψ of the form (10.4.9), inequality (10.4.11) holds. Consequently, proceeding as we did for the case $p = 1$, using the triangle inequality for $\mathbf{L}^p$ norms and Hölder's inequality together with (10.4.6), (10.4.11), and (10.4.5), we obtain

$$\begin{aligned}
\omega_p^*(1/n, f) &\geq \omega_p^*(1/n, |f|) = \sup_{0 \leq h \leq 1/n} \left(\int_0^{1-h} \big|\,|f(t+h)| - |f(t)|\,\big|^p\,dt \right)^{1/p} \\
&\geq \sup_{0 \leq h \leq 1/n} \left(\int_0^{1-h} \big|\,|\psi(t+h)| - |\psi(t)|\,\big|^p\,dt \right)^{1/p} \\
&\qquad - 2\left(\int_0^1 \big|\,|f(t)| - |\psi(t)|\,\big|^p\,dt \right)^{1/p} \\
&\geq n^{1-\frac{1}{p}} \sup_{\substack{E \subset [0,1] \\ |E|=1/n}} \left(\int_E \psi(t)\,dt - \sup_{\substack{E_1 \subset [0,1]\setminus E \\ |E_1|=1/n}} \int_{E_1} \psi(t)\,dt \right) \\
&\qquad - 8\omega_p^*(1/n, |f|) \\
&\geq n^{1-\frac{1}{p}} \sup_{\substack{E \subset [0,1] \\ |E|=1/n}} \left(\int_E |f(t)|\,dt - \sup_{\substack{E_1 \subset [0,1]\setminus E \\ |E_1|=1/n}} \int_{E_1} |f(t)|\,dt \right) \\
&\qquad - 2n^{1-\frac{1}{p}} \sup_{\substack{A \subset [0,1] \\ |A|=1/n}} \int_A \big|\,|f(t)| - |\psi(t)|\,\big|\,dt - 8\omega_p^*(1/n, |f|) \\
&\geq n^{1-\frac{1}{p}} F_n - 2n^{1-\frac{1}{p}} \left(\int_0^1 \big|\,|f(t)| - |\psi(t)|\,\big|^p\,dt \right)^{1/p} (1/n)^{1-\frac{1}{p}} - 8\omega_p^*(1/n, |f|) \\
&\geq n^{1-\frac{1}{p}} F_n - 16\omega_p^*(1/n, f).
\end{aligned}$$

We conclude that (10.4.8) holds. ∎

We recall the definition of equidistributed functions. Let $f(x)$ be defined and measurable on the interval (0,1). For a fixed y denote the set of points $x \in (0,1)$

which satisfy $f(x) > y$ by $E(f > y)$. The function $m(y) = |E(f > y)|$ is called the *distribution function* of f. Two functions f and g, measurable on (0,1), are called *equidistributed* if they have the same distribution function. It is clear that if one of two equidistributed functions is integrable on (0,1), then the other one is also integrable, and their integrals are the same.

10.4.7. *Given any function f, measurable on $(0,1)$ there is a non-increasing function F on $(0,1)$ which is equidistributed with f.*

PROOF. It is obvious that the function $m(y) = |E(f > y)|$ is non-increasing on $(-\infty, \infty)$, continuous from the right, and satisfies $m(-\infty) = 1$, $m(\infty) = 0$. If this function is strictly decreasing and continuous on $(-\infty, \infty)$ then its inverse function $F(x)$ is also strictly decreasing and continuous on $(0,1)$. It is also clear that F is equidistributed with f on (0,1).

Suppose now that the function $x = m(y)$ is not continuous. Consider the curve $\ell : x = m(y)$, $-\infty < y < \infty$, and let y_0 be one of its points of discontinuity. Associate with this curve ℓ and the point y_0 a horizontal line segment consisting of all points (x, y_0) where $m(y_0+) < x \leq m(y_0-)$. Since the function $m(y)$ is continuous from the right, the point $(m(y_0), y_0) = (m(y_0+), y_0)$ must belong to the curve ℓ. Repeat this procedure for each point of discontinuity of the function $x = m(y)$. Let ℓ^* denote the curve generated by combining ℓ with all these horizontal line segments. It is clear that each line $x = x_0$, for $0 < x_0 \leq 1$, intersects the curve ℓ^* at at least one point. The ordinate of this point will be denoted by $F(x_0, f) \equiv F(x_0)$. The function $F(x)$ is uniquely determined for all $0 < x \leq 1$ except those x's which correspond to an interval where $m(y)$ is constant. But such x's constitute no more than a countable set M. For each $x \in M$ let $F(x)$ be any value which does not prevent F from being monotone. From the construction of F it is clear that $|E(F > y_0)| = |E(f > y_0)| = m(y_0)$ for any y_0. Thus f and F are equidistributed. Moreover, it is also clear that the function F is non-increasing[2]. ∎

10.4.8. *Suppose that $0 \leq f(x) \in \mathbf{L}[0,1]$ and that $F(x,f)$ is a non-increasing rearrangement of f. If $E \subset [0,1]$ is any Lebesgue measurable set then*

$$\int_E f(x)\,dx \leq \int_0^{|E|} F(x,f)\,dx. \tag{10.4.14}$$

PROOF. Let $g(x) = f(x)$ for $x \in E$ and $g(x) = 0$ for $x \in [0,1) \setminus E$. Then $g(x) \leq f(x)$ for $x \in [0,1)$ and thus $F(x,g) \leq F(x,f)$. Consequently,

$$\int_E f(x)\,dx = \int_E g(x)\,dx = \int_0^1 F(x,g)\,dx = \int_0^{|E|} F(x,g)\,dx \leq \int_0^{|E|} F(x,f)\,dx. \quad ∎$$

[2] The function F is sometimes called a non-increasing rearrangement of f.

10.4.9. *If* $0 \le f(x) \in \mathbf{L}[0,1]$ *and* $\alpha \in [0,1]$, *then*

$$\left\{ \begin{aligned} \overline{I}_\alpha &\equiv \sup_{\substack{M\subset[0,1]\\ |M|=\alpha}} \int_M f(x)\,dx = \int_0^\alpha F(u,f)\,du, \\ \underline{I}_\alpha &\equiv \inf_{\substack{M\subset[0,1]\\ |M|=\alpha}} \int_M f(x)\,dx = \int_{1-\alpha}^1 F(u,f)\,du. \end{aligned} \right. \tag{10.4.15}$$

PROOF. The identities are easily verified if $\alpha = 0$ or $\alpha = 1$. Indeed, if $\alpha = 0$ then the expressions all are identically zero, and if $\alpha = 1$ then the expressions reduce to

$$\int_0^1 f(x)\,dx = \int_0^1 F(u,f)\,du,$$

which surely holds since f and F are equidistributed.

Thus we may suppose that $0 < \alpha < 1$. Suppose further that there is a point y_0 such that $m(y_0) = \alpha$. Define a function ϕ by $\phi(x) = 0$ if $f(x) \le y_0$ and $\phi(x) = f(x)$ if $f(x) > y_0$. Then it is clear that $0 \le \phi(x) \le f(x)$ holds for any $x \in [0,1]$. Moreover, $F(u,\phi) = F(u,f)$ for $0 \le u < \alpha$ and $F(u,\phi) = 0$ for $\alpha < u \le 1$. Set $M_0 = \{t : t \in [0,1],\ \phi(t) > y_0\}$. It follows that

(10.4.16)

$$\begin{aligned} \overline{I}_\alpha &\ge \sup_{|M|=\alpha} \int_M \phi(x)\,dx \ge \int_{M_0} \phi(x)\,dx = \int_0^1 \phi(x)\,dx \\ &= \int_0^1 F(u,\phi)\,du = \int_0^\alpha F(u,\phi)\,du = \int_0^\alpha F(u,f)\,du. \end{aligned}$$

But (10.4.14) implies

$$\overline{I}_\alpha \le \int_0^\alpha F(u,f)\,du. \tag{10.4.17}$$

Hence by (10.4.16) and (10.4.17) we obtain the first identity in (10.4.15) for the case $\alpha = m(y_0)$.

Suppose now that α is a number for which $\alpha \ne m(y)$ for all real y. This means that α belongs to an interval where $m(y)$ jumps (along the X axis). Let y_0 be the point where $m(y)$ jumps. Set $\beta = \lim_{y\to y_0+} m(y) = m(y_0)$. It is clear that $\beta < \alpha \le m(y_0-)$ and

$$|E(f > y_0)| \equiv |B| = m(y_0) = \beta, \qquad |E(f > y_0 - \frac{1}{n})| \equiv |A_n| \ge \alpha.$$

Thus

$$|\{x : y_0 - \frac{1}{n} < f(x) \le y_0\}| = |A_n \setminus B| \ge \alpha - \beta,$$

and consequently,

$$|D| \geq \alpha - \beta, \qquad D = |\{x : f(x) = y_0\}| = \bigcap_{n=1}^{\infty} (A_n \setminus B).$$

Choose a measurable set $E \subset D$ such that $|E| = \alpha - \beta$. If $M_0 = B \bigcup E$ then we have

$$\begin{aligned} \overline{I}_\alpha &\geq \int_{M_0} f(x)\,dx = \int_B f(x)\,dx + \int_E f(x)\,dx \\ &= \int_B f(x)\,dx + y_0(\alpha - \beta) = \int_B f(x)\,dx + \int_\beta^\alpha y_0\,du. \end{aligned}$$

But since $m(y_0) = \beta = |B|$, we can repeat the proof of (10.4.16) to verify

$$\int_B f(x)\,dx \geq \int_0^{|B|} F(u,f)\,du.$$

Consequently,

$$\overline{I}_\alpha \geq \int_0^\beta F(u,f)\,du + \int_\beta^\alpha y_0\,du = \int_0^\alpha F(u,f)\,du.$$

Thus it follows from (10.4.17) that the first identity in (10.4.15) holds in the case $\alpha \neq m(y)$, $-\infty < y < \infty$. This completes the proof of the first identity in (10.4.15). The second identity is proved in exactly the same way. ∎

In the previous proof we established the fact that the supremum of the numbers $\int_M f(x)\,dx$, as M ranges over the subsets of [0,1] which satisfy $|M| = \alpha$, is attained by some set M_0 of measure $|M_0| = \alpha$.

10.4.10. *Let $0 \leq f(x) \in \mathbf{L}[0,1]$, $\alpha \in [0, 1/2]$ and M_0 be a set which satisfies*

$$\sup_{\{M:|M|=\alpha\}} \int_M f(x)\,dx = \int_{M_0} f(x)\,dx.$$

Then

$$\sup_{\substack{E \subset [0,1]\setminus M_0 \\ |E|=\alpha}} \int_E f(x)\,dx = \int_\alpha^{2\alpha} F(u,f)\,du.$$

PROOF. This proof is exactly like the one for **10.4.8.** ∎

By combining **10.4.9** and **10.4.10**, we obtain the following result.

10.4.11. *Let $f \in \mathbf{L}^p[0,1]$, $1 \le p < \infty$.*
1) If $p = 1$ then

$$\omega_1^*(\frac{1}{n}, f) \ge \frac{1}{17}\left(\int_0^{1/n} F(x,|f|)\,dx - \int_{1-1/n}^{1} F(x,|f|)\,dx\right) \tag{10.4.18}$$

for $n = 1, 2, \ldots$.
2) If $1 < p < \infty$ then

$$\omega_p^*(\frac{1}{n}, f) \ge \frac{n^{1-\frac{1}{p}}}{17}\left(\int_0^{1/n} F(x,|f|)\,dx - \int_{1/n}^{2/n} F(x,|f|)\,dx\right) \tag{10.4.19}$$

for $n = 2, 3, \ldots$.

10.4.12. *Let $f \in \mathbf{L}^p[0,1]$, $1 \le p < \infty$.*
1) If $p = 1$ then

$$n\omega_1^*(\frac{1}{n}, f) \ge \frac{1}{17}\left(F(\frac{1}{n}, |f|) - F(1 - \frac{1}{n}, |f|)\right) \tag{10.4.20}$$

for $n = 1, 2, \ldots$.
2) If $1 < p < \infty$ then

$$n^{1/p}\omega_p^*(\frac{1}{n}, f) \ge \frac{1}{34}\left(F(\frac{1}{2n}, |f|) - F(\frac{1}{n}, |f|)\right) \tag{10.4.21}$$

for $n = 2, 3, \ldots$.

PROOF. 1) Let $p = 1$. Since the function $F(x, |f|)$ is non-increasing, we have by (10.4.18) that

$$\omega_1^*(\frac{1}{n}, f) \ge \frac{1}{17n}\left(F(\frac{1}{n}, |f|) - F(1 - \frac{1}{n}, |f|)\right)$$

and (10.4.20) follows at once.

2) Suppose $p > 1$. Then we have by (10.4.19) that

$$\omega_p^*(\frac{1}{n}, f) \ge \frac{n^{1-\frac{1}{p}}}{17}\left(\int_0^{1/n} F(x,|f|)\,dx - \int_{1/n}^{2/n} F(x,|f|)\,dx\right),$$

and

$$\omega_p^*(\frac{1}{2n}, f) \ge \frac{(2n)^{1-\frac{1}{p}}}{17}\left(\int_0^{1/(2n)} F(x,|f|)\,dx - \int_{1/(2n)}^{1/n} F(x,|f|)\,dx\right).$$

Combining these inequalities we obtain

$$2\omega_p^*(\frac{1}{n},f) \geq \frac{n^{1-\frac{1}{p}}}{17}\left(2\int_0^{1/(2n)} F(x,|f|)\,dx - \int_{1/n}^{2/n} F(x,|f|)\,dx\right)$$
$$\geq \frac{n^{1-\frac{1}{p}}}{17}\left(\frac{F(1/2n,|f|)}{n} - \frac{F(1/n,|f|)}{n}\right)$$
$$= \frac{n^{-\frac{1}{p}}}{17}\left(F(1/2n,|f|) - F(1/n,|f|)\right),$$

i.e., inequality (10.4.21) holds. ∎

Inequality (10.4.21) remains true if $p = 1$. Indeed, for $n \geq 2$ we have by (10.4.20) that

$$n\omega_1^*(\frac{1}{n},f) \geq \frac{1}{2}\cdot 2n\omega_1^*(\frac{1}{2n},f)$$
$$\geq \frac{1}{34}\left(F(\frac{1}{2n},|f|) - F(1-\frac{1}{2n},|f|)\right)$$
$$\geq \frac{1}{34}\left(F(\frac{1}{2n},|f|) - F(\frac{1}{n},|f|)\right).$$

10.4.13. *Let $1 \leq p < q < \infty$ and suppose*

$$D \equiv \sum_{n=1}^{\infty} n^{\frac{q}{p}-2}\left[\omega_p^*(\frac{1}{n},f)\right]^q < \infty. \tag{10.4.22}$$

Then

$$f \in \mathbf{L}^q[0,1] \quad \text{and} \quad \|f\|_q \leq C(q)(\|f\|_1 + D^{1/q}). \tag{10.4.23}$$

PROOF. Notice first that

$$\|f\|_1 = \int_0^1 |f(x)|\,dx = \int_0^1 F(x,f)\,dx \geq \int_0^{1/2} F(x,f)\,dx \geq \frac{1}{2}F(\frac{1}{2},f),$$

i.e.,

$$F(\frac{1}{2},f) \leq 2\|f\|_1. \tag{10.4.24}$$

Next, use Hölder's inequality

$$|\sum_{i=0}^{n} a_i b_i| \leq \left(\sum_{i=0}^{n}|a_i|^p\right)^{1/p}\left(\sum_{i=0}^{n}|b_i|^{p'}\right)^{1/p'}, \qquad \frac{1}{p}+\frac{1}{p'} = 1,$$

to verify

(10.4.25)
$$\begin{aligned}\left(\|f\|_1 + \sum_{k=1}^{n} 2^{k/p}\omega_p^*(2^{-k}, f)\right)^q &= \left(\|f\|_1 + \sum_{k=1}^{n} 2^{k(\frac{1}{p}-\frac{1}{2q})}\omega_p^*(2^{-k}, f)2^{k/(2q)}\right)^q \\ &\le \left(\|f\|_1^q + \sum_{k=1}^{n} 2^{k(\frac{q}{p}-\frac{1}{2})}[\omega_p^*(2^{-k}, f)]^q\right) \times \\ &\quad \times \left(\sum_{k=1}^{n} 2^{\frac{k}{2q}\frac{q}{q-1}}\right)^{q-1} \\ &\le C_1(q)2^{n/2}\left(\|f\|_1^q + \sum_{k=1}^{\infty} 2^{k(\frac{q}{p}-\frac{1}{2})}[\omega_p^*(2^{-k}, f)]^q\right).\end{aligned}$$

Use **10.4.12** to see

$$F(2^{-k-1}, |f|) - F(2^{-k}, |f|) \le 34 \cdot 2^{k/p}\omega_p^*(2^{-k}, f).$$

Adding and telescoping, we obtain

$$F(2^{-n-1}, |f|) - F(\frac{1}{2}, |f|) \le 34 \sum_{k=1}^{\infty} 2^{k/p}\omega_p^*(2^{-k}, f)$$

for $n = 1, 2, \ldots$. Hence it follows from (10.4.24) that

$$F(2^{-n-1}, |f|) \le 2\|f\|_1 + 34 \sum_{k=1}^{\infty} 2^{k/p}\omega_p^*(2^{-k}, f). \tag{10.4.26}$$

Combining (10.4.15) and (10.4.26) we obtain

$$\begin{aligned}
\int_0^1 (F(x,|f|))^q\,dx &= \sum_{n=1}^{\infty}\int_{2^{-n}}^{2^{-n+1}} (F(x,|f|))^q\,dx \\
&\le \sum_{n=1}^{\infty} 2^{-n}(F(2^{-n},|f|))^q \\
&\le 34^q \sum_{n=1}^{\infty} 2^{-n}\left(\|f\|_1 + \sum_{k=1}^{n} 2^{k/p}\omega_p^*(2^{-k},f)\right)^q \\
&\le C_2(q)\sum_{n=1}^{\infty} 2^{-n/2}\left(\sum_{k=1}^{n} 2^{k(\frac{q}{p}-\frac{1}{2})}\omega_p^*(2^{-k},f) + \|f\|_1^q\right) \\
&\le C_3(q)\left(\sum_{k=1}^{\infty} 2^{k(\frac{q}{p}-\frac{1}{2})}[\omega_p^*(2^{-k},f)]^q \sum_{n=k}^{\infty} 2^{-n/2} + \|f\|_1^q\right) \\
&\le C_4(q)\left(\sum_{k=1}^{\infty} 2^{k(\frac{q}{p}-\frac{1}{2})}[\omega_p^*(2^{-k},f)]^q + \|f\|_1^q\right) \\
&\le C_5(q)\left(\sum_{k=1}^{\infty}\sum_{m=2^{k-1}+1}^{2^k} m^{\frac{q}{p}-2}[\omega_p^*(\frac{1}{m},f)]^q + \|f\|_1^q\right) \\
&\le C_5(q)\left(\sum_{m=1}^{\infty} m^{\frac{q}{p}-2}[\omega_p^*(\frac{1}{m},f)]^q + \|f\|_1^q\right).
\end{aligned}$$

Hence we conclude by (10.4.22) that

$$\begin{aligned}
\|f\|_q &= \|F\|_q \\
&\le C_6(q)\left(\|f\|_1^q + \left[\sum_{m=1}^{\infty} m^{\frac{q}{p}-2}[\omega_p^*(\frac{1}{m},f)]^q\right]^{1/q}\right) \\
&= C_6(q)(\|f\|_1 + D^{1/q}). \quad \blacksquare
\end{aligned}$$

The next inequality concerning real numbers is a result of Hardy and Littlewood.

10.4.14. *If* $1 < \alpha, p < \infty$ *and* $a_n \ge 0$ *for* $n = 1, 2, \ldots,$ *then*

$$\sum_{n=1}^{\infty} n^{-\alpha}\left(\sum_{k=1}^{n} a_k\right)^p \le C(\alpha,p)\sum_{n=1}^{\infty} n^{-\alpha}(na_n)^p. \tag{10.4.27}$$

PROOF. Let $\phi_n = n^{-\alpha} + (n+1)^{-\alpha} + \ldots$. Then $\phi_n \le C(\alpha)n^{1-\alpha}$. Set $s_0 = 0$ and

$s_n = \sum_{k=1}^{n} a_k$. We obtain

$$\begin{aligned}
\sum_{n=1}^{m} n^{-\alpha} s_n^p &= \sum_{n=1}^{m} (\phi_n - \phi_{n+1}) s_n^p \\
&= \sum_{n=1}^{m} \phi_n (s_n^p - s_{n-1}^p) - \phi_{m+1} s_m^p \\
&\leq \sum_{n=1}^{m} \phi_n (s_n^p - s_{n-1}^p) \\
&\leq C(\alpha) \sum_{n=1}^{m} n^{1-\alpha} (s_n^p - s_{n-1}^p)
\end{aligned}$$

But $s_n^p - s_{n-1}^p \leq C(p) s_n^{p-1} a_n$. Consequently,

$$\begin{aligned}
\sum_{n=1}^{m} n^{-\alpha} s_n^p &\leq C(\alpha, p) \sum_{n=1}^{m} n^{1-\alpha} s_n^{p-1} a_n \\
&= C(\alpha, p) \sum_{n=1}^{m} \left[(n a_n) n^{-\frac{\alpha}{p}} \right] \left[n^{-\frac{\alpha}{p'}} \right] s_n^{p-1},
\end{aligned}$$

where $\frac{1}{p} + \frac{1}{p'} = 1$. Applying Hölder's inequality for sums

$$\left| \sum_{n=1}^{m} \alpha_n \beta_n \right| \leq \left(\sum_{n=0}^{m} |\alpha_n|^p \right)^{1/p} \left(\sum_{n=0}^{m} |\beta_n|^{p'} \right)^{1/p'},$$

where $\alpha_n = (n a_n) n^{-\alpha/p}$ and $\beta_n = s_n^{p-1} n^{-\alpha/p'}$, and using the fact that $p'(p-1) = p$, we obtain

$$\sum_{n=1}^{m} n^{-\alpha} s_n^p \leq C(\alpha, p) \left(\sum_{n=1}^{m} (n a_n)^p n^{-\alpha} \right)^{1/p} \left(\sum_{n=1}^{m} n^{-\alpha} s_n^p \right)^{1/p'}.$$

Consequently,

$$\left(\sum_{n=1}^{m} n^{-\alpha} s_n^p \right)^{1-\frac{1}{p'}} \leq C(\alpha, p) \left(\sum_{n=1}^{m} n^{-\alpha} (n a_n)^p \right)^{1/p}.$$

Since $1 - \frac{1}{p'} = \frac{1}{p}$ this inequality implies (10.4.27). ∎

10.4.15. *Let $1 \le p < q < \infty$ and suppose $f \in \mathbf{L}^p[0,1]$. Then*

$$\|f\|_q \le C(p,q)\left(\|f\|_p + \left[\sum_{k=1}^{\infty} k^{\frac{q}{p}-2}(E_k^{(p)}(f))^q\right]^{1/q}\right), \tag{10.4.28}$$

and

$$E_n^{(q)}(f) \le C(p,q)\left(E_n^{(p)}(f)n^{\frac{1}{p}-\frac{1}{q}} + \left[\sum_{k=n+1}^{\infty} k^{\frac{q}{p}-2}(E_k^{(p)}(f))^q\right]^{1/q}\right), \tag{10.4.29}$$

where

$$E_n^{(p)}(f) = \inf_{\{a_k\}} \|f(x) - \sum_{k=0}^{n-1} a_k w_k(x)\|_p$$

represents the best approximations to the function f, in $\mathbf{L}^p$ norm, by Walsh polynomials of order n. These inequalities also hold for best approximations by Haar polynomials, namely when $E_n^{(p)}(f)$ is replaced by $E_n^{(p)}(f)_h$.

PROOF. By **10.4.13** we have

$$\|f\|_q \le C(q)\left(\|f\|_1 + \left[\sum_{n=1}^{\infty} n^{\frac{q}{p}-2}\left(\omega_p^*(\frac{1}{n},f)\right)^q\right]^{1/q}\right) \tag{10.4.30}$$

for $f \in \mathbf{L}^p[0,1]$, $1 \le p < q < \infty$ and $\omega_p^*(\delta,f)$ is the modulus of continuity defined in (10.4.1). Since $\omega_p^*(\delta,f) \le \omega_p(\delta,f)$, we have by **10.2.8** that

$$\omega_p^*(\frac{1}{n},f) \le \frac{96}{n^{1/p}}\sum_{k=1}^{n} k^{\frac{1}{p}-1}E_k^{(p)}(f). \tag{10.4.31}$$

Thus it follows from (10.4.30) and the inequality $\|f\|_1 \le \|f\|_p$ that

$$\|f\|_q \le 96C(q)\left(\|f\|_p + \left[\sum_{n=1}^{\infty} n^{-2}(\sum_{k=1}^{\infty} k^{\frac{1}{p}-1}E_k^{(p)}(f))^q\right]^{1/q}\right).$$

Substitute estimate (10.4.27) into the right side for $a_k = k^{\frac{1}{p}-1}E_k^{(p)}(f)$, $k = 1,2,\ldots$. We obtain

$$\|f\|_q \le C(p,q)\left(\|f\|_p + \left[\sum_{n=1}^{\infty} n^{\frac{q}{p}-2}(E_n^{(p)}(f))^q\right]^{1/q}\right).$$

This verifies (10.4.28). Since by **10.2.8** inequality (10.4.31) remains true if we replace $E_n^{(p)}(f)$ with $E_n^{(p)}(f)_h$, a similar proof establishes the following inequality:

$$\|f\|_q \leq C(p,q)\left(\|f\|_p + \left[\sum_{n=1}^{\infty} n^{\frac{q}{p}-2}(E_n^{(p)}(f)_h)^q\right]^{1/q}\right).$$

It remains to prove (10.4.29). For this let $S_n(x) = \sum_{k=0}^{n-1} a_k w_k(x)$ denote a Walsh polynomial which yields the best approximation to the function f in the $\mathbf{L}^p$ norm, i.e., $E_n^{(p)}(f) = \|f - S_n\|_p$. It is obvious that

$$E_k^{(p)}(f - S_n) \leq \|f - S_n\|_p = E_n^{(p)}(f), \qquad k = 1, 2, \ldots,$$

and $E_k^{(p)}(f) \geq E_k^{(p)}(f - S_n)$ for $k \geq n$. Consequently, applying the inequality (10.4.28) to the function $f - S_n$ we obtain

$$\begin{aligned}
E_k^{(q)}(f) &\leq \|f - S_n\|_q \\
&\leq C(p,q)\times \\
&\times \left(\|f - S_n\|_p + \left[\sum_{k=1}^{n} k^{\frac{q}{p}-2}(E_k^{(p)}(f))^q + \sum_{k=n+1}^{\infty} k^{\frac{q}{p}-2}(E_k^{(p)}(f))^q\right]^{1/q}\right) \\
&\leq C_1(p,q)\left(E_n^{(p)}(f)\left(1 + \sum_{k=1}^{n} k^{\frac{q}{p}-2}\right)^{1/q} + \left[\sum_{k=n+1}^{\infty} k^{\frac{q}{p}-2}(E_k^{(p)}(f))^q\right]^{1/q}\right) \\
&\leq C_2(p,q)\left(n^{\frac{1}{p}-\frac{1}{q}} E_n^{(p)}(f) + \left[\sum_{k=n+1}^{\infty} k^{\frac{q}{p}-2}(E_k^{(p)}(f))^q\right]^{1/q}\right).
\end{aligned}$$

This verifies (10.4.29). A similar proof establishes the analogous statement for best approximations by Haar polynomials. ∎

We shall show for $1 \leq p < q < \infty$ and $C(p,q)$ in place of 128, that Theorems **10.3.7** and **10.3.8** are corollaries of Theorem **10.4.15**. To accomplish this we need the following lemma:

10.4.16. *Let $\{a_n\}_{n=1}^{\infty}$ be a sequence of numbers which satisfy $a_n \geq 0$ and $a_n \geq a_{n+1}$ for $n = 1, 2, \ldots$. Then*

$$(10.4.32) \quad \left(\sum_{k=n}^{\infty} k^{\alpha} a_k^{\nu}\right)^{1/\nu} \leq (2 + 2^{(1+\alpha)/\nu})^2 \left(n^{(1+\alpha)/\nu} a_n + \sum_{k=n+1}^{\infty} k^{(1+\alpha-\nu)/\nu} a_k\right)$$

for $n = 1, 2, \ldots$, $\alpha \in (-\infty, \infty)$, and $\nu \geq 0$.

PROOF. Since

$$\sum_{k=n2^m}^{n2^{m+1}-1} k^\alpha \leq (n2^m)(n2^{m+1})^\alpha \leq (2^{\alpha+1}+1)(n2^m)^{1+\alpha}, \qquad \alpha \geq 0,$$

and

$$\sum_{k=n2^m}^{n2^{m+1}-1} k^\alpha \leq (n2^m)(n2^m)^\alpha \leq (2^\alpha+1)(n2^m)^{1+\alpha}, \qquad \alpha < 0,$$

we have

$$\sum_{k=n2^m}^{n2^{m+1}-1} k^\alpha \leq (2^\alpha+1)(n2^m)^{1+\alpha} \tag{10.4.33}$$

for $m = 0, 1, \ldots$, $n = 1, 2, \ldots$, and all real α. Moreover, it is clear that

$$(n2^m)^\beta \leq (1+2^\beta)k^\beta \tag{10.4.34}$$

for $n2^{m-1} < k \leq n2^m$, $n, m = 1, 2, \ldots$, and all real β. Since $\nu \geq 1$ and the sequence $\{a_k\}$ is non-negative and non-increasing, it follows from (10.4.33) and (10.4.34) that

$$\begin{aligned}
\left(\sum_{k=n}^{\infty} k^\alpha a_k^\nu\right)^{1/\nu} &= \left(\sum_{m=0}^{\infty}\sum_{k=n2^m}^{n2^{m+1}-1} k^\alpha a_k^\nu\right)^{1/\nu} \\
&\leq \left(\sum_{m=0}^{\infty} a_{n2^m}^\nu \sum_{k=n2^m}^{n2^{m+1}-1} k^\alpha\right)^{1/\nu} \\
&\leq (1+2^\alpha)^{1/\nu}\left(\sum_{m=0}^{\infty}(n2^m)^{1+\alpha}a_{n2^m}^\nu\right)^{1/\nu} \\
&\leq (1+2^\alpha)^{1/\nu}\sum_{m=0}^{\infty}(n2^m)^{(1+\alpha)/\nu}a_{n2^m} \\
&\leq (1+2^\alpha)^{1/\nu}\left(n^{(1+\alpha)/\nu}a_n + \sum_{m=1}^{\infty}\frac{(n2^m)^{(1+\alpha)/\nu}}{n2^{m-1}}\sum_{k=n2^{m-1}+1}^{n2^m} a_k\right) \\
&\leq (1+2^\alpha)^{1/\nu}\left(n^{(1+\alpha)/\nu}a_n + 2\sum_{m=1}^{\infty}(1+2^{\frac{1+\alpha}{\nu}-1})\sum_{k=n2^{m-1}+1}^{n2^m} k^{\frac{1+\alpha}{\nu}-1}a_k\right) \\
&\leq (1+2^\alpha)^{1/\nu}(2+2^{(1+\alpha)/\nu})\left(n^{(1+\alpha)/\nu}a_n + \sum_{k=n+1}^{\infty}k^{(1+\alpha-\nu)/\nu}a_k\right) \\
&\leq (1+2^{\alpha/\nu})(2+2^{(1+\alpha)/\nu})\left(n^{(1+\alpha)/\nu}a_n + \sum_{k=n+1}^{\infty}k^{(1+\alpha-\nu)/\nu}a_k\right).
\end{aligned}$$

Thus (10.4.32) holds. ∎

If $1 \le p < q < \infty$, then we have by (10.4.32) that

$$n^{\frac{1}{p}-\frac{1}{q}} E_n^{(p)}(f) + \left(\sum_{k=n+1}^{\infty} k^{\frac{q}{p}-2} [E_k^{(p)}(f)]^q \right)^{1/q}$$
$$\le 9 n^{\frac{1}{p}-\frac{1}{q}} E_n^{(p)}(f) + 8 \sum_{k=n+2}^{\infty} k^{\frac{1}{p}-\frac{1}{q}-1} E_k^{(p)}(f).$$

The analogous inequality for best approximations $E_n^{(p)}(f)_h$ by Haar polynomials is also true. Thus if 128 is replaced by a constant $c(p,q)$ then Theorems **10.3.7** and **10.3.8** are corollaries of Theorem **10.4.15**.

Notice also that if

$$E_k^{(p)}(f)_h = O\left(\frac{k^{\frac{1}{q}-\frac{1}{p}}}{\ln(k+1)} \right), \quad \text{or} \quad E_k^{(p)}(f) = O\left(\frac{k^{\frac{1}{q}-\frac{1}{p}}}{\ln(k+1)} \right)$$

for some $1 \le p < q < \infty$ then by Theorems **10.3.7** and **10.3.8**, the function f must belong to $\mathbf{L}^q[0,1]$, since in this case the series on the right side of (10.3.5) and (10.3.9) obviously converges. Furthermore, in this case we have by Theorem **10.4.15** that

$$E_k^{(q)}(f) = O(\ln^{\frac{1}{q}-1}(n+1)), \qquad n \to \infty,$$

and a similar estimate for best approximations $E_n^{(p)}(f)_h$ by Haar polynomials.

§10.5. Best approximations by means of multiplicative and step functions.

As we saw in §2.3 and §2.8, series with respect to multiplicative orthonormal systems share some approximation properties with the trigonometric system and have some approximation properties quite different from the trigonometric system. In this section, we shall obtain one more property which does not have an analogue for the trigonometric system.

Let $\{\chi_n(x)\}_{n=0}^{\infty}$, $x \in [0,1)$, be the multiplicative system defined in §1.5 relative to the sequence $P = (p_1, p_2, \dots, p_n, \dots)$. For each $1 \le p \le \infty$ and each natural number n, let $E_n^{(p)}(f)$ denote the best approximation in the $\mathbf{L}^p[0,1)$ norm to the function $f \in \mathbf{L}^p[0,1)$ by polynomials in the system $\{\chi_n(x)\}_{n=0}^{\infty}$ of order no greater than $n-1$, i.e.,

$$E_n^{(p)}(f) = \inf_{\{c_k\}} \|f(x) - \sum_{k=0}^{n-1} c_k \chi_k(x)\|_p.$$

Let $Q_n(f,x) = \sum_{k=0}^{n-1} \alpha_k \chi_k(x)$ be a polynomial which yields the best approximation to f in $\mathbf{L}^p$ norm, i.e.,

$$\|f(x) - Q_n(f,x)\|_p = E_n^{(p)}(f).$$

Such polynomials exist for each natural number n since $\|f(x) - \sum_{k=0}^{n-1} c_k\chi_k(x)\|_p$ is a continuous, non-negative function on the collection of coefficients c_k.

Using the metric $\rho^*(x,t)$, we shall define an $\mathbf{L}^p$- modulus of continuity, $\overset{*}{\omega}{}^{(p)}(\delta, f)$, analogous to (2.5.1), and for $\delta = 1/m_r$, $m_r = p_r m_{r-1}$, we shall denote it by $\omega_r^{(p)}(f)$, i.e.,

$$\omega_r^{(p)}(f) = \sup_{0\le h<1/m_r} \|f(x\oplus h) - f(x)\|_p, \qquad 1 \le p \le \infty.$$

The following result is true (compare with **10.2.5**).

10.5.1. *If* $1 \le p \le \infty$ *and* $f \in \mathbf{L}^p[0,1)$, *then*

$$E_{m_r}^{(p)}(f) \le \omega_r^{(p)}(f) \le 2E_{m_r}^{(p)}(f).$$

PROOF. Write the partial sums of the Fourier series of a function f with respect to the system $\{\chi_n(x)\}_{n=0}^{\infty}$ in the form (2.8.3):

$$S_n(f,x) = \int_0^1 f(t)D_n(x\ominus t)\,dt.$$

Make the change of variables $u = x \ominus t$, i.e., $t = x \ominus u$, and use the fact that the integral is invariant with respect to translation by $\oplus$. Thus obtain

$$S_n(f,x) = \int_0^1 f(x\ominus u)D_n(u)\,du.$$

Applying (1.5.21), we find that

$$S_{m_r}(f,x) = m_r \int_0^{1/m_r} f(x\ominus u)\,du. \tag{10.5.1}$$

Consider the case $1 \le p < \infty$. Apply Minkowski's inequality to (10.5.1) (see **A5.2.3**) to obtain

$$\begin{aligned}
E_{m_r}^{(p)}(f) &\le \|f(x) - S_{m_r}(f,x)\|_p \\
&= \left(\int_0^1 \left| f(x) - m_r \int_0^{1/m_r} f(x\ominus u)\,du \right|^p dx\right)^{1/p} \\
&\le \left(\int_0^1 \left[m_r \int_0^{1/m_r} |f(x) - f(x\ominus u)|\,du\right]^p dx\right)^{1/p} \\
&= m_r \left[\int_0^1 \left(\int_0^{1/m_r} |f(x) - f(x\ominus u)|\,du\right)^p dx\right]^{1/p} \\
&\le m_r \int_0^{1/m_r} \left(\int_0^1 |f(x) - f(x\ominus u)|^p\,dx\right)^{1/p} du \\
&\le m_r \int_0^{1/m_r} \sup_{0\le u<1/m_r} \left(\int_0^1 |f(x) - f(x\ominus u)|^p\,dx\right)^{1/p} du = \omega_r^{(p)}(f).
\end{aligned}$$

If $p = \infty$ then

$$\begin{aligned} E_{m_r}^{(\infty)}(f) &\le \max_{0\le x\le 1} |f(x) - S_{m_r}(f,x)| \\ &\le \max_{0\le x\le 1} \left(m_r \int_0^{1/m_r} |f(x) - f(x \ominus u)|^p \, du \right) \le \omega_r^{(\infty)}(f). \end{aligned}$$

Thus the left hand inequality of **10.5.1** is established.

To prove the right hand inequality, notice that

$$\chi_k(x \oplus h) = \chi_k(x)$$

for all $0 \le h < 1/m_r$ and $k = 0, 1, \ldots, m_r - 1$. Consequently,

$$Q_{m_r}(f,x) = \sum_{k=0}^{m_r-1} \alpha_k \chi_k(x) = Q_{m_r}(f, x \oplus h).$$

Thus for each $0 \le h < 1/m_r$ we have

$$\begin{aligned} \|f(x) - f(x \oplus h)\|_p &= \|f(x) - Q_{m_r}(f,x) + Q_{m_r}(f, x \oplus h) - f(x \oplus h)\|_p \\ &\le \|f(x) - Q_{m_r}(f,x)\|_p + \|Q_{m_r}(f, x \oplus h) - f(x \oplus h)\|_p \\ &= 2E_{m_r}^{(p)}(f). \end{aligned}$$

Therefore, we find that

$$\omega_r^{(p)}(f) = \sup_{0 \le h < 1/m_r} \|f(x) - f(x \oplus h)\|_p \le 2E_{m_r}^{(p)}(f),$$

i.e., the proof of **10.5.1** is complete. ∎

We shall now consider approximation of a function defined on the positive real axis $0 \le x < \infty$. When approximating functions on the whole axis, entire functions of exponential type are used (instead of polynomials). Moreover by the Paley-Wiener Theorem (see [30], Vol. II., p. 272), there is a 1-1 correspondence between entire functions of exponential type and certain Fourier transforms of functions $f(t) \in \mathbf{L}^2(-\infty, \infty)$. Namely, the Fourier transform

$$\widehat{f}(\nu) = \int_{-\infty}^{\infty} f(t) e^{-2\pi i \nu t} \, dt$$

of a function $f(t) \in \mathbf{L}^2(-\infty, \infty)$ has compact support if and only if the function $f(t)$ can be extended to the complex plane to a function $f(z)$ which is entire and satisfies $|f(z)| = O(e^{(\sigma+\varepsilon)|z|})$, $\varepsilon > 0$, as $|z| \to \infty$.

For multiplicative transformations on the positive, real axis $[0,\infty)$, we see by **6.2.12** and **6.2.13** that an analogue for the collection of entire functions of exponential type is the class of step functions which are constant on intervals of a certain rank. Therefore, we shall study approximations by this class of functions.

We shall call a function $g_r(x)$, defined on $[0,\infty)$, P- adic entire of order less than or equal to r, if it is constant on each interval of rank r:

$$\delta_\ell(r) = [\ell/m_r, (\ell+1)/m_r), \qquad \ell = 0, 1, \ldots.$$

The class of all P-adic entire functions of order less than or equal to r will be denoted by $\mathcal{R}_r$, and we shall denote the best approximation to a function $f(t) \in \mathbf{L}^p(0,\infty)$ in the $\mathbf{L}^p(0,\infty)$ norm by P-adic entire functions of order less than or equal to r by $\mathcal{E}_{m_r}^{(p)}(f)$, i.e.,

$$\mathcal{E}_{m_r}^{(p)}(f) = \inf_{g_r \in \mathcal{R}_r} \|f - g_r\|_{\mathbf{L}^p(0,\infty)},$$

where $\|\phi\|_{\mathbf{L}^p(0,\infty)} = \left(\int_0^\infty |\phi(x)|^p\, dx\right)^{1/p}$, $p \neq \infty$, and $\|\phi\|_{\mathbf{L}^\infty(0,\infty)} = \sup\limits_{x \in [0,\infty)} |\phi(x)|$. We shall also denote the P-adic $\mathbf{L}^p$-modulus of a function $f(x) \in \mathbf{L}^p(0,\infty)$ by

$$\omega_r^{(p)}(f) = \sup_{0 \le h < 1/m_r} \|f(x \oplus h) - f(x)\|_{\mathbf{L}^p(0,\infty)}.$$

The following result is an analogue of Theorem **10.5.1**:

10.5.2. *If $1 \le p \le \infty$ and $f \in \mathbf{L}^1[0,\infty) \bigcap \mathbf{L}^p[0,\infty)$, then*

$$\mathcal{E}_{m_r}^{(p)}(f) \le \omega_r^{(p)}(f) \le 2\mathcal{E}_{m_r}^{(p)}(f).$$

PROOF. Consider the function

$$J_{m_r}(f,x) = m_r \int_0^{1/m_r} f(x \oplus u)\, du,$$

which is constant on each interval $\delta_\ell(r) = [\ell/m_r, (\ell+1)/m_r)$, $\ell = 0, 1, \ldots$. By definition $J_{m_r}(f,x) \in \mathcal{R}_r$. Consequently,

$$\begin{aligned}
\mathcal{E}_{m_r}^{(p)}(f) &= \inf_{g_r \in \mathcal{R}_r} \|f - g_r\|_{\mathbf{L}^p(0,\infty)} \\
&\le \|f - J_{m_r}(f,x)\|_{\mathbf{L}^p(0,\infty)} \\
&= \|f - m_r \int_0^{1/m_r} f(x \oplus u)\, du\|_{\mathbf{L}^p(0,\infty)} \\
&= \|m_r \int_0^{1/m_r} [f(x) - f(x \oplus u)]\, du\|_{\mathbf{L}^p(0,\infty)}.
\end{aligned}$$

By repeating the proof of **10.5.1**, we continue this estimate, obtaining

$$\mathcal{E}_{m_r}^{(p)}(f) \leq m_r \int_0^{1/m_r} \|f(x) - f(x \oplus u)]\,\|_{\mathbf{L}^p(0,\infty)}\, du \leq \omega_r^{(p)}(f),$$

i.e., the left hand inequality of **10.5.2** is established.

Suppose now that $g_r^*(x) = g_r^*(f,x)$ is a step function from $\mathcal{R}_r$ which satisfies

$$\mathcal{E}_{m_r}^{(p)}(f) = \inf_{g_r \in \mathcal{R}_r} \|f - g_r\|_{\mathbf{L}^p(0,\infty)} = \|f - g_r^*\|_{\mathbf{L}^p(0,\infty)}.$$

(Such a function in $\mathcal{R}_r$ exists and is unique.) Since $0 \leq h < 1/m_r$ implies $g_r^*(x) = g_r^*(x \oplus h)$, we can write

$$\begin{aligned}\|f(x) - f(x \oplus h)\,\|_{\mathbf{L}^p(0,\infty)} &= \|f(x) - g_r^*(x) + g_r^*(x \oplus h) - f(x \oplus h)\,\|_{\mathbf{L}^p(0,\infty)} \\ &\leq \|f(x) - g_r^*(x)\|_{\mathbf{L}^p(0,\infty)} + \|g_r^*(x \oplus h) - f(x \oplus h)\,\|_{\mathbf{L}^p(0,\infty)}.\end{aligned}$$

Consequently,

$$\sup_{0 \leq h < 1/m_r} \|f(x) - f(x \oplus h)\,\|_{\mathbf{L}^p(0,\infty)} = \omega_r^{(p)}(f) \leq 2\mathcal{E}_{m_r}^{(p)}(f). \quad \blacksquare$$

Chapter 11

APPLICATIONS OF MULTIPLICATIVE SERIES AND TRANSFORMS TO DIGITAL INFORMATION PROCESSING

In the last decade interest has increased significantly in applications of the Walsh system and its generalizations, especially applications to digital information processing. This interest stems from a peculiarity of the Walsh functions, namely, that each one of them takes on only two values $+1$ and -1. A consequence of this peculiarity is that multiplication can be avoided when utilizing high speed computers for certain problems. In such cases a discrete Hadamard transform can be computed almost 10 times faster than the corresponding discrete Fourier transform. Moreover, with the discrete Hadamard transform (DHT) and the discrete transform with respect to multiplicative systems (DMT), one can perform parallel calculations obtaining output of simultaneous calculations on single instruction processors better than can be done using of the discrete Fourier transform (DFT). This allows one to carry out basic calculations using only addition thereby avoiding the more costly operation of multiplication.

In this chapter we shall discuss discrete multiplicative transforms and practical methods for computing them. In this chapter and the next, we shall restrict our attention to systems $\chi(x,y)$ which are generated by symmetric sequences $\mathcal{P} = (\dots, p_{-j}, \dots, p_{-1}, p_1, \dots, p_j, \dots)$, i.e., sequences where $p_{-j} = p_j$ for $j = 1, 2, \dots$.

§11.1. Discrete multiplicative transforms.

By the discretization of the multiplicative integral

$$\hat{f}(y) = \int_0^\infty f(t)\overline{\chi(t,y)}\,dt \tag{11.1.1}$$

we shall mean the procedure of approximating this integral using Riemann sums; specifically, we shall pass from the integral (11.1.1) to the sum

$$\tilde{f}(y) = \sum_{k=0}^{N-1} f(t_k)\overline{\chi(t_k,y)}\Delta t_k. \tag{11.1.2}$$

It is well-known (see, for example [6], p. 316), that if the spectrum

$$\Phi(\nu) = \int_{-\infty}^\infty \phi(t)\exp(\sqrt{2}\pi i\nu t)\,dt \tag{11.1.3}$$

of a function $\phi(t) \in \mathbf{L}^1(-\infty,\infty)$ has compact support, then discretization of the integral (11.1.3) leads to a "spreading" of peaks in the spectrum $\Phi(\nu)$ and the appearance of false peaks, but if the spectrum $\Phi(\nu)$ does not have compact support, then discretization leads to a "superposition of frequencies". It turns out that these defects are lacking for discretization of the integral (11.1.1).

Before we state and prove some fundamental theorems we establish a number of preliminary results.

11.1.1. *Let n be a natural number and $y \in [0, m_n)$. Then*

$$L_{m_n}(y) = \sum_{k=0}^{m_n-1} \chi\left(\frac{k}{m_n}, y\right) = \begin{cases} m_n & \text{if } [y] = 0, \\ 0 & \text{if } 1 \le [y] \le m_n - 1. \end{cases}$$

PROOF. If $[y] = 0$ then the result is obvious, since $\chi(k/m_n, y) = \chi(k/m_n, 0) = 1$ (see **1.5.3** and **1.5.1**). Suppose $[y] \ge 1$. Write the numbers $k/m_n < 1$ and $[y] \in [1, m_n]$ in the form

$$\frac{k}{m_n} = \frac{\ell_1}{m_1} + \frac{\ell_2}{m_2} + \cdots + \frac{\ell_n}{m_n}, \qquad 0 \le \ell_j \le p_j - 1,\ j = 1, 2, \ldots, n,$$

and

$$[y] = \sum_{j=1}^{q} y_{-j} m_{j-1}, \qquad 0 \le y_{-j} \le p_j - 1,$$

where $y_{-q} \ne 0$ for some $q = q(y) \in [1, n]$. Using definition (1.5.33) and the multiplicative property of the function $\chi(x, y)$, we have

$$\begin{aligned} L_{m_n}(y) &= \sum_{\ell_1=0}^{p_1-1} \sum_{\ell_2=0}^{p_2-1} \cdots \sum_{\ell_n=0}^{p_n-1} \chi\left(\frac{\ell_1}{m_1} + \frac{\ell_2}{m_2} + \cdots + \frac{\ell_n}{m_n}, y\right) \\ &= \prod_{j=1}^{n} \sum_{\ell_j=0}^{p_j-1} \chi\left(\frac{\ell_j}{m_j}, \sum_{\nu=1}^{q} y_{-\nu} m_{\nu-1}\right) \\ &= \prod_{j=1}^{n} \sum_{\ell_j=0}^{p_j-1} \exp \frac{2\pi i \ell_j y_{-j}}{p_j}. \end{aligned}$$

But since $y_{-q} \ne 0$, $q \le n$, we have

$$\sum_{\ell_q=0}^{p_q-1} \exp \frac{2\pi i \ell_q y_{-q}}{p_q} = \sum_{\ell_q=0}^{p_q-1} \left(\exp \frac{2\pi i \ell_q}{p_q}\right)^{y_{-q}} = 0,$$

since this is the sum of all p_q-th roots of unity raised to a fixed power y_{-q} which satisfies $1 \le y_{-q} \le p_q - 1$. Consequently, $L_{m_n}(y) = 0$ for $1 \le [y] \le m_n - 1$. ∎

11.1.2. *Let $f_1(x)$ and $f_2(x)$ be absolutely integrable on $[0,\infty)$, and $\hat{f}_1(y)$, $\hat{f}_2(y)$ represent their respective multiplicative transforms. Suppose that $f_2(x)$ is bounded on $[0,\infty)$ and that its transform $\hat{f}_2(y) \in \mathbf{L}^1(0,\infty)$. Then the multiplicative transform of the product $f(x) = f_1(x)f_2(x)$ is the p-adic convolution of the transforms $\hat{f}_1(y)$ and $\hat{f}_2(y)$, i.e.,*

$$\hat{f}(y) = \mathcal{F}[f_1 \cdot f_2](y) = \int_0^\infty \hat{f}_1(v)\hat{f}_2(y \ominus v)\,dv = \int_0^\infty \hat{f}_1(y \ominus v)\hat{f}_2(v)\,dv.$$

PROOF. Since $f_2(x)$ is bounded and $f_1(x)$ is absolutely integrable, it is clear that $\mathcal{F}[f_1 \cdot f_2](y)$ exists. By **6.2.2** and Fubini's Theorem, we find that

$$\begin{aligned}
\hat{f}(y) &= \int_0^\infty f_1(x)f_2(x)\overline{\chi(x,y)}\,dx \\
&= \int_0^\infty f_1(x)\int_0^\infty \hat{f}_2(v)\chi(x,v)\,dv\overline{\chi(x,y)}\,dx \\
&= \int_0^\infty \hat{f}_2(v)\int_0^\infty f_1(x)\overline{\chi(x,y\ominus v)}\,dx\,dv \\
&= \int_0^\infty \hat{f}_2(v)\hat{f}_1(y\ominus v)\,dv.
\end{aligned}$$

Using the change of variables $v = y \ominus u$, we conclude that

$$\hat{f}(y) = \int_0^\infty \hat{f}_1(u)\hat{f}_2(y\ominus u)\,du. \quad \blacksquare$$

11.1.3. *Let ℓ be a natural number. Then*

$$\int_0^{1/m_\ell} \chi(v,y)\,dv = \begin{cases} 1/m_\ell & \text{for } 0 \le y < m_\ell, \\ 0 & \text{for } m_\ell \le y < \infty. \end{cases}$$

PROOF. Since $v \in [0, 1/m_\ell]$ implies $v < 1$, we see by **1.5.3** and **1.5.1** that

$$\chi(v,y) = \chi(v,[y]).$$

Let $n = [y]$ and write this number in the form

$$n = \sum_{k=1}^\infty n_{-k}m_{k-1}.$$

If $n < m_\ell$ then $n_{-k} = 0$ for $k \ge \ell$. Thus we have by definition (1.5.38) that

$$\chi(v,n) = \chi_n(v) = 1, \qquad n < m_\ell, \quad v \in [0, 1/m_\ell]. \tag{11.1.4}$$

The first part of **11.1.1** follows directly from this identity.

Suppose now that $n \geq m_\ell$. Then

$$n = \sum_{k=1}^{\ell} n_{-k} m_{k-1} + \sum_{k=\ell+1}^{\nu} n_{-k} m_{k-1} = (n)_\ell + \sum_{k=\ell+1}^{\nu} n_{-k} m_{k-1},$$

where $n_{-\nu} \neq 0$, i.e., $n_{-\nu} \in [1, p_\nu - 1]$, $\nu \geq \ell + 1$. We have

$$\begin{aligned}\int_0^{1/m_\ell} \chi(v,y)\,dv &= \int_0^{1/m_\ell} \chi_n(v)\,dv \\ &= \int_0^{1/m_\ell} \chi_{(n)_\ell + \sum_{k=\ell+1}^{\nu} n_{-k} m_{k-1}}(v)\,dv \\ &= \int_0^{1/m_\ell} \chi_{(n)_\ell}(v) \prod_{k=\ell+1}^{\nu} \chi_{n_{-k} m_{k-1}}(v)\,dv.\end{aligned}$$

But $\chi_{(n)_\ell}(v) = 1$ for $v \in [0, 1/m_\ell)$ by (11.1.4). Thus

$$\begin{aligned}\int_0^{1/m_\ell} \chi(v,y)\,dv &= \int_0^{1/m_\ell} \prod_{k=\ell+1}^{\nu} \chi_{n_{-k} m_{k-1}}(v)\,dv \\ &= \sum_{r=0}^{(m_{\nu-1}/m_\ell)-1} \int_{r/m_{\nu-1}}^{(r+1)/m_{\nu-1}} \left(\prod_{k=\ell+1}^{\nu-1} \chi_{n_{-k} m_{k-1}}(v) \right) \chi_{n_{-\nu} m_{\nu-1}}(v)\,dv.\end{aligned}$$

Since the functions

$$\psi_{r,\nu}(v) = \prod_{k=\ell+1}^{\nu-1} \chi_{n_{-k} m_{k-1}}(v)$$

are constant on the intervals $\delta_r(\nu-1)$, $0 \leq r \leq \dfrac{m_{\nu-1}}{m_\ell} - 1$, and

$$\chi_{n_{-\nu} m_{\nu-1}}(v) = \exp \frac{2\pi i n_{-\nu} v_\nu}{p_\nu}, \qquad v \in \delta_r(\nu - 1),$$

we have

$$\int_0^{1/m_\ell} \chi(v,y)\,dv = \sum_{r=0}^{(m_{\nu-1}/m_\ell)-1} \psi_{r,\nu}(\delta_r(\nu-1)) \sum_{v_\nu=0}^{p_\nu-1} \exp \frac{2\pi i n_{-\nu} v_\nu}{p_\nu} = 0.$$

(This last sum is the sum of all p_ν-th roots of unity raised to the power $n_{-\nu}$, where $1 \leq n_{-\nu} \leq p_\nu - 1$). ∎

11.1.4. *Let n and r be natural numbers, $N = m_n m_r$ and*

$$D(u,\zeta) = \int_0^\zeta \chi(x,u)\,dx.$$

Then the function

$$\phi_{n,r}(t) = \frac{1}{m_n}\sum_{k=0}^{N-1} D\left(\frac{k}{m_r}\ominus t, m_r\right), \qquad t \in [0,\infty), \tag{11.1.5}$$

is bounded and its multiplicative transform $\hat{\phi}_{n,r}(y)$ belongs to $\mathbf{L}^1(0,\infty)$ and is given by

$$\hat{\phi}_{n,r}(y) = \begin{cases} m_r & \text{for } 0 \le y < 1/m_n, \\ 0 & \text{for } 1/m_n \le y < \infty. \end{cases} \tag{11.1.6}$$

PROOF. By (1.5.38) we have

$$D\left(\frac{k}{m_r}\ominus t, m_r\right) = \begin{cases} m_r & \text{for } 0 \le (k/m_r)\ominus t < 1/m_r, \\ 0 & \text{for } 1/m_r \le (k/m_r)\ominus t < 1. \end{cases}$$

Thus for each fixed t there are only a fixed number of terms in the sum (11.1.5) which are different from zero. Thus the function $\phi_{n,r}(t)$ is bounded for $0 \le t < m_n$ and identically zero for $t \ge m_r + 1$ (because $[(k/m_r)\ominus t] > 1$ for $t \ge m_n + 1$ and for $0 \le k \le N-1$).

We shall compute $\hat{\phi}_{n,r}(y)$. First, by definition

$$\begin{aligned}
\hat{\phi}_{n,r}(y) &= \int_0^\infty \phi_{n,r}(t)\overline{\chi(t,y)}\,dt \\
&= \frac{1}{m_n}\int_0^\infty \sum_{k=0}^{N-1} D\left(\frac{k}{m_r}\ominus t, m_r\right)\overline{\chi(t,y)}\,dt \\
&= \frac{1}{m_n}\sum_{k=0}^{N-1}\int_0^\infty D\left(\frac{k}{m_r}\ominus t, m_r\right)\overline{\chi(t,y)}\,dt \\
&= \frac{1}{m_n}\sum_{k=0}^{N-1}\int_{0\le(k/m_r)\ominus t<1/m_r} m_r\overline{\chi(t,y)}\,dt.
\end{aligned}$$

Making the change of variables $t = (k/m_r) \ominus v$, $0 \le v < 1/m_r$ we obtain

$$\begin{aligned}
\hat{\phi}_{n,r}(y) &= \frac{m_r}{m_n} \sum_{k=0}^{N-1} \int_0^{1/m_r} \overline{\chi\left(\frac{k}{m_r} \ominus v, y\right)}\, dv \\
&= \frac{m_r}{m_n} \sum_{k=0}^{m_n m_r - 1} \overline{\chi\left(\frac{k}{m_r}, y\right)} \int_0^{1/m_r} \chi(v, y)\, dv \\
&= \frac{m_r}{m_n} \sum_{j=0}^{m_n - 1} \sum_{k=jm_r}^{(j+1)m_r - 1} \overline{\chi\left(\frac{k}{m_r}, y\right)} \int_0^{1/m_r} \chi(v, y)\, dv \\
&= \frac{m_r}{m_n} \int_0^{1/m_r} \chi(v, y)\, dv \sum_{j=0}^{m_n - 1} \overline{\chi(j, y)} \sum_{k=0}^{m_r - 1} \overline{\chi\left(\frac{k}{m_r}, y\right)} \\
&= \frac{m_r}{m_n} \overline{D_{m_n}(y)}\, \overline{L_{m_r}(y)} \int_0^{1/m_r} \chi(v, y)\, dv,
\end{aligned}$$

where $L_{m_r}(y)$ was defined in (11.1.1). Using **11.1.3** we find that

$$\hat{\phi}_{n,r}(y) = \begin{cases} \dfrac{1}{m_n} \overline{D_{m_n}(y)}\, \overline{L_{m_r}(y)} & \text{for } 0 \le y < m_r, \\ 0 & \text{for } m_r \le y < \infty. \end{cases}$$

Applying **11.1.1** we obtain

$$\hat{\phi}_{n,r}(y) = \begin{cases} \dfrac{1}{m_n} \overline{D_{m_n}(y)} m_r & \text{for } 0 \le y < 1, \\ 0 & \text{for } 1 \le y < \infty. \end{cases}$$

Substituting the value of $D_{m_n}(y)$ into this expression we conclude that (11.1.6) holds as promised. In particular, integrability of $\hat{\phi}_{n,r}(y)$ is obvious. ∎

We now come to the discrete integral (11.1.1).

11.1.5. *Let $f(t) \in \mathbf{L}^1(0, \infty)$ be $\mathcal{P}$-continuous on $[0, \infty)$ and $\hat{f}(y)$ have compact support, namely, suppose $\hat{f}(y) = 0$ for $y \ge m_r$. Then for the knot discretization $t_k = k/m_r$, $k = 0, 1, \ldots, m_r m_n - 1$, the step function*

$$\tilde{f}(y) = \begin{cases} \dfrac{1}{m_n} \displaystyle\sum_{k=0}^{m_n m_r - 1} f\left(\frac{k}{m_r}\right) \overline{\chi\left(\frac{k}{m_r}, y\right)} & \text{for } 0 \le y < m_r, \\ 0 & \text{for } m_r \le y < \infty \end{cases} \tag{11.1.7}$$

can be obtained from the transform $\hat{f}(y)$ by the formula

$$\tilde{f}(y) = m_r \int_0^{1/m_n} \hat{f}(y \ominus u)\, du, \tag{11.1.8}$$

i.e., $\tilde{f}(y)$ is an average of $\hat{f}(y)$ over the intervals $\delta_\mu(n) = [\mu/m_n, (\mu+1)/m_n)$, $\mu = 0, 1, \ldots$.

PROOF. Consider the function

$$f_1(t) = \phi_{n,r}(t) f(t),$$

where $\phi_{n,r}(t)$ is defined by (11.1.5). Since $\phi_{n,r}(t)$ is bounded, it is clear that $f_1(t) \in \mathbf{L}^1(0,\infty)$ and $\hat{f}_1(y)$ exists. We compute the function $\hat{f}_1(y)$:

$$\begin{aligned}
\hat{f}_1(y) &= \int_0^\infty f_1(t)\overline{\chi(t,y)}\,dt \\
&= \int_0^\infty f(t)\frac{1}{m_n}\sum_{k=0}^{m_n m_r - 1} D\left(\frac{k}{m_r}\ominus t, m_r\right)\overline{\chi(t,y)}\,dt \\
&= \frac{1}{m_n}\sum_{k=0}^{m_n m_r - 1} m_r \int_{0 \le (k/m_r)\ominus t < 1/m_r} f(t)\overline{\chi(t,y)}\,dt.
\end{aligned}$$

Making the change of variables $t = (k/m_r)\ominus u$ and using properties of the function $\chi(x,y)$, we obtain

(11.1.9)

$$\begin{aligned}
\hat{f}_1(y) &= \frac{m_r}{m_n}\sum_{k=0}^{m_n m_r - 1}\int_0^{1/m_r} f\left(\frac{k}{m_r}\ominus u\right)\overline{\chi\left(\frac{k}{m_r}\ominus u, y\right)}\,du \\
&= \frac{m_r}{m_n}\sum_{k=0}^{m_n m_r - 1}\overline{\chi\left(\frac{k}{m_r}, y\right)}\int_0^{1/m_r} f\left(\frac{k}{m_r}\ominus u\right)\chi(u,y)\,du.
\end{aligned}$$

By hypothesis $\hat{f}(y) = 0$ for $y \ge m_r$. By **6.2.12**, $f(t)$ is constant on the intervals $[\nu/m_r, (\nu+1)/m_r)$ and consequently, $f((k/m_r)\ominus u) = f(k/m_r)$ for $0 \le u < 1/m_r$. Hence it follows from **11.1.3** that

$$\begin{aligned}
\hat{f}_1(y) &= \frac{m_r}{m_n}\sum_{k=0}^{m_n m_r - 1}\overline{\chi\left(\frac{k}{m_r}, y\right)} f\left(\frac{k}{m_r}\right)\int_0^{1/m_r}\chi(u,y)\,du \\
&= \begin{cases} \dfrac{1}{m_n}\displaystyle\sum_{k=0}^{m_n m_r - 1} f\left(\frac{k}{m_r}\right)\overline{\chi\left(\frac{k}{m_r}, y\right)} & \text{for } \ 0 \le y < m_r, \\ 0 & \text{for } \ m_r \le y < \infty, \end{cases}
\end{aligned}$$

i.e., (see (11.1.7)) $\tilde{f}(y) = \hat{f}_1(y)$. On the other hand, by **11.1.2** we have

$$\hat{f}_1(y) = \mathcal{F}[\phi_{n,r}\cdot f] = \int_0^\infty \hat{f}(y\ominus v)\hat{\phi}_{n,r}(v)\,dv.$$

We conclude by (11.1.6) and (11.1.7) that

$$\tilde{f}(y) = \hat{f}_1(y) = \begin{cases} m_r \int_0^{1/m_n} \hat{f}(y \ominus v)\, dv & \text{for} \quad 0 \le y < m_r, \\ 0 & \text{for} \quad m_r \le y < \infty. \end{cases}$$ ∎

11.1.6. *Let $f(t) \in \mathbf{L}^1(0,\infty)$ be $\mathcal{P}$-continuous on $[0,\infty)$. If $\hat{f}(y)$ does not have compact support then the function $\tilde{f}(y)$, determined by the knot discretization $t_k = k/m_r$, $k = 0, 1, \ldots, m_r m_n - 1$, can be written in the form*

$$\tag{11.1.10} \tilde{f}(y) = \begin{cases} \dfrac{1}{m_n m_r} \displaystyle\sum_{k=0}^{m_n m_r - 1} f^*\left(\frac{k}{m_r}\right) \overline{\chi\left(\frac{k}{m_r}, y\right)} & \text{for} \quad 0 \le y < m_r, \\ \dfrac{1}{m_n m_r} \displaystyle\sum_{k=0}^{m_n m_r - 1} f_y^*\left(\frac{k}{m_r}\right) \overline{\chi\left(\frac{k}{m_r}, y\right)} & \text{for} \quad m_r \le y < \infty \end{cases}$$

where

$$f^*\left(\frac{k}{m_r}\right) = m_r \int_0^{1/m_r} f\left(\frac{k}{m_r} \ominus u\right) du,$$

and

$$f_y^*\left(\frac{k}{m_r}\right) = m_r \int_0^{1/m_r} f\left(\frac{k}{m_r} \ominus u\right) \prod_{\nu=r+1}^{q(y)} \chi_{y_{-\nu} m_{\nu-1}}(u)\, du,$$

for $[y] = \sum_{\nu=1}^{q(y)} y_{-\nu} m_{\nu-1}$. Moreover, the function $\tilde{f}(y)$ also satisfies the formula (11.1.8).

PROOF. As in the proof of **11.1.5**, we consider the function $f_1(t) = \phi_{n,r}(t) f(t)$, and show that (11.1.9) holds for $\hat{f}_1(y)$. We consider the cases $0 \le y < m_r$ and $m_r \le y < \infty$ separately.

If $0 \le y < m_r$ then by **11.1.3** we have $\chi(u,y) = 1$ for $0 \le u < 1/m_r$. Thus

$$\int_0^{1/m_r} f\left(\frac{k}{m_r} \ominus u\right) \chi(u,y)\, du = \frac{1}{m_r} m_r \int_0^{1/m_r} f\left(\frac{k}{m_r} \ominus u\right) du = \frac{1}{m_r} f^*\left(\frac{k}{m_r}\right).$$

Substituting this identity into (11.1.9), we establish the first part of **11.1.6**.

If $m_r \le y < \infty$ then we can write

$$[y] = \sum_{\nu=1}^{q(y)} y_{-\nu} m_{\nu-1} = \sum_{\nu=1}^{r} y_{-\nu} m_{\nu-1} + \sum_{\nu=r+1}^{q(y)} y_{-\nu} m_{\nu-1} = [y]^* + [y]_r.$$

Since $[y]^* < m_r$, it follows from the multiplicative property of the function $\chi(x,y)$ that

$$\chi(u,y) = \chi(u,[y]^* + [y]_r) = \chi(u,[y]^*)\chi(u,[y]_r) = \chi(u,[y]_r)$$

for $0 \le u < 1/m_r$, i.e., for $m_r \le y < \infty$ and $[y] = [y]^* + [y]_r$ the function $\chi(x,y)$ is identical with

$$\chi\left(u, \sum_{\nu=r+1}^{q(y)} y_{-\nu} m_{\nu-1}\right) = \prod_{\nu=r+1}^{q(y)} \chi(u, y_{-\nu} m_{\nu-1}) = \prod_{\nu=r+1}^{q(y)} \chi_{y_{-\nu} m_{\nu-1}}(u).$$

Consequently, for $m_r \le y < \infty$ we have

$$\begin{aligned}\int_0^{1/m_r} f\left(\frac{k}{m_r} \ominus u\right) \chi(u,y)\,du &= \int_0^{1/m_r} f\left(\frac{k}{m_r} \ominus u\right) \prod_{\nu=r+1}^{q(y)} \chi_{y_{-\nu} m_{\nu-1}}(u)\,du \\ &= \frac{1}{m_r} f_y^*\left(\frac{k}{m_r}\right).\end{aligned}$$

The proof of (11.1.8) in this situation is similar to the proof of **11.1.5.** ∎

A consequence of **6.2.4**, (11.1.8) and (11.1.10) is that

11.1.7. *If $f(t) \in \mathbf{L}^1(0,\infty)$ is bounded and $\mathcal{P}$-continuous on $[0,\infty)$, then*

$$(11.1.11) \qquad \int_{m_r}^{\infty} \left| \frac{1}{m_n m_r} \sum_{k=0}^{m_n m_r - 1} f_y^*\left(\frac{k}{m_r}\right) \overline{\chi\left(\frac{k}{m_r}, y\right)} \right|^2 dy = \int_{m_r}^{\infty} |\hat{f}(y)|^2\,dy.$$

Combining **6.2.6** and (11.1.11) we obtain the estimate

11.1.8. *Let $m_q \le y < m_{q+1}$ for some $q \ge r$, i.e.,*

$$[y] = \sum_{\nu=1}^{q(y)} y_{-\nu} m_{\nu-1}, \qquad 1 \le y_{-q} < p_q - 1, \quad q = q(y) \ge r.$$

If $f(t) \in \mathbf{L}^1(0,\infty)$ is bounded and $\mathcal{P}$- continuous on $[0,\infty)$, then

$$|\tilde{f}(y)| \le \omega_{q(y)}(f).$$

Theorems **11.1.5** and **11.1.6** give a method for approximating the transform $\tilde{f}(y)$ at any point y. However, in questions of spectral processing, of pattern recognition and in a number of other applications the inverse problem is also significant. By the inverse problem we mean the problem of recapturing the values of $f(t_k)$ from the values of $\tilde{f}(y_\nu)$. Since by (11.1.7), $\tilde{f}(y) = 0$ for $y \ge m_r$, we see that the knot discretization y_ν of the function $\tilde{f}(y)$ need only be chosen from the interval

$0 \leq y < m_r$. Furthermore, it follows from (11.1.7) that $\tilde{f}(y)$ is a step function on $[0, m_r]$ which is constant on the intervals $\delta_\ell(n) = [\ell/m_n, (\ell+1)/m_n)$. Partition the interval $[0, m_r)$ into $m_n m_r$ subintervals $\delta_\ell(n)$ and choose some point from each subinterval, for example $y_\ell = \ell/m_n$. Thus we may write

$$\tilde{f}\left(\frac{\ell}{m_n}\right) = \frac{1}{m_n} \sum_{k=0}^{m_n m_r - 1} f\left(\frac{k}{m_r}\right) \overline{\chi\left(\frac{k}{m_r}, \frac{\ell}{m_n}\right)} \tag{11.1.12}$$

for $\ell = 0, 1, \ldots, m_n m_r - 1$

We shall call the transform (11.1.12) the *direct discrete multiplicative transform* (DDMT).

To obtain the inverse transform of (11.1.12), multiply both sides of (11.1.12) by $\chi(\ell/m_n, q/m_r)$ where $q = 0, 1, \ldots, m_n m_r - 1$ and sum over ℓ. Thus

$$\sum_{\ell=0}^{m_n m_r - 1} \tilde{f}\left(\frac{\ell}{m_n}\right) \chi\left(\frac{\ell}{m_n}, \frac{q}{m_r}\right)$$
$$= \frac{1}{m_n} \sum_{\ell=0}^{m_n m_r - 1} \sum_{k=0}^{m_n m_r - 1} f\left(\frac{k}{m_r}\right) \overline{\chi\left(\frac{k}{m_r}, \frac{\ell}{m_n}\right)} \chi\left(\frac{\ell}{m_n}, \frac{q}{m_r}\right).$$

But

$$\frac{1}{m_n} \sum_{\ell=0}^{m_n m_r - 1} \sum_{k=0}^{m_n m_r - 1} f\left(\frac{k}{m_r}\right) \overline{\chi\left(\frac{k}{m_r}, \frac{\ell}{m_n}\right)} \chi\left(\frac{\ell}{m_n}, \frac{q}{m_r}\right)$$
$$= \frac{1}{m_n} \sum_{k=0}^{m_n m_r - 1} f\left(\frac{k}{m_r}\right) \sum_{\ell=0}^{m_n m_r - 1} \chi\left(\frac{\ell}{m_n}, \frac{q}{m_r} \ominus \frac{k}{m_r}\right)$$
$$= \frac{1}{m_n} \sum_{k=0}^{m_n m_r - 1} f\left(\frac{k}{m_r}\right) \sum_{j=0}^{m_r - 1} \sum_{\ell=j m_n}^{(j+1)m_n - 1} \chi\left(\frac{\ell}{m_n}, \frac{q}{m_r} \ominus \frac{k}{m_r}\right)$$
$$= \frac{1}{m_n} \sum_{k=0}^{m_n m_r - 1} f\left(\frac{k}{m_r}\right) \sum_{j=0}^{m_r - 1} \chi\left(j, \frac{q}{m_r} \ominus \frac{k}{m_r}\right) \sum_{\ell=0}^{m_n - 1} \chi\left(\frac{\ell}{m_n}, \frac{q}{m_r} \ominus \frac{k}{m_r}\right).$$

By **11.1.1**

$$\sum_{\ell=0}^{m_n - 1} \chi\left(\frac{\ell}{m_n}, \frac{q}{m_r} \ominus \frac{k}{m_r}\right) = \begin{cases} m_n, & \text{if } \left[\frac{q}{m_r} \ominus \frac{k}{m_r}\right] = 0, \\ 0, & \text{if } \left[\frac{q}{m_r} \ominus \frac{k}{m_r}\right] \geq 1. \end{cases}$$

Moreover,

$$D_{m_r}\left(\frac{q}{m_r}\ominus\frac{k}{m_r}\right)=\sum_{j=0}^{m_r-1}\chi\left(j,\frac{q}{m_r}\ominus\frac{k}{m_r}\right)$$
$$=\begin{cases} m_r, & \text{if } 0<\dfrac{q}{m_r}\ominus\dfrac{k}{m_r}<\dfrac{1}{m_r}, \\ 0, & \text{if } \dfrac{1}{m_r}\le\dfrac{q}{m_r}\ominus\dfrac{k}{m_r}<1, \end{cases}$$

i.e.,

$$D_{m_r}\left(\frac{q}{m_r}\ominus\frac{k}{m_r}\right)=\begin{cases} m_r, & \text{if } k=q, \\ 0, & \text{if } k\neq q. \end{cases}$$

Therefore,

$$\sum_{\ell=0}^{m_nm_r-1}\tilde{f}\left(\frac{\ell}{m_n}\right)\chi\left(\frac{\ell}{m_n},\frac{q}{m_r}\right)=\frac{1}{m_n}f\left(\frac{q}{m_r}\right)m_nm_r.$$

We conclude that

$$f\left(\frac{q}{m_r}\right)=\frac{1}{m_r}\sum_{\ell=0}^{m_nm_r-1}\tilde{f}\left(\frac{\ell}{m_n}\right)\chi\left(\frac{\ell}{m_n},\frac{q}{m_r}\right) \tag{11.1.13}$$

for $q=0,1,\dots m_nm_r-1$.

We shall call the transform (11.1.13) the *inverse discrete multiplicative transform* (IDMT).

§11.2. Computation of the discrete multiplicative transform.

We shall consider the question of efficiently computing the multiplicative transforms (11.1.12) and (11.1.13). By an arithmetic calculation we shall mean a complex multiplication followed by complex addition. Clearly, the values $\tilde{f}(\ell/m_n)$, $\ell=0,1,\dots,m_nm_r-1$, can be computed using $(m_nm_r)^2$ arithmetic calculations. However, analogous to the discrete Fourier transform case, it is possible to use what is called a *fast algorithm* to compute these values using fewer arithmetic calculations. The idea behind this comes from a special way of representing the quantities $\tilde{f}(\ell/m_n)$.

11.2.1. *Let j and ν be integers with $j\in[0,m_n-1]$, $\nu\in[0,m_r-1]$ written in the form*

$$j=\sum_{\mu=1}^{n}j_{-\mu}m_{\mu-1},\quad \nu=\sum_{\mu=1}^{r}\nu_{-\mu}m_{\mu-1},\qquad 0\le j_{-\mu},\nu_{-\mu}\le p_\mu-1.$$

Then

(11.2.1)

$$\tilde{f}\left(\frac{\ell}{m_n}\right) = \frac{1}{m_n}\sum_{\nu_{-r}=0}^{p_r-1} \overline{\chi}_{[\ell/m_n]}\left(\frac{\nu_{-r}}{p_r}\right)\sum_{\nu_{-r+1}=0}^{p_{r-1}-1} \overline{\chi}_{[\ell/m_n]}\left(\frac{\nu_{-r+1}}{p_r p_{r-1}}\right)\cdots$$

$$\cdots \sum_{\nu_{-2}=0}^{p_2-1} \overline{\chi}_{[\ell/m_n]}\left(\frac{\nu_{-2}}{p_r\dots p_2}\right)\sum_{\nu_{-1}=0}^{p_1-1} \overline{\chi}_{[\ell/m_n]}\left(\frac{\nu_{-1}}{p_r\dots p_1}\right)\times$$

$$\times \sum_{j_{-n}=0}^{p_n-1} \overline{\chi}_{j_{-n}m_{n-1}}\left(\left\{\frac{\ell}{m_n}\right\}\right)\sum_{j_{-n+1}=0}^{p_{n-1}-1} \overline{\chi}_{j_{-n+1}m_{n-2}}\left(\left\{\frac{\ell}{m_n}\right\}\right)\cdots$$

$$\cdots \sum_{j_{-2}=0}^{p_2-1} \overline{\chi}_{j_{-1}m_1}\left(\left\{\frac{\ell}{m_n}\right\}\right)\times$$

$$\times \sum_{j_{-1}=0}^{p_1-1} f\left(\frac{\nu_{-1}+\cdots+\nu_{-r}m_{r-1}+(j_{-1}+\cdots+j_{-n}m_{n-1})m_r}{m_r}\right)\overline{\chi}_{j_{-1}}\left(\left\{\frac{\ell}{m_n}\right\}\right).$$

PROOF. Set $k = jm_r + \nu$ for $0 \le j \le m_n - 1$, $0 \le \nu \le m_r - 1$ and use the multiplicative property of the function $\chi(x, y)$ to see that

$$\begin{aligned}
\tilde{f}\left(\frac{\ell}{m_n}\right) &= \frac{1}{m_n}\sum_{j=0}^{m_n-1}\sum_{k=jm_r}^{(j+1)m_r-1} f\left(\frac{k}{m_r}\right)\overline{\chi}\left(\frac{k}{m_r},\frac{\ell}{m_n}\right)\\
&= \frac{1}{m_n}\sum_{j=0}^{m_n-1}\sum_{\nu=0}^{m_r-1} f\left(j+\frac{\nu}{m_r}\right)\overline{\chi}\left(j+\frac{\nu}{m_r},\frac{\ell}{m_n}\right)\\
&= \frac{1}{m_n}\sum_{\nu=0}^{m_r-1}\overline{\chi}\left(\frac{\nu}{m_r},\frac{\ell}{m_n}\right)\sum_{j=0}^{m_n-1} f\left(j+\frac{\nu}{m_r}\right)\overline{\chi}_j\left(\left\{\frac{\ell}{m_n}\right\}\right)\\
&= \frac{1}{m_n}\sum_{\nu_{-r}=0}^{p_r-1}\cdots\sum_{\nu_{-2}=0}^{p_2-1}\sum_{\nu_{-1}=0}^{p_1-1}\overline{\chi}_{[\ell/m_n]}\left(\frac{\nu_{-1}+\cdots+\nu_{-r}m_{r-1}}{m_r}\right)\times\\
&\quad\times\sum_{j_{-n}=0}^{p_n-1}\cdots\sum_{j_{-2}=0}^{p_2-1}\sum_{j_{-1}=0}^{p_1-1} f\left(\frac{\cdots}{m_r}\right)\overline{\chi}_{j_{-1}+\cdots+j_{-n}m_{n-1}}\left(\left\{\frac{\ell}{m_n}\right\}\right)\\
&= \frac{1}{m_n}\sum_{\nu_{-r}=0}^{p_r-1}\overline{\chi}_{[\ell/m_n]}\left(\frac{\nu_{-r}}{p_r}\right)\cdots\sum_{\nu_{-2}=0}^{p_2-1}\overline{\chi}_{[\ell/m_n]}\left(\frac{\nu_{-2}}{p_r\dots p_2}\right)\times\\
&\quad\times\sum_{\nu_{-1}=0}^{p_1-1}\overline{\chi}_{[\ell/m_n]}\left(\frac{\nu_{-1}}{p_r\dots p_1}\right)\sum_{j_{-n}=0}^{p_n-1}\overline{\chi}_{j_{-n}m_{n-1}}\left(\left\{\frac{\ell}{m_n}\right\}\right)\cdots\\
&\quad\cdots\sum_{j_{-2}=0}^{p_2-1}\overline{\chi}_{j_{-2}m_1}\left(\left\{\frac{\ell}{m_n}\right\}\right)\sum_{j_{-1}=0}^{p_1-1} f\left(\frac{\cdots}{m_r}\right)\overline{\chi}_{j_{-1}}\left(\left\{\frac{\ell}{m_n}\right\}\right),
\end{aligned}$$

where

$$f\left(\frac{\cdots}{m_r}\right) = f\left(\frac{\nu_{-1} + \cdots + \nu_{-r}m_{r-1} + (j_{-1} + \cdots + j_{-n}m_{n-1})m_r}{m_r}\right). \quad \blacksquare$$

For each fixed $\ell = 0, 1, \ldots, m_n m_r - 1$, the sums in (11.2.1) (beginning with the sum over j_{-1}) take respectively $p_1, p_2, \ldots p_n, p_1, \ldots, p_r$ arithmetic operations. Thus we have proved the following result:

11.2.2. *The values $\tilde{f}(\ell/m_n)$, $\ell = 0, 1, \ldots, m_n m_r - 1$ of the discrete multiplicative transform (11.1.12) can be computed in*

$$K = (p_1 + p_2 + \ldots p_n + p_1 + \cdots + p_r)m_n m_r \simeq (\ln m_n + \ln m_r)m_n m_r$$

arithmetic operations.

A similar result holds for the inverse discrete multiplicative transform (11.1.13).

The transforms (11.1.12) and (11.1.13) are more conveniently written in matrix form. To see this we introduce the notation

$$x_k = f\left(\frac{k}{m_r}\right), \quad y_\ell = \tilde{f}\left(\frac{\ell}{m_n}\right), \qquad k, \ell = 0, 1, \ldots m_n m_r - 1, \tag{11.2.2}$$

$$X^T = (x_0, x_1, \ldots, x_{m_n m_r - 1}),$$
$$Y^T = (y_0, y_1, \ldots, y_{m_n m_r - 1}),$$

and $W = (W_{\ell,k})$, $k, \ell = 0, 1, \ldots m_n m_r - 1$, where $W_{\ell,k} = \overline{\chi}(k/m_r, \ell/m_n)$. We shall call the vector Y the ***spectrum of the input vector*** X.

Using this notation we can write the transform (11.1.12) in the form

$$Y = \frac{1}{m_n} WX,$$

and the inverse transform (11.1.13) in the form

$$X = \frac{1}{m_r} W^* X,$$

where $W^* = (W^*_{q,\ell})$, $W^*_{q,\ell} = \overline{\chi}(\ell/m_n, k/m_r)$, $q, \ell = 0, 1, \ldots m_n m_r - 1$.

Writing (11.1.12) in the form (11.2.1) means that the $m_n m_r \times m_n m_r$ matrix W is a product of $n + r$ sparse matrices $W^{(\nu)}$, each of order $m_n m_r \times m_n m_r$. Namely,

$$W = W^{(n+r)} W^{(n+r-1)} \ldots W^{(n+1)} W^{(n)} \ldots W^{(1)},$$

where for $1 \le j \le n$ the matrices $W^{(j)}$ contain only p_j non-zero elements in each row (each a power of $q = \exp(-2\pi i/p_j)$), and for $n + 1 \le j \le n + r$ each row contains p_{j-n} non-zero entries (each a power of $q = \exp(-2\pi i/p_{j-n})$).

We shall presently give a detailed account of the DMT (11.1.12) for the case when $p_k = p \ge 2$ for all $k = 1, 2, \ldots$.

We begin by establishing the following result:

11.2.3. *The identity*

$$\chi(k/p^r, \ell/p^n) = \chi_k(\ell/p^{n+r}) = \chi_\ell(k/p^{n+r})$$

holds for any choice of $k = 0, 1, \dots p^{r+n} - 1$ *and* $\ell = 0, 1, \dots p^{r+n} - 1$.

PROOF. This identity is obvious for $k = 0$ or $\ell = 0$. Let

$$k = \sum_{\nu=1}^{n+r} k_{-\nu} p^{\nu-1}, \qquad k_{-\nu} = 0 \quad \text{for} \quad \nu \geq n + r + 1,$$

$$\ell = \sum_{\nu=1}^{n+r} \ell_{-\nu} p^{\nu-1}, \qquad \ell_{-\nu} = 0 \quad \text{for} \quad \nu \geq n + r + 1,$$

and

$$k_{-\nu} \equiv \left[\frac{k}{p^{\nu-1}}\right] \pmod p, \quad \ell_{-\nu} \equiv \left[\frac{\ell}{p^{\nu-1}}\right] \pmod p.$$

Furthermore, set

$$\frac{k}{p^r} = \sum_{\nu=1}^{\infty} k^*_{-\nu} p^{\nu-1} + \sum_{\nu=1}^{\infty} \frac{k^*_\nu}{p^\nu},$$

$$\frac{\ell}{p^n} = \sum_{\nu=1}^{\infty} \ell^*_{-\nu} p^{\nu-1} + \sum_{\nu=1}^{\infty} \frac{\ell^*_\nu}{p^\nu},$$

$$\frac{\ell}{p^{n+r}} = \sum_{\nu=1}^{\infty} \tilde{\ell}_{-\nu} p^{\nu-1} + \sum_{\nu=1}^{\infty} \frac{\tilde{\ell}_\nu}{p^\nu},$$

$$(\tilde{\ell}_{-\nu} = 0 \quad \text{for} \quad \nu = 1, 2, \dots),$$

$$\frac{k}{p^{n+r}} = \sum_{\nu=1}^{\infty} \tilde{k}_{-\nu} p^{\nu-1} + \sum_{\nu=1}^{\infty} \frac{\tilde{k}_\nu}{p^\nu},$$

$$(\tilde{k}_{-\nu} = 0 \quad \text{for} \quad \nu = 1, 2, \dots).$$

We have

$$k^*_{-\nu} \equiv \left[\frac{k}{p^r p^{\nu-1}}\right] \pmod p = k_{r+\nu} \quad \text{for} \quad \nu \leq n,$$

$$k^*_{-\nu} = 0 \quad \text{for} \quad \nu \geq n + 1,$$

$$k^*_\nu \equiv \left[\frac{k}{p^r} p^\nu\right] \pmod p = \begin{cases} 0 & \text{for} \quad \nu \geq r + 1, \\ \left[\dfrac{k}{p^{r-\nu+1-1}}\right] \pmod p = k_{-(r-\nu+1)} & \text{for} \quad 1 \leq \nu \leq r, \end{cases}$$

i.e.,

$$\frac{k}{p^r} = \sum_{\nu=1}^{n} k^*_{-\nu} p^{\nu-1} + \sum_{\nu=1}^{r} \frac{k^*_\nu}{p^\nu} = \sum_{\nu=1}^{n} k_{-(r+\nu)} p^{\nu-1} + \sum_{\nu=1}^{r} \frac{k_{-(r-\nu+1)}}{p^\nu}.$$

Similarly,

$$\frac{\ell}{p^n} = \sum_{\nu=1}^{r} \ell^*_{-\nu} p^{\nu-1} + \sum_{\nu=1}^{n} \frac{\ell^*_\nu}{p^\nu} = \sum_{\nu=1}^{r} \ell_{-(n+\nu)} p^{\nu-1} + \sum_{\nu=1}^{n} \frac{\ell_{-(n-\nu+1)}}{p^\nu},$$

$$\frac{k}{p^{n+r}} = \sum_{\nu=1}^{n+r} \frac{\tilde{k}_\nu}{p^\nu} = \sum_{\nu=1}^{n+r} \frac{k_{-(n+r-\nu+1)}}{p^\nu},$$

$$\frac{\ell}{p^{n+r}} = \sum_{\nu=1}^{n+r} \frac{\tilde{\ell}_\nu}{p^\nu} = \sum_{\nu=1}^{n+r} \frac{\ell_{-(n+r-\nu+1)}}{p^\nu}.$$

Using the definition (1.5.33) of the function $\chi(x, y)$, we obtain

(11.2.3)

$$\chi\left(\frac{k}{p^r}, \frac{\ell}{p^n}\right) = \exp \frac{2\pi i}{p} \sum_{\nu=1}^{\infty} (k^*_{-\nu} \ell^*_\nu + k^*_\nu \ell^*_{-\nu})$$

$$= \exp \frac{2\pi i}{p} \left(\sum_{\nu=1}^{n} k_{-(r+\nu)} \ell_{-(n-\nu+1)} + \sum_{\nu=1}^{r} k_{-(r-\nu+1)} \ell_{-(n+\nu)} \right),$$

$$\chi\left(k, \frac{\ell}{p^{n+r}}\right) = \exp \frac{2\pi i}{p} \sum_{\nu=1}^{\infty} k_{-\nu} \tilde{\ell}_\nu = \exp \frac{2\pi i}{p} \sum_{\nu=1}^{n+r} k_{-\nu} \ell_{-(n+r-\nu+1)}, \tag{11.2.4}$$

and

$$\chi\left(\frac{k}{p^{n+r}}, \ell\right) = \exp \frac{2\pi i}{p} \sum_{\nu=1}^{n+r} k_{-(n+r-\nu+1)} \ell_{-\nu}. \tag{11.2.5}$$

The sum on the right side of (11.2.5) differs from the one in (11.2.4) only in the order of summation. This can be seen by changing variables with $\nu = n + r - \nu' + 1$. Similarly, changes of variables $\nu = \nu' - r$ in the second sum in (11.2.3) and $\nu = r + 1 - \nu'$ in the third sum transform the right side of (11.2.3) into the right side of (11.2.4). This proves **11.2.3.** ∎

Using the notation (11.2.2) and applying **11.2.3**, we see that when $p_1 = p_2 = \cdots = p \geq 2$, the transforms (11.2.12) and (11.1.13) can be written in the form:

$$y_\ell = \frac{1}{p^n} \sum_{k=0}^{p^{n+r}-1} x_k \overline{\chi}_k \left(\frac{\ell}{p^{n+r}}\right), \qquad \ell = 0, 1, \ldots, p^{n+r} - 1,$$

$$x_k = \frac{1}{p^r} \sum_{\ell=0}^{p^{n+r}-1} y_\ell \chi_\ell\left(\frac{k}{p^{n+r}}\right), \qquad k = 0, 1, \dots, p^{n+r} - 1.$$

Since n and r are arbitrary natural numbers, replacing $n + r$ by n and transferring the factor in front of the second sum to the first one leads us to the formulae

$$(11.2.6) \qquad y_\ell = \frac{1}{p^n} \sum_{k=0}^{p^n-1} x_k \overline{\chi}_k\left(\frac{\ell}{p^n}\right), \qquad \ell = 0, 1, \dots, p^n - 1,$$

$$(11.2.7) \qquad x_k = \sum_{\ell=0}^{p^n-1} y_\ell \chi_\ell\left(\frac{k}{p^n}\right), \qquad k = 0, 1, \dots, p^n - 1.$$

We conclude that if $W = \left(\overline{\chi}_k\left(\frac{\ell}{p^n}\right)\right)$, $\ell, k = 0, 1, \dots, p^n - 1$ is the DDMT matrix, then the IDMT matrix W^* is the complex conjugate of W, i.e., $W^* = \overline{W}$. In particular, computing the DDMT is essentially the same as computing the IDMT and we shall confine our attention to one of these, namely the DDMT (11.2.6).

As we remarked above, the matrix $W = \left(\overline{\chi}_k\left(\frac{\ell}{p^n}\right)\right)$, $\ell, k = 0, 1, \dots, p^n - 1$ of the transform (11.2.6) can be written as a product of n matrices

$$W^{(n)}W^{(n-1)} \dots W^{(1)}$$

of order $p^n \times p^n$ in which each row and each column contains exactly p entries different from zero, and each of these entries is a power of the quantity $q = \exp(-2\pi i/p)$. Although in the general case $W^{(j)} \neq W^{(\nu)}$ for $j \neq \nu$, this representation of W as a product of sparse matrices is not unique. In fact, each different representation generates its own algorithm for computing the DMT. Following the work of Žukov [1], we shall show that the matrix W can be represented as $W = CB^n$, where C is a permutation matrix and B is a sparse block-type matrix. (Recall that a permutation matrix is a matrix whose rows have only one non- zero entry and that every non-zero entry is 1. Multiplication of a column vector by such a matrix results in a permutation of the coordinates of the vector.)

11.2.4. *For $n \geq 2$ there exists a sparse matrix B of order $p^n \times p^n$ such that the matrix $W = \left(\overline{\chi}_k\left(\frac{\ell}{p^n}\right)\right)$, $\ell, k = 0, 1, \dots, p^n - 1$ of the discrete transform (11.2.6) can be written in the form $W = CB^n$, where C is a permutation matrix of order $p^n \times p^n$.*

PROOF. For $\ell, k = 0, 1, \dots, p^n - 1$, denote the elements of the unknown matrix B by $b_{\ell,k}$. Write ℓ and k in the form

$$\ell = \sum_{\nu=1}^{n} \ell_\nu^* p^{\nu-1} = \ell_1^* + p \sum_{\nu=2}^{n} \ell_\nu^* p^{\nu-2} = \ell_1^* + p\gamma_\ell,$$

$$k = \sum_{\nu=1}^{n} k_\nu^* p^{\nu-1} = \sum_{\nu=1}^{n-1} k_\nu^* p^{\nu-1} + k_n^* p^{n-1} = \beta_k + k_n^* p^{n-1}.$$

Here, $k_\nu^* = k_{-\nu}$, $\ell_\nu^* = \ell_{-\nu}$, and $\gamma_\ell = \beta_k = 0$ for $n = 1$.

Set

$$b_{\ell,k} = \begin{cases} \delta_{\gamma_\ell,\beta_k} q^{\ell_1^* k_n^*} & \text{for} \quad n \geq 2, \\ q^{\ell_1^* k_1^*} & \text{for} \quad n = 1, \end{cases} \tag{11.2.8}$$

where

$$q = \exp\frac{-2\pi i}{p}, \qquad \delta_{\mu,\nu} = \begin{cases} 1 & \text{for} \quad \nu = \mu, \\ 0 & \text{for} \quad \nu \neq \mu. \end{cases}$$

Since

$$\begin{aligned} \delta_{\gamma_\ell,\beta_k} &= \delta_{\sum_{\nu=2}^{n} \ell_\nu^* p^{\nu-2}, \sum_{\nu=1}^{n-1} k_\nu^* p^{\nu-1}} \\ &= \begin{cases} \prod_{\nu=2}^{n} \delta_{\ell_\nu^*, k_{\nu-1}^*} & \text{for} \quad n \geq 2, \\ 1 & \text{for} \quad n = 1, \end{cases} \end{aligned} \tag{11.2.9}$$

we have

$$\delta_{\gamma_\ell,\beta_k} = \begin{cases} 0, & \text{if } \ell_\nu^* \neq k_{\nu-1}^* \text{ for some } \nu = 2, 3, \ldots, n, \\ 1, & \text{if } \ell_\nu^* = k_{\nu-1}^* \text{ for all } \nu = 2, 3, \ldots, n. \end{cases}$$

Therefore, for the case $n \geq 2$,

$$b_{\ell,k} = \begin{cases} q^{\ell_1^* k_n^*}, & \text{if} \quad \ell = \ell_1^* + \sum_{\nu=2}^{n} k_{\nu-1}^* p^{\nu-1}, \\ 0 & \text{for all other pairs } (\ell, k). \end{cases}$$

We shall prove by induction that the elements $b_{\ell,k}^{(m)}$ of the matrices B^m can be written in the form

$$b_{\ell,k}^{(m)} = \begin{cases} \prod_{j=m+1}^{n} \delta_{\ell_j^*, k_{j-m}^*} q^{\sum_{\nu=1}^{m} \ell_\nu^* k_{n-m+\nu}^*} & \text{for} \quad m \leq n-1, \\ q^{\sum_{\nu=1}^{n} \ell_\nu^* k_\nu^*} & \text{for} \quad m = n. \end{cases} \tag{11.2.10}$$

The proof of (11.2.10) begins with the case $m = 2$. Notice that $B^2 = B \cdot B$ implies

$$b_{\ell,k}^{(2)} = \sum_{\nu=1}^{p^n-1} b_{\ell,\nu} b_{\nu,k}, \qquad \ell, k = 0, 1, \ldots, p^n - 1.$$

Thus in the case $n \geq 3$ it follows from (11.2.8) and (11.2.9) that

$$b_{\ell,k}^{(2)} = \sum_{\nu_n^*=0}^{p-1} \cdots \sum_{\nu_1^*=0}^{p-1} \left(b_{\ell, \sum_{j=1}^{n} \nu_j^* p^{j-1}} \right) \prod_{j=2}^{n} \delta_{\nu_j^*, k_{j-1}^*} q^{\nu_1^* k_n^*},$$

$$\nu = \sum_{j=1}^{n} \nu_j^* p^{j-1}.$$

In the case $n = 2$ we have

$$b_{\ell,k}^{(2)} = \sum_{\nu_2^*=0}^{p-1} \sum_{\nu_1^*=0}^{p-1} (b_{\ell,\nu_1^*+p\nu_2^*} \delta_{\nu_2^*,k_1^*}) q^{\nu_1^* k_2^*}.$$

Since ℓ and k are fixed, the product of Kronecker delta symbols above is different from zero only when $\nu_j^* = k_{j-1}^*$ for all $j = 2, 3, \ldots, n$. Thus we see that

$$b_{\ell,k}^{(2)} = \begin{cases} \sum_{\nu_1^*=0}^{p-1} \left(b_{\ell,\nu_1^*+\sum_{j=2}^{n} k_{j-1}^* p^{j-1}} \right) q^{\nu_1^* k_n^*} & \text{for} \quad n \geq 3, \\ \sum_{\nu_1^*=0}^{p-1} b_{\ell,\nu_1^*+k_1^* p} q^{\nu_1^* k_2^*} & \text{for} \quad n = 2. \end{cases}$$

Applying (11.2.8) and (11.2.9) again, we obtain

$$\begin{aligned} b_{\ell,k}^{(2)} &= \sum_{\nu_1^*=0}^{p-1} \left(b_{\ell_1^*+p\sum_{j=2}^{n} \ell_j^* p^{j-2},\nu_1^*+\sum_{j=2}^{n} k_{j-1}^* p^{j-1}} \right) q^{\nu_1^* k_n^*} \\ &= \sum_{\nu_1^*=0}^{p-1} \left(\delta_{\sum_{j=2}^{n} \ell_j^* p^{j-2},\nu_1^*+\sum_{j=2}^{n} k_{j-1}^* p^{j-1}} \right) q^{\ell_1^* k_{n-1}^*+\nu_1^* k_n^*}. \end{aligned}$$

Unless $\nu_1^* = \ell_2^*$, the symbol $\delta_{\ell_2^*,\nu_1^*} = 0$. Thus

$$b_{\ell,k}^{(2)} = \begin{cases} \prod_{j=3}^{n} \delta_{\ell_j^*,k_{j-2}^*} q^{\ell_1^* k_{n-1}^*+\ell_2^* k_n^*} & \text{for} \quad n \geq 3, \\ q^{\ell_1^* k_1^*+\ell_2^* k_2^*} & \text{for} \quad n = 2, \end{cases}$$

i.e., (11.2.10) holds for $m = 2$.

Suppose now that (11.2.10) holds for all $m \leq n - 1$. Thus

$$b_{\ell,k}^{(m+1)} = \sum_{\nu=0}^{p^n-1} b_{\ell,\nu} b_{\nu,k}^{(m)} = \sum_{\nu_n^*=0}^{p-1} \cdots \sum_{\nu_1^*=0}^{p-1} b_{\ell,\nu} \prod_{j=m+1}^{n} \delta_{\nu_j^*,k_{j-m}^*} q^{\sum_{i=1}^{m} \nu_i^* k_{n-m+i}^*}.$$

Since this product of Kronecker delta symbols is different from zero only when $\nu_j^* = k_{j-m}^*$ for all $j = m + 1, \ldots, n$, it is evident that

$$\begin{aligned} b_{\ell,k}^{(m+1)} &= \sum_{\nu_m^*=0}^{p-1} \cdots \sum_{\nu_1^*=0}^{p-1} \left(b_{\ell_1^*+p\alpha,\beta} \right) q^{\sum_{i=1}^{m} \nu_i^* k_{n-m+i}^*} \\ &= \sum_{\nu_m^*=0}^{p-1} \cdots \sum_{\nu_1^*=0}^{p-1} \delta_{\alpha,\beta} q^{\ell_1^* k_{n-m}^*+\sum_{i=1}^{m} \nu_i^* k_{n-m+i}^*}, \end{aligned}$$

where

$$\alpha = \sum_{j=2}^{n} \ell_j^* p^{j-2}, \qquad \beta = \sum_{j=1}^{m} \nu_j^* p^{j-1} + \sum_{j=m+1}^{n-1} k_{j-m}^* p^{j-1}$$

and the empty sum (i.e., when $m+1=n$) is interpreted to be zero. The factors involving the quantities ν_j^* will be different from zero only when $\nu_j^* = \ell_{j+1}^*$ for all $j = 1, 2, \dots, m$. Consequently,

$$b_{\ell,k}^{(m+1)} = \begin{cases} \prod_{j=m+2}^{n} \delta_{\ell_j^*, k_{j-m-1}^*} q^{\ell_1^* k_{n-m}^* + \sum_{i=1}^{m} \ell_{i+1}^* k_{n-m+i}^*} & \text{for} \quad m < n-1, \\ q^{\sum_{i=1}^{n} \ell_i^* k_i^*} & \text{for} \quad m = n-1 \end{cases}$$

$$= \begin{cases} \prod_{j=m+2}^{n} \delta_{\ell_j^*, k_{j-(m+1)}^*} q^{\sum_{i=1}^{m+1} \ell_i^* k_{n-(m+1)+i}^*} & \text{for} \quad m < n-1, \\ q^{\sum_{i=1}^{n} \ell_i^* k_i^*} & \text{for} \quad m = n-1, \end{cases}$$

which establishes (11.2.10).

Recall that

$$W_{\ell,k} = \overline{\chi}_k \left(\frac{\ell}{p^n} \right) = \exp \frac{-2\pi i}{p} \sum_{\nu=1}^{n} k_{-\nu} \left(\frac{\ell}{p^n} \right)_\nu,$$

and use the relationship

$$\left(\frac{\ell}{p^n} \right)_\nu \equiv \left[\frac{\ell}{p^n} p^\nu \right] \pmod p = \ell_{-(n-\nu+1)}$$

to write

$$W_{\ell,k} = q^{\sum_{\nu=1}^{n} k_{-\nu} \ell_{-(n-\nu+1)}} = q^{\sum_{\nu=1}^{n} k_\nu^* \ell_{n-\nu+1}^*} \tag{11.2.11}$$

for $\ell, k = 0, 1, \dots, p^n - 1$. Comparing these expressions with the identities

$$b_{\ell,k}^{(n)} = q^{\sum_{\nu=1}^{n} k_\nu^* \ell_\nu^*}, \qquad \ell, k = 0, 1, \dots, p^n - 1$$

we see that the elements of the matrix B^n can be obtained from the elements of the matrix W by a $\mathcal{P}$-adic permutation of the entries of each row, i.e., we can write

$$W = CB^n$$

where $C = (c_{\ell,k})$, $\ell, k = 0, 1, \dots, p^n - 1$, and

$$c_{\ell,k} = \prod_{j=1}^{n} \delta_{\ell_j^*, k_{n-j+1}^*}.$$

In fact,

$$\begin{aligned} W_{\ell,k} &= \sum_{\nu=0}^{p^n-1} c_{\ell,\nu} b_{\nu,k}^{(n)} \\ &= \sum_{\nu_n^*=0}^{p-1} \cdots \sum_{\nu_1^*=0}^{p-1} c_{\ell,\nu} q^{\sum_{j=1}^n \nu_j^* k_j^*} \\ &= \sum_{\nu_n^*=0}^{p-1} \cdots \sum_{\nu_1^*=0}^{p-1} \prod_{i=0}^{n} \delta_{\ell_i^*, \nu_{n-i+1}^*} q^{\sum_{j=1}^n \nu_j^* k_j^*}. \end{aligned}$$

Since $\delta_{\ell_i^*, \nu_{n-i+1}^*} \neq 0$ only when $\nu_{n-i+1}^* = \ell_i^*$, it follows that

$$W_{\ell,k} = q^{\sum_{j=1}^n \nu_{n-j+1}^* k_{n-j+1}^*} = q^{\sum_{j=1}^n \ell_j^* k_{n-j+1}^*},$$

and the proof of (11.2.11) is complete. ∎

Writing the matrix W of the transform (11.2.6) in the form $W = CB^n$ has several advantages over the factorization $W = W^{(n)}W^{(n-1)} \dots W^{(1)}$ in which $W^{(j)} \neq W^{(\nu)}$:

a) The product of the matrix W with a vector X, i.e., the computation $WX = CB^nX$, can be accomplished by n iterations of a single instruction calculation, namely, successive multiplications by the matrix B of the vectors $Z_\nu = B^\nu X$, $\nu = 1, 2, \dots, n-1$, followed by a $\mathcal{P}$-adic permutation. This significantly simplifies the algorithm and programming necessary to perform this product.

b) Since each of the p rows of the matrix B:

$$kp \leq \ell \leq (k+1)p - 1, \qquad k = 0, 1, \dots, p^{n-1} - 1,$$

can be obtained by stretching out the matrix

$$B_1 = \begin{pmatrix} 1 & 1 & 1 & \dots & 1 \\ 1 & q & q^2 & \dots & q^{p-1} \\ 1 & q^2 & (q^4)_* & \dots & (q^{2(p-1)})_* \\ \dots & \dots & \dots & \dots & \dots \\ 1 & q^{p-1} & (q^{2(p-1)})_* & \dots & (q^{(p-1)2})_* \end{pmatrix}$$

$$q = \exp \frac{-2\pi i}{p}, \qquad (q^m)_* = q^{m_1},$$

$$m_1 \equiv m \pmod{p}, \qquad 0 \leq m_1 \leq p-1$$

by judiciously placing zero columns in it, then the computation can be executed in parallel by single instruction processors, the number of which is p^j where $j = 1, 2, \dots, p^{n-1}$.

c) Multiplication of the elements of B_1 by corresponding components of the vector X (or the vectors Z_ν) can be done in the complex basis $(1, i)$ using complex multiplication by the quantity $q^\nu = \exp(-2\pi i\nu/p)$, $\nu = 0, 1, \ldots, p-1$, followed by addition. Each complex multiplication can be accomplished by four real multiplications and two additions. However, it is more appropriate to carry out these computations in the pseudo-complex basis $(1, q, \ldots, q^{p-2})$. In this case, we use the fact that $1+q+\ldots q^{p-1} = 0$ or $q^{p-1} = -(1+q+\ldots q^{p-2})$ to verify the relationships

$$(a_1 + a_2 q + \cdots + a_{p-2}q^{p-2})q = -a_{p-2} + (a_0 - a_{p-2})q + \cdots + (a_{p-3} - a_{p-2})q^{p-2},$$

$$(a_1 + a_2 q + \cdots + a_{p-2}q^{p-2})q^2 = (a_{p-2} - a_{p-3}) - a_{p-3}q + \cdots + (a_{p-4} - a_{p-3})q^{p-2},$$

$$\cdots$$

$$\begin{aligned}(a_1 &+ a_2 q + \cdots + a_{p-2}q^{p-2})q^{p-1} \\ &= -a_1(1 + q + \ldots q^{p-2}) + a_2 + a_3 q + \cdots + a_{p-2}q^{p-3} \\ &= (a_2 - a_1) + (a_3 - a_1)q + \cdots + (a_{p-2} - a_1)q^{p-3} - a_1 q^{p-2}.\end{aligned}$$

Observe that calculating coordinates in the pseudo-complex basis $(1, q, \ldots q^{p-2})$ requires only the addition operation. Moreover, notice that if p is a composite number, say $p = p_1 p_2 = 2 \cdot 3$, then components of the basis which have different signs are considered different from one another ($q^3 = \exp(-2\pi i \cdot 3/6) = -1$).

We examine this computation in the special case when $p = 3$, i.e., in the pseudo-complex basis $(1, q)$, where $q = \exp(-2\pi i/3)$. Let

$$B_1 = \begin{pmatrix} 1 & 1 & 1 \\ 1 & q & q^2 \\ 1 & q^2 & q \end{pmatrix}$$

represent the corresponding block which operates on the components of X or Z_ν, and represent the components of Z_ν in the pseudo-basis $(1, q)$ by $z_1 = \alpha_1 \cdot 1 + \beta_1 q$, $z_2 = \alpha_2 \cdot 1 + \beta_2 q$, and $z_3 = \alpha_3 \cdot 1 + \beta_3 q$. Using the relationships

$$zq = (\alpha \cdot 1 + \beta q)q = \alpha q + \beta q^2 = \alpha q - \beta(1 + q) = -\beta + (\alpha - \beta)q,$$

$$zq^2 = (\alpha \cdot 1 + \beta q)q^2 = \alpha q^2 + \beta q^3 = \beta - \alpha(1 + q) = (\beta - \alpha) - \alpha q,$$

we obtain

$$\begin{aligned} B_1 \begin{pmatrix} z_1 \\ z_2 \\ z_3 \end{pmatrix} &= \begin{pmatrix} z_1 + z_2 + z_3 \\ z_1 + z_2 q + z_3 q^2 \\ z_1 + z_2 q^2 + z_3 q \end{pmatrix} \\ &= \begin{pmatrix} (\alpha_1 + \alpha_2 + \alpha_3) + (\beta_1 + \beta_2 + \beta_3)q \\ (\alpha_1 - \beta_2 + \beta_3 - \alpha_3) + (\beta_1 + \alpha_2 - \beta_2 - \alpha_3)q \\ (\alpha_1 + \beta_2 - \alpha_2 - \beta_3) + (\beta_1 - \alpha_2 + \alpha_3 - \beta_3)q \end{pmatrix}. \end{aligned}$$

Consequently, all the pseudo-components of the vector B_1Z in the basis $(1,q)$ can be computed by using 16 additions followed by 16 multiplications and 12 additions in the basis $(1,i)$.

In addition to eliminating some multiplications, we notice that using the pseudo-basis $(1,q,\ldots,q^{p-2})$ to compute the DMT allows us to simultaneously compute the discrete transform of the data in n steps from only $p^n(p-1)$ real accounts. To do this, we need to consider each component of the vector $X=(x_0,x_1,\ldots,x_{p^n-1})^T$ as a $(p-1)$-dimensional pseudo-vector

$$x_\nu=(x_\nu^{(0)},x_\nu^{(1)},\ldots,x_\nu^{(p-2)})=x_\nu^{(0)}+x_\nu^{(1)}q+\cdots+x_\nu^{(p-2)}q^{p-2},$$

for $\nu=0,1,\ldots,p^n-1$, in which the numbers $x_\nu^{(j)}$, $j=0,1,\ldots,p-2$ are determined in a natural way from the original signal in $p^n(p-1)$ readings. In this way the time to compute the DMT can be reduced still further by a factor of $p-1$.

§11.3. Applications of discrete multiplicative transforms to information compression.

Thanks to the growing availability of high speed computers, the discrete Fourier transform (DFT) and the discrete Walsh transform (DWT) have been used in a wide variety of applications for both theoretical and practical problems. The latest achievements in radio electronics, micro electronics, and scientific computing (see, for example [9]) have fostered applications of general discrete multiplicative transforms (11.1.12), (11.1.13) (DMT), in particular, the DFT and the DWT. This is especially true for digital information processing, including compression of information, and coding theory. Thus research into the properties of the DMT is both practical and timely.

Mathematically, information compression by means of discrete orthogonal transforms can be described as follows. Let T be a non-singular $N\times N$ matrix. A given vector X of dimension N is "transformed" to a vector $Y=TX$ which is altered further by some method to produce a "smaller" vector $\tilde{Y}$. An approximation to the input vector X is reconstructed by $\tilde{X}=T^{-1}\tilde{Y}$ which results in a reconstruction error $\delta=\|X-\tilde{X}\|$. We define the *compression coefficient of the information* to be $\tau=|Y|/|\tilde{Y}|$, where $|Y|$ is some measure of the "size" of the vector Y. We shall call that method which minimizes δ for a given τ the *optimal method of compression of information.* The final objective of information compression is data reduction.

In one method of information compression, which we shall call *zone coding* [9], the vector $\tilde{Y}$ is obtained from the vector Y by replacing some of the coordinates of the vector Y by zero. "Compression" of the vector $\tilde{Y}$ is measured by the number of coordinates not replaced by zero (evidently, $|Y|=N$).

For the DFT, the "transformed" vector Y takes on the form

$$y_n=\frac{1}{N}\sum_{k=0}^{N-1}x_k\exp\left\{-\frac{2\pi ikn}{N}\right\},\qquad n=0,1,\ldots,N-1.$$

For the class of vectors

$$\Xi_\Delta = \{X^T = (x_0, x_1, \ldots, x_{N-1}) : x_k \in \mathbf{R}, \max_{1 \le k \le N-1} |x_{k-1} - x_k| \le \Delta\},$$

Efimov [3] obtained an estimate

$$|y_n| \le \Delta / \sin \frac{\pi n}{N}, \qquad n = 1, 2, \ldots, N-1, \tag{11.3.1}$$

which was improved by him in [5] for integers $n = N/2$ and $n = N/4$. Based on this estimate, optimal methods of zone coding by means of the DFT involve replacing the central coordinates of the vector $Y^T = (y_0, y_1, \ldots, y_{N-1})$ by zero. Earlier [4], he also estimated the coordinate restoration error $\delta = |x_k - \tilde{x}_k|$ for this method of zone coding.

Similar results for the DMT are contained in the work of Kanygin [1], where an analogue of (11.3.1) was obtained. More detailed investigations of the DMT in the uniform and integral norms and for various restrictions on the class of input vectors were conducted by Bespalov in [3], some of whose results will be given below.

We shall restrict ourselves to the case $p_k = p$ for $k = 1, 2, \ldots$ and consider zone coding by means of the discrete transforms (11.2.6) and (11.2.7). The DMT (11.2.6) takes a vector $X^T = (x_0, x_1, \ldots, x_{p^n-1})$ to a vector $Y^T = (y_0, y_1, \ldots, y_{p^n-1})$ by the formula

$$Y = \frac{1}{p^n} W X,$$

i.e.,

$$y_\ell = \frac{1}{p^n} \sum_{k=0}^{p^n-1} x_k \overline{\chi}_k \left(\frac{\ell}{p^n} \right), \qquad \ell = 0, 1, \ldots, p^n - 1. \tag{11.3.2}$$

The inverse transform

$$x_k = \sum_{\ell=0}^{p^n-1} y_\ell \chi_\ell \left(\frac{k}{p^n} \right), \qquad k = 0, 1, \ldots, p^n - 1, \tag{11.3.3}$$

takes the vector Y to the vector X, i.e., $X = W^{-1} Y$.

Let $\tilde{Y}$ be a vector obtained from Y by replacing k of its coordinates by zero. Then the compression coefficient τ equals $p^n/(p^n - k)$, and the reconstruction error is $\delta = \|X - \tilde{X}\|$, where $\tilde{X} = W^{-1}\tilde{Y}$.

For a given compression coefficient, the reconstruction error of a vector X depends on the method of zone coding and on the norm chosen to measure the error. The strategy is to replace by zero those coordinates of the vector Y which are small in absolute value, because these coordinates make smaller contributions to the reconstructed input vector. At this point it is important to recall (see §11.2) that there is a program for computing the DMT which uses only the internal structure of the transformation W. Namely, the matrix W is divided into single-type blocks, which lend themselves readily to parallel computing. Therefore, it is appropriate to replace by zero coordinates of the vector Y which correspond to an entire block (or several blocks) of the matrix W. From the calculations below it will be evident that this strategy brings about the highest reconstruction accuracy. We shall also consider the case when the last coordinates y_ℓ are replaced by zero, which is very convenient for calculation on a computer.

In connection with the preceding remarks we introduce the following notation: we shall call the vector

$$Y_s^T = (\overbrace{0,\ldots,0}^{p^s}, y_{p^s}, y_{p^s+1},\ldots,y_{p^{s+1}-1}, \overbrace{0,\ldots,0}^{p^n-p^{s+1}})$$

the *s-th packet* of the vector Y (in the sequel we usually omit the superscript T); for $j=0,1,\ldots,p-1$, we shall call the vectors

$$Y_{j,s}^T = (0,\ldots,0, y_{jp^s}, y_{jp^s+1},\ldots,y_{(j+1)p^s-1}, 0,\ldots,0)$$

the *j-th subpacket of the s-th packet* of the vector Y. We shall let $V^{(j)} = W^{-1}Y_{j,n-1}$ represent the vector of dimension p^n reconstructed from the j-th subpacket of the last packet of the vector Y, and let

$$X^{(j)} = W^{-1}\left(Y - \sum_{\nu=j}^{p-1} Y_{\nu,n-1}\right)$$

represent the vector reconstructed from the vector $(y_0, y_1, \ldots, y_{jp^{n-1}-1}, 0, \ldots, 0)$ (during the reconstruction which does not replace the first jp^{n-1} coordinates of Y by zero).

Also we will use the notation

$$z_\mu^{(\nu)} = \frac{1}{p}\sum_{r=0}^{p-1} x_{\mu p+r}\overline{\chi}_{\nu p^{n-1}}\left(\frac{r}{p^n}\right), \tag{11.3.4}$$

where $\mu = 0,1,\ldots,p^{n-1}$ and $0 \le \nu \le p-1$.

11.3.1. *For $k=0,1,\ldots,p^n-1$, the coordinates $v_k^{(j)}$ of the vector $V^{(j)}$ satisfy*

$$v_k^{(j)} = z_\mu^{(j)}\chi_{jp^{n-1}}\left(\frac{k}{p^n}\right)$$

where $\mu p \le k \le (\mu+1)p-1$ and $z_\mu^{(j)}$ are defined by (11.3.4).

PROOF. Substitute the formula (11.3.2) for y_ℓ into the expression

$$v_k^{(j)} = \sum_{\ell=jp^{n-1}}^{(j+1)p^{n-1}-1} y_\ell \chi_\ell\left(\frac{k}{p^n}\right)$$

to verify

$$v_k^{(j)} = \frac{1}{p^n} \sum_{\ell=jp^{n-1}}^{(j+1)p^{n-1}-1} \left(\sum_{r=0}^{p^n-1} x_r \overline{\chi}_r\left(\frac{\ell}{p^n}\right)\right) \chi_\ell\left(\frac{k}{p^n}\right).$$

Using **11.2.3** and changing the order of summation, we obtain

$$\begin{aligned} v_k^{(j)} &= \frac{k}{p^n} \sum_{r=0}^{p^n-1} x_r \sum_{\ell=0}^{p^{n-1}-1} \chi_{jp^n+\ell}\left(\frac{k}{p^n}\right) \overline{\chi}_{jp^{n-1}+\ell}\left(\frac{r}{p^n}\right) \\ &= \frac{1}{p^n} \sum_{r=0}^{p^n-1} x_r \chi_{jp^{n-1}}\left(\frac{k}{p^n} \ominus \frac{r}{p^n}\right) \sum_{\ell=0}^{p^{n-1}-1} \chi_\ell\left(\frac{k}{p^n} \ominus \frac{r}{p^n}\right). \end{aligned}$$

But

$$\begin{aligned} \sum_{\ell=0}^{p^{n-1}-1} \chi_\ell\left(\frac{k}{p^n} \ominus \frac{r}{p^n}\right) &= D_{p^{n-1}}\left(\frac{k}{p^n} \ominus \frac{r}{p^n}\right) \\ &= \begin{cases} p^{n-1}, & \text{if } \ 0 \le \dfrac{k}{p^n} \ominus \dfrac{r}{p^n} < \dfrac{1}{p^{n-1}}, \\ 0, & \text{if } \ \dfrac{1}{p^{n-1}} \le \dfrac{k}{p^n} \ominus \dfrac{r}{p^n} < 1. \end{cases} \end{aligned}$$

Since $k \in [\mu p, (\mu+1)p-1]$, the inequality $\dfrac{k}{p^n} \ominus \dfrac{r}{p^n} < \dfrac{1}{p^{n-1}}$ holds only for those $r \in [\mu p, (\mu+1)p-1]$. Consequently,

$$\begin{aligned} v_k^{(j)} &= \frac{1}{p^n} \sum_{r=\mu p}^{(\mu+1)p^n-1} x_r \chi_{jp^{n-1}}\left(\frac{k}{p^n} \ominus \frac{r}{p^n}\right) p^{n-1} \\ &= \frac{1}{p} \sum_{r=0}^{p-1} x_{\mu p+r} \chi_{jp^{n-1}}\left(\frac{k}{p^n} \ominus \frac{\mu p+r}{p^n}\right) \\ &= \frac{1}{p} \sum_{r=0}^{p-1} x_{\mu p+r} \chi_{jp^{n-1}}\left(\frac{k}{p^n}\right) \overline{\chi}_{jp^{n-1}}\left(\frac{\mu}{p^{n-1}} + \frac{r}{p^n}\right). \end{aligned}$$

We notice for $\mu = \sum_{\nu=1}^{n-1} \mu_{-\nu} p^{\nu-1}$ that $\mu_{-n} = 0$ and

$$\chi_{p^{n-1}}\left(\frac{\mu}{p^{n-1}}\right) = \exp\frac{2\pi i}{p} 1 \cdot \left(\frac{\mu}{p^{n-1}}\right)_n = \exp\frac{2\pi i \mu_{-n}}{p} = 1.$$

Therefore,

$$v_k^{(j)} = \frac{1}{p}\chi_{jp^{n-1}}\left(\frac{k}{p^n}\right)\sum_{r=0}^{p-1} x_{\mu p+r}\overline{\chi}_{jp^{n-1}}\left(\frac{r}{p^n}\right) = z_\mu^{(j)}\chi_{jp^{n-1}}\left(\frac{k}{p^n}\right). \quad \blacksquare$$

From the definition of $V^{(j)}$ and $X^{(j)}$ it follows that

$$X^{(j)} = V^{(0)} + V^{(1)} + \cdots + V^{(j-1)}.$$

Consequently, the following result is a corollary of **11.3.1**.

11.3.2. *The coordinates of the vector $X^{(j)}$ satisfy*

$$x_k^{(j)} = \sum_{\nu=0}^{j-1} z_\mu^{(\nu)} \chi_{k_1 p^{n-1}}\left(\frac{\nu}{p^n}\right),$$

where $\mu p \le k \le (\mu+1)p - 1$ and $k_1 = k - \mu p$.

Notation (11.3.4) and this last relationship will be used for the case $s = n-1$. We shall introduce further notation to be used in the cases $s \le n-1$.

Write an arbitrary k, $0 \le k \le p^n - 1$, in the form

$$k = k(\mu, r, \eta) = \mu p^{n-s} + rp^{n-s-1} + \eta, \tag{11.3.5}$$

where $0 \le \mu < p^s$, $0 \le r \le p-1$, and $0 \le \eta \le p^{n-s-1} - 1$. As we saw in (11.2.2), the vector X can be considered as values of a step function $f(t)$ which takes the value x_k on the k-th interval $\delta_k(n)$ of rank n, and moreover, that the t_k's can be distributed on the interval $[0,1]$ across even subintervals. The coordinates of the vector X can be divided into collections of p^{n-s-1} elements whose coordinates lie in the packets $J_{s+1}(k)$ of rank $s+1$. We shall denote the arithmetic means of these collections of coordinates by

$$a(\mu, r) = \frac{1}{p^{n-s-1}} \sum_{\eta=0}^{p^{n-s-1}-1} x_{k(\mu,r,\eta)}, \tag{11.3.6}$$

calling them the *mean values (of the vector X) over packets* of rank $s+1$. Thus

$$a(\mu) = \frac{1}{p}\sum_{r=0}^{p-1} a(\mu, r)$$

is the mean value over the μ-th packet $J_s(\mu)$ of s-th rank. Similar to (11.3.4), we shall also use the notation

$$A(\mu, j, s) = \frac{1}{p} \sum_{r=0}^{p-1} a(\mu, r) \overline{\chi}_{jp^{s-1}}\left(\frac{r}{p^s}\right) \tag{11.3.7}$$

Since

$$\sum_{r=0}^{p-1} \overline{\chi}_{jp^{s-1}}\left(\frac{r}{p^s}\right) = 0,$$

it is easy to see by Abel's transformation that the right side of (11.3.7) can be written in the form

$$\begin{aligned} A(\mu, j, s) &= \frac{1}{p} \sum_{r=0}^{p-1} a(\mu, r) \left(\sum_{\nu=0}^{r} \overline{\chi}_{jp^{s-1}}\left(\frac{\nu}{p^s}\right) - \sum_{\nu=0}^{r-1} \overline{\chi}_{jp^{s-1}}\left(\frac{\nu}{p^s}\right) \right) \\ &= \frac{1}{p} \sum_{r=0}^{p-2} (a(\mu, r) - a(\mu, r+1)) \sum_{\nu=0}^{r} \overline{\chi}_{jp^{s-1}}\left(\frac{\nu}{p^s}\right), \end{aligned}$$

i.e.,

$$A(\mu, j, s) = \frac{1}{p} \sum_{r=0}^{p-2} \Delta a(\mu, r) \sum_{\nu=0}^{r} \overline{\chi}_{jp^{s-1}}\left(\frac{\nu}{p^s}\right). \tag{11.3.8}$$

The following result, which we call the *localization principle*, is a simple consequence of formula (11.3.8).

11.3.3. *For all $1 \leq j \leq p-1$ the values of the quantity $A(\mu, j, s)$ depend only on the "jumps" of the mean values over packets of rank $s+1$, equal to $\Delta a(\mu, r) = a(\mu, r) - a(\mu, r+1)$ for indices $0 \leq r \leq p-2$, and depend neither on the distribution of the individual coordinates x_k nor on the mean values $a(\mu)$ over packets of rank s.*

PROOF. Indeed, if for some μ one adds ε to all quantities $a(\mu, r)$, where $0 \leq r \leq p-1$, then $a(\mu)$ will be changed by ε but the quantities $\Delta a(\mu, r)$ do not change for all $1 \leq r \leq p-2$ and thus neither does $A(\mu, j, s)$. ∎

Let Ξ_Δ be the class of vectors $a = (a_0, a_1, \ldots, a_{p-1})$ with real coordinates which satisfy the condition

$$\max_{1 \leq k \leq p-1} |a_{k-1} - a_k| \leq \Delta,$$

and let $R(a, \nu) = \sum_{k=0}^{p-1} a_k \omega^{\nu k}$, where $\omega = \exp(2\pi i/p)$.

11.3.4. *For any $p \geq 2$ and $1 \leq \nu \leq p-1$ we have*

$$\max_{a \in \Xi_1} |R(a,\nu)| \leq p / \sin\frac{\pi\nu}{p}.$$

PROOF. By Abel's transformation,

$$\begin{aligned} R(a,\nu) &= \sum_{k=1}^{p-1}(a_{k-1}-a_k)\sum_{s=0}^{k-1}\omega^{\nu s} + a_{p-1}\sum_{s=0}^{p-1}\omega^{\nu s} \\ &= \frac{1}{1-\omega^\nu}\sum_{k=1}^{p-1}(a_{k-1}-a_k)(1-\omega^{\nu k}). \end{aligned}$$

Estimating this expression, we find for any $a \in \Xi_1$ that

$$|R(a,\nu)| \leq \frac{1}{|1-\omega^\nu|}\sum_{k=1}^{p-1}|1-\omega^{\nu k}| \leq \frac{2p}{2\sin(\pi\nu)/p} = \frac{p}{\sin(\pi\nu)/p}. \quad \blacksquare$$

Now we shall estimate the spectrum and reconstruction error which these methods of zone coding generate. We will use the following notation: if $a = (a_0, a_1, \ldots, a_{N-1})$ is an N dimensional vector then

$$\|a\|_{\ell^q} = \|a\|_q = \left(\sum_{k=0}^{N-1}|a_k|^q\right)^{1/q}, \qquad 1 \leq q < \infty,$$

and

$$\|a\|_{(\infty)} = \max_{0 \leq k \leq N-1}|a_k|.$$

11.3.5. *For packets Y_s, $s = 1,2,\ldots,n-1$, and subpackets $Y_{j,s}$, $j = 1,2,\ldots,p-1$, of the spectrum of a p^n dimensional vector $X^T = (x_0, x_1, \ldots, x_{p^n-1}) \in \Xi_\Delta$ the following estimates hold:*

$$\max_{X \in \Xi_\Delta}\|Y_{j,s}\|_q \leq p^{n-s-1}\Delta / \sin\frac{\pi j}{p}, \qquad 2 \leq q \leq \infty, \tag{11.3.9}$$

$$\max_{X \in \Xi_\Delta}\|Y_s\|_2 \leq p^{n-s-1}\Delta\sqrt{(p-1)^2 + \frac{1}{12}(p^2-1)}. \tag{11.3.10}$$

PROOF. It suffices to establish estimate (11.3.9) for the case $q = 2$ because for $2 < q \leq \infty$ the ℓ_q norm of a vector is less than or equal to its ℓ_2 norm. Let $0 \leq \nu < p^s$ and $s \leq n-1$. Consider the quantity

$$y_{jp^s+\nu} = \frac{1}{p^n}\sum_{k=0}^{p^n-1} x_k \overline{\chi}_k\left(\frac{jp^s+\nu}{p^n}\right),$$

which by **11.2.3** can be written in the form

$$y_{jp^s+\nu} = \frac{1}{p^n} \sum_{k=0}^{p^n-1} x_k \overline{\chi}_{jp^s+\nu}\left(\frac{k}{p^n}\right).$$

Using notation (11.3.5), we have

$$\begin{aligned} y_{jp^s+\nu} &= \frac{1}{p^n} \sum_{\mu=0}^{p^s-1} \sum_{r=0}^{p-1} \sum_{\eta=0}^{p^{n-s-1}-1} x_{k(\mu,r,\eta)} \overline{\chi}_{jp^s+\nu}\left(\frac{\mu}{p^s} + \frac{r}{p^{s+1}} + \frac{\eta}{p^n}\right) \\ &= \frac{1}{p^n} \sum_{\mu=0}^{p^s-1} \overline{\chi}_{jp^s+\nu}\left(\frac{\mu}{p^s}\right) \sum_{r=0}^{p-1} \overline{\chi}_{jp^s+\nu}\left(\frac{r}{p^{s+1}}\right) \sum_{\eta=0}^{p^{n-s-1}-1} x_{k(\mu,r,\eta)} \overline{\chi}_{jp^s+\nu}\left(\frac{\eta}{p^n}\right). \end{aligned}$$

Since $jp^s + \nu < p^{s+1}$, $\eta/p^n < p^{n-s-1}/p^n = 1/p^{s+1}$, we have $\overline{\chi}_{jp^s+\nu}(\eta/p^n) = 1$, and since $\nu < p^s$, $r/p^{s+1} < 1/p^s$, we have $\overline{\chi}_\nu(r/p^{s+1}) = 1$. Moreover, for $\mu < p^s$ we have

$$(\mu/p^s)_{s+1} = 0,$$

and thus

$$\overline{\chi}_{jp^s}(\mu/p^s) = \exp\frac{-2\pi i}{p} j(\mu/p^s)_{s+1} = 1.$$

Hence, it follows from the notation (11.3.6) and (11.3.7) that

$$\begin{aligned} y_{jp^s+\nu} &= \frac{1}{p^s} \sum_{\mu=0}^{p^s-1} \overline{\chi}_{jp^s+\nu}\left(\frac{\mu}{p^s}\right) \sum_{r=0}^{p-1} \overline{\chi}_{jp^s}\left(\frac{r}{p^{s+1}}\right) p^{n-s-1} a(\mu, r) \\ &= \frac{1}{p^{s+1}} \sum_{\mu=0}^{p^s-1} \overline{\chi}_\nu\left(\frac{\mu}{p^s}\right) pA(\mu, j, s+1) \\ &= \frac{1}{p^s} \sum_{\mu=0}^{p^s-1} A(\mu, j, s+1) \overline{\chi}_\nu\left(\frac{\mu}{p^s}\right) \end{aligned}$$

i.e.,

$$y_{jp^s+\nu} = \frac{1}{p^s} \sum_{\mu=0}^{p^s-1} A(\mu, j, s+1) \overline{\chi}_\nu\left(\frac{\mu}{p^s}\right). \tag{11.3.11}$$

Using properties of the function $\chi_\nu(x)$, we see that

$$\begin{aligned} |y_{jp^s+\nu}|^2 &= y_{jp^s+\nu} \overline{y_{jp^s+\nu}} \\ &= \frac{1}{p^{2s}} \sum_{\mu=0}^{p^s-1} A(\mu, j, s+1) \overline{\chi}_\nu\left(\frac{\mu}{p^s}\right) \cdot \sum_{\ell=0}^{p^s-1} \overline{A}(\ell, j, s+1) \chi_\nu\left(\frac{\ell}{p^s}\right) \\ &= \frac{1}{p^{2s}} \sum_{\mu=0}^{p^s-1} |A(\mu, j, s+1)|^2 + \sum_{k=1}^{p^s-1} \overline{\chi}_\nu\left(\frac{k}{p^s}\right) B(k, j, s+1), \end{aligned}$$

where the quantities $B(k,j,s+1)$ can be expressed by the $A(\mu,j,s+1)$'s and their complex conjugates. Using this representation and the identity

$$\sum_{\nu=0}^{p^s-1} \overline{\chi}_\nu \left(\frac{k}{p^s}\right) = 0, \qquad 1 \le k \le p^s - 1,$$

we have for $2 \le q \le \infty$ that

(11.3.12)

$$\|Y_{j,s}\|_q \le \|Y_{j,s}\|_2 = \left(\sum_{k=0}^{p^n-1} |y_k^*|^2\right)^{1/2} = \left(\sum_{\nu=0}^{p^s-1} |y_{jp^s+\nu}|^2\right)^{1/2}$$

$$= \left[\sum_{\nu=0}^{p^s-1} \frac{1}{p^{2s}} \sum_{\mu=0}^{p^s-1} |A(\mu,j,s+1)|^2 + \sum_{k=1}^{p^s-1} B(k,j,s+1) \sum_{\nu=0}^{p^s-1} \overline{\chi}_\nu \left(\frac{k}{p^s}\right)\right]^{1/2}$$

$$\le \left[\sum_{\nu=0}^{p^s-1} \frac{1}{p^{2s}} \max_{\substack{X\in\Xi_\Delta \\ 0\le\mu\le p^s-1}} |A(\mu,j,s+1)|^2 p^s\right]^{1/2}$$

$$= \max_{\substack{X\in\Xi_\Delta \\ 0\le\mu\le p^s-1}} |A(\mu,j,s+1)|.$$

Since (11.3.6) implies

$$\Delta a(\mu,r) = \frac{1}{p^{n-s-1}} \sum_{\eta=0}^{p^{n-s-1}-1} (x_{\mu p^{n-s}+rp^{n-s-1}+\eta} - x_{\mu p^{n-s}+(r+1)p^{n-s-1}+\eta}),$$

it follows from the restriction $X \in \Xi_\Delta$ that

(11.3.13)

$$|\Delta a(\mu,r)|$$

$$\le \frac{1}{p^{n-s-1}} \sum_{\eta=0}^{p^{n-s-1}-1} |x_{\mu p^{n-s}+rp^{n-s-1}+\eta} - x_{\mu p^{n-s}+(r+1)p^{n-s-1}+\eta}|$$

$$\le \frac{1}{p^{n-s+1}} \sum_{\eta=0}^{p^{n-s-1}-1} p^{n-s-1}\Delta = p^{n-s-1}\Delta.$$

Suppose $\overline{\chi}_{jp^{s-1}}(\nu/p^s) = \omega^{j\nu}$, where $\omega = \exp(-2\pi i/p)$. Notice that

$$\sum_{\nu=0}^{r} \overline{\chi}_{jp^{s-1}}(\nu/p^s) = \frac{1-\omega^{j(r+1)}}{1-\omega^j}$$

for $0 \le r \le p-2$. Thus it follows from **11.3.4** and (11.3.13) that

$$\max_{\substack{X \in \Xi_\Delta \\ 0 \le \mu \le p^s - 1}} |A(\mu, j, s+1)| = \max_{\substack{X \in \Xi_\Delta \\ 0 \le \mu \le p^s - 1}} \frac{1}{p} \Big| \sum_{r=0}^{p-2} \Delta a(\mu, r) \sum_{\nu=0}^{r} \omega^{j\nu} \Big|$$

$$= \max_{\Delta a(\mu,r) \in \Xi_{p^{n-s-1}\Delta}} \frac{1}{p} \Big| \sum_{r=0}^{p-2} \Delta a(\mu, r) \sum_{\nu=0}^{r} \omega^{j\nu} \Big|$$

$$= \max_{a \in \Xi_{p^{n-s-1}\Delta}} \frac{1}{p} |R(a, j)| \le \frac{1}{p} p^{n-s-1} \Delta \frac{p}{\sin(\pi j/p)}.$$

Putting this estimate into (11.3.12), we obtain (11.3.9):

$$\|Y_{j,s}\|_q \le \|Y_{j,s}\|_2 \le p^{n-s-1} \frac{\Delta}{\sin(\pi j/p)}.$$

To obtain estimate (11.3.10), consider the vector X^* whose coordinates x_k^* are obtained from the coordinates x_k of the vector X in the following way:

$$x_k^* = x^*_{k(\mu,r,\eta)} = a(\mu, r) - a(\mu).$$

Clearly, we have packets of p^{n-s-1} identical coordinates arranged in order. Let $Y^* = p^{-n} W X^*$. Using (11.3.11) and the localization principle **11.3.3**, we conclude that $Y^* = Y_s$.

Use Parseval's identity $\|f\|_{\ell_2^*} = \|\hat{f}\|_{\ell_2}$ and take into account the form of the vector Y_s and the fact that its preimage is the vector X^*. We have ($\|X\|^2_{\ell_2^*} = p^{-n} \sum_{k=0}^{p^n - 1} |x_k|^2$)

$$\|Y_s\|^2_{\ell_2} = \|X^*\|^2_{\ell_2^*} = p^{-n} \sum_{k=0}^{p^n-1} |x_k^*|^2 \tag{11.3.14}$$

$$= \frac{1}{p^{s+1}} \sum_{\mu=0}^{p^s - 1} \sum_{r=0}^{p-1} |a(\mu, r) - a(\mu)|^2.$$

Fix μ, r, n and s and temporarily set

$$x(\nu, \eta) = x_{\mu p^{n-s} + \nu p^{n-s-1} + \eta}$$

for $\nu = 0, \ldots, p-1$ and $\eta = 0, 1, \ldots, p^{n-s-1}$. Applying (11.3.6) we find

$$|a(\mu, r) - a(\mu)|$$
$$= \left| \frac{1}{p^{n-s-1}} \sum_{\eta=0}^{p^{n-s-1}-1} x(r,\eta) - \frac{1}{p} \sum_{\nu=0}^{p-1} \frac{1}{p^{n-s-1}} \sum_{\eta=0}^{p^{n-s-1}-1} x(\nu,\eta) \right|$$
$$= \frac{1}{p^{n-s-1}} \left| \sum_{\eta=0}^{p^{n-s-1}-1} \left(x(r,\eta) - \frac{1}{p} \sum_{\nu=0}^{p-1} x(\nu,\eta) \right) \right|$$
$$= \frac{1}{p^{n-s-1}} \times$$
$$\times \left| \sum_{\eta=0}^{p^{n-s-1}-1} \left[(x(r,\eta) - x(0,\eta) - \frac{1}{p} \sum_{\nu=0}^{p-1} (x(\nu,\eta) - x(0,\eta)) \right] \right|.$$

Since $X \in \Xi_\Delta$ we have

$$|a(\mu, r) - a(\mu)|$$
$$\leq \frac{1}{p^{n-s-1}} \sum_{\eta=0}^{p^{n-s-1}-1} \left(r p^{n-s-1} \Delta + \frac{1}{p} \sum_{\nu=0}^{p-1} \nu p^{n-s-1} \Delta \right)$$
$$= \sum_{\eta=0}^{p^{n-s-1}-1} \left(r + \frac{p(p-1)}{2p} \right) \Delta = \left(r + \frac{p-1}{2} \right) p^{n-s-1} \Delta.$$

Putting this estimate into (11.3.14), we conclude that

$$\|Y_s\|_{\ell_2}^2 \leq \frac{1}{p^{s+1}} \sum_{\mu=0}^{p^s-1} \sum_{r=0}^{p-1} \left(r + \frac{p-1}{2} \right)^2 p^{2(n-s-1)} \Delta^2$$
$$= \frac{1}{p} p^{2(n-s-1)} \Delta^2 \sum_{r=0}^{p-1} \left(r^2 - r(p-1) + \frac{(p-1)^2}{4} \right)$$
$$= p^{2(n-s-1)} \Delta^2 \frac{1}{p} \left(\frac{p(p-1)(2p-1)}{6} + \frac{p(p-1)^2}{2} + \frac{p(p-1)^2}{4} \right)$$
$$= p^{2(n-s-1)} \Delta^2 \frac{(p-1)(13p-11)}{12} = p^{2(n-s-1)} \Delta^2 \left((p-1)^2 + \frac{p^2-1}{12} \right),$$

i.e., (11.3.10) is true. ∎

Inequalities (11.3.9) and (11.3.10) show that the smallest contribution to the inverse transform (11.3.3) comes from the last packet ($s = n-1$) of the spectrum, and the smallest contributing subpacket is the one whose index j is nearest to $[p/2]$. We shall now estimate the maximal contribution this j-th subpacket makes to reconstructing the input vector by using formula (11.7.3), i.e., we shall estimate the quantity $V^{(j)} = W^{-1} Y_{j,n-1}$.

11.3.6. *For all $1 \le j \le p-1$ and $1 \le q \le \infty$, the following inequality holds*

$$\max_{X \in \Xi_\Delta} \|V^{(j)}\|_{\ell_q^*} \le \frac{\Delta}{\sin(\pi j/p)}. \tag{11.3.15}$$

PROOF. Let $0 \le \mu \le p^{n-1}-1$ and $\mu p \le k \le (\mu+1)p-1$. Using the representation of the vector $V^{(j)}$ given in **11.3.1** and the norm

$$\|X\|_{\ell_q^*} = \left(\frac{1}{p^n}\sum_{k=0}^{p^n-1} |x_k|^q\right)^{1/q},$$

we have

$$\begin{aligned}\|V^{(j)}\|_{\ell_q^*} &= \left(\frac{1}{p^n}\sum_{k=0}^{p^n-1} |v_k^{(j)}|^q\right)^{1/q} \\ &= \left(\frac{1}{p^n}\sum_{k=0}^{p^n-1} |z_\mu^{(j)}|^q\right)^{1/q} \\ &\le \max_{0\le\mu\le p^{n-1}-1} |z_\mu^{(j)}| = \|V^{(j)}\|_{(\infty)}.\end{aligned}$$

Combining representation (11.3.4) with estimate **11.3.4**, we obtain

$$\begin{aligned}\max_{X\in\Xi_\Delta} \|V^{(j)}\|_{\ell_q^*} &= \max_{0\le\mu\le p^{n-1}-1}\max_{X\in\Xi_\Delta} \frac{1}{p}\left|\sum_{r=0}^{p-1} x_{\mu p+r}\overline{X}_{jp^{n-1}}\left(\frac{r}{p^n}\right)\right| \\ &\le \frac{1}{p}\frac{p\Delta}{\sin(\pi j/p)} = \frac{\Delta}{\sin(\pi j/p)}. \quad \blacksquare\end{aligned}$$

We shall now estimate the error induced by replacing the vector X with the vector $X^{(j)}$, which has been reconstructed from the vector $(y_0, y_1, \ldots y_{jp^{n-1}-1}, 0, \ldots, 0)$.

11.3.7. *The estimate*

$$\max_{X\in\Xi_\Delta} \|X - X^{(j)}\|_{\ell_q^*} \le \Delta \sum_{\nu=1}^{p-j} \frac{1}{\sin(\pi\nu/p)} \tag{11.3.16}$$

holds for any $1 \le j \le p-1$ and $1 \le q \le \infty$.

PROOF. By the definitions, we have

$$\begin{aligned}X - X^{(j)} &= W^{-1}Y - W^{-1}\left(Y - \sum_{\nu=j}^{p-1} Y_{\nu,n-1}\right) \\ &= \sum_{\nu=j}^{p-1} W^{-1}Y_{\nu,n-1} = \sum_{\nu=j}^{p-1} V^{(\nu)}.\end{aligned}$$

Applying **11.3.6** for $1 \le q \le \infty$, we have

$$\max_{X \in \Xi_\Delta} \|X - X^{(j)}\|_{\ell_q^*} \le \sum_{\nu=j}^{p-1} \max_{X \in \Xi_\Delta} \|V^{(\nu)}\|_{\ell_q^*}$$

$$\le \Delta \sum_{\nu=j}^{p-1} \frac{1}{\sin(\pi\nu/p)} = \Delta \sum_{\nu=1}^{p-j} \frac{1}{\sin(\pi\nu/p)}. \quad \blacksquare$$

Thus we have obtained estimates, in the ℓ_q norm, of the coordinates of the vector Y and the error which comes from reconstructing the input data X. Since many applications involve quadratic estimates, we shall examine these results in the ℓ_2 norm.

The choice of the optimal method of zone coding has great practical significance. For a given compression coefficient τ it is easy to determine the number of coordinates of the vector Y which should be replaced by zero. Depending on which coordinates of the vector Y are replaced by zero (i.e., depending on the method of zone coding), we obtain various values of the reconstruction error δ. As was mentioned earlier, the method of zone coding which guarantees the minimal error is called optimal. We shall compare the methods of zone coding generated by the DFT and the DMT.

We begin by examining compression of information by means of the DFT. The spectrum in this case is defined by the formula

$$y_\ell = \frac{1}{N} \sum_{k=0}^{N-1} x_k \exp \frac{-2\pi i k \ell}{N}.$$

For $1 \le \ell \le N-1$, we have by (11.3.1) that

$$\max_{X \in \Xi_\Delta} |y_\ell| \le \frac{\Delta}{\sin(\pi\ell/p)}.$$

Hence it follows from Parseval's identity $\|Y\|_{\ell_2} = \|X\|_{\ell_2^*}$ that the optimal method of zone coding by the DFT is that in which the central coordinates of the vector Y are replaced by zeros.

To estimate the ℓ_2^* norm of the reconstruction error generated by this method, let N be an even integer (the difference between the even and odd cases is insignificant). Replace the $2s+1$ central coordinates of the vector Y by zero, i.e., the coordinates

$$y_{\frac{N}{2}-s}, y_{\frac{N}{2}-s+1}, \ldots, y_{\frac{N}{2}}, \ldots, y_{\frac{N}{2}+s}, \qquad s < \frac{N}{2}.$$

Then the reconstruction error δ is the ℓ_2 norm of the vector formed by these coordinates. Hence it follows that the maximal error generated by the DFT on the class of vectors Ξ_Δ is exactly

$$\delta_F = \Delta \left(1 + \sum_{\ell=(N/2)-s}^{(N/2)+s} \frac{1}{\sin^2(\pi\ell/N)} \right)^{1/2}.$$

Since the function $y = 1/\cos^2 x$ is monotone and the identities $\sin(\frac{\pi}{2} \pm \frac{\pi s}{N}) = \cos\frac{\pi s}{N}$ are well-known, we obtain

$$\frac{N}{\pi}\int_0^{\pi s/N} \frac{dx}{\cos^2 x} \le \frac{N}{\pi}\frac{\pi}{N}\sum_{\ell=1}^{s} \frac{1}{\cos^2(\pi\ell/N)} \le \frac{N}{\pi}\int_0^{\pi(s+1)/N} \frac{dx}{\cos^2 x}.$$

Thus we see that

$$\Delta\sqrt{1 + \frac{2N}{\pi}\tan\frac{\pi s}{N}} \le \delta_F \le \Delta\sqrt{1 + \frac{2N}{\pi}\left(\tan\frac{\pi(s+1)}{N} - \tan\frac{\pi}{N}\right)}.$$

In particular, an approximation to the maximal error of the optimal method of zone coding by means of the DFT is given by

$$\delta_F \simeq \Delta\sqrt{\frac{2N}{\pi}\tan\frac{\pi s}{N}} \le \Delta\sqrt{2s}. \tag{11.3.17}$$

In contrast to zone coding by means of the DFT, when compressing information by means of the DMT it is appropriate to replace the coordinates of the last packets of the vector Y by zero. In this case the basic factor p^{n-s-1} in estimates (11.3.9), (11.3.10) reduces to 1. Based on the results of Theorem **11.3.7**, optimal methods of zone coding by means of the DMT can be obtained by three different methods. Choosing one in a specific problem often depends on the compression coefficient τ and the fixed number p.

THE FIRST METHOD. We replace every element in the last packet by zero, i.e., $p^n - p^{n-1}$ coordinates of the vector Y. Then $\tau = p^n/p^{n-1} = p$. The maximal reconstruction error in the DMT case as measured by the ℓ_2^* norm can be estimated as follows:

$$\delta_M \le \Delta \sum_{\nu=1}^{p-j} \frac{1}{\sin(\pi\nu/p)}$$
$$\simeq \begin{cases} \frac{p\Delta}{2}\ln(p-j+1) & \text{for } \frac{p}{2} \le j \le p-1, \\ \frac{p\Delta}{2}\left(\ln(\frac{p}{2}+1) + \ln\frac{p+1}{2j}\right) & \text{for } 1 \le j < \frac{p}{2}. \end{cases}$$

For the optimal method of zone coding in the DFT case with initial information of size $N = p^n$ and compression coefficient $\tau = p$, it is necessary to replace the $2s+1 = p^n - p^{n-1}$ central coordinates of the vector X by zero, i.e., $s \simeq \frac{p^{n-1}}{2}(p-1)$. Thus, it follows from (11.3.17) that

$$\delta_F \simeq \Delta\sqrt{2\frac{p^{n-1}}{2}(p-1)} \simeq p\sqrt{p^{n-2}}\Delta.$$

THE SECOND METHOD. We replace every element in the central subpacket of the last packet by zero, i.e., p^{n-1} coordinates of the vector Y. Suppose that p is even. Then

$$\tau = \frac{p^n}{p^n - p^{n-1}} = 1 + \frac{1}{p-1}$$

and by (11.3.15), $\delta_M \leq \Delta$.

In the DFT case with the same N and τ we have

$$s \simeq \frac{p^{n-1}}{2}, \quad \delta_F = \Delta\sqrt{2\frac{p^{n-1}}{2}} = p\sqrt{p^{n-3}}\Delta.$$

THE THIRD METHOD. We replace some subsequence lying in the central subpacket of the last packet of Y by zero. This method is used for large p, $p \geq 6$, because a sharp increase in the estimate (11.3.9) is observed only for j approaching 1 or $p-1$.

Unlike the DFT case, notice that for all three methods the quantity δ_M does not depend on the size N of the given input vector, and for large N is significantly smaller than δ_F. Thus for compression of information by means of orthogonal transformations, the method of zone coding which uses the discrete multiplicative transform is more accurate for restoration of the input vector than that which uses the discrete Fourier transform.

§11.4. Practicalities of processing two-dimensional numerical problems with discrete multiplicative transforms.

We saw in the previous section that estimates of the spectrum can be employed for zone coding, i.e., for compression of information. However, for the problem of pattern recognition, the spectrum of the original signal is of primary importance. Moreover for multidimensional signals, pattern recognition uses the multidimensional spectrum. Since multi-dimensional signals can be transformed into one-dimensional signals at the expense of extending the size of the system, it is natural to consider the problem of establishing a correspondence between the spectrum of multidimensional numerical signal and its one-dimensional counterpart. In this context, it is important to notice that the transformation from a multi-dimensional signal to its one- dimensional counterpart is not unique. For example, a two- dimensional signal $\{x_{m,k}\}$, $m, k = 0, 1, \ldots, N-1$, can be converted to a one-dimensional signal by means of row scanning while preserving the order in each row:

$$x_{m,k} = z_{mN+k}, \qquad m, k = 0, 1, \ldots, N-1.$$

Since the difference $z_{(m+1)N} - z_{mN+N-1}$ corresponds to the difference of the value at the left end of the $m+1$-st row and the right end of the m-th row, the one dimensional signal obtained in this way can have large jumps for $k = N-1$. Thus the inverse Fourier transform of the one-dimensional signal (z_{mN+k}) can exhibit Gibbs phenomenon at points of large jumps. Consequently, when converting signals from two-dimensions it is necessary to choose a scanning method which preserves "continuity" when passing from point to point, i.e., choose a method in which distances between data points which are near in the two-dimensional sense remain relatively so in the converted one-dimensional signal. However in this case it is practically impossible to establish some kind of dependence between the two-dimensional spectrum of the initial signal and the one-dimensional spectrum of the converted signal. Moreover, one must keep in mind that computing the DFT of a one-dimensional digital signal of size N^2 takes more calculations than computing the two-dimensional DFT of an $N \times N$ array, including approximately $N(N-1)$ values of the trigonometric functions at points of multiples π/N^2. Therefore, from the point of view of efficiency, the two- dimensional DFT of a system is preferred over the one-dimensional DFT of that same system.

We shall examine the situation in the discrete multiplicative transform case. Existence of dependence between the multiplicative spectrum of a two-dimensional signal and the multiplicative spectrum of its one-dimensional counterpart (obtained through row by row scanning) was discovered by Žukov [2]. We shall define the spectrum $\{y_{m,\mu}\}$, $m = 0,1,\dots,p^n-1$, $\mu = 0,1,\dots,p^r-1$ of a two-dimensional signal $\{a_{k,\nu}\}$, $k = 0,1,\dots,p^n-1$, $\nu = 0,1,\dots,p^r-1$ by

$$
(11.4.1) \qquad
\begin{aligned}
y_{m,\mu} &= \frac{1}{p^{n+r}} \sum_{k=0}^{p^n-1} \sum_{\nu=0}^{p^r-1} a_{k,\nu} \overline{\chi}_\mu\left(\frac{\nu}{p^r}\right) \overline{\chi}_m\left(\frac{k}{p^n}\right) \\
&= \frac{1}{p^{n+r}} \sum_{\nu=0}^{p^r-1} \sum_{k=0}^{p^n-1} a_{k,\nu} \overline{\chi}_m\left(\frac{k}{p^n}\right) \overline{\chi}_\mu\left(\frac{\nu}{p^r}\right).
\end{aligned}
$$

11.4.1. *Let $\{y_{m,\mu}\}$, $m = 0,1,\dots,p^n-1$, $\mu = 0,1,\dots,p^r-1$ be the spectrum of a two-dimensional signal $\{a_{k,\nu}\}$, $k = 0,1,\dots,p^n-1$, $\nu = 0,1,\dots,p^r-1$, and let $\{z_\ell\}$, $\ell = 0,1,\dots,p^{n+r}-1$ be the spectrum of the corresponding one-dimensional signal $\{x_\eta\}$, $\eta = 0,1,\dots,p^{n+r}-1$ obtained by row scanning, i.e., $x_\eta = x_{kp^r+\nu} = a_{k,\nu}$ for $\eta = kp^r+\nu$, $0 \le k \le p^n-1$, and $0 \le \nu \le p^r-1$. Then*

$$z_\ell = z_{\mu p^n + m} = z_{\mu,m} = y_{m,\mu}$$

for $\ell = \mu p^n + m$, $0 \le \mu \le p^r-1$, $0 \le m \le p^n-1$, i.e., the two-dimensional spectrum $\{z_{\mu,m}\}$, obtained from the one-dimensional spectrum $\{z_\ell\}$ by placing its elements one by one in p^n rows (p^r elements to each row) is the transpose of the initial spectrum $\{y_{m,\mu}\}$.

PROOF. By formula (11.3.2) we have

$$z_\ell = \frac{1}{p^{n+r}} \sum_{\eta=0}^{p^{n+r}-1} x_\eta \overline{\chi}_\eta \left(\frac{\ell}{p^{n+r}}\right), \qquad \ell = 0, 1, \ldots, p^{n+r} - 1. \tag{11.4.2}$$

Using property **11.2.3** and the fact that the function $\chi_k(x)$ is multiplicative, we obtain

$$\begin{aligned} z_\ell &= \frac{1}{p^{n+r}} \sum_{\eta=0}^{p^{n+r}-1} x_\eta \overline{\chi}_\ell \left(\frac{\eta}{p^{n+r}}\right) \\ &= \frac{1}{p^{n+r}} \sum_{k=0}^{p^n-1} \sum_{\nu=0}^{p^r-1} x_{kp^r+\nu} \overline{\chi}_\ell \left(\frac{kp^r+\nu}{p^{n+r}}\right) \\ &= \frac{1}{p^{n+r}} \sum_{k=0}^{p^n-1} \sum_{\nu=0}^{p^r-1} a_{k,\nu} \overline{\chi}_\ell \left(\frac{k}{p^n}\right) \overline{\chi}_\ell \left(\frac{\nu}{p^{n+r}}\right). \end{aligned}$$

Setting $\ell = \mu p^n + m$, $0 \le \mu \le p^r - 1$, $0 \le m \le p^n - 1$, we find that

$$\begin{aligned} z_\ell &= z_{\mu p^n+m} = z_{\mu,m} \\ &= \frac{1}{p^{n+r}} \sum_{k=0}^{p^n-1} \sum_{\nu=0}^{p^r-1} a_{k,\nu} \overline{\chi}_{\mu p^n+m} \left(\frac{k}{p^n}\right) \overline{\chi}_{\mu p^n+m} \left(\frac{\nu}{p^{n+r}}\right) \\ &= \frac{1}{p^{n+r}} \sum_{k=0}^{p^n-1} \sum_{\nu=0}^{p^r-1} a_{k,\nu} \overline{\chi}_{\mu p^n} \left(\frac{k}{p^n}\right) \overline{\chi}_m \left(\frac{k}{p^n}\right) \overline{\chi}_{\mu p^n} \left(\frac{\nu}{p^{n+r}}\right) \overline{\chi}_m \left(\frac{\nu}{p^{n+r}}\right). \end{aligned}$$

Since $0 \le \nu \le p^r - 1$ implies $\nu/p^{n+r} < 1/p^n$, we can write

$$\frac{\nu}{p^{n+r}} = \frac{\nu_1}{p^{n+1}} + \frac{\nu_2}{p^{n+2}} \cdots,$$

where $0 \le \nu_j \le p - 1$ for all j, and thus see that

$$\overline{\chi}_m \left(\frac{\nu}{p^{n+r}}\right) = 1$$

for $0 \le m \le p^n - 1$. Using **11.2.3** we obtain the identity

$$\chi_{\mu p^n} \left(\frac{\nu}{p^{n+r}}\right) = \chi_\nu \left(\frac{\mu p^n}{p^{n+r}}\right) = \chi_\nu \left(\frac{\mu}{p^r}\right) = \chi_\mu \left(\frac{\nu}{p^r}\right).$$

Moreover, since the p-adic coordinates of the numbers μp^n and k/p^n satisfy

$$(\mu p^n)_{-q} \equiv \left[\frac{\mu p^n}{p^{q-1}}\right] \pmod p = \mu_{-(n-q)} = \begin{cases} 0 & \text{for } q \le n, \\ \mu_{-(q-n)} & \text{for } q \ge n+1, \end{cases}$$

and

$$\left(\frac{k}{p^n}\right)_{-q} \equiv \left[\frac{k}{p^n} p^q\right] \pmod p \equiv [kp^{q-n}] \pmod p$$
$$= \begin{cases} 0 & \text{for } q \ge n+1, \\ k_{-(n-q+1)} & \text{for } q \le n, \end{cases}$$

we have

$$\chi_{\mu p^n}\left(\frac{k}{p^n}\right) = \exp \frac{2\pi i}{p} \sum_{q=1}^{\infty} (\mu p^n)_{-q} \left(\frac{k}{p^n}\right)_q = 1.$$

Therefore,

$$z_{\mu,m} = \frac{1}{p^{n+r}} \sum_{k=0}^{p^n-1} \sum_{\nu=0}^{p^r-1} a_{k,\nu} \overline{\chi}_m \left(\frac{k}{p^n}\right) \overline{\chi}_\mu \left(\frac{\nu}{p^r}\right). \tag{11.4.3}$$

In particular, comparing (11.4.1) with (11.4.3) we conclude that $z_{\mu,m} = y_{m,\mu}$. ∎

Notice that although the transforms (11.4.1) and (11.4.2) can be computed in roughly the same number of arithmetic calculations, nevertheless in practice it takes considerably more time to compute the system $\{y_{m,\mu}\}$ by formula (11.4.1) than by formula (11.4.2). This is due to the fact that computation of the transform (11.4.1) additional reorganization of all the calculations when one moves from calculating the inner sum to calculating the outer sum.

§11.5. A description of classes of discrete transforms which allow fast algorithms.

In the last decade interest in the study of discrete orthogonal transforms has grown significantly. This interest has been fueled by the appearance of a whole series of algorithms, similar to the fast Fourier transform (FFT) and to the algorithm we looked at in §11.2 for computing the DMT, which allowed these transforms to be computed quickly and efficiently (in real time) on high speed computers, even for problems which involved a large number of calculations. Investigations of the various properties of orthogonal transforms have been influenced by their applications to image processing and speech signals, by identification of tests for pattern recognition, by analyzing and designing communication systems, by generalized Wiener filtering, and by a number of other applied questions.

Because of this, there is keen interest in the problem of determining conditions on an $n \times n$ matrix so that its product with a column vector uses roughly $n \log n$

calculations instead of the n^2 operations used in the general case. Solutions to this problem involve creating an algorithm of type FFT for the given matrix.

In this section we shall find necessary and sufficient conditions for which a square matrix A of order $n \times n$, $n = pq$, can be represented as a product of two sparse matrices, each having a similar sparse structure.

We shall say that a square matrix A of order $n \times n$, $n = pq$, is (p,q)-*factorable* if it can be written as a product $A = BC$ where the elements of the matrices

$$B = (b_{i,j}) = (b_{i_1+i_2p,j_1+j_2q}), \qquad 0 \le i_1, j_2 \le p-1, \quad 0 \le i_2, j_1 \le q-1,$$

$$C = (c_{k,m}) = (c_{k_1+k_2q,m_1+m_2p}), \qquad 0 \le k_2, m_1 \le p-1, \quad 0 \le k_1, m_2 \le q-1$$

satisfy the conditions

$$\text{(11.5.1)} \qquad \begin{cases} b_{i_1+i_2p,j_1+j_2q} = 0 & \text{for} \quad j_1 \neq i_2, \\ c_{k_1+k_2q,m_1+m_2p} = 0 & \text{for} \quad m_1 \neq k_2. \end{cases}$$

The structure of the matrices B and C are shown in Figures 2 and 3, where all elements except those denoted by stars are zero.

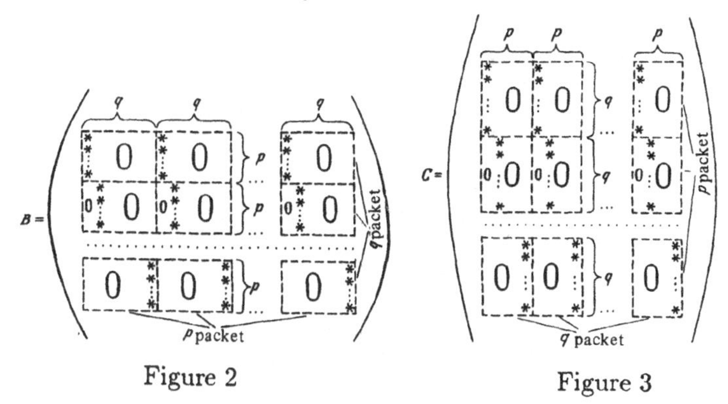

Figure 2 Figure 3

11.5.1. *Elements of a (p,q)-factorable matrix A satisfy the relationship*

$$\text{(11.5.2)} \qquad a_{i_1+i_2p,j_1+j_2q} = b_{i_1+i_2p,i_2+j_1q}c_{i_2+j_1q,j_1+j_2p} = b_{ik}c_{kj}$$

for $k = i_2 + j_1q$.

PROOF. Indeed, by (11.5.1) we find

$$\begin{aligned} a_i &= \sum_{\nu=0}^{n-1} b_{i,\nu}c_{\nu,j} \\ &= \sum_{\nu_1=0}^{q-1} \sum_{\nu_2=0}^{p-1} b_{i_1+i_2p,\nu_1+\nu_2q}c_{\nu_1+\nu_2q,j_1+j_2p} \\ &= b_{i_1+i_2p,i_2+j_1q}c_{i_2+j_1q,j_1+j_2p} = b_{ik}c_{kj}. \quad \blacksquare \end{aligned}$$

11.5.2. *Let A_{ij}, $0 \le i \le q-1$, $0 \le j \le p-1$, be submatrices of order $p \times q$ of a matrix $A = (a_{m,k})$ organized in the following way:*

$$A_{ij} = (a_{r+ip,j+sp}), \qquad 0 \le r \le p-1, \quad 0 \le s \le q-1. \tag{11.5.3}$$

Then the matrix A is (p,q)-factorable if and only if

$$\text{Rank } A_{ij} \le 1 \tag{11.5.4}$$

for all $0 \le i \le q-1$, $0 \le j \le p-1$.

PROOF OF NECESSITY. Let A be (p,q)-factorable. By **11.5.1** we have

$$a_{r+ip,j+sp} = b_{r+ip,i+jq}c_{i+jq,j+sp}.$$

If for fixed i and j we denote

$$\alpha_r = b_{r+ip,i+jq}, \qquad r = 0,1,\ldots,p-1,$$

$$\beta_s = c_{i+jq,j+sp}, \qquad s = 0,1,\ldots,q-1,$$

then $a_{r+ip,j+sp} = \alpha_r\beta_s$. In particular, it follows that Rank $A_{ij} \le 1$.

PROOF OF SUFFICIENCY. Suppose that (11.5.4) holds for every $0 \le \nu \le q-1$, and $0 \le \mu \le p-1$. Then choose numbers $\alpha_{r+\nu p}$ and $\beta_{\mu+sp}$ so that the elements of the matrix $A_{\nu,\mu}$ from (11.5.3) are written in the form

$$a_{r+\nu p,\mu+sp} = \alpha_{r+\nu p}\beta_{\mu+sp}. \tag{11.5.5}$$

Set

$$\begin{aligned} b_{i,j} &= b_{i_1+i_2p,j_1+j_2q} \\ &= \begin{cases} 0 & \text{for } j_1 \ne i_2,\ j_2 = 0,1,\ldots,p-1, \\ \alpha_{i_1+i_2p} & \text{for } j_1 = i_2, \end{cases} \end{aligned} \tag{11.5.6}$$

and

$$\begin{aligned} c_{k,m} &= c_{k_1+k_2q,m_1+m_2p} \\ &= \begin{cases} 0 & \text{for } m_1 \ne k_2,\ k_1 = 0,1,\ldots,q-1, \\ \beta_{m_1+m_2p} & \text{for } m_1 = k_2. \end{cases} \end{aligned} \tag{11.5.7}$$

It is obvious by construction that the matrices $B = (b_{i,j})$ and $C = (c_{k,m})$ are sparse as defined in (11.5.1). It remains to verify that their product coincides with the matrix A. But by (11.5.6), (11.5.7), and (11.5.5) we find that

$$\begin{aligned}\sum_{\nu=0}^{n-1} b_{i,\nu}c_{\nu,m} &= \sum_{\nu_1=0}^{q-1}\sum_{\nu_2=0}^{p-1} b_{i_1+i_2p,\nu_1+\nu_2q}c_{\nu_1+\nu_2q,m_1+m_2p} \\ &= b_{i_1+i_2p,i_2+m_1q}c_{i_2+m_1q,m_1+m_2p} \\ &= \alpha_{i_1+i_2p}\beta_{m_1+m_2p} = a_{i_1+i_2p,m_1+m_2p}. \ \blacksquare\end{aligned}$$

It is fairly difficult to verify condition (11.5.4) for matrices of large order. Because of this the condition of (p,q)- factorability of a matrix will be described in other terms.

We shall associate two bases of the linear space $\mathbf{C}^n$ with each non-singular matrix A : the canonical basis $\{\mathbf{e}_k\}$, $k = 0,1,\dots,n-1$, where the k-th component of $\mathbf{e}_k$ is 1 but all other components of $\mathbf{e}_k$ are zero, and the basis $\{\mathbf{a}_k\}$, $k = 0,1,\dots,n-1$, where $\mathbf{a}_k$ is the k-th column of the matrix A. We shall denote the linear hulls (in $\mathbf{C}^n$) generated by certain collections of these vectors in the following way:

$$\boldsymbol{L}_i = \mathcal{L}(\boldsymbol{e}_{ip}, \boldsymbol{e}_{ip+1}, \dots, \boldsymbol{e}_{ip+p-1}), \qquad i = 0,1,\dots,q-1,$$

$$\boldsymbol{L}'_j = \mathcal{L}(\boldsymbol{a}_j, \boldsymbol{a}_{j+p}, \dots, \boldsymbol{a}_{j+(q-1)p}), \qquad j = 0,1,\dots,p-1.$$

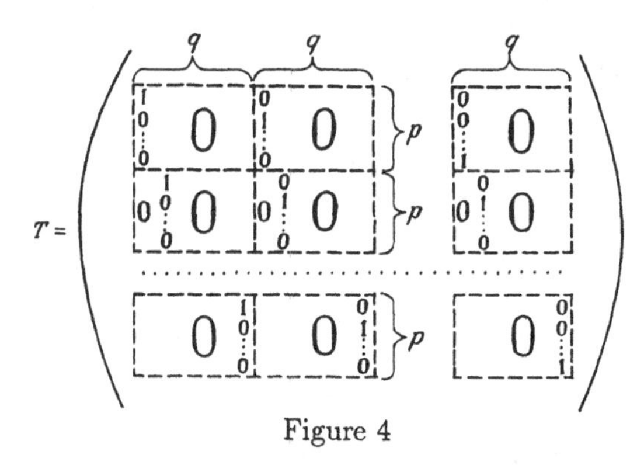

Figure 4

11.5.3. *If A is a non-singular (p,q)-factorable matrix, then*

$$\text{(11.5.8)} \qquad dim\ (\boldsymbol{L}_i \bigcap \boldsymbol{L}'_j) \geq 1$$

for $i = 0,1,\dots,q-1$, and $j = 0,1,\dots,p-1$.

PROOF. For each $i = 0,1,\dots,q-1$ and $j = 0,1,\dots,p-1$ it suffices to show that there exists at least one non-zero vector which belongs simultaneously to the spaces

$\boldsymbol{L}_i$ and $\boldsymbol{L}'_j$. Consider the vector $\boldsymbol{b}_{i+jq}$, i.e., the vector which occupies the $i+jq$-th column of the matrix B. Since A is (p,q) factorable, we have $b_{\nu_1+\nu_2 p, i+jq} = 0$ for $\nu_2 \neq i$. Consequently,

$$\boldsymbol{b}_{i+jq} = \sum_{\nu_1=0}^{p-1} b_{\nu_1+\nu_2 p, i+jq} \boldsymbol{e}_{\nu_1+ip} \in \boldsymbol{L}_i.$$

On the other hand, since the matrix A is non-singular we can write $B = AC^{-1}$. We shall show that the matrix $C^{-1} = (\tilde{c}_{\alpha,\beta})$ is sparse with

(11.5.9) $$\tilde{c}_{\alpha_1+\alpha_2 p, \beta_1+\beta_2 q} = 0, \qquad \alpha_1 \neq \beta_2,$$

where $0 \leq \alpha_1, \beta_2 \leq p-1$, and $0 \leq \alpha_2, \beta_1 \leq q-1$. Indeed, set $T = (t_{i,j})$ where

(11.5.10)
$$\begin{aligned} t_{i,j} &= t_{i_1+i_2 p, j_1+j_2 q} \\ &= \begin{cases} 1 & \text{for } \ i_1 = j_2 \ \text{ and } \ i_2 = j_1, \\ & \qquad 0 \leq i_1, j_2 \leq p-1, \quad 0 \leq i_2, j_1 \leq q-1, \\ 0 & \text{in the remaining cases,} \end{cases} \end{aligned}$$

are elements of a permutation matrix whose structure is shown in Figure 4. Set $C' = CT$ and notice by (11.5.10) that

$$\begin{aligned} c'_{k,m} &= c'_{k_1+k_2 q, m_1+m_2 q} = \sum_{\nu=0}^{n-1} c_{k,\nu} t_{\nu,m} \\ &= \sum_{\nu_1=0}^{p-1} \sum_{\nu_2=0}^{q-1} c_{k_1+k_2 q, \nu_1+\nu_2 p} t_{\nu_1+\nu_2 p, m_1+m_2 q} = c_{k_1+k_2 q, m_1+m_2 q}. \end{aligned}$$

By (11.5.1) we obtain

$$c'_{k_1+k_2 q, m_1+m_2 q} = 0, \qquad m_2 \neq k_2,$$

i.e., the only elements different from zero have the form

$$c'_{k_1+\alpha q, m_1+\alpha q}, \qquad \alpha = 0, 1, \ldots, p-1.$$

Thus the matrix C' is block diagonal and its structure is shown in Figure 5.

Since C' can be obtained from C by a permutation of columns it must also be non-singular. It follows that each block of C is non-singular, since det C is the product of the determinants of these diagonal $q \times q$ blocks. Hence the inverse of C' is also block diagonal (see Figure 6). Since $C' = CT$ we have $(C')^{-1} = (CT)^{-1} = T^{-1}C^{-1}$, i.e.,

$$T(C')^{-1} = TT^{-1}C^{-1} = EC^{-1} = C^{-1}$$

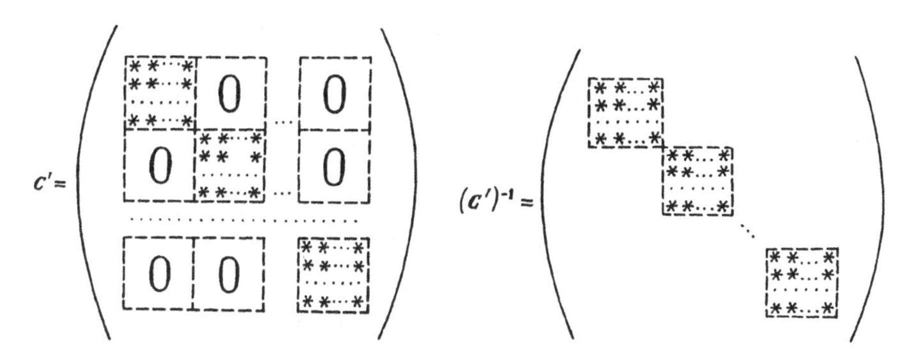

Figure 5 Figure 6

where E is the identity matrix. Consequently, $C^{-1} = T(C')^{-1}$. Since $(C')^{-1}$ has a similar block structure, the product $T(C')^{-1}$, obtained from $(C')^{-1}$ by a permutation of columns, is also sparse with

$$\tilde{c}_{\alpha,\beta} = \tilde{c}_{\alpha_1+\alpha_2 p,\beta_1+\beta_2 q} = \sum_{\nu=0}^{n-1} t_{\alpha_1+\alpha_2 p,\nu} c^*_{\nu,\beta_1+\beta_2 q},$$

where $c^*_{\nu,j}$ are the elements of the matrix $(C')^{-1}$. Using the definition (11.5.10) of the matrix T, we obtain

$$\tilde{c}_{\alpha,\beta} = \sum_{\nu_1=0}^{q-1}\sum_{\nu_2=0}^{p-1} t_{\alpha_1+\alpha_2 p,\nu_1+\nu_2 q} c^*_{\nu_1+\nu_2 q,\beta_1+\beta_2 q} = c^*_{\alpha_1+\alpha_2 q,\beta_1+\beta_2 q}.$$

Since the elements of this block diagonal matrix vanish if the second indices are different, i.e., if $\alpha_1 \neq \beta_2$, we see that (11.5.9) holds as promised.

It remains to show that along with $\boldsymbol{b}_{i+jq} \in \boldsymbol{L}_i$ we also have $\boldsymbol{b}_{i+jq} \in \boldsymbol{L}'_j$. But using (11.5.9) we find

$$\begin{aligned}\boldsymbol{b}_{i+jq} &= \sum_{\nu=0}^{n-1} \tilde{c}_{\nu,i+jq}\boldsymbol{a}_\nu \\ &= \sum_{\nu_1=0}^{p-1}\sum_{\nu_2=0}^{q-1} \tilde{c}_{\nu_1+\nu_2 p,i+jq}\boldsymbol{a}_{\nu_1+\nu_2 p} \\ &= \sum_{\nu_2=0}^{q-1} \tilde{c}_{j+\nu_2 p,i+jq}\boldsymbol{a}_{j+\nu_2 p} \in \boldsymbol{L}'_j. \quad \blacksquare\end{aligned}$$

11.5.4. *Let A be an $n \times n$ non-singular matrix. Suppose that the subspaces $\boldsymbol{L}_i$, $i = 0, 1, \ldots, q-1$, and $\boldsymbol{L}'_j$, $j = 0, 1, \ldots, p-1$, satisfy the conditions*

$$\dim\,(\boldsymbol{L}_i \bigcap \boldsymbol{L}'_j) = 1 \tag{11.5.11}$$

for all i, j. If $\boldsymbol{f}_{i,j} \in \boldsymbol{L}_i \bigcap \boldsymbol{L}'_j$ are non-zero vectors, then the system $\{\boldsymbol{f}_{i,j}\}$, $i = 0, 1, \ldots, q-1$, $j = 0, 1, \ldots, p-1$, is a basis for the space $\mathbf{C}^n$.

PROOF. Since the number of vectors $\boldsymbol{f}_{i,j}$, $i = 0, 1, \ldots, q-1$, $j = 0, 1, \ldots, p-1$, is exactly $pq = n$, it suffices to prove that they are linearly independent. Let

$$\sum_{i=0}^{q-1} \sum_{j=0}^{p-1} \lambda_{i,j} \boldsymbol{f}_{i,j} = 0.$$

Since $\sum_{j=0}^{p-1} \lambda_{i,j} \boldsymbol{f}_{i,j} \in \boldsymbol{L}_i$ then $\sum_{j=0}^{p-1} \lambda_{i,j} \boldsymbol{f}_{i,j} = 0$. But for each fixed i the vector $\boldsymbol{f}_{i,j}$ belongs to $\boldsymbol{L}'_j$. Therefore, the previous relationship implies that $\lambda_{i,j} = 0$ for all i, j. ∎

11.5.5. *Let A be an $n \times n$ non-singular matrix. Then A is (p, q)-factorable if and only if (11.5.11) holds for $i = 0, 1, \ldots, q-1$, $j = 0, 1, \ldots, p-1$.*

PROOF OF NECESSITY. Let A be a non-singular, (p, q)- factorable matrix. Then by relationship (11.5.8) we have

$$\dim \mathbf{C}^n = \sum_{i=0}^{q-1} \dim \boldsymbol{L}_i = \sum_{i=0}^{q-1} \sum_{j=0}^{p-1} \dim (\boldsymbol{L}_i \bigcap \boldsymbol{L}'_j) \geq \sum_{i=0}^{q-1} \sum_{j=0}^{p-1} 1 = pq = n.$$

Consequently, we must have equality in (11.5.8), i.e., $\dim (\boldsymbol{L}_i \bigcap \boldsymbol{L}'_j) = 1$.

PROOF OF SUFFICIENCY. Let A be a non-singular matrix which satisfies condition (11.5.11). Then by **11.5.4** the system of vectors $\{\boldsymbol{f}_{i,j}\}$, $i = 0, 1, \ldots, q-1$, $j = 0, 1, \ldots, p-1$, is a basis for the space $\mathbf{C}^n$.

We introduce the notation

$$\mathcal{B} = \{\boldsymbol{e}_0, \boldsymbol{e}_1, \ldots, \boldsymbol{e}_{n-1}\},$$

$$\mathcal{B}' = \{\boldsymbol{a}_0, \boldsymbol{a}_1, \ldots, \boldsymbol{a}_{n-1}\}, \quad \mathcal{B}'' = \{\boldsymbol{f}_{0,0}, \boldsymbol{f}_{1,1}, \ldots, \boldsymbol{f}_{q-1,p-1}\}.$$

Then $S_{\mathcal{B} \to \mathcal{B}'} = S_{\mathcal{B} \to \mathcal{B}''} S_{\mathcal{B}'' \to \mathcal{B}'}$, where $S_{\cdot \to \cdot}$ represents the matrix which takes one basis into another, for example $S_{\mathcal{B} \to \mathcal{B}'}$ represents the matrix which takes the basis $\mathcal{B}$ to the basis $\mathcal{B}'$. By definition the columns of this matrix are the coordinates of the basis $\mathcal{B}'$ as vectors in the basis $\mathcal{B}$. But $\mathcal{B}$ is the canonical basis. Thus the corresponding coordinates of the vectors from $\mathcal{B}'$ coincide with their components. But by construction, the vectors of the basis $\mathcal{B}'$ are just the columns of the matrix A. Consequently, the coordinates of any vector in $\mathcal{B}'$ with respect to the basis $\mathcal{B}$ is just a corresponding column of A. In particular, $S_{\mathcal{B} \to \mathcal{B}'} = A$.

Set $B = S_{\mathcal{B} \to \mathcal{B}''}$ and $C = S_{\mathcal{B}'' \to \mathcal{B}'}$. Then $A = BC$ and it remains to show that the matrices B and C are sparse matrices as defined by (11.5.1).

We notice that the vectors $\boldsymbol{f}_{i,j}$, $j = 0, 1, \ldots, p-1$, belong to the space $\boldsymbol{L}_i$ and consequently have the form

$$\boldsymbol{f}_{i,j} = \sum_{\nu=0}^{p-1} f_{\nu+ip,i+jp} \cdot \boldsymbol{e}_{\nu+ip}.$$

Thus if we assume $\boldsymbol{b}_k = \boldsymbol{b}_{i+jq} = \boldsymbol{f}_{i,j}$ we find that the components

$$b_{m,k} = b_{m_1+m_2p,k_1+k_2q},$$

$$0 \le m_1, k_2 \le p-1, \quad 0 \le m_2, k_1 \le q-1,$$

of the vector $\boldsymbol{b}_k$ satisfy the relationship $b_{m,k} = 0$ for $m_2 \ne k_1$. Consequently, the matrix $B = S_{B \to B''}$, whose columns are the vectors $\{\boldsymbol{b}_k,\ k = 0, 1, \ldots, n-1\}$, satisfies (11.5.1).

Similarly, the vectors $\boldsymbol{f}_{i,j}$, $i = 0, 1, \ldots, q-1$, belong to the space $\boldsymbol{L}'_j$. Thus the vectors $\boldsymbol{b}_k$, $k = jq, jq+1, \ldots, jq+q-1$ form a basis of this subspace. From this we obtain the following decomposition:

$$\boldsymbol{a}_{j+ip} = \sum_{\nu=0}^{q-1} c_{\nu+jq,j+ip} \cdot \boldsymbol{b}_{\nu+jq} = \sum_{\nu=0}^{q-1} c_{\nu+jq,j+ip} \cdot \boldsymbol{f}_{\nu,j}.$$

Introducing the notation

$$\boldsymbol{c}_{j+ip} = (0, \ldots, 0, c_{iq,j+ip}, c_{iq+1,j+ip}, \ldots, c_{jq+q-1,j+ip}, 0, \ldots, 0),$$

we find that $c_{\nu,\mu} = c_{\nu_1+\nu_2q,\mu_1+\mu_2p} = 0$ if $\mu_1 \ne \nu_2$. Therefore, the matrix $C = S_{B'' \to B'}$ whose columns are the vectors $\boldsymbol{c}_k$, $k = 0, 1, \ldots, n-1$, also satisfies (11.5.1). In particular, the matrix A, as a product of B and C, is (p, q)-factorable. ∎

For applications, it is frequently sufficient to use matrices whose elements are obtained by a permutation of rows and columns of some (p, q)-factorable matrix. Conforming to the terminology in digital methods of information processing, we shall call the set of matrices $\hat{A}$ of the form $\hat{A} = \{A' : A' = T_1 A T_2\}$, where T_1 and T_2 are permutation matrices, the *discrete transformation* $\hat{A} = \hat{A}(A)$ *generated by* the matrix A.

It is evident that a discrete transformation $\hat{A}$ is generated by any matrix it contains, i.e., if $\hat{A} = \hat{A}(A)$ and $A' = T_1 A T_2$ for some permutation matrices T_1 and T_2, then $\hat{A} = \hat{A}(A')$.

It is clear by definition that any discrete transformation $\hat{A}$ is a union of matrices which differ from one another by permutations of rows and columns.

A discrete transformation $\hat{A}$ is called (p, q)-factorable if at least one of its generators is (p, q)-factorable.

Suppose that $\hat{A}$ is the discrete transformation generated by a matrix A'. We shall use this matrix and certain kinds of partitions to generate special subspaces and triangular matrices in the following way.

Suppose $n = pq$. Let $\tau = \{\Delta_i : i = 0, 1, \ldots, q-1\}$ be a partition of the set of numbers $\{0, 1, \ldots, n-1\}$ such that each Δ_i contains exactly p elements chosen from the collection of indices $\{0, 1, \ldots, n-1\}$. Clearly, $\Delta_{i_1} \bigcap \Delta_{i_2} = \emptyset$. Similarly, let $\tau' = \{\Delta'_j : j = 0, 1, \ldots, p-1\}$ be a partition of the set of numbers $\{0, 1, \ldots, n-1\}$ where each Δ'_j contains exactly q elements. Again, $\Delta'_{j_1} \bigcap \Delta'_{j_2} = \emptyset$. Let $\tilde{\boldsymbol{L}}_i = \boldsymbol{L}(\Delta_i)$ be the linear hull of the collection of vectors $\mathbf{e}_\nu$, $\nu \in \Delta_i$, and let $\tilde{\boldsymbol{L}}'_j = \boldsymbol{L}(\Delta'_j)$ be the linear hull of the collection vectors which form the μ-th columns of the matrix A' (which generates the discrete transformation $\hat{A}$), where $\mu \in \Delta'_j$. And, let $A^{(i,j)}$ be the triangular $p \times q$ matrix made up of the elements of the matrix A' which lie in the intersections of the ν-th rows, $\nu \in \Delta_i$, and the μ-th columns, $\mu \in \Delta'_j$, $i = 0, 1, \ldots, q-1$, $j = 0, 1, \ldots, p-1$.

11.5.6. *A discrete transformation $\hat{A} = \hat{A}(A')$ is (p,q)-factorable if and only if there exist partitions τ and τ' of the set $\{0, 1, \ldots, n-1\}$ such that*

$$\text{Rank } A^{(i,j)} \le 1, \qquad i = 0, 1, \ldots, q-1, \quad j = 0, 1, \ldots, p-1. \tag{11.5.12}$$

Proof of Sufficiency. Let $\hat{A}$ be a (p,q)-factorable discrete transformation. Thus there is a (p,q)-factorable matrix A which generates the transformation $\hat{A}$ (we may suppose that the matrix A is different from the matrix A'). Hence by definition there exist permutation matrices T_1 and T_2 such that $A' = T_1 A T_2$. The matrices T_1 and T_2 induce permutations on the set of indices $\{0, 1, \ldots, n-1\}$. For each $i = 0, 1, \ldots, q-1$, let Δ_i represent the subset which results from applying the permutation T_1 to the set $\{ip, ip+1, \ldots, ip+(p-1)\}$. Similarly, let Δ'_j represent the image of $\{j, j+p, \ldots, j+(q-1)p\}$ under the permutation T_2. Then $A^{(i,j)} = A_{ij}$ and we conclude by Theorem **11.5.2** that

$$\text{Rank } A^{(i,j)} = \text{Rank } A_{ij} \le 1.$$

Proof of Necessity. Suppose that the discrete transformation $\hat{A}$ is generated by a matrix A' for which there exist partitions $\tau = \{\Delta_i : i = 0, 1, \ldots, q-1\}$ and $\tau' = \{\Delta'_j : j = 0, 1, \ldots, p-1\}$ of the set $\{0, 1, \ldots, n-1\}$ such that the submatrices $A^{(i,j)}$ satisfy (11.15.12). Let $\tilde{T}_1$ and $\tilde{T}_2$ represent the matrices induced by the permutations on the set $\{0, 1, \ldots, n-1\}$ which take, respectively, the subsets Δ_i and Δ'_j to the sets $\{ip, ip+1, \ldots, ip+(p-1)\}$ and $\{j, j+p, \ldots, j+(q-1)p\}$. The matrix $A = \tilde{T}_1 A' \tilde{T}_2$ also generates the discrete transformation $\hat{A}$ and is (p,q)-factorable by **11.5.3**. Indeed, $A_{ij} = A^{(i,j)}$ and consequently, $\text{Rank} A_{ij} \le 1$. In particular, the discrete transformation $\hat{A}$ is (p,q)-factorable. ∎

A discrete transformation will be called *non-singular* if it is generated by a non-singular matrix. Evidently, all the generators of a non-singular discrete transformation are non- singular since the determinant of a permutation matrix equals ± 1.

11.5.7. *A non-singular discrete transformation $\hat{A} = \hat{A}(A')$ is (p,q)-factorable if and only if there exist partitions τ and τ' of the set $\{0,1,\ldots,n-1\}$ such that*

$$\text{(11.5.13)} \qquad \dim(\tilde{\boldsymbol{L}}_i \bigcap \tilde{\boldsymbol{L}}'_j) = 1, \qquad i = 0,1,\ldots,q-1, \quad j = 0,1,\ldots,p-1.$$

PROOF. The proof is similar to that of **11.5.6** but is based instead on **11.5.5**. Necessity of condition (11.5.13) is obtained from construction of the permutation matrices T_1 and T_2 as in **11.5.6** and necessity of (11.5.13) from construction of the matrices $\tilde{T}_1$ and $\tilde{T}_2$. We notice that if the matrices T_1 and T_2 are already chosen then one can suppose that $\tilde{T}_1 = (T_1)^{-1}$ and $\tilde{T}_2 = (T_2)^{-1}$. ∎

The difference between factorability of discrete transformations and matrices can be illustrated by the following example. Let $\hat{W}$ be the discrete Walsh transformation generated by the matrix $W = (w_{n,k})$, $n,k = 0,1,\ldots,2^r-1$, where if

$$n = n_0 + n_1 2 + \cdots + n_{r-1}2^{r-1}, \quad k = k_0 + k_1 2 + \cdots + k_{r-1}2^{r-1},$$

then $w_{n,k} = \exp\left(\pi i \sum_{\nu=0}^{r-1} n_\nu k_{r-\nu-1}\right)$. We shall show that $\hat{W}$ is $(2^{r-1},2)$-factorable but that the matrix W is not $(2^{r-1},2)$- factorable. Let

$$\boldsymbol{a}_k = (w_{0,k}, w_{1,k}, \ldots, w_{2^r-1,k})^T, \qquad k = 0,1,\ldots,2^r-1$$

denote the column vectors of the matrix W. Below, we shall construct two partitions $\tau = \{\Delta_i : i = 0,1\}$ and $\tau' = \{\Delta'_j : j = 0,1,\ldots,2^r-1\}$ of the set of numbers $\{0,1,\ldots,n-1\}$. Using these partitions, we define subspaces $\tilde{\boldsymbol{L}}_i = \boldsymbol{L}(\Delta_i)$ and $\tilde{\boldsymbol{L}}'_j = \boldsymbol{L}(\Delta'_i)$ which are linear hulls of vectors of the canonical basis and vector columns of the matrix W whose indices belong, respectively, to the sets Δ_i and Δ'_j. We must verify that

$$\text{(11.5.14)} \qquad \dim(\tilde{\boldsymbol{L}}_i \bigcap \tilde{\boldsymbol{L}}'_j) = 1, \qquad i = 0,1, \quad j = 0,1,\ldots,2^r-1.$$

Let the partition τ be determined in the following way: $\Delta_0 = \{0,1,\ldots,2^{r-1}-1\}$, and $\Delta_1 = \{2^{r-1}-1,\ldots,2^r-1\}$. We shall construct the partition τ' so that (11.5.14) holds. Each Δ'_j consists of two elements $j^{(1)}$ and $j^{(2)}$. For any fixed j we will have

$$\tilde{\boldsymbol{L}}'_j = \boldsymbol{L}(\Delta'_j) = \mathcal{L}(\boldsymbol{a}_{j^{(1)}}, \boldsymbol{a}_{j^{(2)}}).$$

Consequently, each vector $\boldsymbol{x}$ of the space $\tilde{\boldsymbol{L}}_j'$ has the form $\boldsymbol{x} = \lambda_1 \boldsymbol{a}_{j^{(1)}} + \lambda_2 \boldsymbol{a}_{j^{(2)}}$, where λ_1 and λ_2 are real numbers. If $\boldsymbol{x} \in \tilde{\boldsymbol{L}}_i$ then $\boldsymbol{x} = \gamma_0 e_0 + \gamma_1 e_1 + \cdots + \gamma_{2^{r-1}-1} e_{2^{r-1}-1}$. In particular, if $\boldsymbol{x} \in \tilde{\boldsymbol{L}}_i \bigcap \tilde{\boldsymbol{L}}_j'$ is a non-zero vector then

$$\lambda_1 \boldsymbol{a}_{j^{(1)}} + \lambda_2 \boldsymbol{a}_{j^{(2)}} = \gamma_0 e_0 + \gamma_1 e_1 + \cdots + \gamma_{2^{r-1}-1} e_{2^{r-1}-1} \tag{11.5.15}$$

and not all of these coefficients are identically zero. The right side of the vector equation (11.5.15) has the form

$$\gamma_0 e_0 + \gamma_1 e_1 + \cdots + \gamma_{2^{r-1}-1} e_{2^{r-1}-1} = \begin{pmatrix} \gamma_0 \\ \vdots \\ \gamma_{2^{r-1}-1} \\ 0 \\ \vdots \\ 0 \end{pmatrix}.$$

Thus (11.5.15) will be satisfied if and only if we can find coefficients λ_1 and λ_2, not both zero, such that the last 2^{r-1} coordinates of the vector $\lambda_1 \boldsymbol{a}_{j^{(1)}} + \lambda_2 \boldsymbol{a}_{j^{(2)}}$ are identically zero. Since $\boldsymbol{a}_{j^{(1)}} = (w_{0,j^{(1)}}, \ldots, w_{2^r-1,j^{(1)}})$ and $\boldsymbol{a}_{j^{(2)}} = (w_{0,j^{(2)}}, \ldots, w_{2^r-1,j^{(2)}})$, we obtain the following system:

$$\lambda_1 w_{\nu,j^{(1)}} + \lambda_2 w_{\nu,j^{(2)}} = 0, \qquad \nu = 2^{r-1}, \ldots, 2^r - 1. \tag{11.5.16}$$

If $\nu \in \{2^{r-1}, \ldots, 2^r - 1\}$ then in the decomposition $\nu = \nu_0 + 2\nu_1 + \cdots + 2^{r-1}\nu_{r-1}$ we have $\nu_{r-1} = 1$. Consequently,

$$w_{\nu,\mu} = \exp\left(\pi i \sum_{s=0}^{r-1} \nu_s \mu_{r-s-1}\right) = \exp\left(\pi i \sum_{s=0}^{r-2} \nu_s \mu_{r-s-1} + \mu_0\right)$$

for $\mu = j^{(1)}, j^{(2)}$. If we set

$$j^{(1)} = j_0^{(1)} + j_1 2 + \cdots + j_{r-1} 2^{r-1} \quad \text{and} \quad j^{(2)} = j_0^{(2)} + j_1 2 + \cdots + j_{r-1} 2^{r-1}$$

then (11.5.16) can be written in the form

$$\begin{aligned} \lambda_1 (-1)^{j_0^{(1)}} \exp\left(\pi i \sum_{s=0}^{r-2} \nu_s j_{r-s-1}\right) &+ \lambda_2 (-1)^{j_0^{(2)}} \exp\left(\pi i \sum_{s=0}^{r-2} \nu_s j_{r-s-1}\right) \\ &= \exp\left(\pi i \sum_{s=0}^{r-2} \nu_s j_{r-s-1}\right) \left[\lambda_1 (-1)^{j_0^{(1)}} + \lambda_2 (-1)^{j_0^{(2)}}\right] = 0 \end{aligned}$$

for $\nu_s = 0, 1$, $s = 0, 1, \ldots, r-2$. Since $j^{(1)}$ and $j^{(2)}$ are distinct, the numbers $(-1)^{j_0^{(1)}}$ and $(-1)^{j_0^{(2)}}$ have different signs. Therefore, (11.5.16) will be satisfied when $\lambda_2 = \lambda_1$.

We have shown that if we take the partition $\tau' = \{\Delta'_j : j = 0, 1, \ldots, 2^{r-1} - 1\}$ where $\Delta'_j = \{2j, 2j+1\}$ for each j, then (11.5.14) holds for $i = 0$, i.e.,

$$\dim\,(\tilde{\boldsymbol{L}}_0 \bigcap \tilde{\boldsymbol{L}}'_j) = 1.$$

A similar construction shows that $\dim\,(\tilde{\boldsymbol{L}}_1 \bigcap \tilde{\boldsymbol{L}}'_j) = 1$. Hence by **11.5.7** the discrete Walsh transformation is $(2^{r-1}, 2)$-factorable. However, we notice by construction that the subspaces $\tilde{\boldsymbol{L}}'_j$ do not coincide with the subspaces $\boldsymbol{L}'_j$ or, equivalently, that the partition τ' does not satisfy the conditions of proposition **11.5.7**. Therefore, we conclude that the matrix W is not $(2^{r-1}, 2)$-factorable.

Chapter 12

OTHER APPLICATIONS OF MULTIPLICATIVE FUNCTIONS AND TRANSFORMATIONS

§12.1. Construction of digital filters based on multiplicative transforms.

Filtering is one of the fundamental techniques of digital signal processing. By a digital filter we shall mean a transformation which takes a sequence of numbers $\{x(n)\}$, called the *input*, in another sequence $\{y(n)\}$, which is called the *output*, or the *filter response* . In general, the relationship between the input sequence and the filter response can be written in the form

$$y(n) = \mathcal{F}(\{y(\nu)\}_{\nu\leq n-1}, \{x(\nu)\}_{\nu\leq n}).$$

During the last decade, rapid expansion of digital methods of signal processing has been a huge stimulus for the development of both a theoretical basis and practical methods to apply digital filtering in various situations. Investigations in this direction are well documented, both in a large number of journal articles and in a series of monographs. Hence instead of dwelling explicitly on the execution of these ideas we shall consider only digital filtering based on spectral analysis of the input signal. Widespread use of such digital filtering began at the end of the sixties with development of the so-called *fast algorithms* for computing the discrete Fourier transform (DFT). The idea behind such filtering is that the DFT can be used to compute the spectrum Y of an input signal X, i.e., the quantity $Y = \mathcal{F}X$. Then by using a diagonal matrix D one can obtain from Y a *filtered* signal $\tilde{X} = \mathcal{F}^{-1}D\mathcal{F}X$. The matrix D compresses, strengthens, weakens, or otherwise affects various spectral components of X.

We shall be especially interested in spectral filtering by means of the discrete multiplicative transform (DMT), and confine ourselves to the case when the DMT is generated by the sequence $p_k = p \geq 2$ for $k = 1, 2, \dots$ and some fixed p. Let $\hat{W}$ denote the direct DMT of size p^n and let $(\hat{W})^{-1}$ denote its inverse. Then the filtered signal generated by spectral filtering using the discrete multiplicative transform can be described by the formula

$$\tilde{X} = (\hat{W})^{-1}D\hat{W}X, \tag{12.1.1}$$

where the transformations $\hat{W}$ and $(\hat{W})^{-1}$ are defined by (11.2.6) and (11.2.7). Let W denote the matrix of the transform (11.2.6) written by **11.2.4** in the form $W = CB^n$, where the elements $b_{\ell,k}$ of the matrix B are defined by (11.2.8), and the elements

$c_{\ell,k}$ of the matrix C are defined by (11.2.12). Then (12.1.1) can be written in the form

$$\tilde{X} = (CB^n)^{-1}DCB^nX.$$

But $(CB^n)^{-1} = (B^n)^{-1}C^{-1} = (B^{-1})^nC^{-1}$ and $C^{-1}DC = D^*$, so

$$\tilde{X} = (B^{-1})^nC^{-1}DCB^nX = (B^{-1})^nD^*B^nX. \tag{12.1.2}$$

We shall show that $B^{-1} = \frac{1}{p}(\overline{B})^T$. Denote the elements of the matrix B by $b_{\ell,k}$ and use identity (11.2.8) for the elements $b^*_{\ell,k}$ of the matrix $(\overline{B})^T$ to write (for $n \geq 2$)

$$\begin{aligned} b^*_{\ell,k} = \overline{b}_{\ell,k} &= \delta_{\sum_{\nu=2}^{n} k_{-\nu}p^{\nu-2}, \sum_{\nu=1}^{n-1} \ell_{-\nu}p^{\nu-1}} (\overline{q})^{k_{-1}\ell_{-n}} \\ &= \prod_{\nu=1}^{n-1} \delta_{k_{-(\nu+1)},\ell_{-\nu}} (\overline{q})^{k_{-1}\ell_{-n}}, \end{aligned} \tag{12.1.3}$$

where $\overline{q} = \exp \frac{2\pi i}{p}$, $k = \sum_{\nu=1}^n k_{-\nu}p^{\nu-1}$ and $\ell = \sum_{\nu=1}^n \ell_{-\nu}p^{\nu-1}$. Let $c_{\ell,k}$ represent the elements of the matrix product $(\overline{B})^TB$, i.e.,

$$c_{\ell,k} = \sum_{m=0}^{p^n-1} b^*_{\ell,m}b_{m,k}.$$

Using (12.1.3) and (11.2.8) we obtain

$$c_{\ell,k} = \sum_{m=0}^{p^n-1} \prod_{\nu=1}^{n-1} \delta_{m_{-(\nu+1)},\ell_{-\nu}} (\overline{q})^{m_{-1}\ell_{-n}} \prod_{\mu=1}^{n-1} \delta_{m_{-(\mu+1)},k_{-\mu}} q^{m_{-1}k_{-n}}.$$

Since for each index m the quantities $\delta_{m_{-(\nu+1)},\ell_{-\nu}}$ are different from zero only when $m_{-(\nu+1)} = \ell_{-\nu}$ for all $\nu = 1, 2, \ldots, n-1$, it follows that

$$\begin{aligned} c_{\ell,k} &= \sum_{m_{-1}=0}^{p-1} \prod_{\mu=1}^{n-1} \delta_{\ell_{-\nu},k_{-\mu}} (\overline{q})^{m_{-1}\ell_{-n}} q^{m_{-1}k_{-n}} \\ &= \prod_{\mu=1}^{n-1} \delta_{\ell_{-\nu},k_{-\mu}} \left(\sum_{m_{-1}=0}^{p-1} (\overline{q})^{m_{-1}\ell_{-n}} q^{m_{-1}k_{-n}} \right). \end{aligned} \tag{12.1.4}$$

Using

$$\overline{q} = \exp \frac{2\pi i}{p} = \exp \frac{-2\pi i(p-1)}{p} = q^{p-1},$$

we can write

$$\sum_{m_{-1}=0}^{p-1} (\overline{q})^{m_{-1}\ell_{-n}} q^{m_{-1}k_{-n}} = \sum_{m_{-1}=0}^{p-1} q^{(k_{-n}+(p-1)\ell_{-n})m_{-1}}$$
$$= \begin{cases} p, & \text{if } k_{-n}+(p-1)\ell_{-n} \equiv 0 \pmod{p}, \\ 0, & \text{if } k_{-n}+(p-1)\ell_{-n} \not\equiv 0 \pmod{p}. \end{cases}$$

But $k_{-n} + (p-1)\ell_{-n} \equiv 0 \pmod{p}$ is equivalent to $k_{-n} - \ell_{-n} \equiv 0 \pmod{p}$, i.e., $k_{-n} = \ell_{-n}$. Hence it follows from (12.1.4) that

$$c_{\ell,k} = \begin{cases} p, & \text{if } k_{-n} = \ell_{-n},\ \nu = 1, 2, \ldots, n, \quad \text{i.e., if } \ell = k, \\ 0, & \text{in all other cases.} \end{cases}$$

Thus we see that $(\overline{B})^T B = pE$, where E is the identity matrix. Consequently, $B^{-1} = \frac{1}{p}(\overline{B})^T$ and the filtering algorithm (12.1.2) becomes

$$\tilde{X} = \frac{1}{p^n}((\overline{B})^T)^n D^* B^n X. \tag{12.1.5}$$

We now consider practical methods for computing the filter (12.1.5). As we noticed in remarks a), b), and c) in §11.2, the computation $B^n X$ can be done using repeated multiplications by the matrix B and can be conducted in the basis $(1, q, \ldots, q^{p-2})$ on single instruction parallel processors using only the addition operation.

We shall find the form of the matrix $D^* = C^{-1}DC$. Since C is a symmetric orthogonal permutation matrix, we have $C^{-1} = C$. Let $c_{\ell,k}$, $\ell, k = 0, 1, \ldots, p^n - 1$, represent the elements of the matrix C. We have by (11.2.12) that

$$c_{\ell,k} = \prod_{j=1}^{n} \delta_{\ell_{-j},k_{-(n-j+1)}}$$

for $\ell = \sum_{j=1}^{n} \ell_{-j} p^{j-1}$, $k = \sum_{j=1}^{n} k_{-j} p^{j-1}$. The elements $d_{i,\ell}$ of the diagonal matrix D can be written in the form $d_{i,\ell} = \delta_{i,\ell}\lambda_i$ where λ_i are the diagonal elements of the matrix D. Consequently, the elements $\tilde{d}_{i,k}$ of the matrix DC have the form

$$\tilde{d}_{i,k} = \sum_{\ell=0}^{p^n-1} d_{i,\ell} c_{\ell,k} = \sum_{\ell=0}^{p^n-1} \delta_{i,\ell}\lambda_i \prod_{\nu=1}^{n} \delta_{\ell_{-\nu},k_{-(n-\nu+1)}}$$
$$= \lambda_i \prod_{\nu=1}^{n} \delta_{i_{-\nu},k_{-(n-\nu+1)}}, \qquad i = \sum_{\nu=1}^{n} i_{-\nu} p^{\nu-1}.$$

Let $d^*_{i,j}$ represent the elements of the matrix $D^* = CDC$. Then

$$d^*_{i,j} = \sum_{k=0}^{p^n-1} c_{i,k}\tilde{d}_{k,j} = \sum_{k=0}^{p^n-1}\prod_{\mu=1}^{n}\delta_{i_{-\mu},k_{-(n-\mu+1)}}\lambda_k\prod_{\nu=1}^{n}\delta_{k_{-\nu},j_{-(n-\nu+1)}}.$$

The first product is different from zero only for those $k \in [0, p^n - 1]$ which satisfy $k_{-(n-\mu+1)} = i_{-\mu}$, i.e., $k_{-\nu} = i_{-(n-\nu+1)}$, $\nu = 1, 2, \ldots, n$. Consequently,

$$d^*_{i,j} = \left(\lambda_{\sum_{\nu=1}^{n} i_{-(n-\nu+1)}p^{\nu-1}}\right)\prod_{\nu=1}^{n}\delta_{i_{-(n-\nu+1)},j_{-(n-\nu+1)}}$$
$$= \left(\lambda_{\sum_{\nu=1}^{n} i_{-\nu}p^{n-\nu}}\right)\prod_{\nu=1}^{n}\delta_{i_{-\nu},j_{-\nu}}.$$

In particular, we see that $d^*_{i,j}$ is different from zero only when $j = i$, in which case $d^*_{i,i} = \lambda_{i^*}$ for $i^* = \sum_{\nu=1}^{n} i_{-\nu}p^{n-\nu}$.

We have shown that by multiplying the i-th element of the column vector B^nX by $d^*_{i,i} = \lambda_{i^*}$, for $i = 0, 1, \ldots, p^n - 1$, we obtain the product D^*B^nX, and furthermore, that the components of this column vector are written in the basis $(1, q, \ldots, q^{p-2})$. It remains to multiply the corresponding column vector by the matrix $(\overline{B})^T$ a total of n times. We note that each p columns of the matrix $(\overline{B})^T$ can be realized by stretching the rows of a matrix $\overline{B}_1$ (see remark b) from §11.2). Thus using the identities

$$(\overline{q})^{\nu} = \exp\frac{2\pi i}{p} = \exp\frac{-2\pi i(p-\nu)}{p} = q^{p-\nu},$$

for $\nu = 1, 2, \ldots, n$, we conclude that multiplication by $(\overline{B})^T$ by a column vector can be carried out in the basis $(1, q, \ldots, q^{p-2})$. This means that to carry out the computations necessary for the digital filtering (12.1.5), the multiplication operation is used only for taking the product of the elements of the diagonal matrix D^* and the elements of the column vector B^nX, the division operation only for dividing the final components by p^n, and all remaining calculations can be performed using only addition and subtraction.

§12.2. Multiplicative holographic transformations for image processing.

At the present, a number of problems concerning storage, transmission and reproduction of information are solved by so- called *digital holography* (see [4] and [14], for example), i.e., analysis and synthesis of wave fields, using the discrete Fourier transform implemented on digital computers. This technique has been used to solve problems of visual information to create optical devices for processing visual signals and for pattern recognition. The use of digital computers for analysis and synthesis of wave fields gives an alternative of analog methods including the methods of physical holography. This alternative brings with it several advantages intrinsic to the digital techniques of signal processing: high accuracy and absolute reducibility of processing, simple algorithms of processing, accessibility of results. Moreover, algorithmic flexibility of digital techniques allow the extension of these techniques to other discrete transforms, in addition to the Fourier transform, which can be realized in optical systems and can be used in digital holography. In other words, there is now the capability to build analog digital holograms on the basis of other linear transforms which preserve all the advantages of Fourier holograms.

The discreet multiplicative transforms (11.1.12) and (11.1.13), which were introduced in §11.1, including the discrete transforms (11.2.6) and (11.2.7), and the special Walsh transform case (when $p = 2$) are of significant interest from this point of view. In this context, we mention the results of Lisovec and Pospelov [1], [2].

Every two-dimensional discrete multiplicative transform is part of a class of spatially separable transformations (see [27], §2.2). In fact, the two-dimensional multiplicative transform generated by a sequence $\{p_k = p\}_{k=1}^{\infty}$ reduces to a one-dimensional transform (see §11.4). Thus construction of holograms based on these transforms can be reduced to the one- dimensional case. Moreover, the two-dimensional extension does not present additional difficulties. Let $p_1, p_2, \ldots, p_s$ be a sequence of whole numbers, $p_\nu \geq 2$, $\nu = 1, 2, \ldots, s$. Let $N = m_s = p_1 \ldots p_s$ and $N_0 = m_{s-1} = p_1 \ldots p_{s-1}$, i.e., $N = p_s N_0$. Using definitions (11.1.12) or (11.1.13), we correspond to each vector $X = (x_0, x_1, \ldots, x_{N-1})^T$ in the real Euclidean space $\mathcal{E}_N$, a vector $Y = (y_0, y_1, \ldots, y_{N-1})^T$, whose components have the form

$$y_k(X) = y_k = \frac{1}{\sqrt{N}} \sum_{\ell=0}^{N-1} x_\ell \overline{\chi}_k\left(\frac{\ell}{N}\right), \qquad k = 0, 1, \ldots, N-1. \tag{12.2.1}$$

The inverse transform is written in the form

$$x_\ell = \frac{1}{\sqrt{N}} \sum_{k=0}^{N-1} y_k \chi_k\left(\frac{\ell}{N}\right), \qquad \ell = 0, 1, \ldots, N-1. \tag{12.2.2}$$

Each component of the vector Y is a complex number of the form

$$y_k = \sqrt{M_k} \exp\{2\pi i \theta_k\}, \qquad M_k = |y_k|^2, \tag{12.2.3}$$

$$\theta_k = \frac{1}{2\pi} \arg y_k, \qquad k = 0, 1, \ldots, N-1.$$

Analogous to the Fourier transform case, the vectors $M = (M_0, M_1, \ldots M_{N-1})^T$ and $\theta = (\theta_0, \theta_1, \ldots, \theta_{N-1})^T$ are called the *spectral capacity* and the *phase spectrum.*

It is well-known that the spectral capacity M by itself does not give enough information to reconstruct the input vector X. Nevertheless, there exists a sufficiently large class of vectors X, whose components can be uniquely determined using only the spectral capacity M and the inverse multiplicative transform. For a description of one of these classes in the vector space $\mathcal{E}_N = \{X : X = (x_0, x_1, \dots, x_{N-1})^T\}$, we shall construct a linear manifold of dimension $N_0 = N/p_s$ in the following way.

Use the components of the vectors $X \in \mathcal{E}_N$ to form vectors

$$X^{(0)} = (x_0, x_1, \dots, x_{N_0-1}, 0, \dots, 0)^T$$

which constitute an N_0-dimensional subspace of $\mathcal{E}_N$. Consider the vector $Z^{(0)}$ whose components $z_k^{(0)}$ are defined by the relationship

$$z_k^{(0)} = \begin{cases} x_\nu & \text{if} \quad k = \nu p_s, \quad \nu = 0, 1, \dots, N_0 - 1, \\ 0 & \text{if} \quad k = \nu p_s + n, \ 1 \le n \le p_s - 1, \end{cases} \tag{12.2.4}$$

and the vector $E = (0, \sqrt{N}, 0, \dots, 0)^T$ which has only one non-zero component, the second one. Then the set of vectors $Z = Z^{(0)} + E$ is a linear manifold in $\mathcal{E}_N$. We shall prove the following result:

12.2.1. *Let $\tilde{X}(Z) = (\tilde{x}_0, \tilde{x}_1, \dots, \tilde{x}_{N-1})^T$ be a vector in $\mathcal{E}_N$ obtained from the vector $Z = Z^{(0)} + E$ by means of the formula*

$$\tilde{x}_n = \frac{1}{\sqrt{N}} \sum_{k=0}^{N-1} M_k(Z) \chi_k\left(\frac{n}{N}\right), \qquad n = 0, 1, \dots, N-1, \tag{12.2.5}$$

where $M_k(Z)$ is the spectral capacity of the vector Z defined by relationships (12.2.2) and (12.2.3). Then for any $\ell = 0, 1, \dots, N_0 - 1$ the coordinates x_ℓ of the vector $X^{(0)}$ are generated by the coordinates $\tilde{x}_n$ in the following way:

$$\begin{cases} \text{either} \quad x_n = \tilde{x}_{np_s+1}, & n = 0, 1, \dots, N_0 - 1, \\ \text{or} \quad x_0 = \tilde{x}_{N-1}, \quad x_{n+1} = \tilde{x}_{(n+1)p_s-1}, & n = 0, 1, \dots, N_0 - 2. \end{cases} \tag{12.2.6}$$

PROOF. We apply the discrete transform (12.2.2) to the vector $Z = Z^{(0)} + E$. Using (12.2.4) and the definition of the vector E we have

$$\begin{aligned} y_k(Z) = y_k &= \frac{1}{\sqrt{N}} \sum_{n=0}^{N-1} z_n \overline{\chi}_k\left(\frac{n}{N}\right) \\ &= \frac{1}{\sqrt{N}} \sum_{\nu=0}^{N_0-1} \left(\sum_{n=\nu p_s}^{(\nu+1)p_s-1} z_n \overline{\chi}_k\left(\frac{n}{N}\right) \right) \\ &= \frac{1}{\sqrt{N}} \sum_{\nu=0}^{N_0-1} \sum_{n=0}^{p_s-1} z_{\nu p_s+n} \overline{\chi}_k\left(\frac{\nu p_s + n}{N}\right) \\ &= \frac{1}{\sqrt{N}} \left\{ \sum_{\nu=0}^{N_0-1} x_\nu \overline{\chi}_k\left(\frac{\nu}{N_0}\right) + \sqrt{N}\, \overline{\chi}_k\left(\frac{1}{N}\right) \right\}. \end{aligned} \tag{12.2.7}$$

Consequently, the spectral capacity of the vector Z satisfies

(12.2.8)
$$\begin{aligned}
M_k(Z) &= M_k = |y_k(Z)|^2 \\
&= \frac{1}{N}\left|\sum_{\nu=0}^{N_0-1} x_\nu \overline{\chi}_k\left(\frac{\nu}{N_0}\right) + \sqrt{N}\,\overline{\chi}_k\left(\frac{1}{N}\right)\right|^2 \\
&= \frac{1}{N}\left(\sum_{\nu=0}^{N_0-1} x_\nu \overline{\chi}_k\left(\frac{\nu}{N_0}\right) + \sqrt{N}\,\overline{\chi}_k\left(\frac{1}{N}\right)\right)\left(\sum_{\nu=0}^{N_0-1} x_\nu \chi_k\left(\frac{\nu}{N_0}\right) + \sqrt{N}\,\chi_k\left(\frac{1}{N}\right)\right) \\
&= \frac{1}{N}\left\{\left|\sum_{\nu=0}^{N_0-1} x_\nu \overline{\chi}_k\left(\frac{\nu}{N_0}\right)\right|^2 + N + 2\mathrm{Re}\left(\sqrt{N}\,\overline{\chi}_k\left(\frac{1}{N}\right)\sum_{\nu=0}^{N_0-1} x_\nu \chi_k\left(\frac{\nu}{N_0}\right)\right)\right\} \\
&= \frac{1}{N}\left|\sum_{\nu=0}^{N_0-1} x_\nu \overline{\chi}_k\left(\frac{\nu}{N_0}\right)\right|^2 + 1 + \frac{2}{\sqrt{N}}\mathrm{Re}\left(\overline{\chi}_k\left(\frac{1}{N}\right)\sum_{\nu=0}^{N_0-1} x_\nu \chi_k\left(\frac{\nu}{N_0}\right)\right) \\
&= M_k^{(1)} + 1 + M_k^{(2)}.
\end{aligned}$$

To compute the components of the vector $\tilde{X}(Z)$ by the formula (12.2.5) write

$$\tilde{x}_\ell = \frac{1}{\sqrt{N}}\sum_{k=0}^{N-1} M_k(Z)\chi_k\left(\frac{\ell}{N}\right) = \sigma_\ell^{(1)} + \sigma_\ell^{(2)} + \sigma_\ell^{(3)}, \tag{12.2.9}$$

where

$$\sigma_\ell^{(1)} = \frac{1}{\sqrt{N}}\sum_{k=0}^{N-1} M_k^{(1)}\chi_k\left(\frac{\ell}{N}\right), \qquad \sigma_\ell^{(2)} = \frac{1}{\sqrt{N}}\sum_{k=0}^{N-1} 1\chi_k\left(\frac{\ell}{N}\right),$$

and

$$\sigma_\ell^{(3)} = \frac{1}{\sqrt{N}}\sum_{k=0}^{N-1} M_k^{(2)}\chi_k\left(\frac{\ell}{N}\right)$$

for $\ell = 0, 1, \ldots, N-1$. Obviously,

$$\begin{aligned}
\sigma_\ell^{(2)} &= \frac{1}{\sqrt{N}}\sum_{k=0}^{N-1} 1\chi_k\left(\frac{\ell}{N}\right) \\
&= \frac{1}{\sqrt{N}}D_N\left(\frac{\ell}{N}\right) \\
&= \begin{cases} \sqrt{N} & \ell = 0, \\ 0 & \ell = 1, 2, \ldots, N-1. \end{cases}
\end{aligned} \tag{12.2.10}$$

Furthermore, for $0 \le \ell \le N-1$ we have

$$\begin{aligned}
\sigma_\ell^{(1)} &= \frac{1}{\sqrt{N}} \sum_{k=0}^{N-1} \left| \frac{1}{\sqrt{N}} \sum_{\nu=0}^{N_0-1} x_\nu \overline{\chi}_k \left(\frac{\nu}{N_0}\right) \right|^2 \chi_k \left(\frac{\ell}{N}\right) \\
&= \frac{1}{\sqrt{N}} \sum_{k_{-s}=0}^{p_s-1} \left(\sum_{k=k_{-s}N_0}^{(k_{-s}+1)N_0-1} \left| \frac{1}{\sqrt{N}} \sum_{\nu=0}^{N_0-1} x_\nu \overline{\chi}_k \left(\frac{\nu}{N_0}\right) \right|^2 \chi_k \left(\frac{\ell}{N}\right) \right) \\
&= \frac{1}{\sqrt{N}} \sum_{k_{-s}=0}^{p_s-1} \sum_{\mu=0}^{N_0-1} \left| \frac{1}{\sqrt{N}} \sum_{\nu=0}^{N_0-1} x_\nu \overline{\chi}_{k_{-s}N_0+\mu} \left(\frac{\nu}{N_0}\right) \right|^2 \chi_{k_{-s}N_0+\mu} \left(\frac{\ell}{N}\right) \\
&= \frac{1}{\sqrt{N}} \sum_{k_{-s}=0}^{p_s-1} \sum_{\mu=0}^{N_0-1} \left| \frac{1}{\sqrt{N}} \sum_{\nu=0}^{N_0-1} x_\nu \overline{\chi}_\mu \left(\frac{\nu}{N_0}\right) \overline{\chi}_{k_{-s}N_0} \left(\frac{\nu}{N_0}\right) \right|^2 \times \\
&\qquad \times \chi_\mu \left(\frac{\ell}{N}\right) \chi_{k_{-s}N_0} \left(\frac{\ell}{N}\right).
\end{aligned}$$

Since $\chi_{k_{-s}N_0}\left(\frac{\nu}{N_0}\right) = \left[\chi_{N_0}\left(\frac{\nu}{N_0}\right)\right]^{k_{-s}} = \left(\exp \frac{2\pi i}{p_s}\left(\frac{\nu}{N_0}\right)_s\right)^{k_s} = 1$ (by the definition and fundamental properties of the function $\chi_m(x)$ and the identity

$$\left(\frac{\nu}{N_0}\right)_s \equiv \left[\frac{\nu}{N_0} N\right] \pmod{p_s} \equiv [\nu p_s] \pmod{p_s} = 0),$$

it follows that

$$\sigma_\ell^{(1)} = \frac{1}{\sqrt{N}} \sum_{\mu=0}^{N_0-1} \left| \frac{1}{\sqrt{N}} \sum_{\nu=0}^{N_0-1} x_\nu \overline{\chi}_\mu \left(\frac{\nu}{N_0}\right) \right|^2 \chi_\mu \left(\frac{\ell}{N}\right) \sum_{k_{-s}=0}^{p_s-1} \chi_{k_{-s}N_0} \left(\frac{\ell}{N}\right).$$

But

$$\begin{aligned}
\sum_{k_{-s}=0}^{p_s-1} \chi_{k_{-s}N_0} \left(\frac{\ell}{N}\right) &= \sum_{k_{-s}=0}^{p_s-1} \left(\chi_{N_0}\left(\frac{\ell}{N_0}\right)\right)^{k_{-s}} \\
&= \sum_{k_{-s}=0}^{p_s-1} \left(\exp \frac{2\pi i}{p_s}\left(\frac{\ell}{N_0}\right)_s\right)^{k_s} \\
&= \begin{cases} p_s & \text{for} \quad (\ell/N)_s \equiv 0 \pmod{p_s}, \\ 0 & \text{for} \quad (\ell/N)_s \not\equiv 0 \pmod{p_s}. \end{cases}
\end{aligned}$$

Consequently, we obtain

$$(12.2.11)\qquad \sigma_\ell^{(1)} = \begin{cases} \dfrac{p_s}{\sqrt{N}} \displaystyle\sum_{\mu=0}^{N_0-1} \left| \sum_{\nu=0}^{N_0-1} x_\nu \bar{\chi}_\mu \left(\frac{\nu}{N_0}\right) \right|^2 \chi_\mu \left(\frac{\ell}{N}\right) \\ \qquad\qquad \text{if } \ell \equiv 0 \pmod{p_s}, \text{ i.e., for } \ell = mp_s\,, \\ \qquad\qquad\qquad m = 0, 1, \ldots, N_0 - 1, \\ 0 \qquad\qquad \text{if } \ell \not\equiv 0 \pmod{p_s}. \end{cases}$$

It remains to evaluate the third expression $\sigma_\ell^{(3)}$. First we write

$$\begin{aligned} \sigma_\ell^{(3)} &= \frac{2}{N} \sum_{k=0}^{N-1} \operatorname{Re} \left(\bar{\chi}_k \left(\frac{1}{N}\right) \sum_{n=0}^{N_0-1} x_n \chi_k \left(\frac{n}{N_0}\right) \right) \chi_k \left(\frac{\ell}{N}\right) \\ &= \frac{1}{N} \sum_{k_{-s}=0}^{p_s-1} \sum_{k=k_{-s}N_0}^{(k_{-s}+1)N_0-1} \sum_{n=0}^{N_0-1} x_n \chi_k \left(\frac{n}{N_0}\right) \chi_k \left(\frac{\ell}{N} \ominus \frac{1}{N}\right) \\ &\quad + \frac{1}{N} \sum_{k_{-s}=0}^{p_s-1} \sum_{k=k_{-s}N_0}^{(k_{-s}+1)N_0-1} \sum_{n=0}^{N_0-1} x_n \bar{\chi}_k \left(\frac{n}{N_0}\right) \chi_k \left(\frac{\ell}{N} \ominus \frac{1}{N}\right) \\ &= \tau_\ell^{(1)} + \tau_\ell^{(2)}. \end{aligned}$$

Since $\chi_{k_{-s}N_0}(n/N_0) = 1$ for $n = 0, 1, \ldots, N_0 - 1$ we have

$$\begin{aligned} \tau_\ell^{(1)} &= \frac{1}{N} \sum_{k_{-s}=0}^{p_s-1} \sum_{k'=0}^{N_0-1} \sum_{n=0}^{N_0-1} x_n \chi_{k'} \left(\frac{n}{N_0}\right) \chi_{k_{-s}N_0+k'} \left(\frac{\ell}{N} \ominus \frac{1}{N}\right) \\ &= \frac{1}{N} \sum_{n=0}^{N_0-1} x_n \sum_{k'=0}^{N_0-1} \chi_{k'} \left(\frac{n}{N_0}\right) \chi_{k'} \left(\frac{\ell}{N} \ominus \frac{1}{N}\right) \sum_{k_{-s}=0}^{p_s-1} \chi_{k_{-s}N_0} \left(\frac{\ell}{N} \ominus \frac{1}{N}\right). \end{aligned}$$

Since

$$\chi_{N_0}(u) = \exp \frac{2\pi i}{p_s} u_s, \qquad u_s \equiv [uN] \pmod{p_s}$$

for $0 \le u < 1$, and

$$\begin{aligned} \left(\frac{\ell}{N} \ominus \frac{1}{N}\right)_s &\equiv \left[\left(\frac{\ell}{N} \ominus \frac{1}{N}\right) N\right] \pmod{p_s} \\ &\equiv \ell \pmod{p_s} \ominus 1 \\ &= \begin{cases} p_s - 1 & \text{for} \quad \ell_{-s}^* = 0, \\ \ell_{-s}^* - 1 & \text{for} \quad 1 \le \ell_{-s}^* \le p_s - 1, \end{cases} \end{aligned}$$

where $\ell^*_{-s} \equiv \ell \pmod{p_s}$, $0 \le \ell^*_{-s} \le p_s - 1$, we also have

$$\sum_{k_{-s}=0}^{p_s-1} \chi_{k_{-s}N_0}\left(\frac{\ell}{N} \ominus \frac{1}{N}\right) = \sum_{k_{-s}=0}^{p_s-1}\left(\chi_{N_0}\left(\frac{\ell}{N} \ominus \frac{1}{N}\right)\right)^{k_{-s}}$$
$$= \sum_{k_{-s}=0}^{p_s-1}\left(\exp\frac{2\pi i}{p_s}\left(\frac{\ell}{N} \ominus \frac{1}{N}\right)\right)^{k_{-s}}$$
$$= \begin{cases} p_s & \text{if } \ell^*_{-s} = 1, \text{ i.e., for } \quad \ell = \gamma p_s + 1, \\ & \qquad \gamma = 0, 1, \ldots, N_0 - 1, \\ 0 & \text{for all other } \ell. \end{cases}$$

Consequently,

$$\tau_\ell^{(1)} = \begin{cases} \dfrac{p_s}{N} \displaystyle\sum_{n=0}^{N_0-1} x_n \sum_{k'=0}^{N_0-1} \chi_{k'}\left(\frac{n}{N_0} \oplus \frac{\gamma p_s + 1}{N}\right) \\ \qquad \text{for} \quad \ell = \gamma p_s + 1, \\ 0 \qquad \text{for all other } \ell. \end{cases}$$

Furthermore, using properties of the Dirichlet kernel we have

$$\sum_{k'=0}^{N_0-1} \chi_{k'}\left(\frac{n}{N_0} \oplus \frac{\gamma p_s + 1}{N}\right) = D_{N_0}\left(\frac{n}{N_0} \oplus \left(\frac{\gamma}{N} + \frac{1}{N}\right)\right)$$
$$= \begin{cases} N_0 & \text{if } \dfrac{n}{N_0} \oplus \left(\dfrac{\gamma}{N} + \dfrac{1}{N}\right) < \dfrac{1}{N_0}, \\ 0 & \text{if } \dfrac{n}{N_0} \oplus \left(\dfrac{\gamma}{N} + \dfrac{1}{N}\right) \ge \dfrac{1}{N_0} \end{cases}$$
$$= \begin{cases} N_0 & \text{for } n = \gamma, \\ 0 & \text{for all remaining } n. \end{cases}$$

Substituting this expression into (12.2.12) we obtain

$$\tau_\ell^{(1)} = \begin{cases} x_\gamma & \text{for} \quad \ell = \gamma p_s + 1, \ \gamma = 0, 1, \ldots, N_0 - 1, \\ 0 & \text{for all other } \ell. \end{cases} \tag{12.2.14}$$

Similarly, we obtain

$$\tau_\ell^{(2)} = \frac{1}{N}\sum_{k_{-s}=0}^{p_s-1} \sum_{k=k_{-s}N_0}^{(k_{-s}+1)N_0-1} \sum_{n=0}^{N_0-1} x_n \overline{\chi}_k\left(\frac{n}{N_0}\right)\chi_k\left(\frac{\ell}{N} \oplus \frac{1}{N}\right)$$
$$= \frac{1}{N}\sum_{k_{-s}=0}^{p_s-1} \sum_{k'=0}^{N_0-1} \sum_{n=0}^{N_0-1} x_n \overline{\chi}_{k'}\left(\frac{n}{N_0}\right)\chi_{k_{-s}N_0+k'}\left(\frac{\ell}{N} \oplus \frac{1}{N}\right)$$
$$= \frac{1}{N}\sum_{n=0}^{N_0-1} x_n \sum_{k'=0}^{N_0-1} \chi_{k'}\left(\frac{\ell}{N} \oplus \frac{1}{N} \ominus \frac{n}{N_0}\right) \sum_{k_{-s}=0}^{p_s-1}\left(\chi_{N_0}\left(\frac{\ell}{N} \oplus \frac{1}{N}\right)\right)^{k_{-s}}.$$

But

$$\sum_{k_{-s}=0}^{p_s-1}\left(\chi_{N_0}\left(\frac{\ell}{N}\oplus\frac{1}{N}\right)\right)^{k_{-s}} =$$

$$= \begin{cases} p_s & \text{if } \ell^*_{-s}+1\equiv 0 \pmod{p_s}, \text{ i.e., for } \ell^*_{-s}=p_s-1 \text{ and } \\ & \ell=\gamma p_s+p_s-1,\ \gamma=0,1,\dots,N_0-1, \\ 0 & \text{for all other } \ell, \end{cases}$$

and

$$\sum_{k'=0}^{N_0-1}\chi_{k'}\left(\frac{\gamma p_s+p_s-1}{N}\oplus\frac{1}{N}\ominus\frac{n}{N_0}\right) = D_{N_0}\left(\frac{\gamma+1}{N_0}-\frac{1}{N}\oplus\frac{1}{N}\ominus\frac{n}{N_0}\right)$$

$$= \begin{cases} N_0 & \text{for } n=\gamma+1, \\ 0 & \text{for all remaining } n. \end{cases}$$

Consequently,

$$\tau_\ell^{(2)} = \begin{cases} x_{\gamma+1} & \text{for } \ell=(\gamma+1)p_s-1,\ \gamma=0,1,\dots,N_0-2, \\ x_0 & \text{for } \ell=N_0-1, \\ 0 & \text{for all other } \ell. \end{cases} \tag{12.2.15}$$

We have an expression for $\sigma_\ell^{(1)}$ from (12.2.11), for $\sigma_\ell^{(3)}$ from (12.2.12), (12.2.14) and (12.2.15), and for $\sigma_\ell^{(2)}$ from (12.2.10). Substituting these expressions into (12.2.9) we obtain

$$\tilde{x}_\ell = \begin{cases} \frac{p_s}{\sqrt{N}}\sum_{\mu=0}^{N_0-1}\left|\frac{1}{\sqrt{N}}\sum_{\nu=0}^{N_0-1}x_\nu\overline{\chi}_\mu\left(\frac{\nu}{N_0}\right)\right|^2\chi_\mu\left(\frac{\ell}{N}\right)+\sqrt{N} \\ \qquad \text{if } \ell=0, \\ \frac{p_s}{\sqrt{N}}\sum_{\mu=0}^{N_0-1}\left|\frac{1}{\sqrt{N}}\sum_{\nu=0}^{N_0-1}x_\nu\overline{\chi}_\mu\left(\frac{\nu}{N_0}\right)\right|^2\chi_\mu\left(\frac{\ell}{N}\right) \\ \qquad \text{if } \ell=mp_s,\quad m=0,1,\dots,N_0-1, \\ x_\gamma & \text{if } \ell=\gamma p_s+1,\ \gamma=0,1,\dots,N_0-1, \\ x_{\gamma+1} & \text{if } \ell=(\gamma+1)p_s-1,\ \gamma=0,1,\dots,N_0-2, \\ x_0 & \text{if } \ell=N-1. \end{cases} \tag{12.2.16}$$

Therefore, either

$$x_n=\tilde{x}_{np_s+1},\qquad n=0,1,\dots,N_0-1,$$

or

$$x_0 = \tilde{x}_{N-1}, \quad x_{n+1} = \tilde{x}_{(n+1)p_s-1}, \qquad n = 0,1,\ldots,N_0-2. \blacksquare$$

The result obtained in **12.2.1** allows us to give a definition of the discrete multiplicative holographic transform which is analogous to the Fourier holographic transform (see [4], §1.4). Let $A = (a_0, a_1, \ldots, a_{N_0-1})^T$ be an information vector, i.e., a vector whose components represent coded information subject to compression. We shall replace it by a corresponding vector $Z^{(0)} = Z^{(0)}(A) \in \mathcal{E}_N$ in the following way:

$$Z_k^{(0)} = \begin{cases} a_n & \text{if } \ k = np_s, \quad n = 0,1,\ldots,N_0-1, \\ 0 & \text{for all remaining } k \in [1, N-1]. \end{cases}$$

This mapping brings about a unique embedding of each information vector A in the Euclidean space $\mathcal{E}_N$. We shall find the spectral capacity M of the sum of the vectors $Z^{(0)}$ and E. The vector M will be called the *hologram* of A and the linear transformation $A \to M$ will be called the *multiplicative holographic transformation.*

We shall establish a connection between the multiplicative holographic transformation and the discrete multiplicative transform of the information vector A. Notice that the sum which appears in (12.2.7) (written for the vector A), namely

$$\sum_{\nu=0}^{N_0-1} a_\nu \overline{\chi}_k(\nu/N_0), \qquad k = 0,1,\ldots,N-1,$$

is periodic in K of period N_0, i.e., for each $k_{-s} = 1,2,\ldots,p_s-1$ and each $k = 0,1,\ldots,N_0-1$ we have

$$\sum_{\nu=0}^{N_0-1} a_\nu \overline{\chi}_{k+k_{-s}N_0}\left(\frac{\nu}{N_0}\right) = \sum_{\nu=0}^{N_0-1} a_\nu \overline{\chi}_k\left(\frac{\nu}{N_0}\right)$$

(see the representation for $\sigma_\ell^{(1)}$ which appears below (12.2.10)). Thus set

$$b_k = \sum_{\nu=0}^{N_0-1} a_\nu \overline{\chi}_k(\nu/N_0), \qquad k = 0,1,\ldots,N_0-1. \tag{12.2.17}$$

Using (12.2.8) and (12.2.17), we see that

(12.2.18)

$$\begin{aligned} M_{k+k_{-s}N_0} &= \frac{1}{N}|b_k|^2 + \frac{2}{\sqrt{N}}\mathrm{Re}\left(b_k\overline{\chi}_{k+k_{-s}N_0}\left(\frac{1}{N_0}\right)\right) + 1 \\ &= \frac{1}{N}|b_k|^2 + \frac{2}{\sqrt{N}}\mathrm{Re}\left\{|b_k|\exp(i \arg b_k)\exp\left(-\frac{2\pi i k_{-s}}{p_s}\right)\right\} + 1 \\ &= \frac{1}{N}|b_k|^2 + \frac{2}{\sqrt{N}}|b_k|\cos\left(\arg b_k - \frac{2\pi k_{-s}}{p_s}\right) + 1 \\ &= \left(\frac{1}{\sqrt{N}}|b_k| + \cos\left(\arg b_k - \frac{2\pi k_{-s}}{p_s}\right)\right)^2 + \sin^2\left(\arg b_k - \frac{2\pi k_{-s}}{p_s}\right) \end{aligned}$$

for $k = 0, 1, \ldots, N_0 - 1$ and $0 \le k_{-s} \le p_s - 1$.

Formulae (12.2.18) and (12.2.17) can be used to write down algorithms to obtain digital holograms. However these algorithms can be simplified considerably in the case when $p_\nu = p$ for $\nu = 1, 2, \ldots, s$. As follows from **11.2.4**, calculation of the quantities b_k from formula (12.2.16), i.e., multiplication of the matrix $W = \left(\overline{\chi}_k\left(\frac{\nu}{N_0}\right)\right)$, $k, \nu = 0, 1, \ldots, N_0 - 1$, and the column vector A, can be obtained by multiplying the $(s-1)$ power ($N_0 = p^{s-1}$) of the sparse matrix B by A, followed by a p-adic permutation using the matrix C. Moreover, if these calculations are carried out in the basis $(1, q, \ldots, q^{p-2})$, $q = \exp(-2\pi i/p)$, then the arithmetic operation of multiplication can be excluded and the final resulting vector can be written using this basis. For example, for $p = 3$ we have

$$b_k = \alpha_k^{(0)} \cdot 1 + \alpha_k^{(1)} q, \qquad k = 0, 1, \ldots, N_0 - 1.$$

Using the relationship

$$\begin{aligned} |b_k|^2 &= \left|\alpha_k^{(0)} + \alpha_k^{(1)}\left(\cos\frac{2\pi}{3} - i\sin\frac{2\pi}{3}\right)\right|^2 \\ &= (\alpha_k^{(0)})^2 + (\alpha_k^{(1)})^2 - \alpha_k^{(0)}\alpha_k^{(1)} = (\alpha_k^{(0)} - \alpha_k^{(1)})^2 + \alpha_k^{(0)}\alpha_k^{(1)} \end{aligned}$$

and the identity

$$\overline{\chi}_{k+k_{-s}, N_0}\left(\frac{1}{N}\right) = \overline{\chi}_{k_{-s}, N_0}\left(\frac{1}{N}\right) = \exp\frac{-2\pi i}{3}k_{-s}, \qquad k \le N_0 - 1,$$

for $k_{-s} = 0, 1, 2$, we have

$$\begin{aligned} &\operatorname{Re}\left(b_k \overline{\chi}_{k+k_{-s}, N_0}\left(\frac{1}{N}\right)\right) \\ &= \operatorname{Re}\left\{\left(\cos\frac{2\pi k_{-s}}{3} - i\sin\frac{2\pi k_{-s}}{3}\right)\left(\alpha_k^{(0)} + \alpha_k^{(1)}\cos\frac{2\pi}{3} - i\alpha_k^{(1)}\sin\frac{2\pi}{3}\right)\right\} \\ &= \alpha_k^{(0)}\cos\frac{2\pi k_{-s}}{3} + \alpha_k^{(1)}\cos\frac{2\pi k_{-s}}{3}\cos\frac{2\pi}{3} - \sin\frac{2\pi k_{-s}}{3}\alpha_k^{(1)}\sin\frac{2\pi}{3} \\ &= \alpha_k^{(0)}\cos\frac{2\pi k_{-s}}{3} + \alpha_k^{(1)}\cos\frac{2\pi}{3}(1 + k_{-s}) \\ &= \begin{cases} \alpha_k^{(0)} - \frac{1}{2}\alpha_k^{(1)} & \text{for } k_{-s} = 0, \\ -\frac{1}{2}\alpha_k^{(0)} - \frac{1}{2}\alpha_k^{(1)} & \text{for } k_{-s} = 1, \\ -\frac{1}{2}\alpha_k^{(0)} + \alpha_k^{(1)} & \text{for } k_{-s} = 2. \end{cases} \end{aligned}$$

Consequently, it follows from (12.2.17) that

$$\begin{aligned}
M_{k+k_{-s}N_0} &= \frac{1}{N}(\alpha_k^{(0)} - \alpha_k^{(1)})^2 + \alpha_k^{(0)}\alpha_k^{(1)} \\
&\quad + \frac{2}{\sqrt{N}}\left(\alpha_k^{(0)} \cos\frac{2\pi k_{-s}}{3} + \alpha_k^{(1)} \cos\frac{2\pi}{3}(1 + k_{-s})\right) + 1 \\
&= 1 + \frac{1}{N}(\alpha_k^{(0)} - \alpha_k^{(1)})^2 + \alpha_k^{(0)}\alpha_k^{(1)} \\
&\quad + \frac{2}{\sqrt{N}} \times \begin{cases} \alpha_k^{(0)} - \frac{1}{2}\alpha_k^{(1)} & \text{for } k_{-s} = 0, \\ -\frac{1}{2}\alpha_k^{(0)} - \frac{1}{2}\alpha_k^{(1)} & \text{for } k_{-s} = 1, \\ -\frac{1}{2}\alpha_k^{(0)} + \alpha_k^{(1)} & \text{for } k_{-s} = 2 \end{cases}
\end{aligned}$$

for $k = 0, 1, \ldots, N_0 - 1$. Substituting these values for M_ν, $\nu = 0, 1, \ldots, N - 1$, into (12.2.5), we can compute the values of $\tilde{x}_n$. Moreover, these calculations again can be carried out in the same way using the basis $(1, q)$ and the $N \times N$ matrix $(\chi_k(n/N))$.

§12.3. Solutions to some optimization problems.

In the previous sections we have examined the questions of applying multiplicative functions to digital processing, i.e., to problems in which both the input and output were discrete. Now we shall consider applications of multiplicative functions to problems where the input data is discrete but the solutions are continuous processes. Such problems can arise in the physical world. For example, let a linear radiator of electromagnetic waves (e.g., an antenna) consist of N elementary dipole radiators, arranged along some straight line. We wish to find, among all practical feasible (i.e., discretely placed) distributions of currents along such a radiator, a distribution whose corresponding radiation pattern is the nearest, in some sense, to a specified radiation pattern.

Another example of such a physical problem is the problem of diffusion of matter. Under circumstances frequently arising in the manufacture of electronic instruments, the concentration of matter before the execution of a diffusion process can only be assigned discretely. The problem we refer to consists in determining, among all possible (i.e., discrete) initial distributions, that distribution which at the end of the diffusion process, i.e., after time T, yields a concentration which is closest, in some sense, to a desired concentration of matter in the sample.

In order to give a precise mathematical formulation of these physical problems and their solutions, we shall introduce several new concepts.

The Chrestenson-Levy system $\chi_n(x)$, defined by (1.5.10) for some integer $p \geq 2$, will be used to generate a system of functions

$$f_n(u) = \begin{cases} \chi_n\left(u+\frac{1}{2}\right) & \text{for} \quad -\frac{1}{2} \le u \le \frac{1}{2}, \\ 0 & \text{for} \quad u \in \left(-\infty, -\frac{1}{2}\right) \cup \left(\frac{1}{2}, \infty\right), \end{cases} \tag{12.3.1}$$

for $n = 0, 1, \ldots$, which like the Chrestenson-Levy system on the interval $[0,1]$, is orthonormal on the interval $[-1/2, 1/2]$ and complete in the space $\mathbf{L}^2(-1/2, 1/2)$. We also form the system of inverse Fourier transforms of this system (12.3.1), i.e., the system

$$R_n(\omega) = \frac{1}{\sqrt{2\pi}} \int_{-\infty}^{\infty} f_n(u) e^{i\omega u}\, du = \frac{1}{\sqrt{2\pi}} \int_{-1/2}^{1/2} f_n(u) e^{i\omega u}\, du, \tag{12.3.2}$$

and establish a number of its properties.

12.3.1. *The system $R_n(\omega)$, $n = 0, 1, \ldots$, defined by (12.3.1) and (12.3.2) is orthonormal in $\mathbf{L}^2(\mathbf{R})$ and complete with respect to the $\mathbf{L}^2(\mathbf{R})$ norm in the class $W^2_{1/2}$ of all functions in $\mathbf{L}^2(\mathbf{R})$ of exponential type with exponent $\le 1/2$.*

PROOF. We shall use the following corollary of Plancherel's Theorem (see [17], p. 442, for example): if ϕ_1, $\phi_2 \in \mathbf{L}^2(\mathbf{R})$ and Φ_1, Φ_2 represent their respective Fourier transforms, then

$$\int_{-\infty}^{\infty} \Phi_1(v)\overline{\Phi_2(v)}\, dv = \int_{-\infty}^{\infty} \phi_1(u)\overline{\phi_2(u)}\, du. \tag{12.3.3}$$

Applying (12.3.3) to the functions $f_n(u)$, $f_m(u)$, which are orthonormal on the interval $[-1/2, 1/2]$, and to their Fourier transforms $R_n(\omega)$, $R_m(\omega)$, we obtain

$$\begin{aligned} \int_{-\infty}^{\infty} R_n(\omega)\overline{R_m(\omega)}\, d\omega &= \int_{-\infty}^{\infty} f_n(u)\overline{f_m(u)}\, du \\ &= \int_{-1/2}^{1/2} f_n(u)\overline{f_m(u)}\, du \\ &= \begin{cases} 1 & \text{for } m = n, \\ 0 & \text{for } m \ne n. \end{cases} \end{aligned}$$

This proves that the system $\{R_n(\omega)\}_{n=0}^{\infty}$ is orthonormal in $\mathbf{L}^2(\mathbf{R})$.

Suppose now that $R(\omega)$ is a function which belongs to the space $W^2_{1/2}$. By a theorem of Paley-Wiener (see [22], p. 26) there exists a function $f(y) \in \mathbf{L}^2(-1/2, 1/2)$ such that

$$R(\omega) = \frac{1}{\sqrt{2\pi}} \int_{-1/2}^{1/2} f(y) e^{i\omega y}\, dy. \tag{12.3.4}$$

Moreover, since the system $\{f_n(y)\}$ is complete in the space $\mathbf{L}^2(-1/2,1/2)$, the Fourier series of the function $f(y)$ in this system must converge to $f(y)$ in the $\mathbf{L}^2(-1/2,1/2)$ norm, i.e.,

$$\lim_{N\to\infty} \|f(y) - \sum_{n=0}^{N-1} c_n f_n(y)\|_{\mathbf{L}^2(-1/2,1/2)} = 0,$$

where

$$c_n = \int_{-1/2}^{1/2} f(y)\overline{f_n(y)}\,dy, \qquad n = 0, 1, \ldots. \tag{12.3.5}$$

But by Plancherel's Theorem,

$$\|R(\omega) - \sum_{n=0}^{N-1} c_n R_n(\omega)\|_{\mathbf{L}^2(\mathbf{R})} = \|f(y) - \sum_{n=0}^{N-1} c_n f_n(y)\|_{\mathbf{L}^2(-1/2,1/2)}. \tag{12.3.6}$$

Since $R(\omega) \in W_{1/2}^2$ was arbitrary, it follows from (12.3.5) and (12.3.6) that the system $\{R_n(\omega)\}$ is complete. ∎

Notice by (12.3.3) that

$$c_n = \int_{-1/2}^{1/2} f(y)\overline{f_n(y)}\,dy = \int_{-\infty}^{\infty} R(\omega)\overline{R_m(\omega)}\,d\omega, \qquad n = 0, 1, \ldots, \tag{12.3.7}$$

i.e., the Fourier coefficients of any function $R(\omega) \in W_{1/2}^2$ with respect to the system $\{R_n(\omega)\}$ coincide with the Fourier coefficients of a function $f(y) \in \mathbf{L}^2(-1/2,1/2)$ with respect to the system $\{f_n(y)\}$, where the functions $R(\omega)$ and $f(y)$ are related by (12.3.4). Consequently, it is possible to use this system as a device for obtaining approximations in various problems connected with the set of functions quadratically integrable on the whole real line which have a finite Fourier transform.

12.3.2. *The Fourier series of any function $R(\omega) \in W_{1/2}^2$ with respect to the system $\{R_n(\omega)\}$ converges to $R(\omega)$ uniformly on the entire real axis and for any $\omega \in \mathbf{R}$,*

$$\left| R(\omega) - \sum_{n=0}^{N-1} c_n R_n(\omega) \right| \le \frac{1}{\sqrt{2\pi}} \|R(\omega) - \sum_{n=0}^{N-1} c_n R_n(\omega)\|_{\mathbf{L}^2(\mathbf{R})}.$$

PROOF. By (12.3.2), (12.3.7), and (12.3.4) we can write

$$\begin{aligned} R(\omega) - \sum_{n=0}^{N-1} c_n R_n(\omega) &= \frac{1}{\sqrt{2\pi}} \left\{ \int_{-1/2}^{1/2} f(y) e^{i\omega y}\,dy - \sum_{n=0}^{N-1} c_n \int_{-1/2}^{1/2} f_n(y) e^{i\omega y}\,dy \right\} \\ &= \frac{1}{\sqrt{2\pi}} \int_{-1/2}^{1/2} \left[f(y) - \sum_{n=0}^{N-1} c_n f_n(y) \right] e^{i\omega y}\,dy. \end{aligned}$$

Hence by the Cauchy-Schwarz inequality we conclude that

$$R(\omega) - \sum_{n=0}^{N-1} c_n R_n(\omega) = \frac{1}{\sqrt{2\pi}} \sqrt{\int_{-1/2}^{1/2} \left| f(y) - \sum_{n=0}^{N-1} c_n f_n(y) \right|^2 dy \int_{-1/2}^{1/2} |e^{i\omega y}|^2 \, dy}$$
$$= \frac{1}{\sqrt{2\pi}} \| f(y) - \sum_{n=0}^{N-1} c_n f_n(y) \|_{\mathbf{L}^2(-1/2,1/2)}$$
$$= \frac{1}{\sqrt{2\pi}} \| R(\omega) - \sum_{n=0}^{N-1} c_n R_n(\omega) \|_{\mathbf{L}^2(\mathbf{R})}. \blacksquare$$

We shall now find a closed form for the function $R_n(\omega)$.

12.3.3. *Let* $n = \sum_{\nu=1}^{\tilde{k}} n_{-\nu} p^{\nu-1}$, $\tilde{k} = k(n)$, $n_{-\tilde{k}} \neq 0$. *Then*

$$R_n(\omega) = \sqrt{\frac{2}{\pi}\frac{1}{\omega}} \sin \frac{\omega}{2p^{\tilde{k}}} \prod_{\nu=1}^{\tilde{k}} \left(\frac{\sin \dfrac{\omega}{2p^{\nu-1}}}{\sin\left(\dfrac{\omega}{2p^{\nu}} + \dfrac{\pi}{p} n_{-\nu}\right)} \right) \exp\left(-\frac{\pi i n_{-\nu}}{p}\right).$$

PROOF. Recall that the function

$$\chi_n(y) = \exp \frac{2\pi i}{p} \sum_{\nu=1}^{\tilde{k}} n_{-\nu} y_\nu$$

is constant on the intervals $\delta_\ell(\tilde{k}) = (\ell/p^{\tilde{k}}, (\ell+1)/p^{\tilde{k}})$, $\ell = 0, 1, \ldots, p^{\tilde{k}} - 1$. Thus

(12.3.8)
$$R_n(\omega) = \frac{1}{\sqrt{2\pi}} \int_{-1/2}^{1/2} \chi_n\left(y + \frac{1}{2}\right) e^{i\omega y} \, dy$$
$$= \frac{e^{-i\omega/2}}{\sqrt{2\pi}} \int_0^1 \chi_n(u) e^{i\omega u} \, du$$
$$= \frac{e^{-i\omega/2}}{\sqrt{2\pi}} \sum_{\ell=0}^{p^{\tilde{k}}-1} \int_{\delta_\ell(\tilde{k})} \chi_n(u) e^{i\omega u} \, du$$
$$= \frac{e^{-i\omega/2}}{\sqrt{2\pi}} \sum_{\ell=0}^{p^{\tilde{k}}-1} \chi_n(\delta_\ell(\tilde{k})) \frac{\exp \dfrac{(\ell+1)i\omega}{p^{\tilde{k}}} - \exp \dfrac{\ell i\omega}{p^{\tilde{k}}}}{i\omega}$$

$$= \frac{2}{\sqrt{2\pi}\omega} \sum_{\ell=0}^{p^{\tilde{k}}-1} \chi_n(\delta_\ell(\tilde{k})) \exp\left\{ \frac{i\omega\left(\ell+\frac{1}{2}-\frac{1}{2}p^{\tilde{k}}\right)}{p^{\tilde{k}}} \frac{\exp\frac{i\omega}{2p^{\tilde{k}}} - \exp\frac{-i\omega}{2p^{\tilde{k}}}}{2i} \right\}$$

$$= \frac{2\sin\frac{\omega}{2p^{\tilde{k}}}}{\sqrt{2\pi}\omega} \sum_{\ell=0}^{p^{\tilde{k}}-1} \chi_n(\delta_\ell(\tilde{k})) \exp\frac{i\omega(2\ell+1-p^{\tilde{k}})}{2p^{\tilde{k}}}$$

$$= \sqrt{\frac{2}{\pi}}\frac{1}{\omega}\sin\frac{\omega}{2p^{\tilde{k}}} S_{\tilde{k}}(\omega),$$

where

$$S_{\tilde{k}}(\omega) = \sum_{\ell=0}^{p^{\tilde{k}}-1} \chi_n(\delta_\ell(\tilde{k})) \exp\frac{i\omega(2\ell+1-p^{\tilde{k}})}{2p^{\tilde{k}}}. \tag{12.3.9}$$

To calculate $S_{\tilde{k}}(\omega)$ consider the sum

$$\sigma_{n,\nu}(\omega) = \sum_{\ell=0}^{p^{\tilde{k}-\nu+1}-1} \chi_{n,\nu}(\delta_\ell(\tilde{k})) \exp\frac{i\omega(2\ell+1-p^{\tilde{k}-\nu+1})}{2p^{\tilde{k}}} \tag{12.3.10}$$

where

$$\chi_{n,\nu}(\delta_\ell(\tilde{k})) = \prod_{m=\nu}^{\tilde{k}} (\chi_{p^{m-1}}(y))^{n-m}, \qquad \nu = 1,2,\ldots,\tilde{k}. \tag{12.3.11}$$

It is clear that

$$\sigma_{n,1}(\omega) = S_{\tilde{k}}(\omega). \tag{12.3.12}$$

We shall establish the recursive relation

$$\sigma_{n,\nu}(\omega) = \sigma_{n,\nu+1}(\omega) \sum_{r=0}^{p-1} (\chi_{p^{\nu-1}}(\delta_{rp^{\tilde{k}-\nu}}))^{n-\nu} \exp\frac{i\omega(2r+1-p)}{2p^\nu}, \tag{12.3.13}$$

for $\nu = 1,2,\ldots,\tilde{k}-1$. Changing the index of summation in (12.3.10) by

$$\ell = rp^{\tilde{k}-\nu} + \ell', \qquad r = 0,1,\ldots,p-1, \quad \ell' = 0,1,\ldots,p^{\tilde{k}-\nu}-1,$$

we obtain

(12.3.14)

$$\sigma_{n,\nu}(\omega) = \sum_{\ell'=0}^{p^{\bar{k}-\nu+1}-1} \sum_{r=0}^{p-1} \chi_{n,\nu}(\delta_{\ell'+rp^{\bar{k}-\nu}}(\tilde{k})) \exp \frac{i\omega(2\ell'+1+2rp^{\bar{k}-\nu}-p^{\bar{k}-\nu+1})}{2p^{\bar{k}}}$$
$$= \sum_{\ell=0}^{p^{\bar{k}-\nu+1}-1} \exp \frac{i\omega(2\ell+1-p^{\bar{k}-\nu+1})}{2p^{\bar{k}}} \sum_{r=0}^{p-1} \chi_{n,\nu}(\delta_{\ell+rp^{\bar{k}-\nu}}) \exp \frac{i\omega r}{p^{\nu}}.$$

But definition (12.3.11) implies

$$\chi_{n,\nu}(\delta_{\ell+rp^{\bar{k}-\nu}}) = (\chi_{p^{\nu-1}}(\delta_{\ell+rp^{\bar{k}-\nu}}))^{n-\nu} \chi_{n,\nu+1}(\delta_{\ell+rp^{\bar{k}-\nu}}).$$

Therefore,

(12.3.15)

$$\sum_{r=0}^{p-1} \chi_{n,\nu}(\delta_{\ell+rp^{\bar{k}-\nu}}) \exp \frac{i\omega r}{p^{\nu}}$$
$$= \sum_{r=0}^{p-1} (\chi_{p^{\nu-1}}(\delta_{\ell+rp^{\bar{k}-\nu}}))^{n-\nu} \chi_{n,\nu+1}(\delta_{\ell+rp^{\bar{k}-\nu}}) \exp \frac{i\omega r}{p^{\nu}}.$$

The function $\chi_{n,\nu+1}(y)$ can be written in the form

$$\chi_{n,\nu+1}(y) = \prod_{m=\nu+1}^{\bar{k}} (\chi_{p^{m-1}}(y))^{n-m} = \exp \frac{2\pi i}{p} \sum_{m=\nu+1}^{\bar{k}} n_{-m} y_m.$$

Since

$$y \in \delta_{\ell+rp^{\bar{k}-\nu}} = \left(\frac{r}{p^{\nu}} + \frac{\ell}{p^{\bar{k}}}, \frac{r}{p^{\nu}} + \frac{\ell+1}{p^{\bar{k}}}\right),$$

then $y = \dfrac{r}{p^{\nu}} + \dfrac{\ell+\theta}{p^{\bar{k}}}$ for some $0 < \theta < 1$ and we have

$$y_m \equiv [yp^m] \pmod{p} \equiv \left[\left(\frac{r}{p^{\nu}} + \frac{\ell+\theta}{p^{\bar{k}}}\right) p^m\right] \pmod{p} \equiv \left[\frac{\ell+\theta}{p^{\bar{k}}} p^m\right] \pmod{p}$$

for any $m \geq \nu + 1$.

This means that for $m \geq \nu+1$ the coordinates y_m of the number $y \in \delta_{\ell+rp^{\bar{k}-\nu}}$ coincide with the coordinates x_m of a number $x \in \delta_{\ell}$. Consequently,

(12.3.16)
$$\chi_{n,\nu+1}(\delta_{\ell+rp^{\bar{k}-\nu}}) = \chi_{n,\nu+1}(\delta_{\ell})$$

for $\ell = 0, 1, \ldots, p^{\bar{k}-\nu} - 1$. Furthermore, since $y \in \delta_{\ell + rp^{\bar{k}-\nu}}$ and $\ell = 0, 1, \ldots, p^{\bar{k}-\nu} - 1$

$$\begin{aligned} y_\nu &\equiv [yp^\nu] \pmod{p} \equiv \left[\left(\frac{r}{p^\nu} + \frac{\ell + \theta}{p^{\bar{k}}} \right) p^\nu \right] \pmod{p} \\ &\equiv r + \left[\frac{\ell + \theta}{p^{\bar{k}-\nu}} \right] \pmod{p} = r, \end{aligned}$$

then

$$\chi_{p^{\nu-1}}(\delta_{\ell + rp^{\bar{k}-\nu}}) = \exp \frac{2\pi i r}{p} = \chi_{p^{\nu-1}}(\delta_{rp^{\bar{k}-\nu}}). \tag{12.3.17}$$

Substituting (12.3.16) and (12.3.17) into (12.3.15) we have

$$\sum_{r=0}^{p-1} \chi_{n,\nu}(\delta_{\ell + rp^{\bar{k}-\nu}}) \exp \frac{i\omega r}{p^\nu} = \sum_{r=0}^{p-1} (\chi_{p^{\nu-1}}(\delta_{rp^{\bar{k}-\nu}}))^{n-\nu} \chi_{n,\nu+1}(\delta_\ell) \exp \frac{i\omega r}{p^\nu}.$$

Combining this last identity with (12.3.14), we obtain

$$\begin{aligned} \sigma_{n,\nu}(\omega) = &\sum_{\ell=0}^{p^{\bar{k}-\nu}-1} \exp \frac{i\omega(2\ell + 1 - p^{\bar{k}-\nu+1})}{2p^{\bar{k}}} \times \\ &\times \sum_{r=0}^{p-1} (\chi_{p^{\nu-1}}(\delta_{rp^{\bar{k}-\nu}}))^{n-\nu} \chi_{n,\nu+1}(\delta_\ell) \exp \frac{i\omega r}{p^\nu} \\ = &\sum_{r=0}^{p-1} (\chi_{p^{\nu-1}}(\delta_{rp^{\bar{k}-\nu}}))^{n-\nu} \exp \frac{i\omega r}{p^\nu} \times \\ &\times \sum_{\ell=0}^{p^{\bar{k}-\nu}-1} \chi_{n,\nu+1}(\delta_\ell) \exp \frac{i\omega(2\ell + 1 - p^{\bar{k}-\nu} + p^{\bar{k}-\nu}(1-p))}{2p^{\bar{k}}} \\ = &\sum_{r=0}^{p-1} (\chi_{p^{\nu-1}}(\delta_{rp^{\bar{k}-\nu}}))^{n-\nu} \exp \left(\frac{i\omega r}{p^\nu} + \frac{i\omega(1-p)}{2p^\nu} \right) \times \\ &\times \sum_{\ell=0}^{p^{\bar{k}-\nu}-1} \chi_{n,\nu+1}(\delta_\ell) \exp \frac{i\omega(2\ell + 1 - p^{\bar{k}-\nu}}{2p^{\bar{k}}} \\ = &\sigma_{n,\nu+1}(\omega) \sum_{r=0}^{p-1} (\chi_{p^{\nu-1}}(\delta_{rp^{\bar{k}-\nu}}))^{n-\nu} \exp \frac{i\omega(2r + 1 - p)}{2p^\nu}. \end{aligned}$$

This completes the proof of (12.3.13).

Write (12.3.13) in the form

$$\sigma_{n,\nu}(\omega) = \sigma_{n,\nu+1}(\omega)\gamma_\nu, \qquad \nu = 1, 2, \ldots, \tilde{k} - 1,$$

and apply (12.3.12). We obtain

$$S_{\tilde{k}}(\omega) = \sigma_{n,1}(\omega) = \sigma_{n,2}(\omega)\gamma_1 = \cdots = \sigma_{n,\tilde{k}-1}(\omega)\prod_{\nu=1}^{\tilde{k}-2}\gamma_\nu = \sigma_{n,\tilde{k}}(\omega)\prod_{\nu=1}^{\tilde{k}-1}\gamma_\nu.$$

Substituting the value for $\sigma_{n,\tilde{k}}(\omega)$ from (12.3.10) and the value of γ_ν we find

$$S_{\tilde{k}}(\omega) = \sum_{\ell=0}^{p-1}\chi_{n,\tilde{k}}(\delta_\ell)\exp\frac{i\omega(2\ell+1-p)}{2p^{\tilde{k}}}\times$$

$$\times\prod_{\nu=1}^{\tilde{k}-1}\sum_{r=0}^{p-1}(\chi_{p^{\nu-1}}(\delta_{rp^{\tilde{k}-\nu}}))^{n_{-\nu}}\exp\frac{i\omega(2r+1-p)}{2p^\nu}$$

$$= \prod_{\nu=1}^{\tilde{k}-1}\sum_{r=0}^{p-1}(\chi_{p^{\nu-1}}(\delta_{rp^{\tilde{k}-\nu}}))^{n_{-\nu}}\exp\frac{i\omega(2r+1-p)}{2p^\nu}.$$

But it is easy to see that

$$\sum_{r=0}^{p-1}\exp\left(\frac{i\omega r}{p^\nu} + \frac{2\pi i n_{-\nu}r}{p}\right) = \frac{1-\exp\left(\dfrac{i\omega}{p^{\nu-1}} + 2\pi i n_{-\nu}\right)}{1-\exp\left(\dfrac{i\omega}{p^\nu} + \dfrac{2\pi i n_{-\nu}}{p}\right)}$$

$$= \frac{\exp\dfrac{i\omega}{2p^{\nu-1}}}{\exp\left(\dfrac{i\omega}{2p^\nu} + \dfrac{\pi i n_{-\nu}}{p}\right)}\,\frac{\sin\dfrac{\omega}{2p^{\nu-1}}}{\sin\left(\dfrac{\omega}{2p^\nu} + \dfrac{\pi n_{-\nu}}{p}\right)}.$$

Consequently, it follows from (12.3.17) and the previous identity that

$$S_{\tilde{k}}(\omega) = \prod_{\nu=1}^{\tilde{k}}\sum_{r=0}^{p-1}\exp\frac{i\omega}{p^\nu}\exp\frac{i\omega(1-p)}{2p^\nu}\exp\frac{2\pi i n_{-\nu}r}{p}$$

$$= \prod_{\nu=1}^{\tilde{k}}\exp\frac{i\omega(1-p)}{2p^\nu}\exp\left(\frac{i\omega}{2p^{\nu-1}} - \frac{i\omega}{2p^\nu}\right)\exp\frac{-\pi i n_{-\nu}}{p}\,\frac{\sin\dfrac{\omega}{2p^{\nu-1}}}{\sin\left(\dfrac{\omega}{2p^\nu} + \dfrac{\pi n_{-\nu}}{p}\right)}$$

$$= \prod_{\nu=1}^{\tilde{k}}\exp\frac{-\pi i n_{-\nu}}{p}\,\frac{\sin\dfrac{\omega}{2p^{\nu-1}}}{\sin\left(\dfrac{\omega}{2p^\nu} + \dfrac{\pi n_{-\nu}}{p}\right)}.$$

Substituting this expression for $S_{\bar{k}}(\omega)$ into (12.3.8), we conclude that

$$R_n(\omega) = \sqrt{\frac{2}{\pi}}\frac{1}{\omega}\sin\frac{\omega}{2p^{\bar{k}}}\prod_{\nu=1}^{\bar{k}}\exp\frac{-\pi i n_{-\nu}}{p}\frac{\sin\dfrac{\omega}{2p^{\nu-1}}}{\sin\left(\dfrac{\omega}{2p^{\nu}}+\dfrac{\pi n_{-\nu}}{p}\right)}. \quad \blacksquare$$

If we set $p = 2$ in the identity for $R_n(\omega)$ found in **12.3.3** and use the fact that

$$\begin{aligned}\sin\frac{\omega}{2^\nu} &= e^{\pi i n_{-\nu}}\sin\left(\frac{\omega}{2^\nu}+\pi n_{-\nu}\right)\\ &= 2e^{\pi i n_{-\nu}}\sin\left(\frac{\omega}{2^{\nu+1}}+\frac{\pi n_{-\nu}}{2}\right)\cos\left(\frac{\omega}{2^{\nu+1}}+\frac{\pi n_{-\nu}}{2}\right), \qquad n_{-\nu}=0 \quad \text{or} \quad 1,\end{aligned}$$

then we see that the function $R_n(\omega)$ which corresponds to the Walsh system is given by

$$R_n(\omega) = \frac{2^{\bar{k}+1}}{\sqrt{2\pi}}\frac{1}{\omega}\sin\frac{\omega}{2^{\bar{k}+1}}\prod_{\nu=1}^{\bar{k}}\exp\frac{\pi i n_{-\nu}}{2}\cos\left(\frac{\omega}{2^{\nu+1}}+\frac{\pi n_{-\nu}}{2}\right).$$

In particular, for the case $p = 2$ the first six functions $R_n(\omega)$ are given by:

$$R_0(\omega) = \frac{2}{\sqrt{2\pi}}\frac{1}{\omega}\sin\frac{\omega}{2},$$

$$R_1(\omega) = \frac{4i}{\sqrt{2\pi}}\frac{1}{\omega}\sin^2\frac{\omega}{4},$$

$$R_2(\omega) = \frac{8i}{\sqrt{2\pi}}\frac{1}{\omega}\sin^2\frac{\omega}{8}\cos\frac{\omega}{4},$$

$$R_3(\omega) = \frac{-8}{\sqrt{2\pi}}\frac{1}{\omega}\sin^2\frac{\omega}{8}\sin\frac{\omega}{4},$$

$$R_4(\omega) = \frac{16i}{\sqrt{2\pi}}\frac{1}{\omega}\sin^2\frac{\omega}{16}\cos\frac{\omega}{8}\cos\frac{\omega}{4},$$

$$R_5(\omega) = \frac{-16}{\sqrt{2\pi}}\frac{1}{\omega}\sin^2\frac{\omega}{16}\cos\frac{\omega}{8}\sin\frac{\omega}{4},$$

$$R_6(\omega) = \frac{-16}{\sqrt{2\pi}}\frac{1}{\omega}\sin^2\frac{\omega}{16}\sin\frac{\omega}{8}\cos\frac{\omega}{4}.$$

We shall give a mathematical formulation of optimization problem concerning discrete radiators of electromagnetic waves, i.e., linear radiators which are the aggregate of elementary dipole radiators, located along the straight line segment $-\ell/2 \le x \le \ell/2$. Suppose that to an elementary radiator of length Δx situated in a neighborhood of the points $x \in [-\ell/2, \ell/2]$, we supply current which changes in time t according to the sinusoidal law

$$f(x)e^{i\omega t}\Delta x, \qquad \omega = \text{const},$$

where $f(x)$ is a complex function interpreted in the following way: $|f(x)|$ represents amplitude and $\arg f(x)$ represents the phase of the current supplied at the point x. In this case, amplitude of the current emitted at an angle ψ depends on the angle ψ in the plane passing across the X axis of the electromagnetic field and is related in a unique way to the radiation pattern of the radiators, written in the form (see [29], p. 13)

$$D(\psi) = \frac{1}{A}\int_{-\ell/2}^{\ell/2} f(x)e^{i(\omega x \sin\psi)/c}\,dx. \tag{12.3.18}$$

Here ψ is the angle between the perpendicular to the X axis and an arbitrary direction ($-\pi/2 \le \psi \le \pi/2$), $c = \text{const}$, A is a normalizing factor which for convenience we shall write in the form $A = 2\ell\sqrt{\pi/2}$.

If we let $u = (\omega\ell \sin\psi)/c$ in (12.3.18) and make the change of variables $y = x/\ell$ then we obtain

$$\begin{aligned} D(\psi) &= D\left(\arcsin\frac{cu}{\omega\ell}\right) = R(u) \\ &= \frac{\ell}{A}\int_{-\ell/2}^{\ell/2} f(\ell y)e^{iuy}\,dy \\ &= \frac{1}{\sqrt{2\pi}}\int_{-\ell/2}^{\ell/2} f^*(y)e^{iuy}\,dy. \end{aligned} \tag{12.3.19}$$

The function $R(u)$ is defined for $u \in [-\omega\ell/c, \omega\ell/c]$. The function $f^*(y) = f(\ell y)$ is piecewise constant for $y \in [-1/2, 1/2]$ since the radiators we are considering are discrete. Moreover, $f^*(y) \in \mathbf{L}^2(-1/2, 1/2)$ and thus the integral (12.3.19) representing $R(u)$ can be analytically continued to the whole complex plane $\mathbf{C}$. Hence it follows from the Paley-Wiener Theorem that a solution $f^*(y)$ of the equation (12.3.19), identically zero outside the interval $[-1/2, 1/2]$, exists if and only if the function $R(u)$ belongs to the class $W^2_{1/2}$. Consequently, we consider the cases $R(u) \in W^2_{1/2}$ and $R(u) \in \mathbf{L}^2(\mathbf{R})$ separately.

12.3.4. *Let $R(u) \in W^2_{1/2}$. There exist piecewise constant solutions $S_n(y)$ to equation (12.3.19) which converge to the solution $f^*(y)$, as $n \to \infty$, in the $\mathbf{L}^2[-1/2, 1/2]$ norm.*

PROOF. To approximate solutions to the equation (12.3.19) we apply the method of partial diagrams (see [29], p. 28). This involves expanding the left side of (12.3.19) in a Fourier series with respect tot the system $\{R_n(u)\}$:

$$R(u) \sim \sum_{n=0}^{\infty} c_n R_n(u), \qquad c_n = \int_{\mathbf{R}} R(x)\overline{R_n(x)}\,dx. \tag{12.3.20}$$

Notice by (12.3.7) that the series in the system $\{f_n(y)\}$ with the same coefficients c_n is the Fourier series of the function $f^*(y)$ with respect to the system $\{f_n(y)\}$. Since the system $\{f_n(y)\} = \{\chi_n(y+1/2)\}$ is complete, it follows that the series

$$f^*(y) \sim \sum_{n=0}^{\infty} c_n f_n(y), \qquad c_n = \int_{\mathbf{R}} R(x) R_n(x)\, dx \tag{12.3.21}$$

converges to $f^*(y)$ in the $\mathbf{L}^2[-1/2, 1/2]$ norm. ∎

The main advantage of considering questions using the system $\{R_n(\omega)\}$ defined in (12.3.2) is that the functions $f_n(u) = \chi_n(u+1/2)$ are piecewise constant and therefore the approximate solutions $S_N(y) = \sum_{n=0}^{N-1} c_n f_n(y)$ to the equation (12.3.19), being a linear combination of these functions, are also piecewise constant. Hence an approximate solution to the problem of distribution of current which is found using the system $\{\chi_n(y+1/2)\}$ can be reproduced in practice, i.e., a solution of equation (12.3.19) by the method of partial diagrams, using the system $\{R_n(\omega)\}$, allows for the discrete structure of the radiators.

Notice also that if a solution to equation (12.3.19) is written as a partial sum of some series involving the functions $\chi_n(u+1/2)$, for some p, consisting of N terms, $p^{\bar{k}-1} \le N < p^{\bar{k}}$, then for an exact reproduction of this distribution of currents the radiator must consist of $p^{\bar{k}}$ elementary radiators of identical length, because this partial sum has $p^{\bar{k}}$ intervals of constancy.

Suppose now that the given radiation pattern $R^*(u) \in \mathbf{L}^2(\mathbf{R})$, i.e., it does not necessarily belong to the class $W^2_{1/2}$. Let S^0_m represent the set of piecewise constant functions which are identically zero outside the segment $[-1/2, 1/2]$ and have m identical intervals of constancy

$$I_k = \left(-\frac{1}{2} + \frac{k-1}{m}, -\frac{1}{2} + \frac{k}{m}\right), \qquad k = 1, 2, \ldots, m.$$

12.3.5. *Suppose a linear radiator of electromagnetic waves consists of p^k elementary radiators of identical length. Among all practical feasible, i.e., those whose distributions of current belong to $S^0_{p^k}$, there is a distribution $\tilde{f}(y)$ which gives the radiation pattern*

$$\tilde{R}(\omega) = \frac{1}{\sqrt{2\pi}} \int_{-\ell/2}^{\ell/2} \tilde{f}(y) e^{i\omega y}\, dy$$

is nearest, in the least squares sense, to a given radiation pattern

$$R^*(\omega) = \frac{1}{\sqrt{2\pi}} \int_{-\ell/2}^{\ell/2} f^*(y) e^{i\omega y}\, dy.$$

Moreover, the distribution $\tilde{f}(y)$ is defined by the formula

$$\tilde{f}(y) = \sum_{n=0}^{p^k-1} c_n \chi_n \left(y + \frac{1}{2}\right), \tag{12.3.22}$$

where

$$c_n = \int_{\mathbf{R}} R(x)\overline{R_n(x)}\,dx, \qquad n = 0, 1, \dots, p^k - 1, \tag{12.3.23}$$

and $R_n(u)$ is defined in **12.3.3**.

PROOF. We need to find a function $\tilde{f}(y) \in S^0_{p^k}$ such that

$$\|R^*(\omega) - \tilde{R}(\omega)\|_{\mathbf{L}^2(\mathbf{R})} = \inf_{f \in S^0_{p^k}} \|R^*(\omega) - R(\omega)\|_{\mathbf{L}^2(\mathbf{R})},$$

where $R(\omega) = \frac{1}{\sqrt{2\pi}} \int_{-\ell/2}^{\ell/2} f(y) e^{i\omega y}\,dy$. Thus this problem about best approximation (in the $\mathbf{L}^2(\mathbf{R})$ norm) reduces to specifying a radiation pattern with the help of a system of p^k elementary radiators.

Let $f^*(y)$ be the inverse Fourier transformation of the given function $R^*(\omega)$. Since any function from the set $S^0_{p^k}$ can be represented exactly as a linear combination of the first p^k Chrestenson-Levy functions $\chi_n(y + 1/2)$, $n = 0, 1, \dots, p^k - 1$, then using the extremal property of partial sums of Fourier series we have

$$\|f^*(y) - \sum_{n=0}^{p^k-1} c_n \chi_n \left(y + \frac{1}{2}\right)\|_{\mathbf{L}^2(\mathbf{R})} = \inf_{f \in S^0_{p^k}} \|f^*(y) - f(y)\|_{\mathbf{L}^2(\mathbf{R})},$$

$$c_n = \int_{-1/2}^{1/2} f^*(y) \overline{\chi}_n \left(y + \frac{1}{2}\right) dy$$

(recall that the functions $\chi_n(y + 1/2)$ have been extended to be zero on the whole real line). Since the Fourier transform is an isometry in $\mathbf{L}^2(\mathbf{R})$, it follows that

$$\begin{aligned}
\|f^*(y) - \sum_{n=0}^{p^k-1} c_n \chi_n \left(y + \frac{1}{2}\right)\|_{\mathbf{L}^2(\mathbf{R})} &= \|R^*(y) - \sum_{n=0}^{p^k-1} c_n R_n(u)\|_{\mathbf{L}^2(\mathbf{R})} \\
&= \inf_{f \in S^0_{p^k}} \|f^*(y) - f(y)\|_{\mathbf{L}^2(\mathbf{R})} \\
&= \inf_{f \in S^0_{p^k}} \|R^*(u) - R(u)\|_{\mathbf{L}^2(\mathbf{R})},
\end{aligned}$$

i.e.,

$$\tilde{f}(y) = \sum_{n=0}^{p^k-1} c_n \chi_n \left(y + \frac{1}{2}\right),$$

where by Plancherel's Theorem the coefficients c_n are determined by the given functions by means of the formula

$$c_n = \int_{\mathbf{R}} R^*(u)\overline{R_n(u)}\,du, \qquad n = 0, 1, \ldots, p^k - 1.$$

Therefore, to define the best distribution (in the indicated sense) of currents $\tilde{f}(y)$ in radiators it is sufficient to find the p^k Fourier coefficients (12.3.23) and put them in the sum (12.3.22). ∎

We shall presently consider the problem of optimizing an initial condition in a diffusion process. First we introduce some more notation.

Using the system $\{f_n(u)\}$, which is given by (12.3.1), we define functions

$$\Phi_{n,k}(u) = f_n(u+k), \qquad k = 0, \pm 1, \pm 2, \ldots, \quad n = 0, 1, \ldots, \tag{12.3.24}$$

which evidently coincide with the functions $\chi_n(u + k + 1/2)$ on the intervals $u \in [k - 1/2, k + 1/2]$, and are identically zero outside these intervals.

We shall represent the Fourier transforms of these functions by

$$\begin{aligned} R_{n,k}(\omega) &= \frac{1}{\sqrt{2\pi}} \int_{k-1/2}^{k+1/2} \Phi_{n,k}(u) e^{i\omega u}\,du \\ &= \frac{1}{\sqrt{2\pi}} \int_{k-1/2}^{k+1/2} f_n(u+k) e^{i\omega u}\,du. \end{aligned} \tag{12.3.25}$$

Making the change of variables $u + k = v$ we obtain

$$R_{n,k}(\omega) = \frac{e^{-i\omega k}}{\sqrt{2\pi}} \int_{-1/2}^{+1/2} \chi_n\left(v + \frac{1}{2}\right) e^{i\omega v}\,dv.$$

Therefore, we have by (12.3.2) that

$$R_{n,k}(\omega) = e^{-i\omega k} R_n(\omega). \tag{12.3.26}$$

12.3.6. *The system $\{R_{n,k}(\omega)\}$ is orthonormal and complete in the space $\mathbf{L}^2(\mathbf{R})$.*

PROOF. Orthonormality of the system $\{R_{n,k}(\omega)\}$ follows from (12.3.3) and the fact that the functions $\Phi_{n,k}(y)$ are orthonormal in $\mathbf{L}^2(\mathbf{R})$.

To prove the system $\{R_{n,k}(\omega)\}$ is complete let $R(\omega) \in \mathbf{L}^2(\mathbf{R})$ and $\varepsilon > 0$ be arbitrary. We shall show that there exists a finite linear combination of the functions $R_{n,k}(\omega)$ which is near the function $R(\omega)$ in the $\mathbf{L}^2(\mathbf{R})$ norm in the following sense:

$$\|R(\omega) - \sum_{k=-M}^{M} \sum_{n=0}^{N_k} c_{n,k} R_{n,k}(\omega)\|_{\mathbf{L}^2(\mathbf{R})} \leq \varepsilon. \tag{12.3.27}$$

Let $\hat{f}(y)$ be the inverse Fourier transform of the function $R(\omega) \in \mathbf{L}^2(\mathbf{R})$. By Plancherel's Theorem, $\hat{f}(y) \in \mathbf{L}^2(\mathbf{R})$, i.e., the integral of the square of this function is convergent. Thus there exists a number M depending on $\varepsilon > 0$ such that

$$\left(\int_{-\infty}^{-M-1/2} + \int_{M+1/2}^{\infty}\right) |\hat{f}(y)|^2\, dy \leq \frac{\varepsilon}{\sqrt{2}}. \tag{12.3.28}$$

Write the function $\hat{f}(y)$ in the following form:

$$\hat{f}(y) = \sum_{k=-\infty}^{\infty} f(y)_k, \qquad \text{where} \quad f(y)_k = \begin{cases} \hat{f}(y) & \text{for } y \in J_k, \\ 0 & \text{for } y \notin J_k. \end{cases}$$

Evidently $f(y)_k \in \mathbf{L}^2(J_k)$ for all k, and thus using the fact that the system $\{\Phi_{n,k}(y)\}$ is complete in the space $\mathbf{L}^2(J_k)$, we can choose coefficients $c_{n,k}$ such that

$$\|f(y)_k - \sum_{n=0}^{N_k} c_{n,k}\Phi_{n,k}(y)\|_{\mathbf{L}^2(J_k)} \leq \frac{\varepsilon}{\sqrt{4M+2}}. \tag{12.3.29}$$

But the functions $\Phi_{n,k}(y)$ vanish off the interval J_k. Consequently,

$$\begin{aligned}
&\|\hat{f}(y) - \sum_{k=-M}^{M}\sum_{n=0}^{N_k} c_{n,k}\Phi_{n,k}(\omega)\|^2_{\mathbf{L}^2(\mathbf{R})} \\
&= \sum_{k=-M}^{M} \|f(y)_k - \sum_{n=0}^{N_k} c_{n,k}\Phi_{n,k}(\omega)\|^2_{\mathbf{L}^2(J_k)} \\
&\quad + \left(\int_{-\infty}^{-M-1/2} + \int_{M+1/2}^{\infty}\right) |\hat{f}(y)|^2\, dy.
\end{aligned}$$

Thus it follows from (12.3.28) and (12.3.29) that

$$\|\hat{f}(y) - \sum_{k=-M}^{M}\sum_{n=0}^{N_k} c_{n,k}\Phi_{n,k}(\omega)\|^2_{\mathbf{L}^2(\mathbf{R})} \leq (2M+1)\frac{\varepsilon^2}{4M+1} + \frac{\varepsilon^2}{2} = \varepsilon^2.$$

Since the Fourier transform is an isometry in $\mathbf{L}^2(\mathbf{R})$, we conclude that

$$\begin{aligned}
\varepsilon &\geq \|\hat{f}(y) - \sum_{k=-M}^{M}\sum_{n=0}^{N_k} c_{n,k}\Phi_{n,k}(\omega)\|^2_{\mathbf{L}^2(\mathbf{R})} \\
&= \|\mathcal{F}^{-1}[\hat{f}(y)] - \sum_{k=-M}^{M}\sum_{n=0}^{N_k} c_{n,k}\mathcal{F}^{-1}[\Phi_{n,k}(y)]\|^2_{\mathbf{L}^2(\mathbf{R})} \\
&= \|R(\omega) - \sum_{k=-M}^{M}\sum_{n=0}^{N_k} c_{n,k}R_{n,k}(\omega)\|^2_{\mathbf{L}^2(\mathbf{R})}
\end{aligned}$$

where $\mathcal{F}^{-1}$ represents the inverse Fourier transform. In particular, inequality (12.3.27) holds and it follows that the system $\{R_{n,k}(\omega)\}$ is complete on $\mathbf{L}^2(\mathbf{R})$. ∎

Let H_γ represent the set of functions $\phi(x) \in \mathbf{L}^2(\mathbf{R})$ whose Fourier transforms have the form

$$\hat{\phi}(u) = f(u)e^{-\gamma u^2} \tag{12.3.30}$$

for some $f(u) \in \mathbf{L}^2(\mathbf{R})$. Let S_m represent the set of piecewise constant functions in $\mathbf{L}^2(\mathbf{R})$ which have m identical intervals of constancy on each of the segments J_k for $k = 0, \pm 1, \pm 2, \ldots$, and let S_m^k represent the set of piecewise constant functions in S_m which vanish outside J_k.

The following theorem gives a solution to the problem mentioned above concerning optimization of an initial condition of a diffusion process.

12.3.7. *There is a function $\phi_0^*(x)$ in the set S_{p^m} which is the best approximation in S_{p^m} with respect to the $\mathbf{L}^2(\mathbf{R})$ norm of the initial condition*

$$u(x,0) = \phi_0(x) \in \mathbf{L}^2(\mathbf{R}), \tag{12.3.31}$$

such that the solution $\tilde{u}(x,t)$ of the differential equation

$$\frac{\partial \tilde{u}(x,t)}{\partial t} = a^2 \frac{\partial^2 \tilde{u}(x,t)}{\partial x^2}, \tag{12.3.32}$$

satisfying the initial condition (12.3.31) coincides with a given function $\phi_T(x) \in H_{a^2T}$ at time $t = T$, i.e.,

$$\tilde{u}(x,0) = \phi_0(x) \quad \text{and} \quad \tilde{u}(x,T) = \phi_T(x).$$

Moreover, the function $\phi_0^*(x)$ can be written in the form

$$\phi_0^*(x) = \sum_{n=0}^{p^m-1} c_{n,k}\Phi_{n,k}(x), \qquad x \in J_k, \quad k = 0, \pm 1, \ldots,$$

where

$$c_{n,k} = \int_{\mathbf{R}} \overline{R_{n,k}(\omega)\mathcal{F}[\phi_T](\omega)e^{a^2\omega^2 T}}\, d\omega,$$

$\Phi_{n,k}(x)$ and $R_{n,k}(\omega)$ are defined by (12.3.24) and (12.3.25), and

$$\mathcal{F}[\phi_T](\omega) = \frac{1}{\sqrt{2\pi}} \int_{\mathbf{R}} \phi_T(u)e^{i\omega u}\, du$$

is the Fourier transform of the function $\phi_T(x)$.

The solution in **12.3.7** is understood in the distributional sense when the function $\phi_0(x)$ satisfies no other condition except $\phi_0(x) \in \mathbf{L}^2(\mathbf{R})$ ([19], p. 202).

PROOF. We shall show that the initial condition $\phi_0(x) \in \mathbf{L}^2(\mathbf{R})$ for which $u(x,T) = \phi_T(x) \in H_{a^2T}$ exists and is unique. For this we choose a solution $u(x,T)$ of equation (12.3.32) which satisfies the initial condition $\phi_0(x)$ by using the Fourier transform ([17], p. 408):

$$\mathcal{F}[u(x,T)](\omega) = \mathcal{F}[\phi_T](\omega) = e^{-a^2T\omega^2}\mathcal{F}[\phi_0](\omega).$$

Since $\phi_T(x) \in H_{a^2T}$, it follows that

$$\mathcal{F}[\phi_0](\omega) = \mathcal{F}[\phi_T](\omega)e^{a^2T\omega^2} \in \mathbf{L}^2(\mathbf{R}), \tag{12.3.33}$$

i.e., the function $\phi_0(x)$ as an element of $\mathbf{L}^2(\mathbf{R})$ can be uniquely determined by using the inverse Fourier transform.

Write the functions $\phi_0(x)$ and $\phi_0^*(x)$ in the form

$$\phi_0(x) = \sum_{k=-\infty}^{\infty} \phi_0(x)_k, \qquad \text{where} \quad \phi_0(x)_k = \begin{cases} \hat{\phi}_0(x) & \text{for } x \in J_k, \\ 0 & \text{for } x \notin J_k, \end{cases}$$

and

$$\phi_0^*(x) = \sum_{k=-\infty}^{\infty} \phi_0^*(x)_k, \qquad \text{where} \quad \phi_0^*(x)_k = \begin{cases} \hat{\phi}_0^*(x) & \text{for } x \in J_k, \\ 0 & \text{for } x \notin J_k. \end{cases}$$

Since

$$\|\phi_0(x) - \phi_0^*(x)\|^2_{\mathbf{L}^2(\mathbf{R})} = \sum_{k=-\infty}^{\infty} \|\phi_0(x)_k - \phi_0^*(x)_k\|^2_{\mathbf{L}^2(\mathbf{R})},$$

the posed problem about finding the best approximation reduces to finding functions $\phi_0^*(x)_k$ which satisfy the property

$$\|\phi_0(x)_k - \phi_0^*(x)_k\|_{\mathbf{L}^2(J_k)} = \inf_{\phi \in S^k_{p^m}} \|\phi_0(x)_k - \phi(x)\|_{\mathbf{L}^2(J_k)}.$$

We notice that any function from the set $S^k_{p^m}$ is by definition a linear combination of the functions $\Phi_{n,k}(u)$ which were defined in (12.3.24). Taking into account the extremal property of the partial sums of a Fourier series, we can write

$$\inf_{\phi \in S^k_{p^m}} \|\phi_0(x)_k - \phi(x)\|_{\mathbf{L}^2(J_k)} = \inf_{\phi \in S^k_{p^m}} \|\phi_0(x)_k - \sum_{n=0}^{p^m-1} c_{n,k}\Phi_{n,k}(x)\|_{\mathbf{L}^2(J_k)}.$$

But this means that

$$\phi_0^*(x)_k = \sum_{n=0}^{p^m-1} c_{n,k}\Phi_{n,k}(x),$$

where

$$c_{n,k} = \int_{J_k} \phi_0(x)_k \overline{\Phi_{n,k}(x)}\, dx,$$

for $k = 0, \pm 1, \ldots$, and $n = 0, 1, \ldots, p^m - 1$. Hence to prove **12.3.7** it suffices to find the coefficients $c_{n,k}$. Using the definition of the function $\phi_0^*(x)_k$ and applying the generalized Parseval identity (12.3.3), we see that

$$\begin{aligned} c_{n,k} &= \int_{J_k} \phi_0(x)_k \overline{\Phi_{n,k}(x)}\, dx \\ &= \int_{\mathbf{R}} \phi_0(x)_k \overline{\Phi_{n,k}(x)}\, dx \\ &= \int_{\mathbf{R}} \mathcal{F}[\phi_0](u) \mathcal{F}[\overline{\Phi}_{n,k}](u)\, du, \end{aligned} \tag{12.3.34}$$

where $\mathcal{F}$ is the Fourier transform operator. But by (12.3.26),

$$\mathcal{F}[\overline{\Phi}_{n,k}](u) = \frac{1}{\sqrt{2\pi}} \int_{\mathbf{R}} \overline{\Phi_{n,k}(x) e^{-iux}}\, dx = \overline{R_{n,k}(u)} = e^{iuk} \overline{R_n(u)}. \tag{12.3.35}$$

Hence we obtain the expression of $c_{n,k}$ introduced above by identities (12.3.33), (12.3.34), and (12.3.35). ▮

APPENDICES

Appendix 1
ABELIAN GROUPS

A1.1. A *group* is a set G together with a binary algebraic operation which satisfies the associative property and for which there exists an inverse operation.

Here we shall consider only groups whose binary operation is commutative. Such groups are called *commutative* or *abelian*. The algebraic operation of a commutative group is usually called *addition* and is denoted by the sign $+$. On occasion we shall use other notation.

Thus a *commutative* (or *abelian*) *group* is a non- empty set G which satisfies the following axioms:

a) There is an *addition operation* defined on G which assigns to each pair a, b in G a third element $a+b$ in G which will be called the sum of the elements a and b.

b) The operation "+" is *associative*, i.e., for any three elements a, b, c in G the identity $(a+b)+c = a+(b+c)$ holds.

c) The operation "+" is *commutative*, i.e., for any pair of elements a, b in G the identity $a+b = b+a$ holds.

d) There is a *zero* element in G, i.e., an element 0 such that $a+0 = a$ for all $a \in G$.

e) For each $a \in G$ there is an *additive inverse*, i.e., an element $(-a)$ such that $a+(-a) = 0$.

It is not difficult to verify that the *zero element of a group is unique*. One can also verify that for each $a \in G$ the *inverse element* $(-a)$ *is uniquely determined* (see [18]).

If A and B are subsets of a group G then we shall represent the set of sums $a+b$ where $a \in A$ and $b \in B$ by $A+B$. If b is a singleton, i.e., $B = \{b\}$, then the set $A+\{b\}$ will be written as $A+b$. We shall also denote the set of elements $(-a)$ where $a \in A$ by $-A$.

One of the simplest examples of a commutative group is the set Z of integers where the usual addition of real numbers serves as the group operation.

Examples of finite groups are given by $Z(n)$, where for each positive integer n $Z(n)$ is the set of integers $0, 1, \ldots, n-1$ and the group operation (which we shall denote by $\oplus$) is defined to be addition modulo n, i.e.,

$$a \oplus b = \begin{cases} a+b & \text{if} \quad a+b < n, \\ a+b-n & \text{if} \quad a+b \geq n. \end{cases}$$

For other examples of groups, see §1.2 and §1.5.

A 1.2. Subgroups. Equivalence classes. Factor groups. A subset H of a group G is called a *subgroup* if it is itself a group with respect to the group operation on G

Let H be a fixed subgroup of a group G. We shall call an element $a \in G$ *equivalent to* an element $b \in G$ *with respect to the subgroup* H if $a-b \in H$. This is an equivalence relation, i.e., it satisfies the following properties:

a) the *reflexive* property: each element $a \in G$ is equivalent to itself (since $a-a = 0 \in H$);

b) the *symmetric* property: if a is equivalent to b then b is equivalent to a (since if $a-b \in H$ then $b-a = -(a-b) \in H$);

c) the *transitive* property: if a is equivalent to b, and b is equivalent to c, then a is equivalent to c (since $a-b \in H$ and $b-c \in H$ imply that $a-c = (a-b)+(b-c) \in H$).

An *equivalence class* in G is a set of points which are all equivalent to a single element in G. It is easy to verify that given two equivalence classes (with respect to the same subgroup H) either these classes are disjoint or they are identical to each other. Consequently, the group G can be decomposed into pairwise disjoint subsets, the *equivalence classes of the group* G with respect to the subgroup H.

It is clear that a given equivalence class is completely determined by any one of its elements a. Indeed, all remaining elements of this equivalence class have the form $a+h$, where h ranges over the subgroup H. For this reason such an equivalence class is denoted by $a+H$ (see **A1.1**). Sometimes the equivalence class $a+H$ is said to be *generated* by the element a. It is clear that an equivalence class is generated by each of its elements.

The set of equivalence classes of a group G with respect to a subgroup H is denoted by G/H. One can define an addition operation on G/H as follows. The *sum* of two equivalence classes $a+H$ and $b+H$ is the class $(a+b)+H$. It is easy to see that this sum does not depend on the choice of the representatives a and b of the class $a+H$ and $b+H$ and that this addition operation satisfies the axioms for a commutative group (see **A1.1**). Thus the set G/H is a commutative group which we shall call the *factor group* of the group G *with respect to* the subgroup H.

A1.3. Isomorphisms and automorphisms. Translations. A map ϕ from one group G to another G_1 is called an *isomorphism* if it is one to one and preserves the group operations, i.e., for any elements a, b in G,

$$\phi(a+b) = \phi(a) + \phi(b),$$

where $\phi(a)+\phi(b)$ is a sum in the group G_1. If $G = G_1$ such a mapping is called an *automorphism* of the group G.

Let a be fixed in some group G. It is easy to verify that the map $x \to x+a$ is an automorphism on G. It is called *translation* of the group G *by the element* a.

For a fixed subgroup H the translation $x \to x+a$ of the group G *corresponds to a translation of the factor group* G/H by the equivalence class $a+H$. Indeed, for the translation $x \to x+a$, all elements of a given equivalence class $b+H$ are carried to elements of the class $a+b+H$. These elements can be written in the form $(b+H)+(a+H)$, i.e., the equivalence class $a+b+H$ can be viewed as the image of the class $b+H$ by the translation of the factor group G/H by the equivalence class $a+H$.

APPENDIX 2
METRIC SPACES. METRIC GROUPS

A2.1. A set X is called a *metric space* if there is a function ρ which takes each pair x and y of X to a non-negative real number $\rho(x,y)$ which satisfies the following conditions:

a) $\rho(x,y) = 0$ if and only if $x = y$;

b) $\rho(x,y) = \rho(y,x)$ (the symmetric property);

c) $\rho(x,y) \leq \rho(x,z) + \rho(z,y)$ (the triangle inequality).

The function $\rho(x,y)$ is called the *distance* between the elements x and y, or the *metric* of the space X.

The set of points

$$U_\epsilon(x_0) = \{x \in X : \rho(x, x_0) < \epsilon\},$$

where x_0 is a fixed point in the space X and $\epsilon > 0$, is called the *open ball of radius* ϵ and *center* x_0 or an *ϵ-neighborhood of the point* x_0.

A point $x \in X$ is called a *cluster point* of a set $m \subset X$ if every neighborhood $U_\epsilon(x)$ contains infinitely many points of M.

A set $M \subset X$ is called *closed* if it contains all of its cluster points.

Denote the union of a set $M \subset X$ and the set of cluster points of M by $\overline{M}$. Then $\overline{M}$ is a closed set and is called the *closure* of the set M.

The set M is called *dense* in X if $\overline{M} = X$.

A point x is said to be *interior* to the set M if there is a neighborhood $U_\epsilon(x)$ which is contained in M.

A set is called *open* if all of its points are interior to it.

The following result is well-known (see [17]):

A2.1.1. *A set M is open if and only if its complement $X \setminus M$ is closed.*

The following theorem describes the situation for open subsets of the real line with respect to the usual metric $\rho(x, y) = |x - y|$ (see [21]):

A2.1.2. *Every open subset of the real line can be written as a finite or countably infinite union of pairwise disjoint open intervals (i.e., sets of the form (a, ∞), $(-\infty, a)$, $(-\infty, \infty)$, or (a, b)).*

Combining this result with **A2.1.1**, we obtain the following corollary:

A2.1.3. *Each closed subset E of the real line can be written as the complement of a finite or countably infinite union of pairwise disjoint open intervals.*

These intervals are called the *contiguous* intervals of the set E.

The collection of sets which can be obtained from a countable (i.e., finite or countably infinite) combination of unions and intersections of open or closed sets is called the collection of *Borel sets.*

A subset M of the space X is called *compact* if given any open covering $\{A_\alpha\}$ (i.e., given any collection of open sets $\{A_\alpha\}$ which satisfy $M \subset \bigcup_\alpha A_\alpha$) there is a finite subcollection $A_{k_1}, A_{k_2}, \ldots, A_{k_N}$ which also covers M.

If the set X is compact then it is called a *compact space.*

A space X is called *locally compact* if each of its points x has a neighborhood $U_\epsilon(x)$ whose closure is compact.

A sequence $\{x_n\}$ of elements of a space X is said to *converge* to an element $x \in X$ if $\rho(x, x_n) \to 0$ as $n \to \infty$.

A sequence $\{x_n\}$ of elements of a space X is called *Cauchy* if given $\epsilon > 0$ there is a number $n_0(\epsilon)$ such that $\rho(x_m, x_n) < \epsilon$ for all $n, m \geq n_0(\epsilon)$.

A metric space is said to be *complete* if each of its Cauchy sequences converge.

A2.2. A group G is called a *metric group* if it is a compact group and a metric space, and the group operations are continuous *with respect to the metric* in the following sense:

a) If $z = x + y$ then given any neighborhood $U(z)$ of z there exists neighborhoods $U(x)$ and $U(y)$ of the points x and y such that

$$U(x) + U(y) \subset U(z).$$

b) Given any neighborhood $U(-x)$ of the element $-x$ there exists a neighborhood $U(x)$ of the point x such that $-U(x) \subset U(-x)$.

The concept of a metric group is a particular case of the more general idea of a *topological group* (see [23]).

APPENDIX 3
MEASURE SPACES

A3.1. A class $\mathcal{A}$ of subsets of a set X is called a *σ-algebra* (or *countably additive*) if

1) $\emptyset \in \mathcal{A}$, where $\emptyset$ represents the empty set;
2) if $A \in \mathcal{A}$ then $X \setminus A \in \mathcal{A}$;
3) if $\{A_n\}$ is any countable collection of sets with $A_n \in \mathcal{A}$ then $\bigcup_n A_n \in \mathcal{A}$.

A set function $\mu(A)$ defined on all sets in some σ- algebra $\mathcal{A}$ is called a *measure* if

a) $0 \leq \mu(A) \leq +\infty$ for all $A \in \mathcal{A}$;
b) $\mu(\emptyset) = 0$;

c) the identity

$$\mu(\bigcup_{n=1}^{\infty} A_n) = \sum_{n=1}^{\infty} \mu(A_n)$$

holds for all sequences $\{A_n\}_{n=1}^{\infty}$ which satisfy $A_n \in \mathcal{A}$ and $A_n \bigcap A_m = \emptyset$ for $n \neq m$.

A *measure space* is a set X together with a σ-algebra $\mathcal{A}$ and a measure μ on $\mathcal{A}$.

A3.2. Measure on the real line. The measure most frequently used on the real line is Lebesgue measure. It can be defined in the following way.

First define the *outer measure* of any subset A of the real line by

$$\text{(A3.2.1)} \qquad \operatorname{mes}^*(A) = \inf\{\sum_n |I_n| : \bigcup_n I_n \supset A\},$$

where each I_n is an interval and $|I_n|$ represents its length. It is clear that $\operatorname{mes}^* I = |I|$ for any interval I.

A3.2.1. A set A is called *Lebesgue measurable* if given any $\epsilon > 0$ there is a closed set F and an open set Q such that $F \subset A \subset Q$ and $\operatorname{mes}^*(Q \setminus F) < \epsilon$.

A3.2.2. *The collection of Lebesgue measurable sets is a σ-algebra which contains the Borel sets and the restriction of mes* to this collection is a measure as defined in* **A3.1**.

This measure is called Lebesgue measure on the real line and will be denoted by mes, i.e., the Lebesgue measure of a measurable set A is defined to be $\operatorname{mes} A = \operatorname{mes}^* A$. (See [21]).

A3.2.3. *If a set A is measurable then*

$$\operatorname{mes} A = \inf \operatorname{mes} Q = \sup \operatorname{mes} F,$$

where the infimum is taken over all open sets Q which satisfy $A \subset Q$ and the supremum is taken over all closed sets F which satisfy $F \subset A$. (See [21]).

A3.3. Lebesgue-Stieltjes measure. Each function $\psi(x)$ which is *continuous from the right* and *non- decreasing* on $(-\infty, \infty)$ gives rise to a generalization of Lebesgue measure on the real line in the following way.

Define a function m on intervals (a, b) by

$$m(a, b) = \psi(b-) - \psi(a).$$

This function generalizes the concept of length since when $\psi(x) = x$ we have $m(a, b) = b-a = |(a, b)|$.

Substituting $m\, I_n$ for $|I_n|$ in formula (A3.2.1), we define an outer measure $\operatorname{mes}^*_\psi$ and as in **A3.2.1**, we form the class of measurable sets with respect to this outer measure and obtain a measure mes_ψ, which is called the *Lebesgue- Stieltjes measure* generated by the function ψ.

We notice by construction that

$$\text{(A3.3.1)} \qquad \begin{cases} \operatorname{mes}_\psi (a, b) = m(a, b) = \psi(b-) - \psi(a), \\ \operatorname{mes}_\psi [a, b] = \psi(b) - \psi(a-), \\ \operatorname{mes}_\psi (a, b] = \psi(b) - \psi(a), \\ \operatorname{mes}_\psi [a, b) = \psi(b-) - \psi(a-). \end{cases}$$

In particular, the measure of a point satisfies $\operatorname{mes}_\psi \{a\} = \psi(a) - \psi(a-)$.

In case the function is continuous from the *left*, instead of continuous from the right, we can define the function m by

$$m(a, b) = \psi(b) - \psi(a+).$$

With slight changes, the program outlined above can be carried out for this case as well. A corresponding analogue of (A3.3.1) holds in this case. For example,

(A3.3.2) $$\text{mes}_\psi[a,b) = \psi(b) - \psi(a).$$

A3.4. *Lebesgue measure on the plane* can be defined as it was on the line with only one change: in formula (A3.2.1) use *two-dimensional intervals* in place of the I_n's, i.e., open rectangles whose sides are parallel to the axes, and let $|I_n|$ represent the *area* of the rectangle I_n.

A3.5. Haar proved the following theorem concerning construction of measure on groups (see [13]).

A3.5.1. *On any locally compact group* (see A2.2) *there exists a measure μ, defined on the σ- algebra of Borel subsets of this group, such that*

a) $\mu(A) > 0$ for any non-empty open set A;

b) $\mu(a + A) = \mu(A)$ for any Borel set A (this property is referred to as the *translation invariance of the measure μ);*

c) $\mu(-A) = \mu(A)$.

Such a measure is called *Haar measure* on the group G.

Lebesgue measure on the real line $\mathbf{R}$ provides a simple example of Haar measure. Here the group structure is provided by the usual addition of real numbers.

APPENDIX 4
MEASURABLE FUNCTIONS. THE LEBESGUE INTEGRAL

A4.1. Let X be a measure space and μ be a measure on some σ-algebra $\mathcal{A}$ of subsets of X. In this case we shall refer to the elements of the class $\mathcal{A}$ as $\mathcal{A}$-*measurable* sets.

A function f defined on X is called $\mathcal{A}$-*measurable* (or μ-*measurable*) if for every real number C the set $\{x : f(x) > C\}$ is measurable. If in place of the collection $\mathcal{A}$ we use the collection of Lebesgue measurable sets (see A3.2), then we obtain the definition of a *Lebesgue measurable* function.

A function f defined on a metric space X is called *lower semi-continuous* on X if for every real number C the set $\{x : f(x) > C\}$ is open.

Two functions f and g defined on the same measurable set E are called *equivalent* if

$$\mu\{x : f(x) \neq g(x)\} = 0.$$

In connection with this recall the following terminology.

If some property holds on the set E, except possibly on some set of measure zero, then we say that this property holds *almost everywhere* (abbreviated as a.e.).

Thus we see that two functions are equivalent if they coincide almost everywhere.

A4.1.1. Egoroff's Theorem. *Let E be a set of finite μ measure and let $\{f_n(x)\}$ be a sequence of measurable functions which converges almost everywhere on E to a function $f(x)$. Then for any $\epsilon > 0$ there exists a measurable set $E_\epsilon \subset E$ such that $\mu(E_\epsilon) > \mu(E) - \epsilon$ and such that the sequence $\{f_n(x)\}$ converges uniformly to $f(x)$ on the set E_ϵ.* (See [17]).

In the case of *Lebesgue measurable* functions, the set E_ϵ can be chosen to be closed.

A4.1.2. Lusin's Theorem. *Let f be a function defined on an interval $[a,b]$. Then f is Lebesgue measurable if and only if for each $\epsilon > 0$ there exists a closed set $F_\epsilon \subset [a,b]$ and a function ϕ continuous on $[a,b]$ such that $\text{mes}\, F_\epsilon > (b-a) - \epsilon$ and $f(x) = \phi(x)$ for $x \in F_\epsilon$.* (See [21]).

A sequence of measurable functions $\{f_n(x)\}$ defined on some measurable set E is said to *converge in measure* to some measurable function $f(x)$ if $\lim_{n\to\infty} \mu\{x : |f_n(x) - f(x)| > \epsilon\} = 0$ for every $\epsilon > 0$.

We notice that every sequence which converges almost everywhere also converges in measure (see [21]).

A4.2. The Lebesgue integral. Let X be a measure space and μ be a measure on some σ-algebra $\mathcal{A}$ of subsets of X. Let $\chi_A(x)$ denote the *characteristic function of the set* A, i.e.,

$$\chi_A(x) = \begin{cases} 1 & \text{for } x \in A, \\ 0 & \text{for } x \notin A. \end{cases}$$

A function of the form

$$\sum_{i=0}^{\infty} a_i \chi_{A_i}(x), \tag{A4.2.1}$$

where $\{A_i\}$ is a sequence of pairwise disjoint sets, is called a *simple function*. It is easy to verify that a simple function is measurable if and only if all the sets A_i which appear in its definition are measurable.

A simple measurable function f of the form (A4.2.1) is called *Lebesgue integrable* (or *summable*) on a set $A = \bigcup_i A_i$ if the series $\sum_i a_i \mu(A_i)$ *converges absolutely*. In this case we define the *Lebesgue integral of the function* f to be the sum of this series, i.e.,

$$\int_A f\, d\mu \equiv \sum_i a_i \mu(A_i).$$

An arbitrary function f is called *Lebesgue integrable* (or *summable*) on a set A of *finite measure* if there exists a sequence $\{f_n\}$ of simple functions summable on A which converge *uniformly* to f. In this case we define

$$\int_A f\, d\mu \equiv \lim_{n\to\infty} \int_A f_n\, d\mu.$$

One can show (see [17]) that the value of the integral so defined does not depend on the choice of the sequence $\{f_n\}$.

Among the collection of *infinite measures* we restrict our attention to σ-finite measures, i.e., those measures whose space X can be written as a *countable* union of sets of *finite* μ-measure:

$$X = \bigcup_n X_n, \qquad \mu(X_n) < \infty. \tag{A4.2.2}$$

A monotone increasing sequence $\{X_n\}$ of measurable subsets of the space X which satisfies condition (A4.2.2) will be called an *exhaustive sequence*.

A function f defined on a space X with σ-finite measure μ is called *summable on* X if it is summable on each measurable subset $A \subset X$ of finite measure and for each exhaustive sequence $\{X_n\}$ the limit

$$\lim_{n\to\infty} \int_{X_n} f\, d\mu$$

exists and does not depend on the choice of this sequence. This limit is called the *integral* of the function f on X and is denoted by $\int_X f\, d\mu$.

Thus we have defined an integral for each space with a σ- finite measure. In the case of a concrete measure, this definition gives rise to a concrete form for the integral. The classical *Lebesgue integral on the real line* or *in the plane* is obtained if the construction carried out above is applied to the measures introduced in **A3.2** and **A3.3**. In this context, the Lebesgue integral corresponding to the measure "mes" is usually written in the form $\int_A f\, dx$ rather than $\int_A f\, d\,\text{mes}$.

A4.3. Functions of bounded variation. The Lebesgue- Stieltjes integral. A function f defined on an interval $[a,b]$ is called a *function of bounded variation* if there exists a constant C such that for any partition $a = x_0 < x_1 < \cdots < x_n = b$ of the interval $[a,b]$ the inequality

$$\sum_{i=1}^{n} |f(x_i) - f(x_{i-1})| < C$$

holds. The quantity

$$V_a^b[f] \equiv \sup \sum_{i=1}^{n} |f(x_i) - f(x_{i-1})|,$$

where the supremum is taken over all partitions of the interval $[a,b]$, is called the *total variation* of the function f on the interval $[a,b]$.

We shall list a number of properties concerning functions of bounded variation and their total variation (see [17]).

A4.3.1. *If $a < b < c$ then $V_a^b[f] + V_b^c[f] = V_a^c[f]$.*

A4.3.2. *For any numbers α and β we have*

$$V_a^b[\alpha f_1 + \beta f_2] \leq |\alpha| V_a^b[f_1] + |\beta| V_a^b[f_2]$$

A4.3.3. *Each monotone function f is a function of bounded variation and $V_a^b[f] = |f(b) - f(a)|$.*

A4.3.4. *If f is a function of bounded variation then the functions $v(x) \equiv V_a^x[f]$ and $v(x) - f(x)$ are monotone non-decreasing. Moreover,*

$$f(x) = v(x) - (v(x) - f(x)),$$

i.e., each function of bounded variation can be written as a difference of two monotone non-decreasing functions[1].

A4.3.5. *Every function of bounded variation has a finite derivative almost everywhere.*

A4.3.6. *If f is a function of bounded variation then*

$$\frac{dV_a^x[f]}{dx} = |f'(x)|$$

holds almost everywhere. (See [25], p. 121.)

A4.3.7. *Let $\psi(x)$ be a function of bounded variation on an interval $[a,b]$. Then*

$$\frac{dV_a^x[\psi(t) - \psi'(x) \cdot t]}{dx} = 0 \tag{A4.3.1}$$

for almost every $x \in [a,b]$.

PROOF. Let A be the set on which the derivative $\psi'(x)$ does not exist. By **A4.3.5**, $\operatorname{mes} A = 0$. Fix a real number c and apply **A4.3.6** to see that

$$\frac{dV_a^x[\psi(t) - ct]}{dx} = |\psi'(x) - c| \tag{A4.3.2}$$

[1] If the function f is continuous at a point x_0 from one side or the other, then the same is true of the functions $v(x)$ and $v(x) - f(x)$.

almost everywhere on $[a,b]$. Let E_c denote the subset of $[a,b]$ on which (A4.3.2) fails to hold. Then we have $\operatorname{mes} E_c = 0$ for each c.

Let $\{r_n\}$ be the set of rational numbers. Set

$$E \equiv \left(\bigcup_n E_{r_n}\right) \bigcup A.$$

Clearly, $\operatorname{mes} E = 0$.We shall prove that (A4.3.1) holds everywhere outside of E.

Indeed, suppose $x \notin E$. Then $\psi'(x)$ exists. Let $\epsilon > 0$ and choose a rational number r_n such that

$$|\psi'(x) - r_n| < \epsilon. \tag{A4.3.3}$$

Then by properties **A4.3.2**, **A4.3.3**, and inequality (A4.3.3) we have

$$\begin{aligned} \frac{1}{h}V_x^{x+h}[\psi(t) - \psi'(x)\cdot t] &\leq \frac{1}{h}V_x^{x+h}[\psi(t) - r_n t] + \frac{1}{h}V_x^{x+h}[(r_n - \psi'(x))t] \\ &\leq \frac{1}{h}V_x^{x+h}[\psi(t) - r_n t] + |r_n - \psi'(x)| \\ &\leq \frac{1}{h}V_x^{x+h}[\psi(t) - r_n t] + \epsilon. \end{aligned} \tag{A4.3.4}$$

Since $x \notin E$ we have $x \notin E_{r_n}$. Thus applying (A4.3.2) for the case $c = r_n$ we see that if h is sufficiently small then

$$\left|\frac{1}{h}V_x^{x+h}[\psi(t) - r_n t] - |\psi'(x) - r_n|\right| < \epsilon.$$

Thus it follows from (A4.3.3) that

$$\frac{1}{h}V_x^{x+h}[\psi(t) - r_n t] \leq 2\epsilon.$$

Putting this estimate into the right side of inequality (A4.3.4), we conclude that

$$\frac{1}{h}V_x^{x+h}[\psi(t) - \psi'(x)\cdot t] < 3\epsilon$$

for h sufficiently small. In particular (A4.3.1) holds. ∎

A4.3.8. Let ψ be a function of bounded variation, continuous (for example) from the right, and written (by **A4.3.4**) as a difference of two monotone non-decreasing functions which are also continuous from the right:

$$\psi(x) = \psi_1(x) - \psi_2(x).$$

Form the Lebesgue-Stieltjes measures (see **A3.3**): $\operatorname{mes}_{\psi_1}$ and $\operatorname{mes}_{\psi_2}$.

Then the *Lebesgue-Stieltjes integral with respect to the function* ψ is defined by the formula

$$\int_a^b f(x)d\psi(x) \equiv \int_a^b f(x)d\operatorname{mes}_{\psi_1} - \int_a^b f(x)d\operatorname{mes}_{\psi_2}.$$

A function f is considered *integrable with respect to* ψ if *both* integrals on the right side of this formula exist and are finite. It is easy to verify that the value of this integral does not depend on the choice of the representation of ψ as a difference of two monotone functions.

A4.3.9. *The Lebesgue-Stieltjes integral satisfies the following inequality* (see [16], p. 206):

$$\left|\int_a^b f(x)d\psi(x)\right| \leq \int_a^b |f(x)|dV_a^x[\psi].$$

A4.4. We shall list several well-known properties of the Lebesgue integral (see [17], [21]).

A4.4.1. The Lebesgue Dominated Convergence Theorem. *If a sequence of measurable functions $\{f_n\}$ converges in measure on a measurable set A (in particular, if it converges a.e.) to a function f and if there exists a function ϕ, summable on A, such that*

$$|f_n(x)| \leq \phi(x)$$

for all n and all $x \in A$, then the limit function f is also summable on A and

$$\lim_{n\to\infty} \int_A f_n\, d\mu = \int_A f\, d\mu.$$

A4.4.2. Levy's Theorem. *Suppose that A is a measurable set and*

$$f_1(x) \leq f_2(x) \leq \cdots \leq f_n(x) \leq \ldots$$

is a sequence of summable functions whose integrals are uniformly bounded on A, i.e.,

$$\int_A f_n\, d\mu \leq K.$$

Then the limit $f(x) = \lim_{n\to\infty} f_n(x)$ is almost everywhere finite on A and

$$\lim_{n\to\infty} \int_A f_n\, d\mu = \int_A f\, d\mu.$$

A4.4.3. A corollary of A4.4.2. *If*

$$\psi_n(x) \geq 0, \quad \text{and} \quad \sum_{n=1}^{\infty} \int_A \psi_n\, d\mu < \infty$$

for almost every $x \in A$, then the series $\sum_{n=1}^{\infty} \psi_n(x)$ converges almost everywhere on A and

$$\int_A \left(\sum_{n=1}^{\infty} \psi_n(x)\right) d\mu = \sum_{n=1}^{\infty} \int_A \psi_n(x)\, d\mu.$$

A4.4.4. Fubini's Theorem. *Let μ represent two- dimensional Lebesgue measure* (see **A3.4**) *and let μ_x, μ_y represent one-dimensional Lebesgue measure on, respectively, the X-axis and the Y-axis. Let $f(x,y)$ be a function integrable with respect to the measure μ on the square $[a,b] \times [a,b]$. Then*

$$\int_{[a,b]\times[a,b]} f(x,y)\, d\mu = \int_a^b \left(\int_a^b f(x,y)\, d\mu_y\right) d\mu_x = \int_a^b \left(\int_a^b f(x,y)\, d\mu_x\right) d\mu_y. \tag{A4.4.1}$$

In this situation, the inner integrals exist for almost all values of the variables and are themselves integrable with respect to the remaining variable.

On the other hand, if even one of the integrals

$$\int_a^b \left(\int_a^b |f(x,y)|\, d\mu_y\right) d\mu_x, \qquad \int_a^b \left(\int_a^b |f(x,y)|\, d\mu_x\right) d\mu_y$$

exist, then $f(x,y)$ is integrable on the square and (A4.4.1) holds.

A4.4.5. *For every function f summable on $[a,b]$ the function $F(x) = \int_a^x f(t)\, dt$ is a function of bounded variation and $F'(x) = f(x)$ almost everywhere on $[a,b]$.*

A proof of the following theorem can be found in [21], p. 461:

A4.4.6. *Let f be summable on $[a,b]$. Then given $\epsilon > 0$ there is a lower semi-continuous function $u(x)$ on $[a,b]$ (see* **A4.1**) *which satisfies the following properties:*

a) $u(x) > -\infty$ everywhere on $[a,b]$;

b) $u(x) \geq f(x)$ everywhere on $[a,b]$;

c) $u(x)$ is summable on $[a,b]$ and

$$\int_a^b u(x)\,dx < \epsilon + \int_a^b f(x)\,dx.$$

APPENDIX 5
NORMED LINEAR SPACES. HILBERT SPACES.

A5.1. A set X is called *linear* (or a *vector space*) if it satisfies the following conditions:

a) X is an *abelian group* (see **A1.1**);

b) there is a *scalar product* which takes a number α and an $x \in X$ to an element $\alpha x \in X$ such that

b1) $\alpha(\beta x) = (\alpha\beta)x$ for all numbers α, β and all $x \in X$;

b2) $\alpha(x+y) = \alpha x + \alpha y$, $(\alpha+\beta)x = \alpha x + \beta x$ for all numbers α, β and all $x, y \in X$;

b3) $1 \cdot x = x$ for all $x \in X$.

If the scalar product is defined for all complex scalars then the space is called a *complex linear space* and if it is defined for all real scalars then the space is called a *real linear space*.

A *normed linear space* is a linear space X together with a function $\|x\|$, called the *norm* of X, which takes the set X into the real line and satisfies the following conditions (the norm axioms):

I. $\|x\| \geq 0$ and $\|x\| = 0$ if and only if $x = 0$.

II. $\|\alpha x\| = |\alpha|\,\|x\|$.

III. $\|x+y\| \leq \|x\| + \|y\|$.

Every normed linear space is a metric space where the distance between two elements $x, y \in X$ is defined to be $\rho(x,y) = \|x-y\|$. Hence all the concepts which were introduced in **A2.1** for metric spaces also make sense for normed linear spaces, in particular, the idea of completeness. A complete normed linear space is called a *Banach space*.

A5.2. An important example of a Banach space is provided by the space of *continuous bounded functions* defined on some metric space X with the usual function addition and scalar multiplication of functions for the linear space structure and

$$\|f\| = \sup_{x \in X} |f(x)|$$

for the norm. This space is denoted by $\mathbf{C}(X)$.

Other examples include the various *integrability spaces of functions* with the integral norms. Let X be some space with a measure μ. The collection of all functions summable on X form a linear space under usual function addition and scalar multiplication of functions. For any $1 \leq p < \infty$ we consider the set of functions f measurable on X such that the power $|f|^p$ is summable. It is easy to see that the quantity

$$\|f\|_p \equiv \left\{ \int_X |f|^p \, d\mu \right\}^{1/p}$$

satisfies axioms II and III of a norm. However, axiom I can fail since from the definition of the Lebesgue integral it easily follows that changing a function on a set of measure zero does not change the value of its integral. Hence in order for axiom I to hold it is necessary that *functions which are equivalent to each other on* X (see **A4.1**) are considered *equal* as elements of these integrability spaces. In particular, we consider any function which is almost everywhere zero to be the zero element of these spaces. We shall let $\mathbf{L}^p(X)$ represent the space determined by the norm $||f||_p$.

In the case $p = \infty$, the norm is defined somewhat differently. Namely, let

$$||f||_\infty = \inf\{\alpha : |f(x)| \leq \alpha \text{ almost everywhere on } X\}$$

and define $\mathbf{L}^\infty(X)$ to be the space of equivalence classes of functions for which the quantity $||f||_\infty$ is finite. This quantity is a norm for this space.

We shall use these spaces only in the case when X is an interval of real numbers, either $[a, b]$ or the positive real axis $[0, \infty)$. Thus we shall denote these spaces by $\mathbf{L}^p[a, b]$ or $\mathbf{L}^p[0, \infty]$.

Here are some of the fundamental properties these spaces enjoy.

A5.2.1. *The space of continuous functions is dense in the spaces* $\mathbf{L}^p[a, b]$ *for any* $1 \leq p < \infty$. (See [2])

A5.2.2. Hölder's inequality. *If* $f \in \mathbf{L}^p[a, b]$, $p > 1$ *and* $g \in \mathbf{L}^q[a, b]$, *where* $\dfrac{1}{p} + \dfrac{1}{q} = 1$ *then the product* fg *is summable on* $[a, b]$ *and*

$$||fg||_1 \leq ||f||_p ||g||_q$$

(see [21], p. 214).

In the case when $p = q = 2$ this inequality is called the *Cauchy- Schwarz integral inequality* to distinguish it from the well-known *Cauchy-Schwarz inequality* for sums:

$$\sum_n |a_n b_n| \leq \left(\sum_n a_n^2\right)^{1/2} \left(\sum_n b_n^2\right)^{1/2}.$$

A5.2.3. For the spaces $\mathbf{L}^p[a, b]$, $p \geq 1$, axiom III of normed linear spaces (see **A5.1**) becomes

$$\left(\int_a^b |f(x) + g(x)|^p \, dx\right)^{1/p} \leq \left(\int_a^b |f(x)|^p \, dx\right)^{1/p} + \left(\int_a^b |g(x)|^p \, dx\right)^{1/p}$$

and is called the *Minkowski inequality*. (See [2].) There is a similar inequality for iterated integrals which is called the *generalized Minkowski inequality* (see [30], Volume I, p. 19). It is

$$\left(\int_a^b \left|\int_c^d f(x, y) \, dy\right|^p dx\right)^{1/p} \leq \int_c^d \left(\int_a^b |f(x, y)|^p \, dx\right)^{1/p} dy,$$

where the function $f(x, y)$ is summable on the rectangle $[a, b] \times [c, d]$ and $p \geq 1$.

A5.2.4. *If a sequence* $\{f_n(x)\}$ *converges in the* $\mathbf{L}[a, b]$ *norm to a function* $f(x)$ *then there is a subsequence* $\{f_{n_k}(x)\}$ *which converges to* $f(x)$ *almost everywhere.* See [17].)

A5.3. A *linear functional* on a linear space X is a real or complex-valued function $F(x)$ which satisfies $F(x + y) = F(x) + F(y)$ for all $x, y \in X$ and $F(\alpha x) = \alpha F(x)$ for all $x \in X$ and all numbers α.

For a normed linear space X we consider only those linear functionals $F(x)$ for which the following quantity is *finite*:

(A5.3.1) $$||F|| = \sup_{||x|| \leq 1} |F(x)|.$$

This quantity is called the norm of the linear functional F. Such functionals are called *continuous* (or *bounded*) since the quantity in (A5.3.1) is finite if and only if the function F is continuous on X (see [17]).

A5.3.1. *If $C > 0$ is a constant and F is a linear functional which satisfies*

$$|F(x)| \leq C\|x\|,$$

then the functional F is continuous and $\|F\| \leq C$. See [17], p. 176.)

A5.3.2. *For each fixed function $f \in \mathbf{L}^p[a,b]$ the map*

$$F(g) \equiv \int_a^b f(x)g(x)\,dx$$

is a linear functional on the space $\mathbf{L}^q[a,b]$ where $\dfrac{1}{p} + \dfrac{1}{q} = 1$. Moreover, $\|F\| = \|f\|_p$.

A5.3.3. The Banach-Steinhaus Theorem. *Suppose that $\{F_n\}$ is a sequence of linear functionals on some Banach space X which is bounded at each point $x \in X$. Then the sequence of norms of these functionals $\{\|F_n\|\}$ is also bounded.*

A5.4. A *Hilbert space* is a complex linear space H together with a map $\langle x, y\rangle$, called an *inner product*, which takes each pair of elements $x, y \in H$ to a complex number and satisfies the following conditions:

a) $\langle x, y\rangle = \overline{\langle y, x\rangle}$ (in particular, $\langle x, x\rangle$ is always a real number);
b) $\langle x_1 + x_2, y\rangle = \langle x_1, y\rangle + \langle x_2, y\rangle$;
c) $\langle \alpha x, y\rangle = \alpha\,\langle x, y\rangle$ for all complex numbers α;
d) $\langle x, x\rangle \geq 0$ and $\langle x, x\rangle = 0$ if and only if $x = 0$.

It is easy to see that the number $\|x\| \equiv \sqrt{\langle x, x\rangle}$ satisfies the *norm axioms* (see **A5.1**). It is also required that H be *complete* with respect to this norm, i.e., that H is in particular a *Banach space*.

Some authors also require that a Hilbert space be *infinite dimensional*.

If the given linear space is *real* then in a similar way we can define a *real Hilbert space*. In this case the scalar product is assumed to be real for all elements.

The space $\mathbf{L}^2(a,b)$ is an important example of a (real) Hilbert space and its inner product is given by

$$\langle f, g\rangle \equiv \int_a^b f(x)g(x)\,dx.$$

The Lebesgue integral of a complex-valued function $\phi = u + iv$ is defined by the identity

$$\int_a^b \phi(x)\,dx \equiv \int_a^b u(x)\,dx + i\int_a^b v(x)\,dx.$$

Using this integral we can define a *complex Hilbert space* $\mathbf{L}^2(a,b)$ by using the inner product

$$\langle f, g\rangle \equiv \int_a^b f(x)\overline{g(x)}\,dx.$$

A system $e_1, e_2, \ldots, e_n, \ldots$ of elements from a Hilbert space H is called *orthonormal* if

$$\langle e_i, e_j\rangle = \begin{cases} 1 & \text{if } i = j, \\ 0 & \text{if } i \neq j. \end{cases}$$

An orthonormal system $\{e_i\}$ is called *complete* if there is no non-zero element $x \in H$ which satisfies $\langle x, e_i\rangle = 0$ for all $i = 1, 2, \ldots$.

An orthonormal system $\{e_i\}$ is called *closed* if the set of all finite linear combinations of the elements of this system is dense in H.

5.4.1. *Let $\{e_i\}$ be an orthonormal system. Then $\{e_i\}$ is complete if and only if it is closed.* (See [15].)

For any element $x \in H$ the numbers $c_i = \langle x, e_i \rangle$ are called the *Fourier coefficients* of the element x with respect to the system $\{e_i\}$ and the series $\sum_{i=1}^{\infty} c_i e_i$ is called the *Fourier series* of the element x in this system.

A5.4.2. *Let $\{e_i\}$ be a complete orthonormal system in a Hilbert space H. Then the Fourier series of any element $x \in H$ converges to it in the norm of H and its Fourier coefficients satisfy Parseval's identity:*

$$\sum_{i=1}^{\infty} c_i^2 = \|x\|^2. \tag{A5.4.1}$$

(See [17]).

A5.4.3. The Riesz-Fischer Theorem. *Let $\{e_i\}$ be any orthonormal system in a Hilbert space H and let $c_1, c_2, \ldots$ be any sequence of numbers such that the series $\sum_{i=1}^{\infty} c_i^2$ converges. Then there is an element $x \in H$ such that the sequence $\{c_i\}$ is the Fourier coefficients of x, i.e., $c_i = \langle x, e_i \rangle$. Moreover, if $\{e_i\}$ is complete, then Parseval's identity (A5.4.1) holds.* (See [17].)

The Hausdorff-Young-F. Riesz Theorem. *Let $\{\phi_n(t)\}$ be a uniformly bounded, orthonormal system in the space $\mathbf{L}^2(a,b)$, i.e., an orthonormal system which satisfies $|\phi_n(t)| \leq M$ for all $t \in [a,b]$, $n = 1, 2, \ldots,$ and some fixed constant M. If $\dfrac{1}{p} + \dfrac{1}{q} = 1$ and $1 < p < 2$ hold, then the following results are true:*

1. *If $f \in \mathbf{L}^p(a,b)$ then the Fourier coefficients $c_n = \int_a^b f(t)\overline{\phi_n(t)}\,dt$ of the function f with respect to he system $\{\phi_n\}$ satisfy the condition*

$$\|c\|_q \equiv \left(\sum_{n=1}^{\infty} |c_n|^q\right)^{1/q} \leq M^{(2/p)-1}\|f\|_p.$$

2. *If a sequence of numbers $\{c_n\}$ satisfies the condition*

$$\|c\|_p \equiv \left(\sum_{n=1}^{\infty} |c_n|^p\right)^{1/p} < \infty$$

then there exists a function $f(t) \in \mathbf{L}^q(a,b)$ such that $\{c_n\}$ are its Fourier coefficients and

$$\|f\|_q \leq M^{(2/p)-1}\|f\|_p.$$

(See [2], Vol. I, p. 218 .)

A5.6. Mercer's Theorem. *If $\{\phi_n(t)\}$ is a uniformly bounded, orthonormal system in the space $\mathbf{L}^2[a,b]$ then the Fourier coefficients of any summable function with respect to this system converge to zero.* (See [2], Vol. I, p. 66 .)

We notice that Theorem **2.7.3** is a special case of this general theorem.

COMMENTARY

In this commentary we include a very brief historical discussion of the themes we have touched upon, some references to scholarly articles from which proofs were adapted which we used in this book, and supplemental information for some of the theorems. For further information about these topics we refer the reader to the articles of Balašov and Rubinšteĭn [1], Wade [4], and the survey found in the monograph [1].

Chapter 1

The Walsh functions were first defined by J.L. Walsh [1]. The enumeration we used here was given by Paley [1]. For other enumerations see Balašov and Rubinšteĭn [1] and Schipp [3]. The idea of viewing the Walsh system as a special case of systems of characters on zero-dimensional groups was first introduced by Vilenkin [1], and somewhat later by Fine [1]. For more about the topological structure of the group G, on which the Walsh functions are defined and continuous, see [1] and [23]. The concept of the modified interval first appeared in the work of Šneĭder [3]. The connection between the Walsh system and the Haar system (see §1.3) was first noticed by Kaczmarz [1]; see also [15]. Estimate (1.4.16) for the Walsh-Dirichlet kernel was obtained by Šneĭder [1] and Fine [1]. The multiplicative systems introduced in §1.5 are a special case of a more general collection of systems which Vilenkin [1] considered. For the case $p_j = p \neq 2$, where $\{p_j\}$ is the sequence (1.5.1), these generalized Rademacher and Walsh functions can be found in the work of Levy [1] and Chrestenson [1]. The systems defined in (1.5.6) and (1.5.10) were considered by Price [1] and Efimov [1].

The Walsh system with continuous index set was introduced by Fine [2]. The groups $G(\mathcal{P})$ introduced in §1.5 are examples of locally compact zero-dimensional groups whose characters were studied by Vilenkin [3] in the general case; see also [1]. Extension of the system of Price to a continuous index set, i.e., the functions (1.5.35), were considered by Efimov and Karakulin [1], under condition (1.5.35), and in the general case by Pospelov [2]. He also considered the functions (1.5.27) on the group $G(\mathcal{P})$. See also Efimov and Pospelov [1]. Relationship (1.5.38) and Proposition **1.5.6** are due to Bespalov.

Chapter 2

Propositions **2.1.1**, **2.1.2** and formula (2.1.9) hold in general for all Fourier transforms on character groups (see [1], [24]). The Lebesgue constants for the Walsh system were in fact estimated by Paley [1]. They were studied further by Fine [1] and Šneĭder [3]. The analogue of the Dini- Lipschitz Theorem for the Walsh system (see **2.3.6**) was proved by Fine [1]. The modulus of continuity (2.3.5) was introduced in the general case of a zero-dimensional group by Vilenkin [1], and the modulus (2.5.1) for the space $\mathbf{L}^p$ by Morgenthaler [1]. Theorem **2.4.1** was established by Onneweer [1]. The analogue of Dini's test **2.5.7** was proved by Fine [1]. Theorem **2.5.12** can be found in the work of Kaczmarz [1], and convergence of the sums $S_{2^k}(x,f)$ at points of continuity of the function f was noticed already in the work of Walsh [1]. For further information concerning completeness and closure of the Walsh system see the more detailed account in [15]. Elementary estimates of Walsh-Fourier coefficients were obtained by Vilenkin [1] and Fine [1]. Concerning **2.7.6** through **2.7.8** see Yoneda [1], also Bljumin and Kotljar [1], McLaughlin [1], Vilenkin and Rubinšteĭn [1]. Theorem **2.7.9** was proved by Fine [1]. We notice that Bočkarev [1] established the finiteness of condition (2.7.11) in Theorem **2.7.10**. Concerning Theorem **2.7.10** see also Vilenkin and Rubinšteĭn [1]. In §2.8 we have given only a very elementary introduction to Fourier series with respect to multiplicative systems. For more details see [1].

Chapter 3

The term by term integral $\psi(x)$ of a Walsh series was used by Fine [1] to study properties of general Walsh series. This series plays the same role for Walsh series that the Riemann functions plays for the study of general trigonometric series (see [2], Vol. I, p. 192).

Propositions **3.1.2** and **3.1.3** can be obtained for Walsh series from corresponding results from the theory of Haar series (see Skvorcov [1]). The idea of using upper and lower derivatives through nets and continuity through nets also is borrowed from the theory of Haar series (see Skvorcov [1]-[3]). The generalized Riemann-Stieltjes integral which was introduced in §3.1 is a very special case of the Kolmogorov integral (see Kolmogorov [1]).

The study of uniqueness for Walsh series is heavily influenced by the well developed theory of uniqueness of trigonometric series (see [2], Chapter XIV). A simple form of uniqueness for Walsh series, namely **3.2.6** in the case that the excluded set E is empty and the limit function $f(x) = 0$ everywhere, was proved by Vilenkin [1]. In the case when $f(x) \in \mathbf{L}[0,1)$ and E is countable, Theorem **3.2.6** was established by Fine [1]. In the case when E is a countable set, it was obtained by Arutunjan and Talaljan [1] as a corollary of an analogous theorem for Haar series, and (by another method) by Crittenden and Shapiro [1]. The proof given here and the more general one for **3.2.5** are due to Skvorcov [4]. For other results concerning uniqueness see Arutunjan [1], Skvorcov [4], [7], [8] and Wade [1], [3], and [5].

The formal product introduced in §3.3 was first mentioned by Šneĭder [2]. He also proved Theorem **3.3.2**. Concerning **3.3.3** see Skvorcov [5]. The first example of a null series in the Walsh system and the first example of an M- set of measure zero were constructed by Šneĭder [2] and Coury [1]. The first example of a perfect M-set of measure zero is contained in the work of Skvorcov [5]. The example which was constructed for the proof of **3.4.2** is a simplified version of an example found in the work of Skvorcov [6]. An uncountable U-set for the Walsh system was constructed by Šneĭder [2]. For further results concerning U-sets and M-sets for the Walsh system see Skvorcov [9], [11], Wade [2], [4], and the survey in the book [1].

Chapter 4

More details about linear methods of summation can be found in [8].

The kernel $K_n(x)$ for the $(C,1)$ method was studies by Fine [1]. In particular, he discovered formula (4.2.10). Formula (4.2.6) and Theorem **4.2.2** were proved by Yano [1], [2]. The proof of the theorems in §4.3 can be found in the article of Skvorcov [10] which deals with more general questions. Theorem **4.3.2** about uniform $(C,1)$ summability of the Walsh- Fourier series of a continuous function was proved by Fine [1] and earlier for character systems on zero-dimensional groups by Vilenkin [1]. Theorem **4.3.3**, for $0 < \alpha < 1$, was proved by Yano [2], and for $\alpha = 1$, by Jastrebova [1]. For analogous results in the multiplicative case see Efimov [1], Bljumin [1], [2]. Theorem **4.4.5** was proved by Fine [3]. Theorem **4.4.4** was formulated without proof in the work of Fine [4] and the complete proof, as well as that of **4.4.3**, appears here for the first time.

Chapter 5

The results about sublinear operators which appear in §5.1, in particular Theorem **5.1.2**, appear in a more general form in the monograph [30], Vol. II, Chapter 12. The dyadic maximal operator (5.2.1) is a special case of maximal operators with respect to differential bases (see [7]). The modified Dirichlet kernel (5.3.1) is in some sense analogous to the conjugate Dirichlet kernel for the trigonometric case. It was introduced and used by Billard [1]. The part of Theorem **5.3.2** which deals with strong type inequalities for the operators S_n was proved by Paley [1]. That weak type inequalities were proved by Watari [1]. For analogous results for multiplicative systems see Young [1].

CHAPTER 6

The Fourier transform for generalized Walsh functions (6.1.1) (which in this case also coincides with (6.1.2)) was introduced by Fine [2], and for locally compact, zero-dimensional abelian groups, by Vilenkin [3].

For symmetric $\mathcal{P}$, i.e., the case when $p_{-j} = p_j$, $j = 1, 2, \ldots$, Propositions **6.1.3**, **6.1.4**, **6.1.5**, **6.2.2**, **6.2.3**, **6.2.4**, and **6.2.9** were formulated in the work of Efimov and Zolotareva [1]. Propositions **6.1.7**, **6.3.1**, **6.3.2** were proved by Bespalov [1]. We notice that using the Plancherel identity **6.2.4**, Proposition **6.1.7** can easily be established using the interpolation theorem of Riesz- Thorin (see [13], pp. 227-229). Propositions **6.2.5** and **6.2.6** were proved in the work of Efimov and Zolotareva [2], Propositions **6.2.10**, **6.2.11**, **6.2.13** were proved by Zolotareva [1], and Propositions **6.2.12**, **6.2.14**, **6.2.15** were proved by Bespalov [3].

CHAPTER 7

Theorem **7.1.1** about uniform convergence on each interval $(\delta, 1)$, $0 < \delta < 1$, of Walsh series with coefficients which decrease monotonically to zero evidently was first proved in the work of Šneĭder [1]. The theorems about integrability, in some sense, of the sum of a Walsh series (including the A-integral and the various classes $\mathbf{L}^p(0,1)$, $0 < p < 1$) were proved in the work of Rubinšteĭn [1]. He also constructed the first example of a Walsh series with coefficients which decrease monotonically to zero whose sum is not Lebesgue integrable on $(0, 1)$. An analogue of Theorem **7.1.3** for the Kaczmarz enumeration was proved by Balašov [1] and Theorem **7.1.4** is due to Moricz [1].

The main result of §7.2 (Theorem **7.2.4**) is due to Yano [1].

The main result of §7.3 (Theorem **7.3.6**) in the general case of character systems is due to Timan (see Timan and Rubinšteĭn [1]). For the Walsh system this was established by Moricz [1].

CHAPTER 8

Theorem **8.1.2** was proved by Rademacher [1], **8.1.3** by Kolmogorov, **8.1.4** and **8.1.5** by Khintchin (see Kolmogorov and Khintchin [1]). We notice that Haagerup [1] found exact values for the constants A_p and B_p in Khintchin's inequality

$$A_p \left(\sum_{k=0}^{n} a_k^2 \right)^{1/2} \le \left(\int_0^1 \left| \sum_{k=0}^{n} a_k r_k(x) \right|^p dx \right)^{1/p} \le B_p \left(\sum_{k=0}^{n} a_k^2 \right)^{1/2}$$

for $0 < p < \infty$.

The results in §§8.2 and 8.3 were proved by Morgenthaler [1], and the special case of Theorem **8.2.5** when $n_k = 2^k$ (i.e., for the Rademacher system) was proved by Kaczmarz and Steinhaus [1].

For more concerning general results about lacunary series see the survey articles of Gapoškin [1] and Wade [4].

CHAPTER 9

Theorem **9.1.2** is a Walsh analogue of a theorem of Kolmogorov, which shows that there exist everywhere divergent trigonometric Fourier series (see [2], Vol. I, pp. 455-464). The example introduced in §9.1 is a simplified version of a construction of Moon [1] which he used for a more general result. The first examples of everywhere divergent Walsh- Fourier series were constructed by Schipp [1], [2], see also Heladze [1]. These examples were preceded by an example of an almost everywhere divergent Walsh-Fourier series (see Stein [1]). For multiplicative systems, examples of everywhere divergent Fourier series were constructed by Heladze [2], in the case when the sequence

$\{p_i\}$ from (1.5.1) is bounded, and by Simon [1], in the general case. For other generalizations of Kolmogorov's theorem see Bočkarev [1] and Ul'janov [2].

Theorem **9.2.1** is due to Billard [1], who transferred the methods of Carleson [1] from the trigonometric case to the Walsh case. The proof presented in §9.2 has been adapted from Hunt [1]. Concerning more general results see Sjölin [1]. For multiplicative systems under the condition that $p_i = p$ for all i, where $\{p_i\}$ is the sequence from (1.5.1), Theorem **9.2.1** appeared in Hunt and Taibleson [1], and for the case when the sequence $\{p_i\}$ is bounded, in Gosselin [1].

Chapter 10

The results in §10.1 are due to Golubov [3]. However, inequality

$$E_n(f)_h \leq 12\omega(1/n, f), \qquad n = 1, 2, \ldots,$$

which is less exact than the left side of inequality (10.1.10), can be obtained as a corollary of a theorem of Nagy [1]. Necessity of the condition in Theorem **10.1.4** can also be obtained as a corollary of a result of Nagy [1]. We notice that the left most inequality in (10.1.10) is exact in the following sense, that

$$E_n(f)_h \leq C\omega(1/n, f)$$

cannot hold for all continuous functions f and all natural numbers n if $C < 1$.

Theorems **10.2.1** through **10.2.6** were proved by Ul'janov [1], and Theorem **10.2.7** by Haar [1]. Theorems **10.2.8-10.2.10** were established by Golubov [3].

Propositions **10.3.1-10.3.3** were proved by Ul'janov [1], and **10.3.4-10.3.11** by Golubov [3].

All the results in §10.4, except **10.4.7**, **10.4.8**, and **10.4.15**, are due to Ul'janov [1]. Propositions **10.4.7** and **10.4.8** are well-known (see, for example, the book of Zygmund [30], Vol. I, pp. 29-31), and **10.4.15** was established by Golubov [3]. For character systems see also Timan and Rubinšteĭn [1].

Proposition **10.5.1**, in the case of approximation by Walsh functions, was established by Watari [2], and in the general case by Efimov [1]. Proposition **10.5.2** was formulated in the work of Efimov [5].

Chapter 11

The proofs of **11.1.1** and **11.1.2** and also the idea behind the proof of **11.1.5** can be found in the work of Efimov and Zolotareva [2]. Propositions **11.1.7**, **11.1.8**, and **11.2.3** appear here for the first time. Theorems **11.2.1** and **11.2.2** were formulated in the work of Efimov and Karakulin [1]. The proof of **11.2.4** was given by Žukov [1], although the representation of the matrix W as a product of matrices of the same type and the passage for calculations to the new basis first appeared in the work of Efimov and Kanygin [1]. Estimates of the spectral coefficients in §11.3 were given by Bespalov [3], [4], but various cases of these estimates are contained in the work of Kanygin [1]. The results of §11.4 are due to Žukov [2], the theorems in §11.5 were formulated in the work of Pospelov [2] (see also Efimov and Pospelov [1]), but the proofs of these theorems, courtesy of A.S. Pospelov, appear here for the first time.

Chapter 12

The foundations of digital filtering are given in an article by Kaiser [1], in the book [6], and the monograph [12]. Construction of digital filters based on the Walsh functions and general multiplicative transforms were considered in the work of Robinson and Grander [1], Harmuth [1] (see also the monographs [9] and [10], the work of Good [1], Tatachar and Prabhakar [1], and Tubol'cev [1], [2]). The algorithm for digital filtering presented here is the information algorithm of Tubol'cev.

Commentary to §12.2 can be found in the text of this section.

The results from §12.3 are due to Lesin [1] (see also Efimov and Lesin [1]).

REFERENCES

Note: For articles and books in Russian, the titles have been translated into English.

I. MONOGRAPHS AND TEXTBOOKS

1. G.N. Agaev, N.Ja. Vilenkin, G.M. Dzafarli, and A.I. Rubinštein, "Multiplicative systems and harmonic analysis on zero- dimensional groups," Baku "ELM", 1981.
2. N.K. Bary, "A Treatise on Trigonometric Series," Pergamon Press, London, 1964.
3. M.M. Džrbašjan, "Integral Transforms and Representation of Functions in the Complex Plane," Moscow "Nauka", 1966.
4. B. F. Fedorov and R.I. El'man, "Digital Holography," Moscow "Nauka", 1976.
5. B. V. Gnenenko, "A Course on the Theory of Probability," Moscow "Nauka", 1969.
6. B. Gol'd and Č Reĭder, "Digital Signal Processing," Moscow "Sov. Radio", 1973.
7. M. Gusman, "Differentiability of Integrals in R^n," Moscow "Mir", 1978.
8. G.H. Hardy, "Divergent Series," Oxford, 1949.
9. H.F. Harmuth, "Transmission of Information by Orthogonal Functions," Springer-Verlag, Berlin, 1972.
10. H.F. Harmuth, "Sequency Theory," Academic Press, New York, N.Y., 1977.
11. R.V. Hemming, "Numerical Methods for Technicians and Engineers," Moscow "Nauka", 1968.
12. R.V. Hemming, "Digital Filters," Moscow "Sov. Radio", 1980.
13. E. Hewitt and K. Ross, "Abstract Harmonic Analysis I," Springer-Verlag, Heidelberg, 1963.
14. L.P. Jaroslavskiĭand H.S. Merzljakov, "Methods of Digital Holography," Moscow "Nauka", 1977.
15. S. Kaczmarz and G. Steinhaus, "Theorie der Orthogonalreihen," Monogr. Mat. Vol. 6, Warsaw, 1935.
16. E. Kamke, "The Lebesgue-Stieltjes Integral," Moscow "Fizmatgiz", 1959.
17. A.N. Kolmogorov and S.V. Fomin, "Elementary Theory of Functions and Functional Analysis," Moscow "Nauka", 1981.
18. A.G. Kuroš, "A Course in Higher Algebra," Moscow "Nauka", 1975.
19. O.A. Ladyženskaja, V.A. Solonnikov, and N.N. Ural'ceva, "Linear and Quasi-linear Equations of Parabolic Type," Moscow "Nauka", 1967.
20. M. Loève, "Probability Theory," Van Nostrand, Princeton, 1963.
21. I.P. Natanson, "Theory of Functions of a Real Variable," 3rd Edition, Moscow "Nauka", 1974.
22. R.E.A.C. Paley and N. Wiener, "Fourier Transforms in the Complex Domain," Amer. Math. Soc., Colloquium Publications, Vol. XIX, 1934.
23. L.S. Pontryagin, "Topological Groups," Princeton Univ. Press, Princeton, N.J., 1946.
24. W. Rudin, "Fourier Analysis on Groups," Interscience Pub., Wiley and Sons, 1967.
25. S. Saks, "Theory of the Integral," Dover Publications, New York, N.Y., 1964.
26. M.H. Taibleson, "Fourier Analysis on Local Fields," Princeton Univ. Press, Princeton, N.J., 1975.
27. A.M. Trahtman and V.A. Trahtman, "Foundations of the Theory of Discrete Signals on Finite Intervals," Moscow "Sov. Radio", 1975.
28. V. S. Vladimirov, "Equations of Mathematical Physics," 3rd Edition, Moscow "Nauka", 1981.
29. E.G. Zelkin, "Construction of Radiation Systems by a Given Diagram of Directions," Moscow "GEI", 1963.
30. A. Zygmund, "Trigonometric Series," Cambridge University Press, New York, N.Y., Vols. I, II, 1959.

II. JOURNAL ARTICLES

F.G. Arutunjan. 1. *Recovery of coefficients of Haar and Walsh series which converge to functions which are Denjoy integrable*, Izv. Akad. Nauk SSSR **30** (1966), 325- -344.

F.G. Arutunjan and A.A. Talaljan. 1. *On uniqueness of Haar and Walsh series*, Izv. Akad. Nauk SSSR **28** (1964), 1391–1408.

L.A. Balašov. 1. *Series with respect to the Walsh system with monotone coefficients*, Sibirsk. Mat. Ž. **12** (1971), 25–39.

L.A. Balašov and A.I. Rubinšteĭn. 1. *Series with respect to the Walsh system and their generalizations*, J. Soviet Math. 1 (1973), 727–763.

M.S. Bespalov. 1. *Multiplicative Fourier transforms in L^p*, in "Itogi Nauki. Mat. Anal.," Bce. Inst. Nauč. Tehn. Inf., No. 100-82, Moscow, 1981.

2. *On Parseval's identity and approximation of functions by Fourier series in multiplicative systems*, in "Applications of Functional Analysis to Approximation Theory," Kalinin: KGU, 1982, pp. 28–42.
3. *Compression of information using discrete multiplicative Fourier transforms*, Sbor. Trud. MIET, Moscow (1982), 91–97.
4. *Multiplicative Fourier transforms*, in "Theory of Functions and Approximation," Proc. Saratov. Zim. Šk., Jan. 24 - Feb. 5, Sarat. Inst., 1982, pp. 39–42.

P. Billard. 1. *Sur la convergence presque partout des series se Fourier-Walsh des fonctions de éspace* $L^2(0,1)$, Studia Math. **28** (1967), 363–388.

S.L. Bljumin. 1. *Linear methods of summation of Fourier series with respect to multiplicative systems*, Sibirsk. Mat. Ž. **9** (1968), 449–455.

2. *Certain properties of a class of multiplicative systems and problems of approximation of functions by polynomials with respect to those systems*, Izv. Vuzov Mat., No. 4 (1968), 13–22.

S.L. Bljumin and B.D. Kotljar. 1. *Hilbert-Schmidt operators and absolute convergence of Fourier series*, Izv. Akad. Nauk SSSR, Ser. Mat. **34** (1970), 209–217.

S.V. Bočkarev. 1. "A method of averaging in the theory of orthogonal series and some problems in the theory of bases," Trudy Mat. Inst. Steklov., No. 146, 1978.

2. *The Walsh Fourier coefficients*, Izv. Akad. SSSR, Ser. Mat. **34** (1970), 203–208.

L. Carleson. 1. *On convergence and growth of partial sums of Fourier series*, Acta Math. **116** (1966), 135–157.

H.E. Chrestenson. 1. *A class of generalized Walsh functions*, Pacific J. Math. **5** (1955), 17–32.

J.E. Coury. 1. *A class of Walsh M-sets of measure zero*, J. Math. Anal. Appl. **31** (1970), 318–320.

2. *A class of Walsh M-sets of measure zero (Corrigendum)*, J. Math. Anal. Appl. **52** (1975), 669–671.

R.B. Crittenden and V.L. Shapiro. 1. *Sets of uniqueness on the group 2^ω*, Annals of Math. **81** (1965), 550–564.

A.V. Efimov. 1. *On some approximation properties of periodic multiplicative orthonormal systems*, Mat. Sbornik **69** (1966), 354–370.

2. *On upper bounds for Walsh-Fourier coefficients*, Mat. Zametki **6** (1969), 725–730.
3. *On summability of the discrete Fourier transform by the Vallee-Pousin method*, in "Proc. Symp. Theory Approx. Funct. Complex Plane," Ufa. "BF", Akad. Nauk SSSR, 1976, p. 33.
4. *On approximation properties of discrete Fourier transforms*, in "Methods of Digital Information Processing," Sbornik Nauč Trud., Moscow "MIET", 1980, pp. 33–45.
5. *Approximation properties of general multiplicative systems*, in "Proc. Inter. Congress Math.," VII, Sect. 9, Part II, Warsaw, 1983, p. 49.

A.V. Efimov and V.S. Kanygin. 1. *An algorithm to compute the discrete transform of Hadamard type*, in "Sbornik Nauč. Trud. Prob. Microelect.," Ser. Fiz.- Mat., Moscow "MIET", 1976, pp. 161–174.

A.V. Efimov and A.F. Karakulin. 1. *A continual analogue of periodic multiplicative orthonormal systems*, Dokl. Akad. Nauk SSSR **218** (1974), 268– 271.

A.V. Efimov and V.V. Lesin. 1. *A solution to an integral equation using the Chrestenson-Levy functions*, in "Applications of the Methods of the Theory of Functions and Functional Analysis to Problems of Mathematical Physics," Trud. Sov. Novosibirsk, 1979, pp. 44–47.

A.V. Efimov and A.S. Pospelov. 1. *Integral transforms in connection with locally compact groups*, in "Proc. Internat. Conf. Theory Approx. Funct.," Kiev, 1983.

A.V. Efimov and S. Ju. Zolotareva. 1. *The multiplicative Fourier integral and some of its applications,* Dokl. Akad. Nauk SSSR **242** (1978), 517–520.
2. *The multiplicative Fourier integral and its discrete analogues,* Analysis Math. **5** (1979), 179–199.

N.J. Fine. 1. *On the Walsh functions,* Trans. Amer. Math. Soc. **65** (1949), 372–414.
2. *The generalized Walsh functions,* Trans. Amer. Math. Soc. **69** (1950), 66–77.
3. *Cesàro summability of Walsh-Fourier series,* Proc. Nat. Acad. Sci. USA **41** (1955), 588–591.
4. *Fourier-Stieltjes series of Walsh functions,* Trans. Amer. Math. Soc. **86** (1957), 246 –255.

V.F. Gapoškin. 1. *Lacunary series and independent functions,* Uspehi Mat. Nauk **21** (1966), 3–82.

S. Ju. Golova. 1. *Discrete multiplicative Fourier transforms and estimates of its spectral density,* in "Methods of Digital Processing of Images," Moscow "MIET", 1982, pp. 45–56.

B. Golubov. 1. *Approximation of functions in the L^p metric by Haar and Walsh polynomials,* Mat. Sbornik **87** (1972), 254–274.
2. *Series in the Haar system,* J. Soviet Math. **1** (1973), 704–726.
3. *Best approximations to functions in the L^p norm by Haar and Walsh polynomials,* Mat. Sbornik **87** (1972), 254–274.

J. Gosselin. 1. *Almost everywhere convergence of Vilenkin- Fourier series,* Trans. Amer. Math. Soc. **185** (1973), 345–370.

J. Good. 1. *The interaction algorithm and practical Fourier analysis,* J. Royal Stat. Soc. (London) **B-20** (1958), 361–372.

V. Haagerup. 1. *Les melleurs constantes de l'inegalite de Khinchine,* Comp. Rend. Acad. Sci. (1978), A259–262.

A. Haar. 1. *Zur Theorie der orthogonalen Funktionen systeme,* Mat. Ann. **69** (1910), 331–371.

H.F. Harmuth. 1. *Sequency filter based on Walsh functions,* IEEE Trans. Electromag. Compatibility **10** (1968), 293–295.

Š.V. Heladze. 1. *Divergence everywhere of Walsh-Fourier series,* Soobšč. Akad. Nauk Gruz. SSR **77** (1975), 305–307.
2. *Divergence everywhere of Fourier series in bounded Vilenkin systems,* Sakharth. SSR Mecn. Akad. Math. Inst. Šrom **58** (1978), 224–242.

R.A. Hunt. 1. *Almost everywhere convergence of Walsh-Fourier series of L^2 functions,* Actes Congr. Int. Math., 1970, Paris, 655–661.

R.A. Hunt and M.H. Taibleson. 1. *Almost everywhere convergence of Fourier series on the ring of integers of a local field,* SIAM J. Math. Anal. **2** (1971), 607–624.

S. Kaczmarz. 1. *Über ein Orthogonal System,* Comp. Rend. Congres Math. (Warsaw, 1929).

S. Kaczmarz. 1. *Le system orthogonal de M. Rademacher,* Studia Math. **2** (1930), 231–247.

J.F. Kaiser. 1. *Digital Filters,* in "Digital Signal Processing," B. Gold and Č Reĭder, Editors, Moscow "Sov. Radio", 1973.

V.S. Kanygin. 1. *Questions of approximation by discrete transforms in finite-dimensional vector spaces,* in "Applications of Functional Analysis to Approximation Theory," Kalinin "KGY", 1975, pp. 37–44.

A.M. Kolmogorov. 1. *Untersuchungen über Integralbegriff,* Math. Ann. **103** (1930), 654–696.

A.M. Kolmogorov and A.Ja. Khintchin. 1. *Konvergenz der Reihen deren Glieder durch den Zufall bestimmt werden,* Mat. Sbornik **32** (1926), 668–677.

V.A. Kotel'nikov. 1. *On the conducting capacity of "ether" and wires in electronic connections,* in "Materials in the All-Union Congress for Questions of Reconstruction Work Connected With the Development of Industry," Moscow "RKKA", 1933.

P. Levy. 1. *Sur une generalisation des fonctions orthogonales de M. Rademacher*, Comment. Math. Helv. **16** (1944), 146–152.

V.V. Lesin. 1. *On an analogue of systems of reference functions corresponding to the Walsh functions*, Mat. Zametki **21** (1977), 485–493.

Ju.P. Lisovec and A.S. Pospelov. 1. *Holographic Walsh transforms*, in "Applications of orthogonal methods to signal processing and systems analysis," Mežbuz. Sbornik, Sverdlovsk "UPI", 1981, pp. 69–73.

2. *Multiplicative holographic transformations for image processing*, in "Methods of digital image processing," Sbornik Nauč. Trud., Moscow "MIET", 1982, pp. 100–109.

J.R. McLaughlin. 1. *Absolute convergence of series of Fourier coefficients*, Trans. Amer. Math. Soc. **184** (1973), 291–316.

K.H. Moon. 1. *An everywhere divergent Fourier-Walsh series of the class* $L(\log^+ \log^+ L)^{1-\epsilon}$, Proc. Amer. Math. Soc. **50** (1975), 309–314.

G.W. Morgenthaler. 1. *Walsh-Fourier series*, Trans. Amer. Math. Soc. **84** (1957), 472–507.

F. Moricz. 1. *On Walsh series with coefficients tending monotonically to zero*, Acta Math. Acad. Sci. Hungar. **38** (1981), 183–189.

C.W. Onneweer. 1. *On uniform convergence of Walsh-Fourier series*, Pacif. J. Math. **34** (1970), 117–122.

R.E.A.C. Paley. 1. *A remarkable system of orthogonal functions*, Proc. Lond. Math. Soc. **34** (1932), 241–279.

A.S. Pospelov. 1. *On a class of factorable matrices*, Abstract, International Congress Math., Warsaw, 1982.

2. *On some properties of P-adic derivatives*, All-Union School on the Theory of Functions in honor of the 100th anniversary of the birth of the academic N.N. Lusin, KemerovskiĭGU, Kemerovo.

J.J. Price. 1. *Certain groups of orthonormal step functions*, Canad. J. Math **9** (1957), 413–425.

H.A. Rademacher. 1. *Einige Sätze Über Reihen von allgemeinen Orthogonalenfunktionen*, Math. Annalen **87** (1922), 112–138.

G.S. Robinson and R. Grander. 1. *A design procedure for nonrecursive digital filters on Walsh functions*, in "Proc. Sympos. Appl. Walsh Functions," Washington, D.C., 1971, pp. 95–98.

A.I. Rubinštein. 1. *The A integral and series with respect to the Walsh system*, Usp. Mat. Nauk **18** (1963), 191–197.

2. *On the moduli of continuity of functions defined on zero-dimensional groups*, Mat. Zametki **23** (1978), 379–388.

3. *On moduli of continuity and best approximations in* $\mathbf{L}^p$ *of functions which are represented by lacunary Walsh series*, Izv. Vuzov. Mat. (1983), 61–68.

F. Schipp. 1. *Über die Größenordnung der Partialsummen der Entwiklung integrierbarer Functionen nach W-systemen*, Acta Sci. Math. (Szeged) **28** (1967), 123–134.

2. *Über die Divergenz der Walsh-Fourierreihen*, Annales Univ. Sci. Budapest Eötvös, Sect. Math. **12** (1969), 49–62.

3. *Certain rearrangements of series in the Walsh series*, Mat. Zametki **18** (1975), 193–201.

P. Simon. 1. *On the divergence of Vilenkin-Fourier series*, Acta Math. Acad. Sci. Hungar. **41** (1983), 359–370.

P. Sjölin. 1. *An inequality of Paley and convergence a.e. of Walsh-Fourier series*, Arkiv för Math. **8** (1969), 551–570.

V.A. Skvorcov. 1. *Differentiation with respect to nets and the Haar series*, Mat. Zametki **4** (1968), 33–40.

2. *Estimating coefficients of everywhere convergent Haar series*, Mat. Sbornik **75** (1968), 349–360.
3. *Haar series with convergent subsequences of partial sums*, Dokl. Akad. Nauk SSSR **183** (1968), 784–786.
4. *Some generalizations of uniqueness theorem for Walsh series*, Mat. Zametki **13** (1973), 367–372.
5. *An example of a Walsh series with a subsequence of partial sums that converges to zero everywhere*, Mat. Sbornik (N.S.) **97** (1975), 517–539.
6. *An example of a zero-series in the Walsh system*, Mat. Zametki **19** (1976), 179–186.
7. *Uniqueness theorems for Walsh series that are summable by the $(C,1)$ method*, Vestnik Mosk. Univ. Ser. Mat. Meh. **31** (1976), 73–80.
8. *Uniqueness conditions for the representation of functions by Walsh series*, Mat. Zametki **21** (1977), 187–197.
9. *The h-measure of M-sets for the Walsh systems*, Mat. Zametki **21** (1977), 335–340.
10. *On Fourier series with respect to the Walsh- Kaczmarz system*, Analysis Math. **7** (1981), 141–150.
11. *An example of a U-set for the Walsh system*, Vestnik Mosk. Univ. Ser. Mat. Meh. (1982), 53–55.

A.A. Šneider. 1. *On series with respect to Walsh functions with monotone coefficients*, Izv. Akad. Nauk SSSR, Ser. Mat. **12** (1948), 179–192.

2. *Uniqueness of expansions with respect to the Walsh system of functions*, Mat. Sbornik **24** (1949), 279–300.
3. *Convergence of Fourier series with respect to the Walsh functions*, Mat. Sbornik **34** (1954), 441–472.

E.M. Stein. 1. *On the limits of sequences operators*, Annals Math. (2) **74** (1961), 140–170.

B. Szökefalvi-Nagy. 1. *Approximation properties of orthogonal expansions*, Acta Sci. Math. **15** (1953/54), 31–37.

M. Tatachar and J.C. Prabhakar. 1. *Design of digital Walsh filters*, Proc. Regional IEEE Conf. Elect. Engineering, New York, N.Y..

M.F. Timan and A.I. Rubinšteĭn. 1. *On embeddings of classes of functions defined on zero- dimensional groups*, Izv. Vuzov. Mat. (1980), 66–76.

M.F. Tubol'cev. 1. *Fast filtering algorithms in the basis of periodic multiplicative functions*, in "Itogi Nauki. Mat. Anal.," Bce. Inst. Nauč. Tehn. Inf., No. 2020, Moscow, 1981, pp..

2. *On the construction of fast filtering algorithms in the basis of the Walsh-Paley functions based on a partial factoring of the discrete transform matrix*, in "Itogi Nauki. Mat. Anal.," Bce. Inst. Nauč. Tehn. Inf., No. 2045, Moscow, 1981, pp..

P.L. Ul'janov. 1. *On series with respect to the Haar system*, Mat. Sbornik **63** (1964), 356–391.

2. *A.N. Kolmogorov and divergence of Fourier series*, Uspehi Mat. Nauk **38** (1983), 51–90.
3. *Embedding of certain classes of functions H_p^ω*, Izv. Akad. Nauk SSSR, Ser. Mat. **32** (1968), 649–686.

N. Ya. Vilenkin. 1. *A class of complete orthonormal series*, Izv. Akad. Nauk SSSR, Ser. Mat. **11** (1947), 363–400.

2. A supplement to, "Theory of orthogonal series," by S. Kaczmarz and G. Steinhaus, Moscow "Fizmatgiz", 1958.
3. *On the theory of Fourier integrals on topological groups*, Mat. Sbornik **30** (1952), 233–244.

N. Ya. Vilenkin and A. Rubinštein. 1. *A theorem of S.B. Stečkin on absolute convergence and character series for a zero dimensional, abelian group*, Izv. Vysš. Učeb. Zaved. (160) No. 9 (1975), 3–9.

N. Ya. Vilenkin and S.V. Zotikov. 1. *On the cross product of orthonormal systems of functions*, Mat. Zametki **13** (1973), 469–480.

W.R. Wade. 1. *A uniqueness theorem for Haar and Walsh series*, Trans. Amer. Math. Soc. **141** (1969), 187- 194.

2. *Summing closed U-sets for Walsh series*, Proc. Amer. Math. Soc. **29** (1971), 123–125.
3. *Growth conditions and uniqueness of Walsh series*, Mich. Math. J. **24** (1977), 153–155.
4. *Recent developments in the theory of Walsh series*, International J. Math. **5** (1982), 625–673.
5. *A unified approach to uniqueness of Walsh series and Haar series*, Proc. Amer. Math. Soc. **99** (1987), 61–65.

J.L. Walsh. 1. *A closed set of normal orthogonal functions*, Amer. J. Math. **45** (1923), 5–24.

C. Watari. 1. *On generalized Walsh-Fourier series, I*, Proc. Japan Acad. **73** (1957), 435–438.

2. *Best approximation by Walsh polynomials*, Tôhoku Math. J. **15** (1963), 1–5.
3. *Mean convergence of Walsh-Fourier series*, Tôhoku Math. J. **16** (1964), 183–188.

Sh. Yano. 1. *On Walsh series*, Tôhoku Math. J. **3** (1951), 223–242.

2. *On approximation by Walsh functions*, Proc. Amer. Math. Soc. **2** (1951), 962–967.

K. Yoneda. 1. *On absolute convergence of Walsh-Fourier series*, Math. Japonica **18** (1973), 71–78.

Wo-Sang Young. 1. *Mean convergence of generalized Walsh-Fourier series*, Trans. Amer. Soc. **218** (1976), 311–320.

S. Ju. Zolotareva. 1. *Frequency properties of the multiplicative Fourier transform*, in "Methods of Digital Information Processing," Moscow "MIET", 1980, pp. 82–93.

D.M. Žukov. 1. *Realizing algorithms of multiplicative discrete Fourier transforms*, in "Methods of Digital Information Processing," Moscow "MIET", 1980, pp. 51–59.

2. *Equivalence of one-dimensional and two-dimensional Chrestenson-Levy transforms*, in "Methods of Digital Information Processing," Moscow "MIET", 1982, pp. 65–70.

SUBJECT INDEX

9 780792 311003